AF820523

Computer Mathematics

Ruyong Feng · Wen-shin Lee
Yosuke Sato
Editors

Computer Mathematics

9th Asian Symposium (ASCM2009), Fukuoka, December 2009, 10th Asian Symposium (ASCM2012), Beijing, October 2012, Contributed Papers and Invited Talks

Springer

Editors
Ruyong Feng
Academy of Mathematics
and Systems Science
Beijing
China

Yosuke Sato
Department of Mathematical
Information Science
Tokyo University of Science
Tokyo
Japan

Wen-shin Lee
Department of Mathematics
and Computer Science
University of Antwerp
Antwerp
Belgium

ISBN 978-3-662-43798-8 ISBN 978-3-662-43799-5 (eBook)
DOI 10.1007/978-3-662-43799-5

Library of Congress Control Number: 2014949487

Mathematics Subject Classification: 68W30, 65Y20, 68U05, 68P05

Springer Heidelberg New York Dordrecht London

© Springer-Verlag Berlin Heidelberg 2014
This work is subject to copyright. All rights are reserved by the Publisher, whether the whole or part of the material is concerned, specifically the rights of translation, reprinting, reuse of illustrations, recitation, broadcasting, reproduction on microfilms or in any other physical way, and transmission or information storage and retrieval, electronic adaptation, computer software, or by similar or dissimilar methodology now known or hereafter developed. Exempted from this legal reservation are brief excerpts in connection with reviews or scholarly analysis or material supplied specifically for the purpose of being entered and executed on a computer system, for exclusive use by the purchaser of the work. Duplication of this publication or parts thereof is permitted only under the provisions of the Copyright Law of the Publisher's location, in its current version, and permission for use must always be obtained from Springer. Permissions for use may be obtained through RightsLink at the Copyright Clearance Center. Violations are liable to prosecution under the respective Copyright Law.
The use of general descriptive names, registered names, trademarks, service marks, etc. in this publication does not imply, even in the absence of a specific statement, that such names are exempt from the relevant protective laws and regulations and therefore free for general use.
While the advice and information in this book are believed to be true and accurate at the date of publication, neither the authors nor the editors nor the publisher can accept any legal responsibility for any errors or omissions that may be made. The publisher makes no warranty, express or implied, with respect to the material contained herein.

Printed on acid-free paper

Springer is part of Springer Science+Business Media (www.springer.com)

Preface

This volume contains the proceedings of the Ninth Asian Symposium on Computer Mathematics (ASCM 2009), held at the JAL Resort Sea Hawk Hotel, Fukuoka, Japan, December 14–17, 2009, and the proceedings of the Tenth Asian Symposium on Computer Mathematics (ASCM 2012), held at the Chinese Academy of Sciences, Beijing, China, October 26–28, 2012. In both conferences, the contributed papers were selected by the Program Committee for presentation at the symposium and went through a standard refereeing process after the symposium. Both Program Committees had strong Asian participation, and the reviewing process was aided by reviewers from around the world.

The ASCM 2009 was jointly held with the Third International Conference on Mathematical Aspects of Computer and Information Sciences (MACIS 2009). The invited speakers at the joint conference were Markus Rosenkranz, Toshinori Oaku, Kokichi Sugihara, and Lihong Zhi. The ASCM 2012 had three excellent plenary talks delivered by Erich Kaltofen, Markus Püschel, and Josef Schicho. This volume also contains the extended abstracts provided by Erich Kaltofen and Josef Schicho after the symposium.

In addition to the regular sessions, both ASCM 2009 and ASCM 2012 had organized sessions presenting research of selected topics. The three organized sessions of ASCM 2009 were "Digitizing Mathematics: From Pen and Paper to Digital Content" organized by Volker Sorge and Alan P. Sexton, "Validated Numerical Computation" by Mitsuhiro Nakao, and "Computational Algebraic Number Theory" by Guenaël Renault. The four organized sessions of ASCM 2012 were "On the Latest Progress in Verified Computation" organized by Lihong Zhi, "Computational Geometry" by Jin-San Cheng, "Parametric Polynomial Computations" by Yosuke Sato, and "Differential and Difference Algebra" by Chun-Ming Yuan.

We would like to express our gratitude to all those who have contributed to the present volume and the organization of ASCM 2009 and ASCM 2012. We thank the authors of the papers for contributing their work and the conference participants for their presence. We also thank the organizers of the organized sessions for presenting selected research topics and the invited speakers for accepting our invitations. We are very grateful to the Program Committee members and the reviewers for their time and efforts in evaluating the submissions before and after each conference. We especially thank both teams for the local arrangements, who made the conferences successful and enjoyable. Last but not least, we thank Ziming Li for his valuable assistance in the conference organization and the publication of the proceedings. ASCM 2009 thanks Math-for-industry and Mathematical Research Center for Industrial Technology at Kyushu University, Ehime Campus Information Service, Cybernet, Japan Society for Symbolic and Algebraic Computation, Maplesoft, and JAL Hotels for their financial support. ACSM 2012 gratefully acknowledges the generous support of the Academy of Mathematics and Systems Science and the Key Laboratory of Mathematics Mechanization at the Chinese Academy of Sciences, the National Natural Science Foundation of China, Maplesoft, Cybernet Systems China.

Previous ASCM meetings were held in Beijing, China (1995), Kobe, Japan (1996), Lanzhou, China (1998), Chiang Mai, Thailand (2000), Matsuyama, Japan (2001), Beijing, China (2003), Seoul, Korea (2005), and Singapore, Singapore (2007). We hope that the ASCM continues to serve as a forum for participants to present original research, learn of research progress and developments, and exchange ideas and views on doing mathematics using computers.

Beijing, May 2014 — Ruyong Feng
Antwerp — Wen-shin Lee
Tokyo — Yosuke Sato

Conference Organization

Organizing Committee of ASCM2009

General Chair

Masakazu Suzuki, Kyushu University, Japan

Program Committee

Xavier Dahan, Kyushu University, Japan
Xiao-Shan Gao, Chinese Academy of Sciences, China
Deepak Kapur, University of New Mexico, USA
Ziming Li, Academy of Mathematics and Systems Science, China
Hirokazu Murao, The University of Electro Communications, Japan
Mitsuhiro Nakao, Kyushu University, Japan
Hyungju Park, KIAS, Korea
Guenael Renault, UPMC, INRIA, France
Ko Sakai, Tsukuba University, Japan
Yosuke Sato (co-chair), Tokyo University of Science, Japan
Alan P. Sexton, University of Birmingham, UK
Volker Sorge, University of Birmingham, UK
Gert Vegter, Groningen University, The Netherlands
Chee Yap (co-chair), New York University, USA

Local Arrangements

Tatsuyoshi Hamada, Fukuoka University/JST CREST, Japan
Koji Nakagawa, Kyushu University, Japan
Hiroshi Yoshida, Kyushu University, Japan

Organizing Committee of ASCM2012

General Chair
Ziming Li, Chinese Academy of Sciences, China

Program Committee
Shaoshi Chen, North Carolina State University, USA
Howard Cheng, University of Lethbridge, Canada
Ruyong Feng (co-chair), Chinese Academy of Sciences, China
Hoon Hong, North Carolina State University, USA
Wen-shin Lee (co-chair), University of Antwerp, Belgium
Kosaku Nagasaka, Kobe University, Japan
Chau Ngo, Quy Nhon University, Vietnam
Tateaki Sasaki, University of Tsukuba, Japan
Yosuke Sato, Tokyo University of Science, Japan
Luu Ba Thang, Hanoi National University of Education, Vietnam
Irem Yaman, Gebze Institute of Technology, Turkey
Zhengfeng Yang, East China Normal University, China
Chee K. Yap, New York University, USA
Jian Zhang, Chinese Academy of Sciences, China
Zhifang Zhang, Chinese Academy of Sciences, China

Local Arrangements
Jinsan Cheng, Chinese Academy of Sciences, China
Lei Huang, Chinese Academy of Sciences, China
Chunming Yuan (chair), Chinese Academy of Sciences, China
Daizhen Zhou, Chinese Academy of Sciences, China

Contents

Contributors

Yoshinori Aono NICT, Koganei, Japan

François Boulier LIFL, Université de Lille 1, Villeneuve D'Ascq Cedex, France

Brice Boyer Department of Mathematics, North Carolina State University, Raleigh, NC, USA

Changbo Chen ORCCA, University of Western Ontario (UWO), London, ON, Canada; Chongqing Key Laboratory of Automated Reasoning and Cognition, Chongqing Institute of Green and Intelligent Technology, Chinese Academy of Sciences, Chongqing, China

Liangyu Chen Shanghai Key Laboratory of Trustworthy Computing, East China Normal University, Shanghai, China

Howard Cheng Department of Mathematics and Computer Science, University of Lethbridge, Lethbridge, Canada

Matthew T. Comer Department of Mathematics, North Carolina State University, Raleigh, NC, USA

Xiao-Shan Gao KLMM, Academy of Mathematics and Systems Science, Chinese Academy of Sciences, Beijing, People's Republic of China

Mark Giesbrecht Cheriton School of Computer Science, University of Waterloo, Waterloo, ON, Canada

Jingjun Han Beijing International Center for Mathematical Research and School of Mathematical Sciences, Peking University, Beijing, China

Hoon Hong Department of Mathematics, North Carolina State University, Raleigh, USA

Akinari Hoshi Department of Mathematics, Rikkyo University, Toshima-ku, Tokyo, Japan

Daiju Inaba Mathematics Certification Institute of Japan, Katsushika-ku, Japan

Shutaro Inoue Tokyo University of Science, Tokyo, Japan

Qingfang Jin Key Laboratory of Mathematics Mechanization, AMSS, Chinese Academy of Science, Beijing, China

Erich L. Kaltofen Department of Mathematics, North Carolina State University, Raleigh, NC, USA

T.W. Kelsey School of Computer Science, University of St Andrews, St Andrews, UK

Heinz Kredel IT-Center, University of Mannheim, Mannheim, Germany

George Labahn Symbolic Computation Group, David R. Cheriton School of Computer Science, University of Waterloo, Waterloo, Canada

Na Lei School of Mathematics, Jilin University, Changchun, China

François Lemaire LIFL, Université de Lille 1, Villeneuve D'Ascq Cedex, France

Tuo Leng School of Computer Engineering and Science, Shanghai University, Shanghai, China

U. Leonhardt School of Physics and Astronomy, University of St Andrews, St Andrews, UK

Yaohui Li Department of Computer Science, Tianjin University of Technology and Education, Tianjin, People's Republic of China

S.A. Linton School of Computer Science, University of St Andrews, St Andrews, UK

Xuefeng Liu Research Institute for Science and Engineering, Waseda University, Shinjuku-ku, Tokyo, Japan

Zhuojun Liu Key Laboratory of Mathematics Mechanization, AMSS, Chinese Academy of Science, Beijing, China

Michael Monagan Department of Mathematics, Simon Fraser University, Burnaby, BC, Canada

Marc Moreno Maza ORCCA, University of Western Ontario (UWO), London, ON, Canada

Akira Nagai NTT Information Sharing Platform Laboratories, Tokyo, Japan

Shin'ichi Oishi Faculty of Science and Engineering, Waseda University, Shinjuku-ku, Tokyo, Japan; CREST/JST, Saitama, Japan

Tomoaki Okayama Graduate School of Economics, Hitotsubashi University, Kunitachi, Tokyo, Japan

Roman Pearce Department of Mathematics, Simon Fraser University, Burnaby, BC, Canada

Nam Pham Cheriton School of Computer Science, University of Waterloo, Waterloo, ON, Canada

Sanzheng Qiao Shanghai Key Laboratory of Contemporary Applied Mathematics, Department of Computing and Software, McMaster University, Hamilton, ON, Canada

Yuxue Ren School of Mathematics, Jilin University, Changchun, China

Tateaki Sasaki Institute of Mathematics, University of Tsukuba, Tsukuba, Japan

Josef Schicho RICAM Linz, Austrian Academy of Sciences, Vienna, Austria

Hiroshi Sekigawa Tokyo University of Science, Tokyo, Japan

Fei Shen LMAM and School of Mathematical Sciences, Peking University, Beijing, China

Liyong Shen School of Mathematical Science, University of Chinese Academy of Sciences, Beijing, China

Petr Sojka Faculty of Informatics, Masaryk University, Brno, Czech Republic

Yuqing Song Department of Computer Science, Tianjin University of Technology and Education, Tianjin, People's Republic of China

Yao Sun SKLOIS, Institute of Information Engineering, CAS, Beijing, China

Joris van der Hoeven LIX, CNRS, École polytechnique, Palaiseau Cedex, France

Dingkang Wang KLMM, Academy of Mathematics and Systems Science, CAS, Beijing, China

Dongming Wang Laboratoire D'Informatique de Paris 6, CNRS—Université Pierre et Marie Curie, Paris, France

Yimin Wei Shanghai Key Laboratory of Contemporary Applied Mathematics, School of Mathematical Sciences, Fudan University, Shanghai, People's Republic of China

Baofeng Wu Key Laboratory of Mathematics Mechanization, AMSS, Chinese Academy of Science, Beijing, China

Min Wu Shanghai Key Laboratory of Trustworthy Computing, East China Normal University, Shanghai, China

Wenyuan Wu Chongqing Institute of Green and Intelligent Technology, Chinese Academy of Sciences, Chongqing, China

Xiaoli Wu The School of Science, Hangzhou Dianzi University, Hangzhou, Zhejiang, China; Key Laboratory of Mathematics Mechanization, Chinese Academy of Sciences, Beijing, China

Zhifeng Wu Department of Computer Science, Tianjin University of Technology and Education, Tianjin, People's Republic of China

Bican Xia LMAM and School of Mathematical Sciences, Peking University, Beijing, China

Rong Xiao University of Western Ontario, London, ON, Canada

C. Xiong School of Computer Science, University of St Andrews, St Andrews, UK

Jing Yang LMIB and School of Mathematics and Systems Science, Beihang University, Beijing, China

Zhengfeng Yang Shanghai Key Laboratory of Trustworthy Computing, East China Normal University, Shanghai, China

Chun-Ming Yuan KLMM, Academy of Mathematics and Systems Science, Chinese Academy of Sciences, Beijing, People's Republic of China

Zhenbing Zeng Department of Mathematics, Shanghai University, Shanghai, China

Wen Zhang School of Mathematics and Physics, Qingdao University of Science and Technology, Qingdao, People's Republic of China

Xiaoming Zhang Key Laboratory of Mathematics Mechanization, AMSS, Chinese Academy of Science, Beijing, China

Yang Zhang Department of Mathematics, University of Manitoba, Winnipeg MB, Canada

Zhi-Yong Zhang KLMM, Academy of Mathematics and Systems Science, Chinese Academy of Sciences, Beijing, People's Republic of China

Xiaopeng Zheng School of Mathematics, Jilin University, Changchun, China

Part I
Invited Talks of ASCM2012

Symbolic Computation and Complexity Theory Transcript of My Talk

Erich L. Kaltofen

Keywords Computational complexity · Exponential-time algorithms · Practicality

1 The Setting

I gave talks at the conference *Alan Turing's Heritage: Logic, Computation & Complexity* in Lyon, France on July 3, 2012, at the Pierre and Marie Curie University (UPMC) Paris 6, France on July 17, 2012, and at the Tenth Asian Symposium on Computer Mathematics (ASCM) in Beijing, China, on October 26, 2012 on the complexity theoretic hardness of many problems that the discipline of symbolic computation tackles. Here is a brief transcript of part of those talks.

1.1 NP-Completeness and Beyond

A fundamental problem of symbolic computation, that of solving systems of polynomial equations, is easily shown to be NP-hard: $x \vee \neg y \quad \equiv \quad (1-x)y = 0, x(x-1) = 0, y(y-1) = 0$, which shows how to encode a clause in a satisfiability problem as polynomial equations.

This material is based on work supported in part by the National Science Foundation under Grants CCF-1115772.

E.L. Kaltofen (✉)
Department of Mathematics, North Carolina State University,
Raleigh, NC 27695-8205, USA
e-mail: kaltofen@math.ncsu.edu

© Springer-Verlag Berlin Heidelberg 2014
R. Feng et al. (eds.), *Computer Mathematics*, DOI 10.1007/978-3-662-43799-5_1

Real geometry, when the solutions of the polynomial systems are restricted to real numbers, is by Tarski's algorithm decidable. However, Fischer and Rabin [1] have shown that the problem requires exponential space, $2^{\Omega(n)}$, where n is the number of variables in the polynomials. Furthermore, Mayr and Meyer [2] have extended the result to Polynomial Ideal Membership over the rationals, that is, they show $2^{\Omega(n)}$-space hardness.

Finally, Fröhlich and Shepherdson [3] shows that there are fields K in which the five arithmetic operations, namely addition, negation, multiplication, division, and equality testing are computable but where factorization in $\mathsf{K}[x]$ is undecidable ("unentscheidbar"). The proof is based on the infinite tower of extensions by squareroots of prime integers and the fact that $\sqrt{2} \notin \mathbb{Q}(\sqrt{3}, \sqrt{5}, \sqrt{7}, \sqrt{11}, \ldots)$, for instance.

These are indeed formidable computational complexity theoretic barriers to the discipline of symbolic computation.

1.2 Early Symbolic Computation Algorithms

Buchberger's famous 1965 Gröbner basis algorithm tackles exactly these hard problems: it decides ideal membership and computes solutions to polynomial systems. Berlekamp's and Zassenhaus's 1968 polynomial factorization algorithms work for coefficients in finite fields and the rational numbers. Collins's 1974 cylindrical algebraic decomposition algorithm performs Tarski's quantifier elimination.

The pursuit of symbolic computation algorithms that solved these computational hard problems in the early 1980s was ridiculed by some theorists as hopeless.

2 Cook's Thesis

In his plenary talk at the ICM in Kyoto Cook [4], three function classes were introduced:
polytime, the functions computable by polynomial-time algorithms,
NAT, functions arising from natural computational problems,
PracSolv, functions computable on an actual computer on **all inputs** of 10,000 bits or less.

Stephen Cook then formalizes his thesis:

Thesis: *PracSolv* $\cap$ *NAT* $=$ *polytime* $\cap$ *NAT*.

The notion that *polytime*captures the domain of efficiently computable functions is ingrained in theoretical computer science. Many reductions in modern theoretical cryptography make use of the device. Stephen Cook refines the notion to natural functions. As an unnatural function he gives the example of $2^{\lceil \log n \rceil^{1000}} \in$ *polytime*, that with high exponent of 1,000, and which he excludes from *NAT*.

As evidence in 1990 there were polynomial-time algorithms for linear programming and for polynomial factorization over rational numbers. Polynomial identity testing via the Zippel-Schwartz lemma was and is random polynomial time, and at that time it was hypothesized that with randomization in polynomial-time algorithms one could not reach beyond the class *polytime*. Today, results of Impagliazio and Kabanets raise more doubts. Indeed, already in 1990 polynomial factorization, even for polynomials in many variables, was known to be in randomized *polytime* Kaltofen and Trager [5]. Supersparse polynomial factorization algorithms would follow 15 years later Kaltofen and Koiran [6].

Richard D. Jenks had expressed some doubt in the thesis at that time, telling me as PhD student: *You prove that problems are hard and I write computer programs that solve them.* In the following I will attempt to challenge Cook's Thesis.

The Thesis can fail in two directions. There may exist an $f \in \mathit{PracSolv} \cap \mathit{NAT}$ but $f \notin \mathit{polytime}$. Many programs in symbolic computation can produce outputs to problems that are, in the worst case, hard.

1. Proofs that a positive semidefinite polynomial is not a sum-of-squares: 462-dimensional linear matrix inequalities (LMI) with 7,546 variables Guo, Kaltofen and Zhi [7]. Semidefinite programming constitutes a far-reaching generalization to linear programming with a limited nonlinearity in its control parameters: the solution must remain a definite matrix.
2. Large Gröbner basis problems, e.g., compact McEliece crypto system: 115 variables, 193,584 equations, degrees $= 2, 3, \ldots, 256$ Faugére et al. [8].
3. Proofs that certain nonlinear polynomial equation problems (LMIs) do not have a rational solution, while they have a real solution Guo, Din and Zhi [9]. I have substituted this diophantine problem to my list for this paper; in my talk I listed a problem in real algebraic geometry with 4 variables.

The above examples do not violate Cook's Thesis. Clearly, a superpolynomial-time algorithm can have polynomial-time running time on a subset of inputs. Many algorithms in symbolic computation, such as Buchberger's algorithm and its modern variant *FGb*, have unpredictable running time. Cook's *NAT* is the class of "natural" functions, not "natural" inputs. It is important for algorithmic infrastructure to know the worst-case behavior of an algorithm: Google returns a list of ranked pages for all queries. Nonetheless, it is the hallmark of the symbolic computation discipline to have greatly enlarged the domain of natural and solvable inputs to hard problems.

The Thesis can also fail in the opposite: We may have an $f \in \mathit{polytime} \cap \mathit{NAT}$ but $f \notin \mathit{PracSolv}$. It is actually not so easy to find a natural problem in symbolic computation that is in *polytime* but whose worst-case complexity is super-quadratic in its input size. I offer three examples:

1. The characteristic polynomial of a sparse matrix $\in \mathbb{Z}_p^{n\times n}$ with $O(n)$ nonzero entries is notoriously difficult to compute with $O(n)$ auxiliary space in the worst case. The best algorithm is of $n^{2+1/2+o(1)}$-time, $O(n)$-space Villard [10]. For $n = 10^6$ this is, ignoring the implied $n^{o(1)}$, $\geq 10^9\times$ input size. We restrict to

$O(n)$ space because by using quadratic space one has $O(n^{2.38})$ time with fast matrix multiplication, although the solution is quite impractical.
2. The Sylvester resultant in x of $f(x, y), g(x, y)$ can be computed by the half-GCD algorithm in $\max\{\deg_x(f), \deg_x(g)\}^{2+o(1)} \times \max\{\deg_y(f), \deg_y(g)\}^{1+o(1)}$ scalar operations.
3. Lattice basis reduction is polynomial-time, but the dependency on the dimension may be superquadratic. I have substituted this diophantine problem to my list for this paper; in my talk I listed sharp estimates for the structured condition numbers of Hankel matrices.

In sparse/structured linear algebra, $O(n \log(n))$ versus $O(n^2)$ running time makes all the difference, for example in discrete Fourier transform algorithms. Polynomial factorization is again a forerunner: polynomials modulo 2 can be factored in subquadratic time since Kaltofen and Shoup [11].

Shaoshi Chen mentioned while he was at NCSU that some of the algorithms for symbolic summation have worst-case performance beyond quadratic. Those and the above are all candidates for polynomial-time problems that are not practically solvable for large inputs, although one may require much more than Cook's original 10,000 bits for the inputs. It is my conclusion that the Thesis fails on that side: $\mathit{PracSolv} \cap \mathit{NAT} \subsetneq \mathit{polytime} \cap \mathit{NAT}$.

Stephen Cook at the Turing Centennial celebration in San Francisco in June 2012 suggested to me to consider in place of *polytime* the class of *logarithmic space* as the practical one (in my talk I stated it as *poly-logarithmic space*).

3 Faugère's Question

After my talk in Paris, Jean-Charles Faugère asked me the following question: Does it make sense to study **and implement** algorithms that have exponential running time? My answer was "no," with some disapproval from the audience. I clarified that one may study algorithms that are worst-case exponential, but that run polynomial-time on the inputs studied. Executing an exponential-time algorithm, say a combinatorial search, constitutes a single computation, not providing an algorithm for general use.

I should add Ludovic Perret's comment to me at NCSU in October 2012: One studies exponential algorithms to know their run times, for example when choosing size of a key in a crypto scheme.

References

1. Fischer, M.J., Rabin, M.O.: Super-exponential complexity of presburger arithmetic. In: R. M. Karp, editor, Complexity of Computation, pp. 27–41. Amer. Math. Soc. (1974)
2. Mayr, E.W., Meyer, A.R.: The complexity of the word problem for commutative semigroups and polynomial ideals. Advances Math. **46**, 305–329 (1982)

3. Fröhlich, A., Shepherdson, J.C.: Effective procedures in field theory. Phil. Trans. Roy. Soc. Ser. A 248, 407–432 (1955/1956)
4. Cook, S.A.: Computational complexity of higher type functions. In: Proceedings of the ICM Kyoto Japan, ICM Series, pp. 55–69 (1990)
5. Kaltofen, E., Trager, B.: Computing with polynomials given by black boxes for their evaluations: Greatest common divisors, factorization, separation of numerators and denominators. J. Symbolic Comput., 9(3):301–320 (1990) URL: http://www.math.ncsu.edu/kaltofen/bibliography/90/KaTr90.pdf
6. Kaltofen, E., Koiran, P.: Finding small degree factors of multivariate supersparse (lacunary) polynomials over algebraic number fields. In: Jean-Guillaume Dumas, editor, ISSAC MMVI Proceedings of the 2006 International Symposium Symbolic Algebraic Computation, pp. 162–168, New York, N. Y., 2006. ACM Press. ISBN 1-59593-276-3. URL: http://www.math.ncsu.edu/kaltofen/bibliography/06/KaKoi06.pdf
7. Guo, F., Kaltofen, E.L., Zhi, L.: Certificates of impossibility of Hilbert-Artin representations of a given degree for definite polynomials and functions. In: Joris van der Hoeven and Mark van Hoeij, editors, ISSAC 2012 Proceedings of the 37th International Symposium Symbolic Algebraic Computation, pp. 195–202, New York, N. Y., July 2012. Association for Computing Machinery. ISBN 978-1-4503-1269. URL: http://www.math.ncsu.edu/kaltofen/bibliography/12/GKZ12.pdf;URL: http://arxiv.org/abs/1203.0253
8. Faugère, J.-C., Otmani, A., Perret, L., Tillich, J.-P.: Algebraic cryptanalysis of McEliece variants with compact keys. In: Proceedings of the Eurocrypt 2010, volume 6110 of Lecture Notes in Computer Science, pp. 279–298, Heidelberg, Germany, 2010. Springer Verlag
9. Guo, Q., Safey El Din, M., Zhi, L.: Computing rational solutions of linear matrix inequalities. In: Manuel Kauers, editor, ISSAC 2013 Proceedings of the 38th International Symposium Symbolic Algebraic Computation, New York, N. Y., 2013. Association for Computing Machinery
10. Villard, G.: Computing the Frobenius normal form of a sparse matrix. In: Proceedings of the Third International Workshop on Computer Algebra in Scientific Computing, pp. 395–407, Heidelberg, Germany, 2000. Springer Verlag
11. Kaltofen, E., Shoup, V.: Subquadratic-time factoring of polynomials over finite fields. Math. Comput., 67(223):1179–1197, July 1998. URL: http://www.math.ncsu.edu/kaltofen/bibliography/98/KaSh98.pdf

Factorization of Motions

Josef Schicho

1 Extended Abstract

We define *motion polynomials* as polynomials with coefficients in the dual quaternions and study their factorizations. The motion polynomials correspond to motions in 3D space, and factoring into linear factors means to compose the motion into translations and rotations. This allows to realize the motion by a linkage with revolute or prismatic joints. This is joint work with G. Hegedüs (Univ. Oboda), Z. Li (RICAM), and H.-P. Schröcker (Univ. Innsbruck). The results are published in [1]. This research has been supported by the Austrian Science Fund (FWF): DK W 1214-N15.

Let $\mathbb{H} = \langle 1, \mathbf{i}, \mathbf{j}, \mathbf{k} \rangle_{\mathbb{R}}$ be the skew field of quaternions. It is well known [2] that $\mathbb{H}$ is algebraically closed, in the sense that every univariate left polynomial $P \in \mathbb{H}[t]$ can be written as a product of linear polynomials. Here, the variable t is supposed to commute with the coefficients. To decompose P, one looks for right zeroes in $\mathbb{H}$: if $P(q) = 0$, then $(t - q)$ is a right factor of P, and the polynomial quotient has degree one less.

In order to find right zeroes, we compute the norm polynomial $N(t) = P(t)\overline{P}(t)$, where $\overline{P}$ is obtained by conjugating all coefficients. It is a real polynomial that does not assume negative values when evaluated at real numbers. Generically, it has no real zeroes, so that it can be written as a product of irreducible quadratic factors. For any such factor Q, there is a unique common right zero of P and Q in $\mathbb{H}$, and this common right zero can be computed by polynomial division: the polynomial remainder of P mod Q is linear.

The factorization algorithm can be extended to skew ring $\mathbb{DH} = \langle 1, \mathbf{i}, \mathbf{j}, \mathbf{k} \rangle_{\mathbb{D}}$ of dual quaternions, where $\mathbb{D} = \mathbb{R} \oplus \mathbb{R}\epsilon$ is the two-dimensional $\mathbb{R}$-algebra generated by $\mathbb{R}$ and ϵ with $\epsilon^2 = 0$. We are especially interested in polynomials with real

J. Schicho (✉)
RICAM Linz, Austrian Academy of Sciences, Vienna, Austria
e-mail: josef.schicho@ricam.oeaw.ac.at

© Springer-Verlag Berlin Heidelberg 2014
R. Feng et al. (eds.), *Computer Mathematics*, DOI 10.1007/978-3-662-43799-5_2

norm polynomials; these polynomials are called *motion polynomials*. Since $\mathbb{D}$ is not a field, the algorithm may sometimes fail, so that there exist polynomials in $\mathbb{DH}[t]$ without factorization into linear ones. For generic motion polynomials of degree d, the algorithm works, and one gets $d!$ different factorizations into linear motion polynomials. (In contrast to the commutative case, it is not allowed to permute the factors.)

The special interest in motion polynomials comes from a well-known isomorphism of the six-dimensional Lie group SE_3 of Euclidean displacements and the multiplicative group of dual quaternions with nonzero real norm modulo multiplication by nonzero real scalars. Motions are curves in SE_3, and in this sense motion polynomials parameterize motions. Conversely, every motion that has a parameterization by rational functions can also be parameterized by a motion polynomial.

Linear motion polynomials parameterize revolutions around a fixed axis or translational pushes in fixed directions. Hence the factorization into linear motion polynomials decomposes the parameterized motion into revolutions or pushes, and the motion can be realized by a chain of revolute or prismatic joints.

A generic quadratic motion has two factorizations into two revolutions. The two chains of revolute joints can be combined to a movable closed chain with four links and four revolute joints. This linkage is called the Bennett linkage after its discoverer Bennett [3].

For $d > 2$, a generic motion of degree d can be decomposed into d revolutions in $d!$ different ways. Again, it is possible to combine the corresponding chains into one linkage. For instance, for $d = 3$ we obtain a movable linkage with 8 links connected by 12 revolute joints. Since the 6 decompositions are in relation to the permutations of the 3 irreducible factors of the norm polynomial of the parameterizing cubic motion polynomial, and the group of permutations is generated by transpositions, one can construct all decompositions by composing two neighboring revolutions and decomposing in the second way, as above. We call this operation *Bennett flip*.

By multiplying linear motion polynomials and applying Bennett flips, one can construct various families of closed overconstrained linkages. For instance, let us multiply two linear motion polynomials parameterizing revolutions around the same axes and a third linear motion polynomial; then we do two Bennett flips and construct a closed 5R linkage. This linkage is called the Goldberg 5R linkage after its discoverer [4]. It was shown in [5] that the Goldberg 5R linkage is the only movable 5R linkage with all 5 joints actually moving that is neither planar nor spherical.

Similar constructions lead to various families of 6R linkages, some well known, and some new. It should be mentioned that the classification of closed 6R linkages is a famous open problem in kinematics. Not all known families have a motion that can be rationally parameterized. An upper bound for the genus of the motion of a 6R linkage is given in [6].

References

1. Hegedüs, G., Schicho, J., Schröcker, H.-P.: Factorization of rational curves in the study quadric and revolute linkages. Mech. Mach. Theory **69**(1), 142–152 (2013)
2. Gordon, B., Motzkin, T.S.: On the zeros of polynomials over division rings. Trans. Amer. Math. Soc. **116**, 218–226 (1965)
3. Bennett, G.T.: The skew isogramm-mechanism. Proc. London Math. Soc. (2nd Series) 13, 151–173 (1913–1914).
4. Goldberg, M.: New five-bar and six-bar linkages in three dimensions. Trans. ASME **65**, 649–656 (1943)
5. Karger, A.: Classification of 5r closed kinematic chains with self mobility. Mech. Mach. Th. **33**, 213–222 (1998)
6. Hegedüs, G., Schicho, J., Schröcker, H.-P.: The theory of bonds II: Closed 6R linkages with maximal genus. Technical Report, arxiv 1309.6558 (2013)

Part II
Contributed Papers of ASCM2009

Simplification of the Lattice Based Attack of Boneh and Durfee for RSA Cryptoanalysis

Yoshinori Aono

Abstract We present a new formulation and its simpler analysis of the lattice-based attack of Boneh and Durfee for the RSA cryptography [1]. We follow the same approach as theirs, however, we propose a new way of defining a lattice with which we can achieve the same solvable key bound $d < N^{0.292}$. Our lattice is represented as a lower triangle matrix, which makes its analysis much simpler than that of [1]. We think that our analysis technique would be useful for considering applications/generalizations of this approach.

1 Introduction

Boneh and Durfee [1] proposed a polynomial time attack by which we can recover the RSA secret key d from the public information (e, N) when $d < N^{0.292}$; which we will refer as the *Boneh-Durfee* bound. The basic idea of the attack is based on the Coppersmith technique [3] by which we can obtain small solutions of a modular equation such as $f(x_1, x_2, \ldots, x_n) \equiv 0 \pmod{W}$. The technique converts the problem of finding small solutions of the equation into the problem of solving an algebraic equation by a lattice reduction algorithm.

Here is a detailed explanation of their approach. The goal is to obtain a small solution (x_0, y_0) of the following target equation to recover the secret key:

$$f_{\mathrm{BD}}(x, y) = -1 + x(y + A) \equiv 1 \pmod{e} \tag{1}$$

Here $A = N + 1$. From this, the following bivariate polynomials are defined:

$$g_{i,j}(x, y) = \begin{cases} x^{i-j}(f_{\mathrm{BD}}(x, y))^j e^{m-j} & \text{for} \quad i \geq j \\ y^{j-i}(f_{\mathrm{BD}}(x, y))^i e^{m-i} & \text{for} \quad i < j \end{cases} \tag{2}$$

for a certain range of (i, j) and a fixed integer m. These polynomials are converted to a lattice represented by a row echelon matrix L_{BD} defined by using the coefficients

Y. Aono (✉)
NICT, Koganei, Japan
e-mail: aono@nict.go.jp

© Springer-Verlag Berlin Heidelberg 2014
R. Feng et al. (eds.), *Computer Mathematics*, DOI 10.1007/978-3-662-43799-5_3

of $g_{i,j}(x, y)$ with some parameters. Then by using a lattice reduction algorithm, we obtain a system of polynomial equations from which we can compute polynomial number of candidates of the solution (x_0, y_0) numerically.

In this approach a technically crucial point is to design a matrix for a lattice with a small determinant. They showed that their matrix has a sufficiently small determinant; however, its analysis is complicated since the technique is of geometrically progressive matrices, and it seems hard to apply for the other situations. The purpose of this paper is to give a new way to construct a lattice with asymptotically the same determinant that is much simpler to analyze.

Since Boneh and Durfee's work, variants of their technique have been proposed. Blömer and May [2] proposed a new lattice-based algorithm for attacking RSA with a short secret key. They constructed a lower triangle lattice by eliminating some columns from the original lattice; this simplifies the determinant analysis. In [7], Jochemsz and May gave an algorithm for finding small roots of a multivariate modular/integer equation based on a generalized lattice construction strategy. Note that both algorithms achieve a slightly weaker solvable key bound than the Boneh–Durfee bound.

In this paper we follow the strategy of Boneh and Durfee to give a new variation of the lattice-based attack with a simpler analysis. We propose a conversion from the polynomials (2) to three-variable polynomials $G_{i,j}(x, y, z)$ when we construct lattice; on the other hand, Boneh and Durfee directly constructed the lattice from $g_{i,j}(x, y)$. Since we obtain a lower triangle matrix representation of our lattice, we can easily compute its determinant. Therefore, we give a new simple algorithm to achieve the Boneh–Durfee bound. We remark that the same idea was independently found by Herrmann and May [6]; they referred their technique as "llinearization" and applied it for analyzing an attack for RSA-CRT.

We carry out our computer experiments to compare the qualities of our lattice and that of Boneh and Durfee. We check the solvable key ranges, the determinants, and the length of obtained vectors by L^2 algorithm [9] on lattice generated by these two algorithms. As shown in Sect. 5, we confirm that the qualities of the two lattice series are equivalent for various parameters in practice. We find the computational time of the L^2 algorithm is reduced by about 30 % from the original attack of Boneh and Durfee.

This paper is organized as follows: In Sect. 2, we give basic symbols, notations, lemmas. We give our formulation of the lattice-based attack in Sect. 3 and its detailed analysis is explained in Sect. 4. The computer experiments to compare our lattice construction and that in [1] are described in Sect. 5.

2 Preliminaries

In this section, for the following discussions, we introduce some notations and state some known facts and key technical lemmas.

We use the standard RSA notations throughout this paper. A given RSA instance is defined by p, q, e, and d, where p and q are large primes, e is a public key, and d is the corresponding secret key. Let $N = p \times q$, and assume that $\gcd(e, (p-1)(q-1)) = 1$. The key relation between e and d is

$$ed \equiv 1 \pmod{(p-1)(q-1)} \tag{3}$$

from which we derive our target Eq. (1) by following the argument in [1].

The basic strategy of the lattice-based attack is to convert the problem of recovering RSA key into the problem of finding small solution of a modular equation within the certain range. In general, solving modular equation is not easy, whereas there are some cases where we may be able to use the standard numerical method for solving this problem. The Howgrave–Graham lemma [5] provides us with one of such cases.

To state the Howgrave–Graham lemma, we introduce the notion of *XY-norm* for a bivariate polynomial $f(x, y) = \sum_{i,j} a_{i,j} x^i y^j$ and integers X and Y by

$$||f(x, y)||_{XY} \stackrel{\text{def}}{=} \sqrt{\sum_{i,j} a_{i,j}^2 X^{2i} Y^{2j}}.$$

Lemma 1 (Howgrave–Graham [5]) *For any positive integers X, Y and W, let $f(x, y)$ be a bivariate polynomial consisting of w terms with integral coefficients such that the following holds:*

$$||f(x, y)||_{XY} < W/\sqrt{w}.$$

Then we have for x and y satisfying $|x| < X$ and $|y| < Y$,

$$f(x, y) \equiv 0 \pmod{W} \Leftrightarrow f(x, y) = 0.$$

Note that $f(x, y) = 0$ clearly implies $f(x, y) \equiv 0 \pmod{W}$. What is important is its converse. This lemma guarantees that we can find all solutions of the modular equation within the range from the integer solutions of $f(x, y) = 0$.

Now we introduce some definitions and lemmas about the lattice; we need to obtain a polynomial having small XY-norm to use Lemma 1, and this problem can be reduced to a problem of finding short vectors in a lattice.

Consider linearly independent vectors $\mathbf{b}_1, \ldots, \mathbf{b}_n$, then the lattice with basis $\mathbf{b}_1, \ldots, \mathbf{b}_n$ is defined as

$$L(\mathbf{b}_1, \ldots, \mathbf{b}_n) = \{a_1\mathbf{b}_1 + a_2\mathbf{b}_2 + \cdots + a_n\mathbf{b}_n | a_i \in \mathbb{Z} \quad \text{for} \quad i = 1, \ldots, n\}. \tag{4}$$

That is, the lattice is the set of integral linear combinations of its basis vectors. We often represent the lattice $L(\mathbf{b}_1, \ldots, \mathbf{b}_n)$ by the matrix $\begin{pmatrix} \mathbf{b}_1 \\ \vdots \\ \mathbf{b}_n \end{pmatrix}$.

To find short vectors in a lattice, it can be used a lattice reduction algorithm; here we use the LLL algorithm [8] in our analysis. The two short vectors in the LLL reduced basis described in the following theorem are important.

Theorem 1 *[1, Fact 3.3] Let* $\mathbf{b}_1, \ldots, \mathbf{b}_n$ *be a lattice basis. Then the LLL algorithm can find linearly independent lattice vectors* $\mathbf{v}_1$ *and* $\mathbf{v}_2$ *such that*

$$|\mathbf{v}_1| \leq 2^{(n-1)/4}|\det(L)|^{1/n} \text{ and } |\mathbf{v}_2| \leq 2^{n/2}|\det(L)|^{1/(n-1)}. \tag{5}$$

Here, $\det(L)$ is the determinant of the lattice which is defined by the determinant of a matrix representation of the lattice; note that suppose we have a lower triangle matrix representation of a lattice, $\det(L)$ is easily calculated by the product of its diagonal elements.

In the lattice-based attack, it needs to convert polynomials to vectors; since a lattice reduction algorithm is designed for vectors, while our targets are polynomials. Then we introduce a mapping.

We divide this mapping into two steps, named a *vectorisation* and an *instantiation* respectively. We introduce a way to map three-variable polynomials to vectors since we consider three-variable polynomials in our construction.

Definition 1 (*Polynomials* $\Rightarrow$ *vectors*)
Let $\mathbf{K}$ be a finite sequence of distinct three-variable monomials; let its order be fixed, and for any t, $x^{i_t}y^{j_t}z^{k_t}$ be the tth monomial in this order. Then for any $f(x,y,z) = \sum_{1\leq t\leq|\mathbf{K}|} a_t x^{i_t}y^{j_t}z^{k_t}$, we map it to the following vector $\mathbf{b}$, which is called the *vectorization* of $f(x,y,z)$ and is denoted as $\mathscr{V}_{\mathbf{K}}(f)$.

$$\begin{array}{llllll} f(x,y,z) = & a_1x^{i_1}y^{j_1}z^{k_1} & + a_2x^{i_2}y^{j_2}z^{k_2} + & \cdots & + & a_{|\mathbf{K}|}x^{i_{|\mathbf{K}|}}y^{j_{|\mathbf{K}|}}z^{k_{|\mathbf{K}|}} \\ & \downarrow & \downarrow & & & \downarrow \\ \mathbf{b} = & (a_1x^{i_1}y^{j_1}z^{k_1}, & a_2x^{i_2}y^{j_2}z^{k_2} & ,\ldots, & & a_{|\mathbf{K}|}x^{i_{|\mathbf{K}|}}y^{j_{|\mathbf{K}|}}z^{k_{|\mathbf{K}|}}). \end{array}$$

For example the polynomial $f(x,y,z) = -3x^3 + 4x^2yz - 2xy^2z^2 + 7xy^3z^3$ is mapped to the vector $(-3x^3, 4x^2yz, -2xy^2z^2, 7xy^3z^2)$.

We introduce a conversion named an instantiation and its inverse; it converts a three-variable monomials to integers by substitution. Our matrix is defined using the vectorizations and hence each element of the matrix is monomial. On the other hand, a lattice reduction algorithm is designed for integer lattices or integer matrices. Thus, for using a lattice reduction algorithm, we need to instantiate our matrix by substituting some integers X, Y and Z to x, y and z, which we call an *instantiation* with X, Y and Z. Conversely, converting an integer vector to a polynomial is called a *deinstantiation*. Note that (since $\mathbf{K}$ and the order of monomials is fixed) we know a monomial $x^{i_t}y^{j_t}z^{k_t}$ corresponding to the tth entry of a given vector; hence, deinstantiation at the tth entry can be achieved by simply dividing its integral value by $X^{i_t}Y^{j_t}Z^{k_t}$.

These vectorization, instantiation, and deinstantiation procedures are essentially the same as those used by some published works such as Boneh and Durfee.

3 A New Lattice-Based Algorithm

In this section, we give a new lattice-based algorithm for RSA with a short secret key; that is, a new lattice construction and its simpler analysis to derive the same Boneh–Durfee bound $d < N^{0.292}$. Our analysis requires only elementary calculations. The detailed analysis is given in the next section. What is different from the original algorithm is to use three-variable polynomials to construct a lattice; while the bivariate polynomials (2) are used directly in the original paper. We first state some definitions and lemma to explain our lattice construction.

3.1 Canonical Replacement

We introduce a way to convert the bivariate polynomials $g_{i,j}(x, y)$ into three-variable polynomials artificially. We first express $g_{i,j}(x, y)$ as a sum of monomials and then replace every xy by $z + 1$ in $g_{i,j}(x, y)$. For example, the polynomial $g_{2,3}(x, y) = e^{m-2}(-1 + xy + Ax)^2 y = e^{m-2}y + A^2 e^{m-2}x^2 y + \cdots$ is converted into the polynomial $G_{3,2}(x, y, z) = e^{m-2}y + A^2 e^{m-2}(1 + z)x + \cdots$. Like this example, we denote the converted polynomial from $g_{i,j}(x, y)$ by $G_{i,j}(x, y, z)$, and we call this conversion a *canonical replacement*. It is clear that $G_{i,j}(x, y, -1 + xy) = g_{i,j}(x, y)$. Though artificial, this canonical replacement allows us to define a lower triangle matrix representation of our lattice.

Here we extend the notion of XY-norm for converted three-variable polynomials and show some useful bound. Though we consider such converted three-variable polynomials, they are essentially bivariate polynomials; hence, we still discuss its XY-norm. Let $F(x, y, z) = \sum_{i,j,k} b_{i,j,k} x^i y^j z^k$ be a three-variable polynomial. For this F, we define a three-variable version of the XY-norm as

$$||F(x, y, z)||_{XY} \stackrel{\text{def}}{=} \sqrt{\sum_{i,j,k} b_{i,j,k}^2 X^{2i} Y^{2j} (X^2 Y^2 + 1)^k}.$$

Again this is somewhat artificial; one motivation is to have the following bound:

Lemma 2 *Let $f(x, y)$ be any bivariate polynomial and $F(x, y, z)$ be its canonical replacement. Let v be the maximum degree of z in $F(x, y, z)$. Then for any nonnegative integers X and Y, we have $||f(x, y)||_{XY} \le (v + 1)||F(x, y, z)||_{XY}$.*

Proof We let $f(x, y) = \sum_{s,t} a_{s,t} x^s y^t$, and let $F(x, y, z) = \sum_{i,j,k} b_{i,j,k} x^i y^j z^k$. Then since F is obtained from f by the canonical replacement, it follows that $f(x, y) = F(x, y, -1 + xy) = \sum_{i,j,k} b_{i,j,k} x^i y^j (-1 + xy)^k$. Thus we have

$$\sum_{i,j,k} b_{i,j,k} x^i y^j (-1 + xy)^k = \sum_{i,j,k} b_{i,j,k} x^i y^j \sum_{c=0}^{k} \binom{k}{c} (-1)^{k-c} x^c y^c$$

$$= \sum_{i,j,k,c} b_{i,j,k} \binom{k}{c} (-1)^{k-c} x^{i+c} y^{j+c}$$
$$= \sum_{s,t,k,c} b_{s-c,t-c,k} \binom{k}{c} (-1)^{k-c} x^s y^t.$$

Here, we let $s = i + c$ and $t = j + c$. Comparing the coefficient of $x^s y^t$, we have

$$a_{s,t} = \sum_{k,c} b_{s-c,t-c,k} (-1)^{k-c} \binom{k}{c}. \tag{6}$$

Now k and c are integers between 0 and v since its range is from the degree of z in $F(x, y, z)$. Hence, the number of terms on the right-hand side of (6) is equal to or less than $(v+1)^2$. Thus we have $|a_{s,t}|^2 \le (v+1)^2 \sum_{k,c} \left| b_{s-c,t-c,k} \binom{k}{c} \right|^2$. Hence, we derive our claim as follows:

$$\begin{aligned}
||f(x,y)||_{XY}^2 &= \sum_{s,t} |a_{s,t}|^2 X^{2s} Y^{2t} \le \sum_{s,t} (v+1)^2 \sum_{k,c} \left| b_{s-c,t-c,k} \binom{k}{c} \right|^2 X^{2s} Y^{2t} \\
&= (v+1)^2 \sum_{s,t,k,c} \left| b_{s-c,t-c,k} \binom{k}{c} \right|^2 X^{2s} Y^{2t} \\
&= (v+1)^2 \sum_{i,j,k,c} \left| b_{i,j,k} \right|^2 \binom{k}{c}^2 X^{2(i+c)} Y^{2(j+c)} \\
&= (v+1)^2 \sum_{i,j,k} \left| b_{i,j,k} \right|^2 X^{2i} Y^{2j} \sum_{c=0}^{k} \binom{k}{c}^2 X^{2c} Y^{2c} \\
&= (v+1)^2 \sum_{i,j,k} \left| b_{i,j,k} \right|^2 X^{2i} Y^{2j} (1 + X^2 Y^2)^k \\
&= (v+1)^2 ||F(x,y,z)||_{XY}^2.
\end{aligned}$$

□

Remark We will assume that $Z = \sqrt{X^2 Y^2 + 1}$ whenever we consider instantiation/deinstantiation with some X and Y to keep its consistent with this extended XY-norm notion. Thus, for any three-variable polynomial $F(x, y, z)$ obtained as a sum of monomials of the deinstantiation of some vector $\mathbf{F}$ w.r.t. X and Y, the following relation is immediate:

$$||F(x,y,z)||_{XY} = |\mathbf{F}|. \tag{7}$$

Step 1: Choose attack parameters m and δ. Step 2: Define an index sequence **I** and a monomial sequence **K** (as explained in Section 4). For each $(i,j) \in \mathbf{I}$, define a polynomial $g_{i,j}(x,y)$ as (2) and a polynomial $G_{i,j}(x,y,z)$ by the canonical replacement of $g_{i,j}(x,y)$. Construct a lattice L using vectors $\mathscr{V}_{\mathbf{K}}(G_{i,j})$ as row vectors in the order of (i,j) following **I**. Step 3: Instantiate L with $X = \lfloor N^{\delta} \rfloor$, $Y = \lfloor N^{0.5} \rfloor$ (and $Z = \sqrt{X^2Y^2+1}$). Then apply a lattice reduction algorithm to it. Step 4: For two short vectors $\mathbf{v}_1'$ and $\mathbf{v}_2'$ computed by the LLL algorithm, compute their deinstantiations $\mathbf{v}_1$ and $\mathbf{v}_2$. Define polynomials $H_1(x,y,z)$ and $H_2(x,y,z)$ by summing up the monomials in $\mathbf{v}_1$ and $\mathbf{v}_2$ respectively. Then define $h_1(x,y) = H_1(x,y,-1+xy)$ and $h_2(x,y) = H_2(x,y,-1+xy)$. Step 5: Enumerate all integral solutions of $h_1(x,y) = h_2(x,y) = 0$. For each of those solutions, compute d by (1) and check whether it is an integer.

Fig. 1 Our version of the lattice-based attack

Now we explain our version of the lattice-based attack for RSA following its outline stated in Fig. 1. This is essentially the same as the one by Boneh and Durfee except for polynomials and a lattice construction.

We first define symbols used in the algorithm. Let δ be the ratio of the bit-length of d to that of N; here we assume that $\delta < 0.5$. Let $m \in \mathbb{N}$ be another parameter; the larger m would yield the better solvable key range but the more computation time is necessary. The Boneh–Durfee bound $\delta < 0.292$ is the approximated value when we take sufficiently large m. Thus, considering available computational resource and δ, an appropriate number should be chosen for m; for our experiment, we set m from 6 to 10.

Then we define the set $I = \{(i, j) \in \mathbb{Z}^2 | 0 \leq i \leq m, 0 \leq j \leq 2(1-\delta)i\}$. The sequence **I** is defined by introducing some order to elements in I; however, we postpone its explanation to the next section. For $(i, j) \in I$, polynomials $g_{i,j}(x, y)$ are defined as (2); thus, they are the same polynomials defined in [1].[1]

We then further extend them to three-variable polynomials $G_{i,j}(x, y, z)$ by the canonical replacement. Consider the set of monomials of type $x^i y^j z^k$ that appear in some $G_{i,j}(x, y, z)$. Again its ordered version **K** is defined in the next section. Now our lattice L is defined by the vectors $\mathscr{V}_{\mathbf{K}}(G_{i,j})$ for $(i, j) \in \mathbf{I}$. One important point here is that we can choose some appropriate ordering for **K** and **I** so that the matrix representation of L becomes lower triangle, which makes us the determinant analysis simply.

Next we carry out the LLL algorithm on L' that is obtained as the instantiation of L with parameters $X = \lfloor N^{\delta} \rfloor$ and $Y = \lfloor 3N^{0.5} \rfloor$ (and $Z = \sqrt{1 + X^2Y^2}$).

From two short vectors $\mathbf{v}_1'$ and $\mathbf{v}_2'$ in the reduced basis, we construct the corresponding polynomials $h_1(x, y)$ and $h_2(x, y)$. For each $c = 1$ and 2, convert the vector $\mathbf{v}_c$ to its deinstantiation $\mathbf{v}_c$. Then define $H_c(x, y, z)$ as the sums of all monomials in $\mathbf{v}_c$, and define $h_c(x, y) = H_c(x, y, -1 + xy)$. Here, it can be shown that each

[1] Our $g_{i,j}(x, y)$ for $i \geq j$ and for $i < j$ correspond to their $g_{i,j}(x, y)$ and $h_{i,j}(x, y)$ respectively.

$H_c(x, y, z)$ is an integer linear combination of $G_{i,j}(x, y, z)$'s since each $\mathbf{v}'_c$ is also an integer linear combination of $\mathscr{V}_{\mathbf{K}}(G_{i,j})$'s.

Finally, in step 5, we solve the simultaneous equation $h_1(x, y) = h_2(x, y) = 0$. For each (x_1, y_1) of the found integral solutions, compute d by $d = (-1 + x_1(y_1 + A))/e$ and check it is indeed the correct secret key, i.e., check whether it is a nonnegative integer. This is the outline of our version of the lattice-based attack.

Now we show relationships between the polynomials computed in the algorithm; our target is to derive the Boneh–Durfee bound via considering a sufficient condition of Lemma 1. We have by construction

$$^{\forall}x, y \left[\begin{array}{l} f_{\mathrm{BD}}(x, y) \equiv 0 \pmod{e} \\ \Rightarrow g_{i,j}(x, y) = G_{i,j}(x, y, -1 + xy) \equiv 0 \pmod{e^m} \text{ for}^{\forall}(i, j) \in I \end{array} \right].$$

Then, since each $H_c(x, y, z)$ is an integral linear combination of $G_{i,j}(x, y, z)$,

$$^{\forall}x, y \left[\begin{array}{l} G_{i,j}(x, y, -1 + xy) \equiv 0 \pmod{e^m} \text{ for } \forall(i, j) \in I \\ \Rightarrow H_c(x, y, -1 + xy) = h_c(x, y) \equiv 0 \pmod{e^m} \text{ for } c = 1, 2 \end{array} \right]$$

holds. Thus, if both $h_1(x, y)$ and $h_2(x, y)$ satisfy the condition of the Howgrave–Graham lemma for X, Y and $W = e^m$, we have

$$^{\forall}x, y, |x| < X, |y| < Y \left[h_c(x, y) \equiv 0 \pmod{e^m} \Leftrightarrow h_c(x, y) = 0 \text{ for } c = 1, 2 \right],$$

and this implies that

$$\forall x, y, |x| < X, |y| < Y \left[f_{\mathrm{BD}}(x, y) \equiv 0 \pmod{e} \Rightarrow h_c(x, y) = 0 \text{ for } c = 1, 2 \right]. \tag{8}$$

Therefore, a small solution of $f_{\mathrm{BD}}(x, y) \equiv 0 \pmod{e}$ must be included in the set of small integer solutions of $h_1(x, y) = h_2(x, y) = 0$ when h_1 and h_2 satisfy the Howgrave–Graham condition; thus, it suffices that the polynomials satisfy

$$||h_c(x, y)||_{XY} < e^m/\sqrt{w}. \tag{9}$$

Here w is the number of terms in $h_c(x, y)$.

Now we consider this condition for each $h_c(x, y)$ to derive the Boneh–Durfee bound. We have by Lemma 2,

$$||h_c(x, y)||_{XY} \leq (v + 1)||H_c(x, y, z)||_{XY} \leq (m + 1)||H_c(x, y, z)||_{XY}.$$

Note that the second inequality is from the degree of z in $H_c(x, y, z)$ is smaller than m. Moreover, by Theorem 1 and (7) we have

$$||H_c(x, y, z)||_{XY} = |\mathbf{v}_c| \leq 2^{n/2} \det(L')^{1/(n-1)}.$$

Rearranging these conditions, we have a sufficient condition for the Howgrave–Graham lemma as $(m+1)2^{n/2}\det(L')^{1/(n-1)} < e^m/\sqrt{1-\delta m}$.

Following the analysis of Boneh and Durfee, we disregard the numbers $(m+1)2^{n/2}$ and $\sqrt{1-\delta m}$ because they are sufficiently small compared to the RSA parameters, and we use $\det(L')^{1/n}$ instead of $\det(L')^{1/(n-1)}$. Hence we have our simplified sufficient condition:

$$\det(L')^{1/n} < e^m. \tag{10}$$

Here from the analysis in the next section, we have both

$$n = (1-\delta)m^2 + \mathrm{o}(m^2) \text{ and } \det(L') = N^{\left(-\frac{1}{3}\delta^2 - \frac{1}{3}\delta + \frac{5}{6}\right)m^3 + \mathrm{o}(m^3)}.$$

Therefore, the condition (10) is equivalent to

$$N^{\left(-\frac{1}{3}\delta^2 - \frac{1}{3}\delta + \frac{5}{6}\right)m^3 + \mathrm{o}(m^3)} = \det(L') < e^{nm} = N^{(1-\delta)m^3 + \mathrm{o}(m^3)}.$$

Hence we have $-\frac{1}{3}\delta^2 - \frac{1}{3}\delta + \frac{5}{6} < 1 - \delta$ for sufficiently large m. Then we have the condition for δ which s the same as the Boneh–Durfee bound [1]:

$$\delta < 1 - 1/\sqrt{2} \approx 0.292. \tag{11}$$

4 Analysis in Detail

In this section we show that L defined in the above section is lower triangle; hence we can easily derive the determinant of the L' defined as the instantiation of L with parameters X, Y and $Z = \sqrt{1 + X^2Y^2}$.

We need to give the detailed construction of L to prove our claim; before this, we define an index sequence $\mathbf{I}$ and a monomial sequence $\mathbf{K}$ to set an order of terms in our matrix. For fixed m and $\delta < 0.5$, we define the set $I \stackrel{\text{def}}{=} \{(i,j) \in \mathbb{Z}^2 \mid 0 \leq i \leq m, 0 \leq j \leq 2(1-\delta)i\}$. We respectively define $\mathbf{I}_1$ and $\mathbf{I}_2$ by the lexicographic order of (i,j) in $\{(i,j) \in I \mid i \geq j\}$ and that of (j,i) in $\{(i,j) \in I \mid i < j\}$; we use these sequences to define the order of the vector $G_{i,j}(x,y,z)$. We further set the index sequence $\mathbf{I}$ as the concatenation of $\mathbf{I}_1$ and $\mathbf{I}_2$. For $\mathbf{I}_1 = ((i_1, j_1), \ldots, (i_u, j_u))$ and $\mathbf{I}_2 = ((i'_1, j'_1), \ldots, (i'_{u'}, j'_{u'}))$, we construct monomial sequences $\mathbf{K}_1$ and $\mathbf{K}_2$; we use these sequences to set the monomial order in vectorization. We define the monomial sequences $\mathbf{K}_1$ and $\mathbf{K}_2$ by $\mathbf{K}_1 = (x^{i_1-j_1}z^{j_1}, \ldots, x^{i_u-j_u}z^{j_u})$ and $\mathbf{K}_2 = (y^{j'_1-i'_1}z^{i'_1}, \ldots, y^{j'_{u'}-i'_{u'}}z^{i'_{u'}})$ respectively. We also set $\mathbf{K}$ by the concatenation of $\mathbf{K}_1$ and $\mathbf{K}_2$. We use these sequences to define our lattice.[2]

[2] For example, for $m = 3$ and $\delta = 0.25$; we have $\mathbf{I}_1 = ((0,0), (1,0), (1,1), (2,0), (2,1), (2,2), (3,0), (3,1), (3,2), (3,3))$ and $\mathbf{I}_2 = ((2,3), (3,4))$. By them, we have the monomial sequence $\mathbf{K}_1 = (1, x, z, x^2, xz, z^2, x^3, x^2z, xz^2, z^3)$ and $\mathbf{K}_2 = (yz^2, yz^3)$.

We state two facts for our analysis; we use them in the proof of Lemma 3 and Lemma 4. We denote a symbol $\prec$ the order in $\mathbf{K}_1$ and $\mathbf{K}_2$.

Fact 1 We have for the elements in $\mathbf{K}_1$, $x^i z^j \prec x^{i'} z^{j'} \Leftrightarrow i+j < i'+j'$ or $[i+j = i'+j'$ and $j < j']$. For the elements in $\mathbf{K}_2$, $y^i z^j \prec y^{i'} z^{j'} \Leftrightarrow i+j < i'+j'$ or $[i+j = i'+j'$ and $j < j']$.

Fact 2 For elements in $\mathbf{K}$, we have $x^j z^i \in \mathbf{K}_1 \Leftrightarrow 0 \le i+j \le m$, and $y^j z^i \in \mathbf{K}_2 \Leftrightarrow$ $[0 \le i \le m$ and $0 < j < (1-2\delta)i]$.

Now we define our lattice L by using the polynomials $G_{i,j}(x, y, z)$ and the defined sequences; here we actually give a matrix representation of L. Our matrix is defined by the row matrix of vectors $\mathscr{V}_{\mathbf{K}}(G_{i,j})$ for $(i, j) \in I$ whose order is from $\mathbf{I}$. We divide L as follows to show its lower triangularity:

$$L = \begin{bmatrix} \mathscr{V}_{\mathbf{K}}(G_{0,0}) \\ \vdots \\ \mathscr{V}_{\mathbf{K}}(G_{m,m'}) \end{bmatrix} = \overbrace{\phantom{L_{00}}}^{\mathbf{K}_1}\overbrace{\phantom{L_{01}}}^{\mathbf{K}_2} \left[\begin{array}{c|c} L_{00} & L_{01} \\ \hline L_{10} & L_{11} \end{array}\right] \begin{matrix} \}\mathbf{I}_1 \\ \}\mathbf{I}_2 \end{matrix} \tag{12}$$

Here, $m' = \lfloor 2(1-\delta)m \rfloor$. Therefore, we need to show that L_{00} and L_{11} are lower triangle matrices and show that L_{01} is the zero matrix to proof the triangularity of L. Moreover, we prove that the monomials in $G_{i,j}(x, y, z)$ are contained in $\mathbf{K}$; this will be shown via the proof of Lemma 3 and Lemma 4.

Lemma 3 *L_{00} and L_{01} are a lower triangle and the zero matrix respectively.*

Proof Let (i_k, j_k) be the kth element in $\mathbf{I}_1$. We first show the triangularity of L_{00}; we need to show that the polynomial $G_{i,j}(x, y, z)$ can be expressed as a linear combination of the first k elements in $\mathbf{K}_1$, and show that the coefficient of $x^{i_k - j_k} z^{j_k}$ in $G_{i_k,j_k}(x, y, z)$ is not zero.

Here, $G_{i,j}(x, y, z)$ is expressed for certain integers a_ℓ by the definition (2) and the canonical replacement:

$$G_{i_k,j_k}(x, y, z) = x^{i_k - j_k}(z + Ax)^{j_k} e^{m - j_k} = \sum_{\ell=0}^{j_k} a_\ell x^{i_k - \ell} z^\ell .$$

Thus, by Fact 1, $x^{i_k - j_k} z^{j_k}$ (the kth element in $\mathbf{K}_1$) is the most right nonzero element in $\mathscr{V}_{\mathbf{K}}(G_{i_k,j_k})$ in the order $\prec$; this also corresponds to the kth diagonal element in our matrix. Hence the nonzero elements in $\mathbf{b}_{i_k,j_k}$ are on the diagonal position or its left; this shows that L_{00} is a lower triangle matrix. It is clear that L_{01} is the zero-matrix since the polynomial $G_{i_k,j_k}(x, y, z)$ for $(i_k, j_k) \in \mathbf{I}_1$ does not have a monomial of type $z^{j'} y^{i'}$. □

Lemma 4 *L_{11} is a lower triangle matrix.*

Proof Let (i_k, j_k) be the kth element in $\mathbf{I}_2$. We carry the proof by showing the monomials $G_{i_k,j_k}(x, y, z)$ are included in $\mathbf{K}_1$ or the first k elements in $\mathbf{K}_2$.

We first give an expression of $G_{i_k,j_k}(x,y,z)$ for $(i_k,j_k)\in \mathbf{I}_2$; we have by (2) and the canonical replacement,

$$\begin{aligned}
G_{i_k,j_k}(x,y,z)/e^{m-i_k} &= y^{j_k-i_k}(-1+xy+Ax)^{i_k} = y^{j_k-i_k}(z+Ax)^{i_k}\\
&=\sum_{t=0}^{i_k}\binom{i_k}{t}(Ax)^t y^{j_k-i_k} z^{i_k-t}\\
&=\sum_{t=0}^{j_k-i_k-1}\binom{i_k}{t}(Ax)^t y^{j_k-i_k} z^{i_k-t}\\
&\quad+\sum_{t=j_k-i_k}^{i_k}\binom{i_k}{t}(Ax)^t y^{j_k-i_k} z^{i_k-t}\\
&=\sum_{t=0}^{j_k-i_k-1}\binom{i_k}{t}A^t(xy)^t y^{j_k-i_k-t} z^{i_k-t}\\
&\quad+\sum_{t=j_k-i_k}^{i_k}\binom{i_k}{t}A^t(xy)^{j_k-i_k} x^{t-j_k+i_k} z^{i_k-t}\\
&=\sum_{t=0}^{j_k-i_k-1}\binom{i_k}{t}A^t(1+z)^t y^{j_k-i_k-t} z^{i_k-t}\\
&\quad+\sum_{t=j_k-i_k}^{i_k}\binom{i_k}{t}A^t(1+z)^{j_k-i_k} x^{t-j_k+i_k} z^{i_k-t}\\
&=\sum_{t=0}^{j_k-i_k-1}\binom{i_k}{t}A^t\sum_{\ell=0}^{t}\binom{t}{\ell} y^{j_k-i_k-t} z^{i_k-t+\ell}\\
&\quad+\sum_{t=j_k-i_k}^{i_k}\binom{i_k}{t}A^t\sum_{\ell=0}^{j_k-i_k}\binom{t}{\ell} x^{t-j_k+i_k} z^{i_k-t+\ell}.
\end{aligned}$$

Therefore, the monomials included in the expression of $G_{i_k,j_k}(x,y,z)$ are

$$x^{t-j_k+i_k}z^{i_k-t+\ell} \text{ for } j_k-i_k\le t\le i_k \text{ and } 0\le \ell\le j_k-i_k, \tag{13}$$

and

$$y^{j_k-i_k-t}z^{i_k-t+\ell} \text{ for } 0\le t\le j_k-i_k-1 \text{ and } 0\le \ell\le t. \tag{14}$$

We show that the terms in (13) and (14) are included in $\mathbf{K}_1$ and $\mathbf{K}_2$ respectively.

First we argue the terms in (13). By Fact 2, we have $x^{t-j_k+i_k}z^{i_k-t+\ell}\in \mathbf{K}_1 \Leftrightarrow 0\le (t-j_k+i_k)+(i_k-t+\ell)=2i_k-j_k+\ell\le m$, thus we need to show

$$0 \le 2i_k - j_k + \ell \le m \text{ for } (i_k, j_k) \in \mathbf{I}_2 \text{ and } 0 \le \ell \le j_k - i_k.$$

Here, $j_k - i_k \le t \le i_k$ implies $0 \le t - j_k + i_k \le 2i_k - j_k$. Then by $0 \le \ell \le j_k - i_k$, we have $0 \le 2i_k - j_k + \ell \le i_k \le m$ and thus the monomials (13) are in $\mathbf{K}_1$.

Next we show the terms (14) are in $\mathbf{K}_2$. By Fact 2, we have

$$y^{j_k-i_k-t}z^{i_k-t+\ell} \in \mathbf{K}_2 \Leftrightarrow [0 \le i_k-t+\ell \le m \text{ and } 0 < j_k-i_k-t < (1-2\delta)(i_k-t+\ell)].$$

Hence we show that these two inequalities satisfy for $(i_k, j_k) \in \mathbf{I}_2, 0 \le t \le j_k-i_k-1$ and $0 \le \ell \le t$. We have $i_k - t + \ell \le i_k \le m$ from $0 \le \ell \le t$. On the other hand,

$$i_k - t + \ell \ge 2i_k - j_k + 1 > 2i_k - 2(1-\delta)i_k + 1 = 2\delta i_k + 1 \ge 0$$

holds from $t \le j_k - i_k - 1$ and $\ell \ge 0$. Thus, $0 \le i_k - t + \ell \le m$ holds. This is the first inequality. Next we show the second inequality. We have that $t \le j_k - i_k - 1$ derives $j_k - i_k - t \ge 1 > 0$. On the other hand, $j_k - i_k - t < (1-2\delta)i_k - t \le (1-2\delta)i_k - t + \ell$ holds. Then we have for $0 \le \delta < 0.5$, $-t + \ell \le (1-2\delta)(-t+\ell)$ since $-t+\ell \le 0$. Hence we have

$$j_k-i_k-t < (1-2\delta)i_k-t+\ell \le (1-2\delta)i_k+(1-2\delta)(-t+\ell) \le (1-2\delta)(i_k-t+\ell).$$

Therefore, the monomials (14) are in $\mathbf{K}_2$.

Finally, we need to show that each maximum element in (14) in the order $\prec$ corresponds to a diagonal element in L_{11}. We have by Fact 1, the maximum element in (14) is a monomial $y^{j_k-i_k-t}z^{i_k-t+\ell}$ such that:

$(j_k - i_k - t) + (i_k - t + \ell) = j_k + \ell - 2t$ is maximum, and $j_k - i_k - t$ is also maximum under maximized $j_k + \ell - 2t$ within the range of (14).

Hence this is the case $\ell = t = 0$; this corresponds to the monomial $y^{j_k-i_k}z^{i_k}$, which is the kth element in $\mathbf{K}_2$; thus; it corresponds to the kth diagonal element in L_{11}. Therefore L_{11} is a lower triangle matrix. □

Now we can easily calculate $\det(L')$; since L and its instantiation L' are lower triangle matrices by combining Lemma 3 and Lemma 4. We have from the expressions, the diagonal elements in L' corresponding to $G_{i,j}(x,y,z)$ are $e^{m-j}X^{i-j}Z^j$ for $(i,j) \in \mathbf{I}_1$, and $e^{m-i}Y^{j-i}Z^i$ for $(i,j) \in \mathbf{I}_2$ respectively. Hence by using the approximations $e \approx N, X \approx N^{\delta}, Y \approx N^{0.5}$ and $Z = \sqrt{X^2Y^2+1} \approx N^{\delta+0.5}$, we have

$$\begin{aligned}
\det(L_{00}) &= e^{m(m+1)(m+2)/3}X^{m(m+1)(m+2)/6}Y^{m(m+1)(m+2)/6} \\
&= N^{\left(\frac{5}{12}+\frac{1}{3}\delta\right)m^3+o(m^3)}, \\
\det(L_{11}) &= e^{(1-2\delta)m^3/6+o(m^3)}Y^{(1-2\delta)^2m^3/6+o(m^3)}Z^{(1-2\delta)m^3/3+o(m^3)} \\
&= N^{\left(-\frac{1}{3}\delta^2-\frac{2}{3}\delta+\frac{5}{12}\right)m^3+o(m^3)}, \text{ and} \\
\det(L') &= \det(L_{00}) \cdot \det(L_{11}) = N^{\left(-\frac{1}{3}\delta^2-\frac{1}{3}\delta+\frac{5}{6}\right)m^3+o(m^3)}.
\end{aligned} \tag{15}$$

On the other hand, the dimension of the matrix is $|\mathbf{I}| = (1-\delta)m^2 + o(m^2)$, thus we have $e^{nm} = N^{(1-\delta)m^3+o(m^3)}$.

Therefore as explained in the previous section, we can derive the bound $\delta < 0.292$ by using these values.

5 Computer Experiments

We carry out our computer experiments to check that our lattice construction is valid for recovering short secret key; for various parameters, we compare the solvable key ranges, the determinants and the computational times of lattices between our lattice L and Boneh and Durfee's L_{BD}. The results of our experiments are shown in Tables 1 and 2. Then we confirm the qualities between two lattice series are equivalent in practice. Moreover, we find the computational time of the L^2 algorithm is reduced by about 30 % from the original attack.

We implement our experiment procedure by the C++ language using Shoup's NTL [10] of version 5.4.2. We carry out the lattice reduction part by the L^2 algorithm [9] with parameters[3] $\delta = 0.99$ and $\eta = 0.51$, and implement the resultant calculation algorithm by [4]. We compile our source code by `gcc-4.1.2` (64 bit version) with `-O6` option. We conduct our computer experiments on the TSUBAME supercomputer.[4]

The procedures of our experiments are shown in Fig. 2. In steps 1-2 and 2-2, the reason for using L^2 algorithm, while we used the LLL algorithm in the analysis, is to speed up the experiment. Moreover, we verify that the L^2 algorithm can find sufficiently short vectors for our propose, see Table 2. In step 2-2, we use a parameter $Z = \lfloor\sqrt{X^2Y^2+1}\rfloor$ for instantiation, while we used $Z = \sqrt{X^2Y^2+1}$ in analysis; this is from our implementation of L^2 algorithm designed for the integer vectors. However, we think that this does not affect the quality of the algorithm. In steps 1-3 and 2-3, the vectors obtained by the L^2 algorithm are sorted by their length; this is because these vectors are approximate ones and we cannot guarantee that $\mathbf{v}'_1$ and $\mathbf{v}'_2$ are the shortest two in the reduced basis $\mathbf{b}'_1, \ldots, \mathbf{b}'_n$. In steps 1-4 and 2-4, we check the algebraic independence of $h_1(x, y)$ and $h_2(x, y)$ by checking $R(x) \neq 0$ holds or not, where $R(x)$ is the resultant of h_1 and h_2. Moreover, we regard an experiment instance is successful if $R(x) \neq 0$ and $R(x_0) = 0$ where x_0 is from the small solution of the target equation.

Input parameters of experiments are ℓ, m, and δ which define respectively the bit length of N, the parameter for constructing the lattice, and the ratio of bit length of d

[3] This δ is L^2 algorithm's parameter and different from δ used for defining RSA instance. See the original paper [9] about this δ.

[4] TSUBAME is a grid type supercomputer at Tokyo Inst. of Tech. A node of the supercomputer which we used contains eight Opteron Dual Core model 880 processors of 2.4 GHz and 32 GB RAM. Note, however, we have not been able to make a parallel version of our algorithm; it was used only for reducing the total experiment time.

Table 1 Key recoverability and computational time for $\ell = 512$ and 1024

Experiment parameters		Lattice	Results		
m	δ		s.	L^2 time	Total time
$\ell = 512$					
6	0.265	L_{BD}	5	31.9 s	49.5 s
		L	5	22.6 s	39.9 s
	0.270	L_{BD}	4	31.0 s	50.2 s
		L	4	21.8 s	40.8 s
8	0.265	L_{BD}	5	360 s	721 s
		L	5	251 s	610 s
	0.270	L_{BD}	4	318 s	613 s
		L	4	218 s	514 s
10	0.270	L_{BD}	5	39 min	159 min
		L	5	28 min	147 min
	0.275	L_{BD}	4	33 min	132 min
		L	4	23 min	121 min
$\ell = 1024$					
6	0.270	L_{BD}	5	123 s	195 s
		L	5	86 s	157 s
	0.275	L_{BD}	0	112 s	120 s
		L	0	75 s	82 s
8	0.275	L_{BD}	5	1322 s	2553 s
		L	5	860 s	2081 s
	0.280	L_{BD}	0	1096 s	1230 s
		L	0	695 s	823 s
10	0.270	L_{BD}	5	150 min	572 min
		L	5	110 min	575 min
	0.275	L_{BD}	4	127 min	493 min
		L	4	91 min	489 min
	0.280	L_{BD}	0	108 min	267 min
		L	0	78 min	250 min

to that of N. We carry out the experiments for $m = 6, 8$ and 10, $\ell = 512$ and 1024, and $\delta = 0.260$–0.280 in 0.005 intervals.

We generate each sample instance for given parameters ℓ and δ as follows. First, randomly choose $\ell/2$-bit primes p and q, and let $N = pq$ (Here we choose the primes as the Euler–Jacobi pseudoprimes to bases 2, 3, 5, 7 and 11.) Next, randomly choose $\lfloor \delta\ell \rfloor$-bit odd integer as the secret key d such that $\gcd(d, (p-1)(q-1)) = 1$. Compute the corresponding public key $e \equiv d^{-1} (\mod (p-1)(q-1))$ and let $A = N + 1$, $y_0 = p + q$ and $x_0 = (1 - ed)/(p-1)(q-1)$. Note that they

Table 2 Determinant and length of vectors

Experiment parameters		RSA parameters			Lattice		Results					
m	δ	$\log_2 N$	$\log_2 e$	$\log_2 d$	Deg.	Type	$\log_2$ det	D	B_1	H_1	B_2	H_1
6	0.265	512.0	511.8	135.0	34	L_{BD}	103989	−2.15	3056.35	3056.35	3056.54	3056.54
						L	103989	−2.15	3056.35	3056.35	3056.54	3056.54
8	0.265	512.0	510.9	135.0	57	L_{BD}	231587	−1.17	4060.96	4060.96	4060.99	4060.99
						L	231587	−1.17	4060.96	4060.96	4060.99	4060.99
10	0.270	512.0	508.2	138.0	86	L_{BD}	436180	1.67	5073.53	5073.53	5073.58	5073.58
						L	436180	1.67	5073.53	5073.53	5073.58	5073.58
6	0.265	1024.0	1022.47	271.0	34	L_{BD}	207700	−5.08	6103.74	6103.74	6103.89	6103.89
						L	207700	−5.08	6103.74	6103.74	6103.89	6103.89
8	0.265	1024.0	1022.65	271.0	57	L_{BD}	463048	−6.74	8116.91	8116.91	8117.11	8117.11
						L	463048	−6.74	8116.91	8116.91	8117.11	8117.11
10	0.270	1024.0	1023.67	276.0	86	L_{BD}	875290	−4.46	10173.33	10173.33	10173.36	10173.36
						L	875290	−4.46	10173.33	10173.33	10173.36	10173.36

The procedure of our computer experiments for L_{BD}: ExpBD(e,d,N,x_0,y_0; δ,m)

Step 1-1: Let $X = \lfloor N^\delta \rfloor$ and $Y = \lfloor 3N^{0.5} \rfloor$. Compute polynomials $g_{i,j}(x,y)$ in (2) for $I = \{(i,j) \in \mathbb{Z}^2 | 0 \le i \le m, 0 \le j \le 2(1-\delta)i\}$. Then construct the lattice L_{BD} by following [2].
Step 1-2: Apply the L^2 algorithm for L_{BD}.
Step 1-3: Sort the vectors in the reduced basis $\mathbf{b}'_1, \dots, \mathbf{b}'_n$ by these length to $\mathbf{v}'_1, \dots, \mathbf{v}'_n$. Compute $h_c(x,y)$ as the sum of the elements in the deinstantiation of $\mathbf{v}'_c$ for $c = 1$ and 2.
Step 1-4: Check $h_1(x_0,y_0) = 0$ and $h_2(x_0,y_0) = 0$. (If $h_1(x_0,y_0) \ne 0$ or $h_2(x_0,y_0) \ne 0$, the experiment is failure.) Compute $R(x) = \mathrm{Res}(h_1,h_2)$ and check $R(x) \ne 0$ and $R(x_0) = 0$ holds or not.

The procedure of our computer experiments for our L: ExpOurs(e,d,N,x_0,y_0; δ,m)

Step 2-1: Let $X = \lfloor N^\delta \rfloor$, $Y = \lfloor 3N^{0.5} \rfloor$. Compute polynomials $g_{i,j}(x,y)$ in (2) and convert them to $G_{i,j}(x,y,z)$ for $I = \{(i,j) \in \mathbb{Z}^2 | 0 \le i \le m, 0 \le j \le 2(1-\delta)i\}$. Then construct the lattice L by the method in Section 3
Step 2-2: Apply the L^2 algorithm for L'. Here, L' is the instantiation of L with parameter X, Y (and $Z = \lfloor \sqrt{X^2Y^2+1} \rfloor$).
Step 2-3: Sort the vectors of reduced basis $\mathbf{b}'_1, \dots, \mathbf{b}'_n$ by these length to $\mathbf{v}'_1, \dots, \mathbf{v}'_n$. Compute $H_c(x,y,z)$ as the sum of the elements in the deinstantiation of $\mathbf{v}'_c$, and $h_c(x,y) = H_c(x,y,-1+xy)$ for $c = 1,2$
Step 2-4: Check $h_1(x_0,y_0) = 0$ and $h_2(x_0,y_0) = 0$. (If $h_1(x_0,y_0) \ne 0$ or $h_2(x_0,y_0) \ne 0$, the experiment is failure.) Compute $R(x) = \mathrm{Res}(h_1,h_2)$ and check $R(x) \ne 0$ and $R(x_0) = 0$ holds or not.

Fig. 2 Our computer experiment procedure for L_{BD} and L

satisfy $f_{BD}(x_0, y_0) \equiv 0 (\mod e)$. Then the generating algorithm outputs the tuple (e, d, N, x_0, y_0) of RSA instances and the solution of the target equation.

The results are shown in Tables 1 and 2. Table 1 shows the solvable key ranges of our lattices and that of [1]; that is, the experiment is successful or not, for each parameter. The computational times on the table is the average of five experiments for each parameter. Table 2 shows some values for comparing the qualities of L and L_{BD} for some experiment instances; for example, $\log_2$ of the determinants, $\log_2$ of the length of short vectors and other values.

5.1 Key Recoverability

We check that the difference in the key recoverable ranges between L_{BD} and L. We regard that a lattice can recover the secret key if the polynomials $h_1(x, y)$ and $h_2(x, y)$ pass the check in step 1-4 or 2-4. We perform our experiments five times for each parameter.

The results of the experiments are shown in Table 1. The column "lattice" and "s." mean the type of lattice used in the procedure and the number of experiments that pass the check respectively. We also give the "L^2 time", which means the CPU time processing step 1-2 or 2-2, and give the "total time", which means the CPU time processing the routine ExpBD or ExpOurs in the table. We conclude that the key recoverable ranges of both algorithms are equivalent in our experiments.

We give some remark on the computational times. We can see the L^2 time is reduced by about 30 % compared with the L^2 time of the previous lattice; we think this is caused as the matrix representation of L is simpler than that of L_{BD}. We remark that the total computational time is approximately the sum of L^2 time and the time for computing the resultant; thus, we can see the time of the resultant computation is longer than that of L^2 algorithm when m is large. We are able to avoid this by improving the source code for computing the resultant polynomial of bivariate polynomials; however, our interest is in the lattices, hence we think that this is not essential for our study.

We further remark on the recoverable range at $\ell = 1024$. We can see the qualities of lattices for $m = 8$ are better than those for $m = 10$ in the table, while we find that the recoverable range expands with larger m. This is caused by an irregular instance; in fact, the fault sample instance for $m = 8$ and $\delta = 0.275$ has the public key $e \approx 2^{1018}$; which is smaller than $N \approx 2^{1024}$ while we assumed that $e \approx N$ in our analysis. Since then, this error is caused by the sample instance is not suitable for our analysis.

5.1.1 Determinant and Obtained Vectors

We compare the determinants, the length of obtained vectors, and some amounts to check the qualities of L and L_{BD}; we pick up some instances in our experiments. These are shown in Table 2.

We explain the columns in Table 2. The column deg. shows the degree of the lattice, that is, the number of vectors in the lattice basis. The column D means the value $\log_2(|\mathbf{v}'_1|/\det^{1/\text{deg}})$. The columns B_1, H_1, B_2 and H_2, respectively, mean the value $\log_2|\mathbf{v}'_1|$, $\log_2||h_1(x,y)||_{XY}$, $\log_2|\mathbf{v}'_2|$ and $\log_2||h_2(x,y)||_{XY}$.

We give some remarks on the results in the table. The values D in the table are sufficiently smaller than $(\deg-1)/4$; this is the upper bound guaranteed by Theorem 1 when we use the LLL algorithm. Hence, we verify the L^2 algorithm finds sufficiently short vectors for our objective. We can see values in "result" are equivalent for L and L_{BD}; in fact, we confirmed that they are equivalent at least 10 most significant digits in practice. On the other hand, we have for $c = 1$ and 2, $|\mathbf{v}'_c| = ||h_c(x,y)||_{XY}$ for the lattice L_{BD} and $|\mathbf{v}'_c| = ||H_c(x,y,z)||_{XY} \geq ||h_c(x,y)||_{XY}/(m+1)$ for the lattice L. That is, the result shows that the inequality is unnecessary pessimistic. In sum, we verified that the obtained polynomials by L are valuable as those by the algorithm of Boneh and Durfee.

6 Conclusion

We study the lattice-based attack for RSA with short secret key. We give the new simple analysis to obtain the Boneh–Durfee bound $\delta < 0.292$. Through computer experiments, we verify that the recoverable ranges of our lattice and that of Boneh and Durfee's are equivalent, and furthermore, by our approach, the computational time of L^2 algorithm is reduced by about 30 % compared with the original lattice. One important advantage of our technique is that it does not require any technical method or involved calculation required in the original technique. We hope that our analysis technique will be applicable in other situations of the lattice-based attack.

Acknowledgments I am grateful to Osamu Watanabe for his advice, careful reading, and for correcting some expressions. The author and this research was supported in part by the JSPS Global COE program "Computationism as a Foundation for the Sciences."

References

1. Boneh, D., Durfee, G.: Cryptanalysis of RSA with private key *d* less than $N^{0.292}$. IEEE Trans. Inf. Theory **46**(4), 1339–1349 (2000)
2. Blömer, J., May, A.: Low secret exponent RSA revisited. In: Proceedings of the CaLC 2001. LNCS, vol. 2146, pp. 4–19 (2001)
3. Coppersmith, D.: Finding a small root of a univariate modular equation. In: Proceedings of the EUROCRYPT 1996. LNCS, vol. 1070, pp. 155–165 (1996)
4. Healy, A.D.: Resultants, resolvents and the computation of Galois Groups. http://www.alexhealy.net/papers/math250a.pdf
5. Howgrave-Graham, N.: Finding small roots of univariate modular equations revisited.In: Proceedings of the Cryptography and Coding. LNCS, vol. 1355, pp. 131–142 (1997)
6. Herrmann, M., May, A.: Maximizing small root bounds by linearization and applications to small secret exponent RSA. In: Proceedings of the PKC 2010. LNCS, vol. 6056, pp. 53–69 (2010)
7. Jochemz, E., May, A.: A strategy for finding roots of multivariate polynomials with new applications in attacking RSA variants. In: Proceedings of the Asiacrypt 2006. LNCS, vol. 4284,pp. 267–282 (2006)
8. Lenstra, A.K., Lenstra Jr, H.W., Lovász, L.: Factoring polynomials with rational coefficients. Mathematische Annalen **261**(4), 515–534 (1982)
9. Nguyen, P., Stehlé, D.: Floating-point LLL (Full version). ftp://ftp.di.ens.fr/pub/users/pnguyen/FullL2.pdf.
10. Shoup, V., NTL: A library for doing number theory. http://www.shoup.net/ntl/index.html

Real Root Isolation of Regular Chains

François Boulier, Changbo Chen, François Lemaire and Marc Moreno Maza

Abstract We present an algorithm RealRootIsolate for isolating the real roots of a polynomial system given by a zerodimensional squarefree regular chain. The output of the algorithm is guaranteed in the sense that all real roots are obtained and are described by boxes of arbitrary precision. Real roots are encoded with a hybrid representation, combining a symbolic object, namely a regular chain and a numerical approximation given by intervals. Our algorithm is a generalization, for regular chains, of the algorithm proposed by Collins and Akritas. We have implemented RealRootIsolate as a command of the module `SemiAlgebraicSetTools` of the `RegularChains` library in MAPLE. Benchmarks are reported.

1 Introduction

Finding real roots for univariate polynomials has been widely studied. Some methods guarantee the number of real roots and isolate each real root in an arbitrary small interval. The algorithm presented in this paper is a generalization to regular chains of the algorithm given by Collins and Akritas [6].

There exist many different approaches for isolating real roots of univariate polynomials by means of Descartes rules of signs [11]. Uspensky [30] rediscovered

F. Boulier · F. Lemaire (✉)
LIFL, Université de Lille 1, 59655 Villeneuve D'Ascq Cedex, France
e-mail: Francois.Boulier@lifl.fr

F. Lemaire
e-mail: Francois.Lemaire@lifl.fr

C. Chen · M. Moreno Maza (✉)
ORCCA, University of Western Ontario (UWO), London, ON, Canada

C. Chen
Chongqing Key Laboratory of Automated Reasoning and Cognition, Chongqing Institute of Green and Intelligent Technology, Chinese Academy of Sciences, Chongqing, China
e-mail: changbo.chen@hotmail.com

M.Moreno Maza
e-mail: moreno@csd.uwo.ca

© Springer-Verlag Berlin Heidelberg 2014
R. Feng et al. (eds.), *Computer Mathematics*, DOI 10.1007/978-3-662-43799-5_4

independently[1] an inefficient version of Vincent's work [1]. More recent algorithms are closer to the original work of Vincent and based on continuous fractions [2, 3]. The approach of [28] is efficient in memory since it avoids the storage of one polynomial at each node of the tree of the recursive calls.

The methods mentioned above are all for univariate polynomials with integral or rational coefficients. In [12], the authors apply Descartes Algorithm for polynomials with bit-stream coefficients. In [7, 14], the authors present algorithms for isolating the real roots of univariate polynomials with real algebraic number coefficients. There exist different approaches for isolating real roots of polynomial systems with finitely many complex solutions. Various constructions are employed to generalize to multivariate systems the techniques known for univariate equations: rational univariate representation [26], polyhedron algebra [22], and triangular decompositions [5, 19, 24, 33].

In this paper, we generalize the Vincent-Collins-Akritas Algorithm to zerodimensional squarefree regular chains; therefore our work falls in the same category as this latter group of papers. Our idea is to build inductively (one variable after another) "boxes" in which one and only one real solution lies. This basically amounts to applying the Vincent-Collins-Akritas Algorithm to polynomials with real algebraic coefficients defined by a regular chain. Our main algorithm RealRootIsolate takes a zerodimensional squarefree regular chain T as an input and returns a list of disjoint boxes (Cartesian products of intervals) such that each box contains exactly one real root of T. We have implemented our algorithm in MAPLE in the module `SemiAlgebraicSetTools` of the `RegularChains` library.

Although rediscovered independently, the techniques presented here share some ideas with those of [24, 25]. However, our algorithm focuses on finding isolation boxes for real solutions of polynomial systems, whereas Rioboo's primary goal is to implement the real closure of an ordered field. Moreover, Rioboo uses Sturm sequences and subresultants for univariate polynomial real root isolation.

Other real root isolation algorithms based on triangular decompositions [5, 19, 33] rely on the so-called "sleeve polynomials", see Sect. 2.5.

We do not report on a comparative implementation with the methods in [5, 7, 19, 24, 33]. In order to ensure a fair comparison, one would need to bring these six real root isolation methods (including ours) in a common implementation framework, which would require a significant amount of work.

As mentioned, the algorithm presented here has been implemented in MAPLE interpreted code. However, it does not rely yet on fast polynomial arithmetic nor modular methods for regular chain computations. As shown in [17], these techniques should speedup our implementation dramatically.

We compare our code with the RootFinding[Isolate] command (available in MAPLE) based on the rational univariate representation [26]. With no surprise, the highly optimized supporting C code allows RootFinding[Isolate] to outperform our modest MAPLE implementation on systems that are in Shape Lemma position [4].

---

[1] Recent investigations of A. Akritas seem to prove that Uspensky only had an incomplete knowledge of Vincent's paper, from [29, pp. 363–368].

However, for different families of examples, corresponding to non-equiprojectable[2] varieties the situation is reversed which demonstrates the interest of our approach, even in this unfair comparative implementation framework.

Another contribution of our work is that it equips MAPLE with a tool for manipulating real numbers exactly. For instance, our code provides a data-type (called a *box*) for encoding a point with n coordinates that are real algebraic numbers, together with a function for deciding whether this point cancels a given n-variate polynomial.

We investigate the impact of different strategies for isolating roots. In particular, we identify a family of examples where the use of normalized regular chains instead of arbitrary (but still zero-dimensional) regular chains can speedup the root isolation even though normalization tends to substantially increase coefficient sizes, as established in [9].

2 Real Root Isolation of a Zerodimensional Regular Chain

After recalling the Vincent-Collins-Akritas algorithm in Sect. 2.1 and introducing definitions in Sects. 2.2 and 2.3, the algorithm RealRootIsolate and its subalgorithms are presented in Sect. 2.4. In Sect. 2.5 we compare our method with other existing approaches.

2.1 The Vincent-Collins-Akritas Algorithm

The Vincent-Collins-Akritas algorithm isolates the real roots of a squarefree polynomial (with rational coefficients) with an arbitrary precision. A basic version (Algorithm 1) is recalled here, before its generalization in Sect. 2.4.

Definition 1 Let V be a finite set of t real numbers. An *interval decomposition* of V is a list $I_1, \ldots, I_t$ such that each I_i is an open rational interval $]a, b[$ or a rational singleton $\{a\}$, each I_i contains one element of V and $I_i \cap I_j = \emptyset$ if $i \neq j$.

In Algorithm 1, there are different ways to compute a strict bound H (in the sense that any root α of p satisfies $|\alpha| < H$). For example, if $p = \sum_{i=0}^{d} a_i x^i$, take the Cauchy bound $H = \frac{1}{|a_d|} \sum_{i=0}^{d} |a_i|$. Sharper bounds are given in [2].

The main arguments for the correctness of Algorithm 1 are the following. Algorithm 1 computes a polynomial $\bar{p}$ whose positive real roots are in bijection with the real roots of p which lie in $]a, b[$. The application of Descartes' rule of signs on $\bar{p}$ thus provides a bound on the number of real roots of p which lie in $]a, b[$. This bound is exact when equal to 0 or 1 [23, Theorem 1.2]. Since p is squarefree, the bound returned by Algorithm 3 will eventually become 0 or 1, by [23, Theorem 2.5] so that the whole method terminates.

[2] The notions of an equiprojectable variety and equiprojectable decomposition are discussed in [8].

Algorithm 1 RootIsolateVCA(p)

Input: p squarefree polynomial of $\mathbb{Q}[x]$
Output: an interval decomposition of the real roots of p
$H \leftarrow$ a strict bound on the roots of p
return RootIsolateAuxVCA($p,]-H, H[$)

Algorithm 2 RootIsolateAuxVCA($p,]a, b[$)

Input: p squarefree polynomial in $\mathbb{Q}[x]$ and $a < b$ rational
Output: an interval decomposition of the real roots of p which lie in $]a, b[$
$nsv \leftarrow$ BoundNumberRootsVCA($p,]a, b[$)
if $nsv = 0$ **then return** $\emptyset$
else if $nsv = 1$ **then return** $]a, b[$
else
  $m \leftarrow (a+b)/2 \quad res \leftarrow \emptyset$
  if $p(m) = 0$ **then** $res \leftarrow \{\{m\}\}$
  {Next line ensures the roots are sorted increasingly}
  return RootIsolateAuxVCA($p,]a, m[$) $\cup\ res \cup$ RootIsolateAuxVCA($p,]m, b[$)

Algorithm 3 BoundNumberRootsVCA($p,]a, b[$)

Input: $p \in \mathbb{Q}[x]$ and $a < b$ rational
Output: a bound on the number of roots of p in the interval $]a, b[$
$\bar{p} \leftarrow (x+1)^d\, p\left(\frac{a\,x+b}{x+1}\right)$ where d is the degree of p, and denote $\bar{p} = \sum_{i=0}^{d} a_i x^i$
$a'_e, \ldots, a'_0 \leftarrow$ the sequence obtained from $a_d, \ldots, a_0$ by removing zero coefficients
return the number of sign variations in the sequence $a'_e, \ldots, a'_0$

2.2 *Regular Chains*

In this paper one only considers zerodimensional squarefree regular chains, denoted *zs-rc*. Roughly speaking, a zerodimensional regular chain is a triangular set[3] of polynomials, with as many equations as variables, and which has a finite number of complex roots (and consequently a finite number of real roots).

Let $x_1 < \cdots < x_s$ be s variables, and denote $\mathbb{Q}_s = \mathbb{Q}[x_1, \ldots, x_s]$. Let $p \in \mathbb{Q}_s$ be a non-constant polynomial. We denote by $\text{mvar}(p)$ the *main variable* of p, by $\text{init}(p)$ the *initial* (or leading coefficient w.r.t. $\text{mvar}(p)$) of p, by $\text{mdeg}(p)$ the degree of p in its main variable and by $\text{sep}(p)$ the *separant* of p, that is, $\partial p/\partial \text{mvar}(p)$. If T is a set of polynomials in $\mathbb{Q}_s$, $\langle T \rangle$ denotes the ideal generated by T and $V(T)$ denotes the set of all complex solutions of the system $T = 0$. For a given x_i, $T_{\le x_i}$ (resp. $T_{>x_i}$) denotes the elements of T whose main variable is less (resp. strictly greater) than x_i.

Definition 2 Let $T = \{p_1, \ldots, p_s\}$ where each p_i lies in $\mathbb{Q}_s$. The set T is a *zerodimensional squarefree regular chain* (or *zs-rc*) of $\mathbb{Q}_s$ if $\text{mvar}(p_i) = x_i$ for $1 \le i \le s$, $\text{init}(p_i)$ does not vanish on $V(\{p_1, \ldots, p_{i-1}\})$ for any $2 \le i \le s$, and $\text{sep}(p_i)$ does not vanish on $V(\{p_1, \ldots, p_i\})$ for any $1 \le i \le s$.

[3] triangular set in the sense that each polynomial introduces exactly one more variable.

Thanks to the first two conditions, it is easy to show that the system $T = 0$ has a finite number of complex solutions (counted with multiplicity), which is equal to the product of the main degrees of the elements of T denoted $\mathrm{DEG}(T)$. The third condition, which forbids multiple roots, is the natural generalization of squarefree polynomials to regular chains. As for the algorithm RootIsolateVCA, this condition is only required to make the isolation algorithms terminate.

In practice, the zs-rc can be computed using the TriangularizeAlgorithm [21] available in the `RegularChains` library shipped with MAPLE. Moreover, the regular chains are not built by checking the conditions of Definition 2 but by using regularity tests of polynomials modulo ideals. A polynomial p is said to be *regular* modulo an ideal I if it is neither zero nor a zero-divisor modulo I. If T is a regular chain, p is said to be regular modulo T if p is regular modulo the ideal $\langle T \rangle$. Thus, the following definition is equivalent to Definition 2.

Definition 3 Let $T = \{p_1, \ldots, p_s\}$ where each p_i lies in $\mathbb{Q}_s$. The set T is a *zerodimensional squarefree regular chain* (or *zs-rc*) of $\mathbb{Q}_s$ if $\mathrm{mvar}(p_i) = x_i$ for any $1 \leq i \leq s$, $\mathrm{init}(p_i)$ is regular modulo the ideal $\langle p_1, \ldots, p_{i-1} \rangle$ for any $2 \leq i \leq s$, and $\mathrm{sep}(p_i)$ is regular modulo the ideal $\langle p_1, \ldots, p_i \rangle$ for any $1 \leq i \leq s$.

The next lemma makes the link between the regularity property of a polynomial q modulo a zs-rc and the fact that q does not vanish on the solutions of a zs-rc. It is implicitly used to check whether or not a polynomial vanishes on a root of a regular chain in the CheckZeroDivisor algorithm.

Lemma 1 *Let T be a zs-rc of $\mathbb{Q}_s$ and q a polynomial of $\mathbb{Q}_s$. Then q is regular modulo T iff q does not vanish on any complex solution of T.*

2.3 Boxes

This section defines the boxes used for isolating solutions of zs-rc, as well as extra definitions needed to specify the algorithms of Sect. 2.4.

Definition 4 An s-box (or box) B is a Cartesian product $B = I_1 \ldots I_s$ where each I_i is either a rational open interval $]a, b[$ (a and b are rational) or a singleton $\{a\}$ with a rational. The width of B, denoted by $|B|$, is defined as the maximum of the $|I_i|$ where $|I_i| = 0$ if it is a singleton and $b - a$ if $I_i =]a, b[$.

The algorithm EvalBox(p, B), where $p \in \mathbb{Q}_s$, and B is a s-box, returns an interval I such that $p(v) \in I$ for any $v \in B$. Different variants for EvalBox(p, B) exist. Any variant for EvalBox(p, B) satisfying the following property can be used: the box EvalBox(p, B) should tend to the singleton $\{p(x_0)\}$ when the width of B tends to zero (by keeping the condition $x_0 \in B$). This simply ensures that the interval EvalBox(p, B) should shrink as the width of the box B decreases.

Definition 5 Let $B = I_1 \dots I_s$ be an s-box and $T = \{p_1, \dots, p_s\}$ be a zs-rc of $\mathbb{Q}_s$. We say (B, T) satisfies the Dichotomy Condition (or **DC**) if

- one and only one real root of T lies in B
- if $I_1 =]a, b[$, $p_1(x_1 = a)$ and $p_1(x_1 = b)$ are nonzero and have opposite signs
- for each $2 \le k \le s$, if $I_k =]a, b[$ then the two intervals $\mathsf{EvalBox}(p_k(x_k = a), B)$ and $\mathsf{EvalBox}(p_k(x_k = b), B)$ do not meet 0 and have opposite signs.[4]

This last condition is the natural generalization of the condition $p(a)$ and $p(b)$ are nonzero and have opposite sign, and p vanishes only once on the interval $]a, b[$ in the univariate case. Condition **DC** allows to refine a box very much like one refines the interval $]a, b[$ by dichotomy.

Definition 6 Let V be a finite set of t points of $\mathbb{R}^s$. A list $B_1, \dots, B_t$ of s-boxes is called a *box-decomposition* of V if each point of V lies in exactly one B_l and $B_i \cap B_j = \emptyset$ if $i \neq j$. If T is a zs-rc, we call box-decomposition of T a box-decomposition of the real roots of $T = 0$.

Definition 7 A task $\mathcal{M} = \text{TASK}(p,]a, b[, B, T)$ is defined as: T is a zs-rc of $\mathbb{Q}_s$, $p \in \mathbb{Q}_{s+1}$, $T \cup \{p\}$ is a zs-rc, B is an s-box, (B, T) satisfies **DC**, and $a < b$ are rational numbers. The solutions of $\mathcal{M}$ denoted by $V_t(\mathcal{M})$ are defined as $V(T \cup \{p\}) \cap (B \times]a, b[)$ (i.e. the real solutions of $T \cup \{p\}$ which prolong the real root in B and whose component x_{s+1} lies in $]a, b[$).

2.4 Algorithms

The main algorithm **RealRootIsolate**, which isolates the real roots of a zerodimensional squarefree regular chain, is presented here. Only elements of proofs are given. However, specifications are stated with details. One assumes $n > 1$.

The algorithms presented here use the mechanism of exceptions which is available in a lot of programming languages. We find it appropriate since doing computations using the D5 principle [10] can be seen as doing computations as if the coefficient ring were a field. When a zero divisor is hit (leading to a splitting), one raises an exception exhibiting the splitting. This exception can then be caught to restart computations. This shortens and makes clearer[5] the algorithms presented here. Only Algorithm 4 throws exceptions.

Algorithm 4 checks whether p is regular modulo T or not. If p is regular modulo T, the algorithm returns normally, otherwise an exception is raised. Algorithm 4 is called whenever one needs to know whether a polynomial vanishes or not, on a real root x^0 of T isolated by a box B. Indeed, if p is regular modulo T, thanks to Lemma 1, p does not vanish on x^0. This allows to refine B until $\mathsf{EvalBox}(p, B)$ does not contain 0, which gives the sign of $p(x^0)$.

[4] the sign of an interval not meeting zero is just the sign of any element of it.

[5] otherwise, splittings need to be handled each time a function returns a value.

Algorithm 4 CheckZeroDivisor(p, T)

Input: T a zs-rc $\mathbb{Q}_s$ and $p \in \mathbb{Q}_s$

Output: If p is regular modulo T, then the algorithm terminates normally. Otherwise, an exception is thrown exhibiting t zs-rc $T_1, \ldots, T_t$ such that **C1** $V(T_1) \cup \cdots \cup V(T_t) = V(T)$, and **C2** $\sum_{i=1}^{t} \mathrm{DEG}(T_i) = \mathrm{DEG}(T)$ hold.

$T_1, \ldots, T_t \leftarrow$ Regularize(p, T)

if p belongs to at least one $\langle T_i \rangle$ **then throw exception**($T_1, \ldots, T_t$)

Algorithm 5 RefineBox(B, T)

Input: T is a zs-rc of $\mathbb{Q}_s$, (B, T) satisfies **DC** and $|B| > 0$

Output: an s-box $\bar{B}$ such that $|\bar{B}| \leq |B|/2$, $\bar{B} \subset B$ and $(\bar{B}, T)$ satisfies the **DC**

Algorithm 6 RealRootIsolate(T)

Input: T is a zs-rc

Output: a box-decomposition $B_1, \ldots, B_p$ of T

1: $I_1, \ldots, I_t \leftarrow$ RootIsolateVCA(T_{x_1})
2: $toDo \leftarrow \{(T_{>x_1}, (I_i, T_{\leq x_1}))\}_{1 \leq i \leq t}$
3: $res \leftarrow \emptyset$
4: **while** $toDo \neq \emptyset$ **do**
5: pick and remove a $(T_{>x_i}, (B, T_{\leq x_i}))$ from $toDo$
6: $B'_1, \ldots, B'_{t'} \leftarrow$ SolveNewVar($T_{x_{i+1}}, B, T_{\leq x_i}$)
7: **if** $x_{i+1} = x_n$ **then** $res \Leftarrow res \cup \{B'_1, \ldots, B'_{t'}\}$
8: **else** $toDo \leftarrow toDo \cup \{(T_{<x_{i+1}}, (B'_j, T_{\geq x_{i+1}}))\}_{1 \leq j \leq t'}$
9: **return** res

The algorithm Regularizeis not recalled here (see [21] for details) but its specification is: if T is a zs-rc, Regularize(p, T) returns a list of zs-rc $T_1, \ldots, T_t$ such that for each T_i, p either belongs to $\langle T_i \rangle$ or is regular modulo T_i. Moreover $T_1, \ldots, T_t$ is (what we call) a splitting of T, which in dimension 0 satisfies the two conditions **C1** and **C2** of the output of Algorithm 4. Due to condition **C2**, splittings cannot occur indefinitely.

Algorithm 5 is able to refine a box containing a real root by dividing its width by 2. It is simply the generalization of the dichotomy process for splitting into two an isolating interval. The algorithm is not detailed here for brevity. The main idea is to divide by two each interval I_i of $B = I_1 \times \cdots \times I_s$ which is larger than $|B|/2$ while keeping the **DC** condition.

Algorithm 6 is a generalization of Algorithm 1 for a zs-rc. Line 1 isolates the real roots of the univariate polynomial T_{x_1}. The variable $toDo$ is a set of $(T_{>x_i}, (B, T_{\leq x_i}))$ such that each $(B, T_{\leq x_i})$ satisfies **DC**. It means that $(B, T_{\leq x_i})$ represents one (and only one) real root of $T_{\leq x_i}$. The set $T_{>x_i}$ simply is the set of polynomials which have not be solved yet. Algorithm 6 calls Algorithm 7 (which allows to solve one new variable) until all variables are solved. Note that Algorithm 6 could be followed by a refinement of each returned box so that the width of each box is smaller than a given precision.

Algorithm 7 SolveNewVar(p, B, T)

Input: T is a zs-rc of $\mathbb{Q}_s$, $p \in \mathbb{Q}_{s+1}$, $T \cup \{p\}$ is a regular chain, (B, T) satisfies **DC**
Output: a box-decomposition of the roots $(x_1^0, \ldots, x_{s+1}^0)$ of $T \cup \{p\}$ such that $(x_1^0, \ldots, x_s^0)$ is the root of T which lies in B

```
1: refine B into a box B' such that 0 ∉ EvalBox(i_p, B')
2: compute a bound H on the roots of p(x_1^0, ..., x_s^0, x_{s+1})
3: toDo ← {TASK(p, ]−H, H[, B', T)}
4: res ← ∅
5: while toDo ≠ ∅ do
6:   pick and remove a task M from toDo
7:   for all e in SolveTask(M) do
8:     if e is a box then res ← res ∪ {e} else toDo ← toDo ∪ {e}
9: return res
```

Algorithm 8 SolveTask$(\mathcal{M})$

Input: a task $\mathcal{M} = \text{TASK}(p,]a, b[, B, T)$ where T is a zs-rc of $\mathbb{Q}_s$
Output: One of the four following cases:
1: $\emptyset$ which means $V_t(\mathcal{M}) = \emptyset$.
2: a box B' such that $(B', T \cup \{p\})$ satisfies **DC** and B' is a box-decomposition of $V_t(\mathcal{M})$, which means $V_t(\mathcal{M})$ is composed of only one point **3**: two tasks $\mathcal{M}_1$ and $\mathcal{M}_2$ such that $V_t(\mathcal{M}_1)$ and $V_t(\mathcal{M}_2)$ forms a partition of $V_t(\mathcal{M})$ **4**: two tasks $\mathcal{M}_1$ and $\mathcal{M}_2$ plus a box B' such that $(B', T \cup \{p\})$ satisfies **DC** and the three sets $V_t(\mathcal{M}_1)$, $V_t(\mathcal{M}_2)$ and $\{x^0\}$ form a partition of $V_t(\mathcal{M})$, where x^0 denotes the only real root of $T \cup \{p\}$ which lies in B'.

```
nsv, B' ← BoundNumberRoots(M)
if nsv = 0 then return ∅
else if nsv = 1 then
  B'' ← B'×]a, b[
  refine B'' until (B'', T ∪ {p}) satisfies DC
  return {B''}
else
  m ← (a + b)/2   res ← ∅   p' ← p(x_{s+1} = m)
  if p' ∈ ⟨T⟩ then res ← {B' × {m}} else CheckZeroDivisor(p', T)
  return res ∪ {TASK(p, ]a, m[, B', T), TASK(p, ]m, b[, B', T)}
```

Also remark that any raised exception will hit Algorithm 6 since none of the algorithms presented here catches any exception. It is however easy to adjust Algorithm 6 so that it would catch exceptions and recall itself on each regular chain returned by the splitting. The recursion would eventually stop because of condition **C2** of Algorithm 4 (i.e. splittings cannot occur indefinitely).

Algorithm 7 finds the real roots of p (seen as univariate in x_{s+1}) that "prolong" the real root which lies in B. Line 1 always terminates, since i_p is regular modulo T, so it does not vanish on any root of T. The bound H at line 2 can be computed in the following way. Denote $p = \sum_{i=0}^{d} a_i x_{s+1}^i$ and $A_i = \text{EvalBox}(a_i, B')$. Then take $H = \frac{1}{\min |A_d|} \sum_{i=0}^{d} (\max |A_i|)$ where $\min |A_i|$ (resp. $\max |A_i|$) denotes the minimum (resp. maximum) of the modulus of the bounds of the interval A_i. The rest of the algorithm is based on Algorithm 8 which transforms tasks into new tasks and boxes.

Algorithm 9 BoundNumberRoots($\mathcal{M}$)

Input: a task $\mathcal{M} = \text{TASK}(p,]a, b[, B, T)$ where T is a zs-rc of $\mathbb{Q}_s$

Output: (nsv, B') such that $B' \subset B$, (B', T) satisfies **DC**, and nsv is a bound on the cardinal of $V_t(\mathcal{M})$. The bound is exact if $nsv = 0$ or 1.

1: $\bar{p} \leftarrow (x_{s+1} + 1)^d \, p\left(x_{s+1} = \frac{a\,x_{s+1}+b}{x_{s+1}+1}\right)$ with $d = \text{mdeg}(p)$

2: denote $\bar{p} = \sum_{i=0}^{d} a_i x_{s+1}^i$

3: $a'_e, \ldots, a'_0 \leftarrow$ the sequence obtained from $a_d, \ldots, a_0$ by removing the a_i belonging to $\langle T \rangle$

4: **for all** a'_i **do** CheckZeroDivisor(a'_i, T)

5: $B' \leftarrow B$

6: **while** there is an a'_i such that $0 \in$ EvalBox(a'_i, B') **do** $B' =$ RefineBox(B', T)

7: **return** the number of sign variations of the sequence
EvalBox(a'_e, B'), EvalBox(a'_{e-1}, B'), ..., EvalBox(a'_0, B')

Algorithm 8 is a generalization of Algorithm 2. The cases $nsv = 0$ or 1 are straightforward. When $nsv > 1$, one needs to split the interval $]a, b[$ into two, yielding two tasks returned on line 10. Lines 8–9 correspond to the lines 5–6 of Algorithm 2. Indeed, checking $p(m) = 0$ is transformed into checking if p' lies in $\langle T \rangle$ or is not a zero divisor modulo T.

Algorithm 9 is a generalization of Algorithm 3. One discards the coefficients of p' which lie in $\langle T \rangle$ because they vanish on the real root v which is in B. One also ensures that the other coefficients (the a'_i) are not zero divisors, so they cannot vanish on v. Thus the loop at line 6 terminates. Moreover, this guarantees that the number of sign variations is correct. Please note that the sequence $a'_e, \ldots, a'_0$ is never empty. Indeed if all a_i's were in $\langle T \rangle$, then all coefficients of p would lie in $\langle T \rangle$ (impossible since i_p is regular modulo T).

2.5 Comparison with Other Methods

In the introduction we provided a comparison of our work with others. More technical details are reported below.

References [24, 25] give algorithmic methods (available in AXIOM) to manipulate real algebraic numbers. These developments were designed for improving *Cylindrical Algebraic Decomposition* (CAD) methods in AXIOM. Although [24] contains all the tools to solve our problem, this paper focuses on the problem of manipulating real algebraic numbers. It does not address directly the problem of isolating the real roots of a given zerodimensional regular chain [25] provides tools to perform univariate polynomial real root isolation by using *quasi Sylvester sequence* which according to [25] can be faster than the techniques based on the Descartes rules.

References [7, 14] present algorithms for isolating real roots of univariate polynomials with algebraic coefficients. Their algorithms require the ideal to be *prime*, and this condition is ensured by performing univariate factorization [20] into irreducible factors for polynomials with algebraic coefficients. Our method does not require such factorizations and only requires the ideal to be squarefree. Thus, our method replaces a decomposition into prime ideals by regularity tests which are often less costly. Please note that the regularity tests we perform are in fact replaced by interval arithmetics, as explained in the paragraph CheckZeroDivisorof Sect. 3.1.

Reference [26] is based on Gröbner basis computations and rational univariate representation. Thus, [26] transforms the initial problem into the problem of isolating the real roots of a univariate polynomial with rational number coefficients

Reference [19] starts from a zerodimensional regular chain (although [19] uses the terminology of characteristic sets) and proceeds variable by variable. Their technique is different from ours. After isolating a real root say x_1^0 for $p_1(x_1) = 0$, they build two univariate polynomials $\overline{p}_2(x_2)$ and $\underline{p}_2(x_2)$ whose real roots will interleave nicely (see [19, Definition 2]) when the precision on x_1^0 is sufficiently low, yielding isolation intervals for the variable x_2 [33] improves techniques of [19] by avoiding restarting the isolation from the beginning when decreasing the precision. Such techniques are also used in [5], where the authors consider general zerodimensional triangular systems (which may not be a regular chain) and treat multiple zeros directly.

Quoting the abstract of [22], the authors use *a powerful reduction strategy based on univariate root finder using Bernstein basis representation and Descartes' rule.* Basically, they reduce the problem to solving univariate polynomials by using the Bernstein basis representation and optimizations based on convex hulls.

3 Implementation

3.1 The SemiAlgebraicSetTools Package

The algorithm RealRootIsolate has been coded using exceptions in MAPLE in the module `SemiAlgebraicSetTools` of the `RegularChains` library [16]. We present some implementation issues and optimizations integrated in our code.

Precision. The user can specify a positive precision so all isolation boxes have a width smaller than the given precision. If an infinite precision is provided, then the algorithm only isolates the real roots by refining the boxes the least possible. We take the precision into account as soon as possible in the algorithm, meaning that the box is refined each time an box is extended with a new variable.

Constraints. The user can restrict the solutions by imposing that some variables lie in a prescribed interval. If the intervals are restrictive (i.e., smaller than the intervals computed using bounds), many useless branches are cut.

The CheckZeroDivisor algorithm is not directly called in our code. Indeed, regularity test can be very expensive and should be avoided as much as possible. When a call CheckZeroDivisor(p, T) returns, one knows that a box B isolating a real root of T can always be refined until the interval EvalBox(p, B) does not meet zero. This is in fact the only reason why we call CheckZeroDivisor. In order to avoid a regularity test, we first try to refine B a few times to see if EvalBox(p, B) still meets zero. If it does not, we do not need to check the regularity.

Refining boxes. In the MAPLE implementation, Algorithm 5 receives an extra parameter x_k. In that case, the box is only refined for the variables smaller than x_k (i.e. the variables x_i with $i \leq k$). This is useful at line 6 of Algorithm BoundNumberRoots. Indeed, if $\mathrm{mvar}(a_i') = x_k$ holds, then it is not necessary to refine the complete box B' to ensure that EvalBox(a_i', B') does not meet 0.

Change of variables. By modifying Algorithms 7 and 8, we call Algorithm 9 with $a = 0$ and $b = 1$. This replaces $(x_{s+1}+1)^d\, p\left(x_{s+1} = \frac{a\,x_{s+1}+b}{x_{s+1}+1}\right)$ by several substitutions $p(x_{s+1} = x_{s+1}/2)$, $p(x_{s+1} = 1/x_{s+1})$ and $p(x_{s+1} = x_{s+1}+1)$ which can be written efficiently, the last one using fast Taylor shift [31].

Refining other branches. Due to the triangular structure of the system, many roots share a common part (i.e., values for some variables are equal). When refining a root, we refine the roots sharing a common part to save computations.

Further refining. After being computed, an isolation box of a real root v can be refined further using the MAPLE command RefineBox. To do so, exceptions have to be caught. Our implementation associates a regular chain T to each box B encoding a real root. Thus, if T is split into $T_1, \ldots, T_s$, one replaces (B, T) by the right (B, T_i) which also defines the real root v as done in [25, p. 528].

EvalPoly. For evaluating EvalBox(p, B), we apply a Hörner scheme to p. For example, the polynomial $p := x_2^3x_1 + 3x_2^2 + x_2x_1^2 + x_1^2 + x_1 + 1$ is rearranged as $1 + (1 + x_1)x_1 + (x_1^2 + (3 + x_1x_2)x_2)x_2$. Assuming $x_2 > x_1$, the interval of B for the variable x_2 tends to be in practice wider than that for the variable x_1, since the intervals for smaller variables tend to be more refined than those for higher variables. On the example, the Hörner form decreases the exponents of x_2.

3.2 Further Development

Using fast polynomial arithmetic and modular methods. The current implementation of the CheckZeroDivisor algorithm can be improved in a significant manner. Indeed, the modular algorithm for regularity test of [17] and implemented with the MODPN library [18] outperform the regularity test used in CheckZeroDivisor by several orders of magnitude.

Computing with algebraic numbers. Using the two algorithms RefineBox and CheckZeroDivisor, one can encode algebraic numbers and check if a multivariate polynomial cancels on some algebraic numbers. This allows computing with algebraic numbers, very much as in [24]. Moreover, inequations and inequalities could be included with almost no work. Indeed they can be handled at the end of RealRootIsolate using CheckZeroDivisor. They can also be treated inside the sub-algorithms as soon as a box in construction involves all the variables of an inequality or inequation, allowing to cut some branches.

Floating-point computations. As suggested by Fabrice Rouillier (private communication), it would speedup the algorithm to use multiple-precision floating-point computations with exact rounding (as in the MPFI library [27]).

Exceptions could be caught sooner to avoid losing already done computations.

Using continuous fractions as in [2, 3] may also be investigated.

Interval arithmetics. The algorithm EvalBox could certainly benefit from techniques for "optimizing" polynomial expressions, as in [15].

Newton's method. Some attempts were made to incorporate a Newton method for system of polynomials in the RefineBox algorithm. Due to the triangular form of the system, the Jacobian is also triangular which eases the method. However, although the convergence was really faster, it was not satisfactory because of the coefficient swell in the isolation intervals. However, we believe that Newton's method should be investigated further.

4 Benchmarks

4.1 Description of the Experimentation

The names of the examples used for benchmarking are listed in Fig. 1. Most of them are classical. The lhlp files tests are taken from [19]. The examples *chemical-reaction*, *geometric-constraints*, *neural-network*, *p3p-special* and *Takeuchi-Lu* appear in [32]. The nld-*d*-*n* and nql-*n*-*d* examples are described in Sect. 4.3. The set of all the examples can be found at www.lifl.fr/~lemaire/BCLM09/BCLM09-systems.txt. Benchmark results are given in Fig. 1. They were run on an AMD Phenom II X4 (4 Gb of mem.) using MAPLE 14 64 bits and our latest development version of the `RegularChains` library. Timings are in seconds. Timeouts are indicated with >. The column *Sys* denotes the name of the system. The column *v/e/s* stands for the number of variables/equations/real solutions.

The MAPLE command RootFinding[Isolate] isolates real roots within the times indicated in the group of columns RF/Is. For multivariate systems, this command relies on Gröbner basis computations [13] and rational univariate representation [26]. In Column 1, the command used is RootFinding [Isolate](sys, variables, digits=10, output=interval). For Column 2 the same command is used with the variable ordering reversed, in case the variable ordering is important. Note that the option digits = 10 ensures that the first ten digits of the results are correct which is not the same as guaranteeing a width less than `1e-10` for the isolation boxes in RealRootIsolate. However, the difficulty in isolating the real roots is comparable since the roots are neither close to zero nor too big.

The other groups of columns correspond to three strategies for isolating real roots using our algorithm RealRootIsolate. In each strategy, the initial system is first decomposed into zs-rc using the Triangularize command together with the option

		RF/Is		Strategy 1		Strategy 2		Strategy 3			
Sys	v/e/s	1	2	Tr	Is/10	Tr/No	Is/10	Tr	Is/∞	∞/5	5/10
4-body-homog	3/3/7	0.16	0.17	0.58	4.1	1.9	4.6	0.58	1.5	1.5	1.5
5-body-homog	3/3/11	0.19	0.2	0.83	11	10	16	0.81	3.5	3.9	3.7
Arnborg-Lazard-rev	3/3/8	<0.1	<0.1	0.35	2.9	0.42	2.8	0.34	0.88	1.1	1
Arnborg-Lazard	3/3/8	<0.1	<0.1	0.36	3	0.42	2.7	0.35	0.91	1.3	1.1
Barry	3/3/2	<0.1	<0.1	0.2	0.64	0.22	1.6	0.19	0.19	0.26	0.2
Caprasse-Li	4/4/18	<0.1	<0.1	0.61	1.4	0.77	0.9	0.63	0.35	0.62	0.46
Caprasse	4/4/18	<0.1	<0.1	0.65	1.4	0.8	0.9	0.62	0.37	0.65	0.5
chemical-reaction	4/4/4	<0.1	<0.1	0.23	1.3	0.26	1	0.23	0.29	0.71	0.45
circles	2/2/22	0.6	0.57	0.32	10	0.41	10	0.32	6.4	1.8	1.6
cyclic-5	5/5/10	0.22	0.22	1.1	1.9	1.6	0.77	1.1	0.4	1.5	0.69
Czapor-Geddes-Wang	5/5/2	<0.1	<0.1	0.98	2.5	5.4	3.3	0.99	1.1	1	0.84
fabfaux	3/3/3	<0.1	<0.1	0.65	2.9	14	3.4	0.66	0.89	1.3	1.3
geometric-constraints	3/3/8	<0.1	<0.1	0.16	0.91	0.18	0.92	0.17	0.23	0.41	0.35
GonzalezGonzalez	3/3/2	<0.1	<0.1	0.21	0.45	0.26	0.42	0.22	0.19	0.16	0.14
Katsura-4	5/5/12	<0.1	<0.1	0.41	5.7	0.51	7.8	0.41	1.4	2.6	2.4
lhlp1	3/3/6	<0.1	<0.1	0.19	0.54	0.21	0.76	0.19	0.19	0.2	0.17
lhlp2	3/3/2	<0.1	<0.1	0.2	0.46	0.24	0.65	0.19	0.18	0.19	0.14
lhlp3	3/3/2	<0.1	<0.1	0.15	0.37	0.18	0.4	0.16	0.14	0.15	<0.1
lhlp4	2/2/4	<0.1	<0.1	0.18	1.3	0.22	2.1	0.18	0.3	0.63	0.42
lhlp5	3/3/4	<0.1	<0.1	0.26	0.97	0.3	1.1	0.27	0.27	0.39	0.34
lhlp6	4/4/4	<0.1	<0.1	0.29	1.1	0.33	0.84	0.28	0.26	0.59	0.34
neural-network	4/4/22	0.57	0.57	0.44	7	0.67	6.1	0.42	1.7	3	2.7
nld-3-4	4/4/27	0.72	0.73	1.1	2.9	1.6	1.9	1.2	0.6	1.5	1.3
nld-3-5	5/5/111	47	47	9.1	23	12	14	8.9	4	12	10
nld-4-5	5/5/?	>2000	>2000	>2000	?	>2000	?	>2000	?	?	?
nld-7-3	3/3/7	58	58	1.7	3.8	2.9	3.1	1.6	4.6	<0.1	<0.1
nld-8-3	3/3/8	275	275	2.1	9.9	11	7	2.1	8.7	1.3	1.2
nld-9-3	3/3/7	1078	1083	7.9	14	32	27	7.9	16	<0.1	<0.1
nld-10-3	3/3/8	>2000	>2000	10	45	341	118	10	44	2.4	2.3
nql-5-4	5/5/2	66	62	0.2	0.59	0.22	0.57	0.2	0.26	0.11	0.14
nql-10-2	10/10/2	144	132	0.25	1.4	0.29	1.4	0.25	0.49	0.31	0.44
nql-10-4	10/10/2	>2000	>2000	0.32	1.4	0.38	1.4	0.31	0.53	0.27	0.37
nql-15-2	15/15/2	>2000	>2000	0.39	2.6	0.45	2.5	0.38	1.5	0.52	0.81
p3p-special	5/5/24	0.22	0.27	0.31	9	0.5	13	0.31	2.8	3.2	3.4
PlateForme2d-easy	6/6/0	<0.1	<0.1	0.57	0.16	0.76	0.17	0.56	0.14	<0.1	<0.1
r-5	5/5/1	1.1	1.1	0.32	0.13	0.36	0.13	0.31	0.12	<0.1	<0.1
r-6	6/6/1	>2000	>2000	0.48	0.11	0.56	0.1	0.48	<0.1	<0.1	<0.1
Rose	3/3/18	0.34	0.37	0.46	15	0.54	22	0.45	2.2	8.5	7.3
simple-nql-20-30	20/20/2	>2000	>2000	0.57	12	0.65	12	0.55	28	1.2	0.13
Takeuchi-Lu	4/4/14	<0.1	<0.1	0.27	3	0.31	3.9	0.27	0.46	1.9	1.2
Trinks-2	6/7/0	<0.1	<0.1	0.18	<0.1	0.19	<0.1	0.18	<0.1	<0.1	<0.1
Trinks-difficult	6/6/2	<0.1	<0.1	0.24	1.2	0.29	1.8	0.24	0.25	0.64	0.53
wilkinson20	1/1/21	<0.1	<0.1	0.11	0.49	0.13	0.5	0.11	0.13	0.21	0.17
wilkinsonxy	2/2/25	<0.1	<0.1	0.17	3.2	0.19	3.2	0.17	1.2	1	1

Fig. 1 Benchmark

radical='yes' ensuring these regular chains are squarefree. In order to keep things simple and uniform, the option probability=xx of Triangularize is not used. Thus the modular algorithm of [8] is not applied even though it can solve all our examples that the non-modular version cannot.

Strategy 1. We build regular chains (column Tr) and call the RealRootIsolate algorithm (column Is/10) on each regular chain with a precision of `1e-10`.

Strategy 2. A variant of Strategy 1 where we compute strongly normalized regular chains (column Tr/No) using the option normalized='strongly' of Triangularize.

Strategy 3. Another variant of Strategy 1. We build regular chains (column Tr) and call the RealRootIsolatealgorithm on each regular chain with an infinite precision (column Is/∞), in the sense that the width of the boxes are not constrained. Thus, only the isolation is performed. Then we call the command RefineListBox to refine the list of boxes with a precision of `1e-5` (column ∞/5). Then we refine again the boxes for a precision of `1e-10` (column 5/10).

4.2 Comparison of Different Strategies

Strategies 1 and 2 are comparable. Strongly normalized regular chains take more time to be computed, since normalization is a post-process for the command Triangularize. The isolation time is roughly the same in general for both types of regular chains. For the nld-d-n (except nld-9-3) family of examples, normalization helps the isolation process. However, for some other examples, such as 5-body-homog, p3p-special and Rose, normalization makes things worse.

Comparing Strategies 1 and 3 shows two things. First, it is usually faster to isolate solutions with an infinite precision rather than with a small precision. Second, it shows that the overall times for Strategies 1 and 3 are comparable.

4.3 Comparison with RootFinding

The RootFinding[Isolate] is obviously a lot faster on many examples. One should keep in mind that this command calls internal routines written in C that have been developed intensively for years. However, the RootFinding[Isolate] has difficulties on some systems such as the nql-n-d and nld-d-n ones.

The nql-n-d (for non quasi linear) example is specific and was suggested by Fabrice Rouillier. It is defined by n equations in n variables $x_1^d - 2 = 0, x_i^d + x_i^{d/2} - x_{i-1} = 0$ for $2 \leq i \leq n$ for some even degree d. This system is already a zs-rc. The algorithm RealRootIsolate solves it easily since the degrees are distributed evenly among the equations. On the other hand, the RootFinding[Isolate] needs to build a rational univariate representation which we believe has a very large degree roughly equal to d^n (i.e., about one million when $d = 4$ and $n = 10$).

A similar example is simple-nql-n-d defined by $x_1^d - 2 = 0, x_i^d - x_{i-1} = 0$ for $2 \leq i \leq n$. The degree of the rational univariate representation is also roughly d^n. For the example simple-nql-20-30, d^n is around 10^{29}.

The second family of systems which causes difficulties to RootFinding[Isolate] are the nld-d-n (for non-leading linear) defined by n equations of the form $x_1 + \cdots + x_{i-1} + x_i^d + x_{i+1} + \cdots + x_n - 1 = 0$ for $1 \leq i \leq n$. On these systems the computations performed by Triangularize tend to split into many branches, even though the equiprojectable decomposition consists of a few components (generally 2). For System nld-9-3, the command Triangularize (used without normalization option) produces 15 components where the largest coefficient has size 20 digits. The command EquiprojectableDecomposition (which requires the use of normalized regular chains) produces 3 components for nld-9-3, where most coefficients have more than 500 digits. Since nld-9-3 has 729 complex solutions, this suggests that the univariate polynomial in the rational univariate representation has degree 729 and coefficients with size at least 500 digits. This makes it difficult to isolate the real roots of such polynomial. Therefore, the nld-d-n examples show that splitting can help in solving some problems.

5 Conclusion

We presented a generalization of the Vincent-Collins-Akritas Algorithm for zerodimensional squarefree regular chains, and its implementation in MAPLE. Each box isolating a root can be refined arbitrarily after being computed. This allows manipulating algebraic numbers (encoded by a isolation box and a regular chain) very much like in [24]. Many improvements in the algorithm RealRootIsolate are possible and should be investigated. Among them, we believe that writing a C library to perform the isolation would improve a lot the timings. Yet for some non-equiprojectable varieties, our algorithm and its MAPLE implementation show favorable performances.

References

1. Akritas, A.G.: There is no Uspensky's method. In: proceedings of ISSAC (1986)
2. Akritas, A.G., Strzeboński, A.W., Vigklas, P.S.: Implementations of a new theorem for computing bounds for positive roots of polynomials. Comput. **78**(4), 355–367 (2006)
3. Akritas, A.G., Vigklas, P.S.: A comparison of various methods for computing bounds for positive roots of polynomials. J. Univ. Comput. Sci. **13**(4), 455–467 (2007)
4. Becker, E., Mora, T., Marinari, M.G., Traverso, C.: The shape of the shape lemma. In: Proceedings of the International Symposium on Symbolic and Algebraic Computation, pp. 129–133. ACM Press, New York, USA (1994)
5. Cheng, J.S., Gao, X.S., Yap, C.K.: Complete numerical isolation of real roots in zero-dimensional triangular systems. Journal of Symbolic Computation **44**(7), 768–785 (2009). http://www.sciencedirect.com/science/article/B6WM7-4TKPVBT-1/2/34927cd53d447d69e3b126d3fa93006a
6. Collins, G.E., Akritas, A.G.: Polynomial real root isolation using Descartes'rule of signs. In: proceedings of ISSAC'76, pp. 272–275. Yorktown Heights NY (1976)
7. Collins, G.E., Johnson, J.R., Krandick, W.: Interval arithmetic in cylindrical algebraic decomposition. J. Symb. Comput. **34**(2), 145–157 (2002)
8. Dahan, X., Moreno Maza, M., Schost, É., Wu, W., Xie, Y.: Lifting techniques for triangular decompositions. In: ISSAC'05, pp. 108–115. ACM Press (2005)
9. Dahan, X., Schost, É.: Sharp estimates for triangular sets. In: ISSAC 04, pp. 103–110. ACM (2004)
10. Della Dora, J., Dicrescenzo, C., Duval, D.: About a new method for computing in algebraic number fields. In: Proceedings of EUROCAL 85 Vol. 2, Lect. Notes in Comp. Sci. vol. 204, pp. 289–290. Springer (1985)
11. Descartes, R.: Géométrie (1636)
12. Eigenwillig, A., Kettner, L., Krandick, W., Mehlhorn, K., Schmitt, S., Wolpert, N.: A descartes algorithm for polynomials with bit-stream coefficients. In: CASC, vo. 3718 of LNCS, pp. 138–149. Springer (2005)
13. Faugère, J.C.: A new efficient algorithm to compute Gröbner bases (F_4). J. Pure Appl. Algebra **139**, 61–88 (1999)
14. Johnson, J.R., Krandick, W.: Polynomial real root isolation using approximate arithmetic. In: Proceedings of the 1997 International Symposium on Symbolic and Algebraic Computation, pp. 225–232. ACM (1997)
15. Leiserson, C.E., Li, L., Moreno Maza, M., Xie, Y.: Efficient evaluation of large polynomials. In: Proceedings of International Congress of Mathematical Software—ICMS 2010. Springer (2010)

16. Lemaire, F., Moreno Maza, M., Xie, Y.: The RegularChains library. In: Ilias S. Kotsireas (ed.) Maple Conference 2005, pp. 355–368 (2005)
17. Li, X., Moreno Maza, M., Pan, W.: Computations modulo regular chains. In: proceedings of ISSAC'09, pp. 239–246. ACM Press, New York, USA (2009)
18. Li, X., Moreno Maza, M., Rasheed, R., Schost, É.: The modpn library: Bringing fast polynomial arithmetic into MAPLE. In: MICA'08 (2008)
19. Lu, Z., He, B., Luo, Y., Pan, L.: An algorithm of real root isolation for polynomial systems. In: Wang, D., Zhi, L. (eds.) Proceedings of Symbolic Numeric Computation 2005, pp. 94–107 (2005)
20. McCallum, S.: An improved projection operation for cylindrical algebraic decomposition. In: Caviness, B.F., Johnson, J.R. (eds.) Quantifier Elimination and Cylindrical Algebraic Decomposition, pp. 242–268. Springer (1988)
21. Moreno Maza, M.: On triangular decompositions of algebraic varieties. Tech. Rep. TR 4, 99, NAG Ltd, Oxford, UK, : presented at the MEGA-2000 Conference. Bath, England (1999)
22. Mourrain, B., Pavone, J.P.: Subdivision methods for solving polynomial equations. Tech. rep., INRIA (Aug 2005), (number RR-5658)
23. Mourrain, B., Rouillier, F., Roy, M.F.: The bernstein basis and real root isolation. Comb. Comput. Geom., MSRI Publ. **52**, 459–478 (2005)
24. Rioboo, R.: Real algebraic closure of an ordered field, implementation in axiom. In: Proceedings of ISSAC'92, pp. 206–215. ISSAC, ACM Press (1992)
25. Rioboo, R.: Towards faster real algebraic numbers. J. Symb. Comput. **36**(3–4), 513–533 (2003)
26. Rouillier, F.: Solving zero-dimensional systems through the rational univariate representation. AAECC **9**, 433–461 (1999)
27. Rouillier, F., Revol, N.: The multiple precision floating-point interval library (MPFI) library. http://gforge.inria.fr/projects/mpfi/
28. Rouillier, F., Zimmermann, P.: Efficient isolation of polynomial real roots. J. Comput. Appl. Math. **162**(1), 33–50 (2003)
29. Serret, J.A.: Cours d'Algèbre Supérieure, 4th edn. Gauthier-Villars, Paris (1877)
30. Uspensky, J.V.: Theory of Equations. McGraw-Hill Co., New York (1948)
31. von zur Gathen, J., Gerhard, J.: Fast algorithms for taylor shifts and certain difference equations. In: ISSAC, pp. 40–47 (1997).
32. Xia, B., Yang, L.: An algorithm for isolating the real solutions of semi-algebraic systems. J. Symb. Comput. **34**(5), 461–477 (2002)
33. Xia, B., Zhang, T.: Real solution isolation using interval arithmetic. Comput. Math. Appl. **52**(6–7), 853–860 (2006)

A Practical Implementation of a Modular Algorithm for Ore Polynomial Matrices

Howard Cheng and George Labahn

Abstract We briefly review a modular algorithm to perform row reduction of a matrix of Ore polynomials with coefficients in $\mathbb{Z}[t]$, and describe a practical implementation in Maple that improves over previous modular and fraction-free versions. The algorithm can be used for finding the rank, left nullspace, and the Popov form of such matrices.

1 Introduction

Ore domains provide a general setting for describing the arithmetic of linear differential, difference, and q-difference operators. Systems of differential, difference, and q-difference equations can then be defined via matrices of Ore operators (polynomials) evaluated at unknown functions. One can then make use of matrix constructions to investigate such systems. For example, performing row reduction on a matrix of Ore polynomials to simpler forms allows one to determine its rank and left nullspace which give the minimum number of equations needed to represent the system of equations [1]. When a transformation is invertible, the normal form gives the matrix representing an equivalent system with a minimum number of equations. When the leading coefficient is triangular (as in the weak Popov form), then the normal form allows one to rewrite high-order operators (e.g. derivatives) in terms of lower ones [2]. These transformations can also be applied to the computation of greatest common right divisors (GCRDs) and least common left multiples (LCLMs) [3–6], which represents the intersection and the union of the solution spaces of systems of equations.

H. Cheng (✉)
Department of Mathematics and Computer Science, University of Lethbridge, Lethbridge, Canada
e-mail: howard.cheng@uleth.ca

G. Labahn
Symbolic Computation Group, David R. Cheriton School of Computer Science, University of Waterloo, Waterloo, Canada
e-mail: glabahn@uwaterloo.ca

© Springer-Verlag Berlin Heidelberg 2014
R. Feng et al. (eds.), *Computer Mathematics*, DOI 10.1007/978-3-662-43799-5_5

The FFreduce algorithm [3] is a procedure for row reducing a matrix of Ore operators which performs row operations in a fraction-free way to reduce to simpler form while still controlling coefficient growth. This algorithm computes the rank and left nullspace of these matrices, and can be used to compute the row-reduced and weak Popov forms of shift polynomial matrices [3], as well as the Popov form of general Ore polynomial matrices [7]. It can also be used to compute a greatest common right divisor (GCRD) and a least common left multiple (LCLM) of such matrices. Besides their general use with systems of equations, LCLMs are also used in nonlinear control theory in order to define the notion of transfer function in some cases [8].

A modular version of the FFreduce algorithm was developed by the authors to reduce the computational complexity [2]. In the modular algorithm, it was observed that the evaluation reduction $\mathbb{Z}_p[t][Z;\sigma,\delta] \to \mathbb{Z}_p[Z;\sigma,\delta]$ is not generally an Ore ring homomorphism [6]. Instead of performing the row operations on the Ore polynomial matrices directly, the problem was converted to one involving a system of linear equations over $\mathbb{Z}_p$. Larger striped Krylov matrices over $\mathbb{Z}_p$ was constructed and row reductions were performed on these matrices. Each Krylov matrix was constructed dynamically—rows were added depending on which row is selected as the pivot in each step. This was needed to ensure that the correct image was computed during the reduction in the presence of potential unlucky homomorphisms, even though unlucky homomorphisms occur rarely in practice. Thus, the modular algorithm was a trade-off between not exploiting polynomial arithmetic (or equivalently, the structure of the matrix) and the improved efficiency of coefficient arithmetic in simpler domains.

One obstacle in obtaining further improvement was that the row operations to reduce the Krylov matrix have to be done one step at a time, because it is not possible to construct the entire Krylov matrix a priori or the wrong system of solutions may have been solved. As a result, the only linear algebra subroutines in the `LinearAlgebra:-Modular` package in Maple used to accelerate the computation were operations on individual rows instead of the entire matrix. The resulting implementation has to switch back and forth between high-level Maple code and low-level compiled linear algebra subroutines that are significantly faster. In practice, the resulting modular algorithm was only faster than the corresponding fraction-free algorithm for very large inputs.

In this work, we investigate the applicability of linear algebra subroutines on blocks of matrices to speedup the computation. Assuming that the first evaluation point is "lucky," the Krylov matrices for the remaining evaluation points can be constructed and the entire matrix can be reduced with a few calls to the appropriate linear algebra subroutines. This allows more sophisticated implementations of linear algebra subroutines to speedup the reduction process (e.g. [9]).

2 Notation and Definitions

The definitions given here are similar to those in our previous works [2, 3].

For any vector of integers (also called *multi-index*) $\boldsymbol{\omega} = (\omega_1, \ldots, \omega_p)$, we denote by $|\boldsymbol{\omega}| = \sum_{i=1}^{p} \omega_i$. The vector $\mathbf{e}_i$ denotes the ith unit vector (of the appropriate dimension) such that $(e_i)_j = \delta_{ij}$; we also have $\mathbf{e} = (1, \ldots, 1)$ (of the appropriate dimension). We denote by $Z^{\boldsymbol{\omega}}$ the diagonal matrix having Z^{ω_i} on the diagonal.

Let k be any field and let $\sigma : k \to k$ be an injective endomorphism of k. Then, a *derivation* $\delta : k \to k$ with respect to σ is an endomorphism of the additive group of k satisfying

$$\delta(rs) = \sigma(r)\delta(s) + \delta(r)s$$

for all $r, s \in k$. In this paper, we examine Ore polynomial rings with coefficients in $\mathbb{Z}[t]$. That is, the ring $\mathbb{Z}[t][Z; \sigma, \delta]$ with σ an automorphism, δ a derivation and with the multiplication rule

$$Z \cdot a = \sigma(a)Z + \delta(a)$$

for all $a \in \mathbb{Z}[t]$. When $\delta \equiv 0$, we call the polynomials *shift polynomials*. For brevity, we use $\mathbb{Z}[t][Z]$ when the specific choices of σ and δ are not important.

Let $\mathbb{Z}[t][Z]^{m\times n}$ be the ring of $m \times n$ Ore polynomial matrices over $\mathbb{Z}[t]$. We adapt the following conventions for the remainder of this paper. Let $\mathbf{F}(Z) \in \mathbb{Z}[t][Z]^{m\times n}$, $N = \deg \mathbf{F}(Z)$, and write

$$\mathbf{F}(Z) = \sum_{j=0}^{N} F^{(j)} Z^j, \text{ with } F^{(j)} \in \mathbb{Z}[t]^{m\times n}.$$

We also write $c_j(\mathbf{F}(Z)) = F^{(j)}$ to denote the jth coefficient matrix. The *row degree* of an Ore polynomial matrix $\mathbf{F}(Z)$ is $\boldsymbol{\nu} = \operatorname{rdeg} \mathbf{F}(Z)$ if the ith row has degree ν_i. Some useful properties of matrices of Ore polynomials, such as linear independence and rank, can be found in [3].

The problem of row reduction of Ore polynomial matrices can be formalized as follows. An Ore polynomial vector $\mathbf{P}(Z) \in \mathbb{Z}[t][Z]^{1\times m}$ is said to have *order*[1] $\boldsymbol{\omega}$ with respect to $\mathbf{F}(Z)$ if

$$\mathbf{P}(Z) \cdot \mathbf{F}(Z) = \mathbf{R}(Z) \cdot Z^{\boldsymbol{\omega}}$$

for some *residual* $\mathbf{R}(Z)$. The set of all vectors of a particular order $\boldsymbol{\omega}$ forms a $\mathbb{Q}[t][Z]$-module. An *order basis* for this module, $\mathbf{M}(Z) \in \mathbb{Z}[t][Z]^{m\times m}$ of row degree $\boldsymbol{\mu}$, is a basis such that

[1] Orders in this paper will be with respect to $\mathbf{F}(Z)$ and it will not be explicitly stated for the remainder of the paper.

1. every row, $\mathbf{M}(Z)_{i,*}$, has order $\boldsymbol{\omega}$ for all $1 \leq i \leq m$;
2. the rows of $\mathbf{M}(Z)$ form a basis of the module of all vectors of order $\boldsymbol{\omega}$. That is, every $\mathbf{P}(Z) \in \mathbb{Q}[t][Z]^{1\times m}$ of order $\boldsymbol{\omega}$ can be written as $\mathbf{P}(Z) = \mathbf{Q}(Z) \cdot \mathbf{M}(Z)$ for some $\mathbf{Q}(Z) \in \mathbb{Q}[t][Z]^{1\times m}$;
3. the leading column coefficient is normalized. That is, there exists a nonzero $d \in \mathbb{Z}[t]$ such that
$$\mathbf{M}(Z) = d \cdot Z^{\boldsymbol{\mu}} + \mathbf{L}(Z)$$
where $\deg \mathbf{L}(Z)_{k,l} \leq \mu_l - 1$.

An order basis represents all row operations to eliminate a specified number of low-order coefficients. An order basis of a particular order and degree, if it exists, is unique up to a constant multiple [3, Theorem 4.4]. When $\boldsymbol{\omega} = (mN + 1) \cdot \mathbf{e}$ and $\mathbf{R}(Z)$ is the corresponding residual, the rows in $\mathbf{M}(Z)$ corresponding to the zero rows in $\mathbf{R}(Z)$ give a basis for the left nullspace of $\mathbf{F}(Z)$. However, it is not known a priori the row degree $\boldsymbol{\mu}$ of the order basis. A row-reduced form and weak Popov form, together with the unimodular transformation matrix, can be extracted from $\mathbf{M}(Z)$ and $\mathbf{R}(Z)$ if $\mathbf{F}(Z)$ is a matrix of shift polynomials [3]. In the general case of matrices of Ore polynomials, the computation of the Popov form can be formulated as a left nullspace computation and can be extracted from the result of an order basis computation [2].

3 The Modular Algorithm

A modular algorithm was given in [2] to compute the order basis and the residual. The fraction-free algorithm [3] can be reduced easily from $\mathbb{Z}[t][Z]$ to $\mathbb{Z}_p[t][Z]$ using Chinese remaindering. The usual issue of normalization of the image, detecting unlucky homomorphisms, and termination can be dealt with as described in [2]. It should be noted that the algorithm is output-sensitive in that the number of primes used is determined by the output size, and there is no need to verify the result (e.g. by trial division).

However, the reduction from $\mathbb{Z}_p[t][Z]$ to $\mathbb{Z}_p[Z]$ was not possible because the evaluation homomorphism $t \leftarrow \alpha$ is generally not an Ore ring homomorphism. Instead, we formulate the order basis problem as a system of linear equations over $\mathbb{Z}_p$ and perform Gaussian elimination on the coefficient matrix. The method we follow is similar to polynomial GCD computation by Gaussian elimination on the well-known Sylvester matrix [10]. It can also be considered an extension to the modular algorithm for Ore polynomial GCRD computation of Li and Nemes [6].

Given row degree $\boldsymbol{\mu}$ and order $\boldsymbol{\omega}$, the coefficients in the order basis $\mathbf{M}(Z)$ can be viewed as a solution to a linear system of equations over the coefficient ring. By equating the coefficients of like powers, each row of the order basis satisfies a system of equations of the form

$$\begin{array}{c} \quad Z^0 \;\cdots\; Z^{\mu_k-1+\delta_{1,k}} \\ \left[\cdots \middle| \; p_k^{(0)} \cdots p_k^{(\mu_k-1+\delta_{1,k})} \middle| \cdots \right] \end{array} \cdot \begin{array}{c} Z^0 \qquad \cdots \qquad Z^{\boldsymbol{\omega}-\mathbf{e}} \\ \left[\begin{array}{c|c|c} & \vdots & \\ \hline \cdots & Z^0 \cdot F_{k,\cdot}(Z) & \cdots \\ & \vdots & \\ \cdots & Z^{\mu_k-1+\delta_{1,k}} \cdot F_{k,\cdot}(Z) & \cdots \\ \hline & \vdots & \end{array}\right] \end{array} = \mathbf{0}.$$

More formally, for any $\mathbf{P}(Z) \in \mathbb{Q}[t][Z]^{m\times n}$ we define

$$\mathbf{P}_{\mathbf{v}} = \left[P_{*,1}^{(0)} \cdots P_{*,1}^{(v_1)} | \cdots | P_{*,n}^{(0)} \cdots P_{*,n}^{(v_n)}\right].$$

We also define (recall that $\boldsymbol{\omega} = \sigma \cdot \mathbf{e}$)

$$K(\boldsymbol{\mu}, \boldsymbol{\omega}) = \begin{bmatrix} c_0(\quad \mathbf{F}(Z)_{1,*}) & \cdots & c_{\sigma-1}(\quad \mathbf{F}(Z)_{1,*}) \\ \vdots & & \vdots \\ c_0(Z^{\mu_1}\cdot \mathbf{F}(Z)_{1,*}) & \cdots & c_{\sigma-1}(Z^{\mu_1}\cdot \mathbf{F}(Z)_{1,*}) \\ \vdots & & \vdots \\ c_0(\quad \mathbf{F}(Z)_{m,*}) & \cdots & c_{\sigma-1}(\quad \mathbf{F}(Z)_{m,*}) \\ \vdots & & \vdots \\ c_0(Z^{\mu_m}\cdot \mathbf{F}(Z)_{m,*}) & \cdots & c_{\sigma-1}(Z^{\mu_m}\cdot \mathbf{F}(Z)_{m,*}) \end{bmatrix}. \tag{1}$$

Then the ith row of the order basis satisfies

$$(\mathbf{M}_{i,*})_{\boldsymbol{\mu}-\mathbf{e}+\mathbf{e}_i} \cdot K(\boldsymbol{\mu} - \mathbf{e} + \mathbf{e}_i, \boldsymbol{\omega}) = \mathbf{0}. \tag{2}$$

The matrix $K(\boldsymbol{\mu}, \boldsymbol{\omega})$ has dimensions $|\boldsymbol{\mu} + \mathbf{e}| \times |\boldsymbol{\omega}|$, and is called a *striped Krylov matrix* (with m stripes). This is a generalization of the well-known Sylvester matrix when $m = 2$ and $n = 1$. We also define $K^*(\boldsymbol{\mu}, \boldsymbol{\omega})$ to be the matrix $K(\boldsymbol{\mu}, \boldsymbol{\omega})$ with linearly dependent columns removed.

Example 1 Let $\boldsymbol{\mu} = (2, 2)$, $\boldsymbol{\omega} = (3, 3)$, and

$$\mathbf{F}(Z) = \begin{bmatrix} 2Z^2 + 3tZ + 6t^2 & Z^2 - Z + 2 \\ (t-1)Z + 3t^3 & 3tZ + t \end{bmatrix} \in \mathbb{Z}[t][Z; \sigma, \delta]^{2\times 2},$$

with $\sigma(a(t)) = a(t)$ and $\delta(a(t)) = \frac{d}{dt}a(t)$. Then

$$K(\mu, \omega) = \left[\begin{array}{cc|cc|cc} 6t^2 & 2 & 3t & -1 & 2 & 1 \\ 12t & 0 & 6t^2+3 & 2 & 3t & -1 \\ 12 & 0 & 24t & 0 & 6t^2+6 & 2 \\ 3t^3 & t & t-1 & 3t & 0 & 0 \\ 9t^2 & 1 & 3t^3+1 & t+3 & t-1 & 3t \\ 18t & 0 & 18t^2 & 2 & 3t^3+2 & t+6 \end{array}\right].$$

□

One way to obtain an order basis of degree μ and order ω is to perform Gaussian elimination on $K(\mu, \omega)$ so that the first ω columns are eliminated. The rows in the submatrix $K(\mu - \mathbf{e}, \omega)$ are used for pivots in the elimination process, and the remaining rows give the residual $\mathbf{R}(Z)$. The order basis can be recovered from the transformation matrix corresponding to these rows.

Example 2 Continuing from Example 1, we perform Gaussian elimination on $K((2, 2), (3, 3))$ using the first two rows of each stripe as pivots. After removing some common factors in each row to reduce the results, the resulting matrix is

$$\begin{bmatrix} 6t^2 & 2 & 3t & -1 & 2 & 1 \\ 0 & -4 & 6t^3-3t & 2t+2 & 3t^2-4 & -t-2 \\ 0 & 0 & 0 & 0 & -252t^5+270t^4-234t^3-22t^2-16t+16 & 882t^4-104t^2-56t-10 \\ 0 & 0 & -3t^2+2t-2 & 7t & -2t & -t \\ 0 & 0 & 0 & 21t^2-14 & 9t^4-12t^3+10t^2-8t+8 & -21t^3+11t^2-14t+2 \\ 0 & 0 & 0 & 0 & -126t^6+135t^5-180t^4-11t^3+118t^2-20t & 441t^5-52t^3-28t^2-103t \end{bmatrix}$$

with the corresponding transformation matrix

$$\begin{bmatrix} 1 & 0 & 0 & 0 & 0 & 0 \\ -2 & t & 0 & 0 & 0 & 0 \\ -6t^2+4 & -126t^4+6t^3-4t^2+4t+14 & 21t^3-14t & -252t^2+24t+34 & 252t^3-12t^2-34t-8 & 0 \\ -t & 0 & 0 & 2 & 0 & 0 \\ -3t^2+2 & 3t^3-2t^2+2t & 0 & 12t-4 & -6t^2+4t-4 & 0 \\ -3t^3+2t & -63t^5+3t^4-2t^3+2t^2+21t & 0 & -126t^3+12t^2+59t & 126t^4-6t^3-59t^2-4t & 21t^3-14t \end{bmatrix}.$$

The order basis $\mathbf{M}(Z)$ of degree $\mu = (2, 2)$ and order $\omega = (3, 3)$ can be easily extracted. The rows of $\mathbf{M}(Z)$ are:

$$\left[(21t^3 - 14t)Z^2 + (-126t^4 + 6t^3 - 4t^2 + 4t + 14)Z - 6t^2 + 4 \quad (252t^3 - 12t^2 - 34t - 8)Z - 252t^2 + 24t + 34\right]$$

and

$$\left[(-63t^4 + 3t^3 - 2t^2 + 2t + 21)Z - 3t^2 + 2 \quad (21t^3 - 14t)Z^2 + (126t^3 - 6t^2 - 59t - 4)Z - 126t^2 + 12t + 59\right].$$

□

Unfortunately, the row degree μ of the order basis $\mathbf{M}(Z)$ of order ω is not known a priori. In practice, one starts with $\mu_0 = \mathbf{0}$ and performs elimination on $K(\mu_0, \omega)$. For any $i \geq 0$, μ_{i+1} is determined by the pivoting needed to reduce $K(\mu_i, \omega)$ by one more column. Thus, each step in the algorithm involves performing Gaussian elimination of one column followed by adding one row to the matrix. Unlucky homomorphisms occur when the determinant of $K^*(\mu, \omega)$ vanishes under the evaluation $t \leftarrow \alpha$. In such case, the pivoting that occurs during the elimination is different. Unlucky homomorphisms can be detected by comparing the different row degrees of the final order basis computed under each evaluation homomorphism.

The `LinearAlgebra:-Modular` package in Maple was used to perform efficient computations over $\mathbb{Z}_p$. The use of Gaussian elimination for solving the system of linear equations instead of working on the Ore polynomial matrices directly means that the modular algorithm is no longer exploiting the structure present in the Krylov matrix. On the other hand, coefficient arithmetic over $\mathbb{Z}[t]$ can be replaced by simpler coefficient arithmetic over $\mathbb{Z}_p$. For larger problems, the gain in simpler coefficient arithmetic more than offsets the loss in efficiency by not exploiting the structure. The algorithm outperforms the fraction-free algorithm [3] for very large problems even though the fraction-free algorithm exploits the structure of the Krylov matrix. However, the modular algorithm is not competitive for small input [2].

4 Improved Implementation

The implementation of the modular algorithm described in [2] has two drawbacks. First, the interleaving between matrix construction and row elimination means that routines such as Gaussian elimination (on an entire matrix) or block matrix multiplication cannot be applied to speedup the computation further. The implementation would have to switch between high-level Maple code and the faster, low-level compiled code in the `LinearAlgebra:-Modular` package. Second, the extra work and bookkeeping required for incremental matrix construction reduce the advantage of the modular algorithm. We would like to make use of low-level compiled linear algebra routines as much as possible without switching to Maple code.

In order to improve the modular algorithm, we note the unpredictability of the final row degree is mostly due to the presence of unlucky homomorphisms, but they occur rarely in practice. Therefore, the incremental elimination algorithm given previously [2] is used on one evaluation point in $\mathbb{Z}_p$. Assuming that the evaluation point (and the prime p) is not unlucky, the order basis computed has the correct degree μ. If μ turns out to be incorrect, it will be detected when combined with the results from other primes. In that case, we perform extra computations in $\mathbb{Z}_p$ that are wasted. However, this does not occur often in practice.

When the correct degree μ of the order basis is known (as assumed), it is relatively straightforward to compute the order basis and the residual:

1. construct $\mathbf{A} = [K(\mu, \omega + (N+1) \cdot \mathbf{e}) \mid \mathbf{I}]$;
2. perform Gaussian elimination on $\mathbf{A}$ to compute a reduced row echelon form to eliminate the first $|\omega|$ columns, using only rows in $K(\mu - \mathbf{e}, \omega)$ as pivots;
3. record the linearly dependent columns J as well as $d = \det K^*(\mu - \mathbf{e}, \omega)$ which is (up to sign) the product of the pivots used;

4. construct $\mathbf{B}$ from $-\mathbf{A}_{*,J}$ after removing the pivot rows and inserting the $m \times m$ identity matrix into the columns corresponding to those rows.
5. compute $\mathbf{C} = (-1)^{\sum_{i=2}^{m} \mu_i} \cdot d \cdot \mathbf{B} \cdot \mathbf{A}$;
6. if $\mathbf{C}$ is not zero in the first $|\boldsymbol{\omega}|$ columns, then the homomorphism is unlucky. Otherwise, extract $\mathbf{R}(Z)$ from the left part and $\mathbf{M}(Z)$ from the right part of $\mathbf{C}$.

The Gaussian elimination in Step 2 can be performed, for example, by calling the `RowReduce` routine in the `LinearAlgebra:-Modular` package of Maple on the entire matrix $\mathbf{A}$. The matrix multiplication in Step 5 can be performed by the `Multiply` routine. As a result, the new implementation can fully take advantage of good low-level implementation of block Gaussian elimination and multiplication (e.g. [9]). Since there is no need to perform incremental matrix construction, both memory management and bookkeeping are reduced. In addition, the control of the program can stay inside the low-level `LinearAlgebra:-Modular` subroutines instead of switching back and forth between them and Maple code.

Example 3 We apply this method to Example 1. We perform our calculations in $\mathbb{Z}_{31}$ and perform the evaluation $t \leftarrow 7$. To conserve space, we only show $\mathbf{A}' = [K(\boldsymbol{\mu}, \boldsymbol{\omega}) \mid \mathbf{I}]$ and compute only $\mathbf{M}(Z)$. Initially,

$$\mathbf{A}' = \begin{bmatrix} 15 & 2 & 21 & 30 & 2 & 1 & 1 & 0 & 0 & 0 & 0 & 0 \\ 22 & 0 & 18 & 2 & 21 & 30 & 0 & 1 & 0 & 0 & 0 & 0 \\ 12 & 0 & 13 & 0 & 21 & 2 & 0 & 0 & 1 & 0 & 0 & 0 \\ 6 & 7 & 6 & 21 & 0 & 0 & 0 & 0 & 0 & 1 & 0 & 0 \\ 7 & 1 & 7 & 10 & 6 & 21 & 0 & 0 & 0 & 0 & 1 & 0 \\ 2 & 0 & 14 & 2 & 8 & 13 & 0 & 0 & 0 & 0 & 0 & 1 \end{bmatrix}.$$

Performing Gaussian elimination on rows 1, 2, 4, and 5, we obtain:

$$\begin{bmatrix} 1 & 0 & 0 & 0 & 14 & 14 & 29 & 5 & 0 & 27 & 1 & 0 \\ 0 & 1 & 0 & 0 & 25 & 9 & 11 & 1 & 0 & 24 & 28 & 0 \\ 12 & 0 & 13 & 0 & 21 & 2 & 0 & 0 & 1 & 0 & 0 & 0 \\ 0 & 0 & 1 & 0 & 14 & 25 & 0 & 23 & 0 & 6 & 20 & 0 \\ 0 & 0 & 0 & 1 & 25 & 8 & 22 & 2 & 0 & 21 & 26 & 0 \\ 2 & 0 & 14 & 2 & 8 & 13 & 0 & 0 & 0 & 0 & 0 & 1 \end{bmatrix}.$$

Here, $J = \{1, 2, 3, 4\}$ and $d = 26$. Thus,

$$\begin{aligned} \mathbf{C} &= 26 \cdot \begin{bmatrix} -12 & 0 & \mathbf{1} & -13 & 0 & \mathbf{0} \\ -2 & 0 & \mathbf{0} & -14 & -2 & \mathbf{1} \end{bmatrix} \cdot \mathbf{A}' \\ &= \begin{bmatrix} 0 & 0 & 0 & 0 & 2 & 6 & 4 & 28 & 26 & 26 & 27 & 0 \\ 0 & 0 & 0 & 0 & 28 & 14 & 14 & 6 & 0 & 1 & 27 & 26 \end{bmatrix}, \end{aligned}$$

where the highlighted entries are the identity matrix inserted to form $\mathbf{B}$. Therefore, the image of the order basis computed is

$$\mathbf{M}(Z) = 5 \cdot \begin{bmatrix} 6Z^2 + 16Z + 20 & 11Z + 6 \\ 30Z + 8 & 6Z^2 + 11Z + 5 \end{bmatrix}.$$

One may easily verify that this is a scalar multiple of the image of the order basis computed in Example 2 under the evaluation $t \leftarrow 7$ in $\mathbb{Z}_{31}$.

The introduction of the scalar multiple is due to the removal of content in Example 2. The implementation given here in fact computes exactly the same result (including the scalar multiple) as the previous fraction-free and modular implementations for the order basis problem [2, 3]. □

5 Experimental Results

Experiments were performed on Ore polynomial matrices in differential case. The results of these experiments are shown in Tables 1 and 2.

The application of block linear algebra routines reduces the running time of the modular algorithm in all cases. The improvement is more significant for smaller problems, where the original modular algorithm is not competitive against the fraction-free problems.

For the larger problems, however, the improvement is less significant. For larger problems, the size of the coefficients in the output becomes larger as well. More time is spent on the other parts of the algorithm such as reconstruction by Chinese

Table 1 Comparison of fraction-free, modular, and the new modular algorithm on random $m \times n$ matrices with $\deg_t = 1$ and integer coefficients having magnitude ≤ 5

m, n	N	FFreduce (s)	Modular (s)	New modular (s)	Improvement (%)
2	1	0.023	0.115	0.069	40
2	2	0.107	0.242	0.192	21
2	4	1.689	2.984	2.301	23
2	8	15.047	28.499	23.232	18
2	16	278.883	279.041	232.877	17
2	32	5447.542	4669.801	3992.689	15
3	1	0.472	1.060	0.723	32
3	2	3.808	6.268	5.416	20
3	4	41.549	51.498	44.253	14
3	8	667.682	599.466	521.571	13
4	2	41.348	47.841	41.765	13
4	4	663.707	554.060	487.000	12
4	6	3850.143	2561.281	2303.989	10
5	2	293.122	260.021	227.314	13
5	4	6179.169	3845.945	3362.376	13
8	1	1660.258	1088.609	998.659	8
10	1	16179.879	8019.137	7524.240	6

Table 2 Comparison of fraction-free, modular, and the new modular algorithm on random $m \times n$ matrices with $\deg_t = 2$ and integer coefficients having magnitude ≤ 5

m, n	N	FFreduce (s)	Modular (s)	New modular (s)	Improvement (%)
2	2	0.470	1.647	1.378	16
2	4	5.920	11.611	10.004	14
2	8	86.214	128.528	109.911	14
2	16	1237.410	1437.492	1300.804	10
3	2	14.718	25.101	21.954	13
3	4	216.214	267.295	238.497	11
3	6	1157.705	1220.524	1114.106	9
3	8	3933.234	3994.955	3735.837	6
4	2	170.561	193.981	174.399	10
4	4	2397.460	2270.272	2096.580	8

remaindering and memory management, and the amount of time spent on actual elimination is proportionally smaller. Since the new implementation improves mainly the elimination process, the improvement is less significant for larger problems. On the other hand, we see that the improved implementation given in this paper increases the advantage of the modular algorithm over the fraction-free algorithm, and allows the modular algorithm to be used beneficially for smaller problems.

References

1. Beckermann, B., Cheng, H., Labahn, G.: Fraction-free row reduction of matrices of skew polynomials. In: Proceedings of the 2002 International Symposium on Symbolic and Algebraic Computation, pp. 8–15. ACM (2002)
2. Cheng, H., Labahn, G.: Modular computation for matrices of ore polynomials. In: Computer Algebra 2006: Latest Advances in Symbolic Algorithms, pp. 43–66 (2007)
3. Beckermann, B., Cheng, H., Labahn, G.: Fraction-free row reduction of matrices of ore polynomials. J. Symb. Comput. **41**(5), 513–543 (2006)
4. Li, Z.: A Subresultant Theory for Linear Differential, Linear Difference and Ore Polynomials, with Applications. PhD thesis, RISC-Linz, Johannes Kepler University, Linz, Austria, 1996
5. Li, Z.: A subresultant theory for ore polynomials with applications. In: Proceedings of the 1998 International Symposium on Symbolic and Algebraic Computation, pp. 132–139. ACM (1998)
6. Li, Z., Nemes, I.: A modular algorithm for computing greatest common right divisors of ore polynomials. In: Proceedings of the 1997 International Symposium on Symbolic and Algebraic Computation, pp. 282–289. ACM (1997)
7. Davies, P., Cheng, H., Labahn, G.: Computing popov form of general ore polynomial matrices. Milest. Comput. Algebra (MICA) 2008, 149–156 (2008)
8. Halas, M., Kotta, U., Li, Z., Wang, H., Yuan, C.: Submersive rational difference systems and formal accessibility. In: Proceedings of the 2009 International Symposium on Symbolic and Algebraic Computation, pp. 175–182. ACM (2009)

9. Dumas, J.-G., Giorgi, P., Pernet, C.: FFPACK: finite field linear algebra package. In: Proceedings of the 2004 International Symposium on Symbolic and Algebraic Computation, pp. 119–126. ACM (2004)
10. Loos, R.: Generalized Polynomial Remainder Sequences. In: Computer Algebra: Symbolic and Algebraic Computation, pp. 115–137. Springer, New York (1982)

Computing Popov Forms of Matrices Over PBW Extensions

Mark Giesbrecht, George Labahn and Yang Zhang

Abstract In this paper we define the Popov and weak Popov forms of matrices over Poincaré–Birkhoff–Witt (PBW) extensions, and exhibit effective algorithms to find them. As applications we give general methods to calculate the ranks of such matrices, and a method to transfer a system of differential equations into a first order equation.

1 Introduction

When seeking to determine left and right equivalence properties of a matrix over a domain, one often seeks to transform that matrix into canonical form, such as the Hermite or Smith form. These forms capture all left and two-sided equivalence properties, respectively, but suffer from substantial growth in the size of entries. For example, for polynomial matrices, their degrees are generally of the order of the size of the matrix, even when the degrees of the entries in the original matrix are small. In order to avoid this increase in degree, [1] introduced another normal form, which has been successfully applied to control theory (see for example, [2]). This form has come to be known as the Popov normal form.

Recently, the Popov form has attracted considerable interest in the computer algebra field for its low degree and correspondingly efficient algorithms. For example, [3, 4] defined and discussed shifted-Popov forms. In [5] the *weak Popov form* (called the quasi-Popov form in [4]) is discussed, which while not canonical, is easier to compute and elicits many of the matrix properties that we are interested in.

M. Giesbrecht (✉) · G. Labahn
Cheriton School of Computer Science, University of Waterloo, Waterloo ON N2L 3G1, Canada
e-mail: mwg@uwaterloo.ca

G. Labahn (✉)
e-mail: glabahn@uwaterloo.ca

Y. Zhang (✉)
Department of Mathematics, University of Manitoba, Winnipeg MB R3T 2N2, Canada
e-mail: zhang39@cc.umanitoba.ca

© Springer-Verlag Berlin Heidelberg 2014
R. Feng et al. (eds.), *Computer Mathematics*, DOI 10.1007/978-3-662-43799-5_6

In the noncommutative case, matrices over *Ore domains* (domains which satisfy the *Ore condition*, which essentially says that any two elements have a nontrivial left (or right) common multiple) have been considered at least since the 1930s (see [6–8]). In [7] the Hermite and Smith forms of matrices over skew polynomial rings are considered. Recent treatments from a ring theoretic perspective can be found in [9]. In the computer algebra field, [10] gave a method to calculate the ranks of matrices over skew polynomial rings $\mathsf{R}[x, x^{-1}; \sigma]$, and their method can be applied to Weyl algebras [11] use fraction-free methods to discuss weak Popov form of matrices over skew polynomial rings. Giesbrecht et al. [12] discuss Popov forms of matrices over valuation domains. Davies et al. and Giesbrecht et al. [13, 14] develop reductions to linear algebra over a (commutative) field to give polynomial-time algorithms for Popov and, respectively, Hermite forms of matrices over an Ore domain.

Matrices over multivariate polynomial rings and rings of differential operators have been extensively used in multidimensional linear systems since the mid-1970s (see, for example [15–19]), and also in other areas such as operator algebras; see for example [20]. This work motivates us not only to consider the univariate case, but also the case of multivariate noncommutative polynomial rings.

In this paper, we discuss matrices over noncommutative rings called PBW extensions, which includes most popular rings with derivations studied in computer algebra. At first, we consider the row spaces of matrices as left modules over base rings and define two kinds of reductions which are used to construct weak Popov forms and Popov forms respectively. The rank of matrices remains invariant under our reductions, and equals the number of nonzero rows of the (weak) Popov forms. Note that the ranks of matrices are independent of term orders. Therefore some term orders may produce ranks very quickly, while some may be considerably slower.

As an application, this gives a direct method to calculate the rank of matrices over Weyl algebras, as compared to the method given by [10], which sets up a bijection between the Weyl algebra and a skew–Laurent polynomial ring.

In this extended abstract, we first recall the definition of PBW extensions, and then outline how to define (weak) Popov forms of matrices over PBW extensions. We then present an algorithm to construct Popov forms. More results will appear in a forthcoming journal version.

2 Definitions and Some Results

The definition of a PBW extension was first given by [21]. This led to a unified treatment of many polynomial-type rings which are currently studied in associative ring theory and computer algebra.

Definition 1 Let R and E be two associative rings with $\mathsf{R} \subseteq \mathsf{E}$. E is called a *(finite) PBW extension* of R if there exist $x_1, x_2, \ldots, x_n \in \mathsf{E}$ such that

(a) monomials $x_1^{i_1} \ldots x_n^{i_n}$ form a basis for E as a free left R-odule, for $i_1, \ldots, i_n \in \mathbb{N}$;
(b) $x_i r - r x_i \in \mathsf{R}$ for each $i = 1, \ldots, n$ and any $r \in \mathsf{R}$;

(c) $x_i x_j - x_j x_i \in \mathsf{R} + \mathsf{R}x_1 + \cdots + \mathsf{R}x_n$ for all $i, j = 1, \ldots, n$.

Let $\mathfrak{R} = \mathsf{R}\langle x_1, \ldots, x_n \rangle$ be a PBW extension of an Ore domain R. One can naturally define a term-ordering on $\mathfrak{R}$ which satisfies the usual multiplicative properties and respects degree; we refer to [22, Section 2] for details. Furthermore, we can extend the term ordering on $\mathfrak{R}$ to the left $\mathfrak{R}$-module $\mathfrak{R}^m$ and define leading monomials, leading coefficients and leading terms in the natural way. Throughout this paper assume that $\prec$ is a *term-over-position* admissible term order on $\mathfrak{R}^m$, i.e., the monomial term order has higher priority than the position in the vector.

We now define the notions of weak reduction and reduction as mechanisms for cancelling terms in vectors via the leading term (designated by $\mathrm{lt}(\cdot)$) of another vector.

Definition 2 Given $\mathbf{a}, \mathbf{b}, \mathbf{c} \in \mathfrak{R}^m$, we say that

(a) $\mathbf{a}$ *weakly reduces* to $\mathbf{c}$ modulo $\mathbf{b}$ in one step if and only if $\mathrm{lt}(\mathbf{b})$ divides $\mathrm{lt}(\mathbf{a})$ and $\mathbf{c} = \mathbf{a} - q_1\mathbf{b}$, where $q_1 \in \mathfrak{R}$ is such that $\mathrm{lt}(\mathbf{a}) = \mathrm{lt}(q_1\mathbf{b})$.
(b) $\mathbf{a}$ *reduces* to $\mathbf{c}$ modulo $\mathbf{b}$ in one step if and only if $\mathrm{lt}(\mathbf{b})$ divides a term $\mathbf{d}$ that appears in $\mathbf{a}$ and $\mathbf{c} = \mathbf{a} - q_2\mathbf{b}$, where $q_2 \in \mathfrak{R}$ is such that $\mathbf{d} = \mathrm{lt}(q_2\mathbf{b})$.

Thus, weak reduction uses the leading term of $\mathbf{b}$ to cancel the leading term of $\mathbf{a}$, whereas full reduction is much stronger, and uses the leading term of $\mathbf{b}$ to cancel *any* possible term in $\mathbf{a}$. The *weak division algorithm* (resp. the *division algorithm*) for $\mathbf{r} \in \mathfrak{R}^m$ by a set $\Gamma \subseteq \mathfrak{R}^m$ is defined correspondingly to reduce the least leading monomials (resp. all possible monomials) in $\mathbf{r}$ by the leading monomials of elements of Γ.

Definition 3 A nonzero vector $\mathbf{a}$ in $\mathfrak{R}^m$ is called *(weakly) reduced* with respect to a set $S = \{\mathbf{s}_1, \ldots, \mathbf{s}_l\}$ of nonzero vectors in $\mathfrak{R}^m$ if no (leading) term that appears in $\mathbf{a}$ is divisible by any one of the $\mathrm{lt}(\mathbf{s}_i)$, $i = 1, \ldots, l$.

Furthermore, a set $S = \{\mathbf{s}_1, \ldots, \mathbf{s}_\ell\}$ of nonzero vectors is called *(weakly) reduced* if each vector $\mathbf{s}_i$ (for $1 \leq i \leq \ell$) is (weakly) reduced with respect to $S \backslash \{\mathbf{s}_i\}$.

We can now define the weak Popov and Popov forms as follows:

Definition 4 Given a matrix $\mathfrak{R}^{m\times n}$ and a term-over-position admissible term order on $\mathfrak{R}^m$, let $\Gamma = \{\mathbf{r}_1, \ldots, \mathbf{r}_m\}$ be its set of its row vectors.

(a) $\mathfrak{R}^{m\times n}$ is in *weak Popov form* with respect to $\prec$ if Γ is a weakly reduced set.
(b) $\mathfrak{R}^{m\times n}$ is in *Popov form* with respect to $\prec$ if

  (i) Γ is a reduced set;
  (ii) the leading coefficients of $\{\mathbf{r}_i\}$ are monic;
  (iii) rows are in a descending chain with respect to $\prec$, that is, $\mathbf{r}_m \prec \cdots \prec \mathbf{r}_2 \prec \mathbf{r}_1$.

The weak reduction and reduction correspond to weak Popov form and Popov form, respectively. In this paper, we first describe two algorithms to present (weak) divisions, and then use them to construct (weak) Popov forms. Here we list one as follows:

Algorithm: `Popov form for` $\mathfrak{R}^{m\times n}$

Input: ▸ row vectors $\mathbf{r}_1, \ldots, \mathbf{r}_m$ of a matrix $A \in \mathfrak{R}^{m\times n}$;

Output: ▸ row vectors $\mathbf{p}_1, \ldots, \mathbf{p}_m$ of Popov form of $A \in \mathfrak{R}^{m\times n}$;

Initialization: $\mathbf{p}_1 := 0, \ldots, \mathbf{p}_m := 0$; changes:=true;
While (changes) do
  changes := false;
  Swap rows so that they are in a descending chain with respect to $\prec$;
  For i from 2 to m do
    If $\mathbf{r}_i$ is reducible modulo $\mathbf{r}_1$ then
      Using the division algorithm, let $\mathbf{r}_i$ be remainder of $\mathbf{r}_i$ by $\mathbf{r}_1$;
      changes := true;
  end do;
end do;
Return: $\mathbf{p}_1 := \mathbf{r}_1, \ldots, \mathbf{p}_m := \mathbf{r}_m$;
Make leading coefficients 1 by multiplying by suitable elements of quotient field of R.

Theorem 1 *The Popov form algorithm terminates after a finite number of steps and produces a Popov form.*

The proof follows relatively easily from the fact that $\mathfrak{R}^m$ is noetherian, and the repeated reduction yields a descending chain of ideals which must be finite. Of course, the number of reduction steps is of great interest in the efficiency of our algorithms.

One of the applications of Popov forms is to provide the rank information efficiently. First, we prove the following theorem which implies that the ranks of matrices are invariant under Popov form transformations.

Theorem 2 *For any matrix $A \in \mathfrak{R}^{m\times n}$ there exists a unimodular matrix U such that UA is in Popov form (similarly for weak Popov form).*

The transformation matrix U and the Popov form can be computed by our algorithms. As an application, we can find the rank of a matrix by counting the number of nonzero rows. Also, given a unimodular matrix $\mathfrak{R}^{n\times n}$ (one whose inverse is also in $\mathfrak{R}^{n\times n}$), one can find its inverse simply by noting that the Popov form of a unimodular matrix is the identity.

We also anticipate that the Popov form will be useful as an intermediate step to efficiently computing the Hermite and Smith/Jacobson forms in an effective manner.

Acknowledgments All the authors would like to thank NSERC Canada for their support of this research.

References

1. Popov, V.M.: Invariant description of linear, time-invariant controllable systems. SIAM J. Control **10**(2), 252–264 (1972)
2. Kailath, T.: Linear Systems. Prentice Hall, Upper Saddle River (1980)
3. Villard. G.: Computing Popov and Hermite forms of polynomial matrices. In: Proceedings of ISSAC'96, pp. 250–258. ACM Press (1996)
4. Beckermann, B., Labahn, G., Villard, G.: Shifted normal forms of polynomial matrices. In: Proceedings of ISSAC'99, pp. 189–196. ACM Press (1999)
5. Mulders, T., Storjohann, A.: On lattice reduction for polynomial matrices. J. Symb. Comput. **35**, 377–401 (2003)
6. Ore, O.: Theory of non-commutative polynomials. Ann. Math. **34**(22), 480–508 (1933)
7. Jacobson, N.: The Theory of Rings. American Math. Soc, New York (1943)
8. Jacobson, N.: Finite-Dimensional Division Algebras over Fields. Springer, Berlin (1996)
9. Cohn, P.M.: Skew Fields. Cambridge University Press, Cambridge (1995)
10. Abramov, S., Bronstein, M.: Linear algebra for skew-polynomial matrices. Technical Report 4420, INRIA. INRIA Rapport de Recherche (2002)
11. Beckermann, B., Cheng, H., Labahn, G.: Fraction-free row reduction of matrices of ore polynomials. J. Symb. Comput. **41**(5), 513–543 (2006)
12. Giesbrecht, M., Labahn, G., Zhang, Y.: Computing valuation Popov forms. In: Proceedings of CASA'2005, Lecture Notes on Computer Science 3516, pp. 619–626. Springer (2005)
13. Davies, P., Cheng, H., Labahn, G.: Computing Popov Form of General Ore Polynomial Matrices. In: Milestones in Computer, Algebra, pp. 149–156 (2008)
14. Giesbrecht, M., Kim, M.: On Computing the Hermite Form of a Matrix of Differential Polynomials. In: Proceedings of CASC'09. (2009)
15. Bose, N.K.: Multidimensional Systems Theory. D. Reidel, Dordrecht (1985)
16. Park, H.: Symbolic computation and signal processing. J. Symb. Comput. **37**, 209–226 (2004)
17. Youla, D.C., Pickel, P.F.: The Quillen-Suslin theorem and the structure of n-dimemsional elementary polynomial matrices. IEEE Trans. Circuits Syst. **31**, 513–518 (1984)
18. Zerz, E.: Topics in Multidimensional Linear Systems Theory. Lecture notes in control and information sciences. Springer, Berlin (2000)
19. Hillebrand, A., Schmale, W.: Towards an effective version of a theorem of Stafford. J. Symb. Comput. **32**, 699–716 (2001)
20. Finkel, F., Kamran, N.: On the equivalence of matrix differential operators to Schrödinger form. Nonlinear Math. Phys. **4**(3–4), 278–286 (1997)
21. Bell, A.D., Goodearl, K.R.: Uniform rank over differential operator rings and Poincaré-Birkhoff-Witt extension. Pac. J. Math. **131**(1), 13–37 (1988)
22. Giesbrecht, M., Reid, G., Zhang, Y.: Non-commutative Gröbner Bases in Poincaré–Birkhoff–Witt extensions. In: Proceedings of Conference on Computer Algebra and Scientific Computation (CASC'02), pp. 97–106 (2002)

On the Simplest Quartic Fields and Related Thue Equations

Akinari Hoshi

Abstract Let K be a field of char $K \neq 2$. For $a \in K$, we give an explicit answer to the field isomorphism problem of the simplest quartic polynomial $X^4 - aX^3 - 6X^2 + aX + 1$ over K as the special case of the field intersection problem via multi-resolvent polynomials. From this result, over an infinite field K, we see that the polynomial gives the same splitting field over K for infinitely many values a of K. We also see by Siegel's theorem for curves of genus zero that only finitely many algebraic integers $a \in \mathcal{O}_K$ in a number field K may give the same splitting field. By applying the result over the field $\mathbb{Q}$ of rational numbers, we establish a correspondence between primitive solutions to the parametric family of quartic Thue equations

$$X^4 = mX^3Y - 6X^2Y^2 + mXY^3 + Y^4 = c,$$

where $m \in \mathbb{Z}$ is a rational integer and c is a divisor of $4(m^2 + 16)$, and isomorphism classes of the simplest quartic fields.

1 Introduction and Main Results

Let K be a field of char $K \neq 2$ and $K(s)$ the rational function field over K with variable s. We take the simplest quartic polynomial

$$f_s(X) := X^4 - sX^3 - 6X^2 + sX + 1 \in K(s)[X]$$

with discriminant $4(s^2 + 16)^3$. The Galois group $\mathrm{Gal}_{K(s)} f_s(X)$ of the polynomial $f_s(X)$ over $K(s)$ is isomorphic to the cyclic group C_4 of order four.

In the case where $K = \mathbb{Q}$, for $a \in \mathbb{Z}\backslash\{0, \pm 3\}$ the polynomials $f_a(X)$ are irreducible over $\mathbb{Q}$ with $\mathrm{Gal}_{\mathbb{Q}} f_a(X) \cong C_4$ and the splitting fields $\mathrm{Spl}_{\mathbb{Q}} f_a(X)$ of $f_a(X)$

A. Hoshi (✉)
Department of Mathematics, Rikkyo University,
3-34-1 Nishi-Ikebukuro, Toshima-ku, Tokyo 171-8501, Japan
e-mail: hoshi@rikkyo.ac.jp

© Springer-Verlag Berlin Heidelberg 2014
R. Feng et al. (eds.), *Computer Mathematics*, DOI 10.1007/978-3-662-43799-5_7

over $\mathbb{Q}$ are totally real cyclic quartic number fields which are called the simplest quartic fields (cf. e.g. [1–3], [4, Sect. 6.2], [5–8]).

For $b = -a \in K$, the polynomials $f_a(X)$ and $f_b(X)$ have the same splitting field over K. In this paper, we consider the field isomorphism problem of $f_s(X)$, i.e., for a fixed $a \in K$, determine whether $b \in K$ gives the same splitting field over K as $\mathrm{Spl}_K f_a(X) = \mathrm{Spl}_K f_b(X)$ or not.

For $n \geq 3$, Rikuna [9] constructed one-parameter families of cyclic polynomials of degree n over K with char $K \nmid n$ and $K \ni \zeta + \zeta^{-1}$ where ζ is a primitive n-th root of unity, and $f_s(X)$ may be obtained the quartic case $n = 4$ of Rikuna's cyclic polynomials (see also [10, 11]). An answer to the field isomorphism problem to Rikuna's cyclic polynomials was given by Komatsu [12] as a generalization of Kummer theory (cf. also [13, 14]).

In Sect. 3, by using multi-resolvent polynomials, we give an explicit form of answer to the field isomorphism problem of $f_s(X)$ over K as the special case of the field intersection problem (cf. the simplest cubic case [12, 14–18]). One of the advantages of using multi-resolvent polynomials is the validity for non-abelian groups (see [19–22]).

Theorem 1.1 *Let K be a field of char $K \neq 2$ and $f_a(X) = X^4 - aX^3 - 6X^2 + aX + 1 \in K[X]$ for $a \in K$. For $a, b \in K$ with $a \neq \pm b$ and $(a^2+16)(b^2+16) \neq 0$, the following three conditions are equivalent:*
(i) *the splitting fields of $f_a(X)$ and of $f_b(X)$ over K coincide;*
(ii) *the polynomial $f_A(X)$ splits completely into four linear factors over K for $A = A_1$ or $A = A_2$ where*

$$A_1 = \frac{ab+16}{-a+b} \quad \text{and} \quad A_2 = \frac{ab-16}{a+b};$$

(iii) *there exists $z \in K$ such that*

$$B = a + \frac{(a^2+16)z(z+1)(z-1)}{f_a(z)}$$

where $B = b$ or $B = -b$.

Moreover, if $\mathrm{Gal}_K f_a(X) \cong C_4$ *(resp.* $\mathrm{Gal}_K f_a(X) \cong C_2$ *or* $\{1\}$*) then* (ii) *occurs for only one of A_1 and A_2 (resp. for both of A_1 and A_2) and* (iii) *occurs for only one of b and $-b$ (resp. for both of b and $-b$).*

Note that the equivalence of the conditions (i) and (iii) is valid also for $a = \pm b$.

By Theorem 1.1, for a fixed $a \in K$ with $a^2 + 16 \neq 0$, we have $\mathrm{Spl}_K f_b(X) = \mathrm{Spl}_K f_a(X)$ where b is given as in Theorem 1.1 (iii) for arbitrary $z \in K$ with $f_a(z) \neq 0$ and $b^2 + 16 \neq 0$. Hence we have the following:

Corollary 1.2 *Let K be an infinite field of char $K \neq 2$. For a fixed $a \in K$ with $a^2 + 16 \neq 0$, there exist infinitely many $b \in K$ such that* $\mathrm{Spl}_K f_b(X) = \mathrm{Spl}_K f_a(X)$.

The following theorem is well known as Siegel's theorem for curves of genus zero (cf. [23, Theorem 6.1], [24, Chap. 8, Sect. 5]).

Theorem (Siegel) *Let K be a number field and $\mathcal{O}_K$ the ring of integers in K. If a rational function $\varphi(s) \in K(s)$ has at least three distinct poles, then there are only finitely many $z \in K$ such that $\varphi(z) \in \mathcal{O}_K$.*

In contrast with Corollary 1.2, by applying Siegel's theorem to Theorem 1.1, we get:

Corollary 1.3 *Let K be a number field and $\mathcal{O}_K$ the ring of integers in K. Assume that $a \in \mathcal{O}_K$ with $a^2 + 16 \neq 0$. Then there exist only finitely many integers $b \in \mathcal{O}_K$ such that $\mathrm{Spl}_K f_b(X) = \mathrm{Spl}_K f_a(X)$. In particular, there exist only finitely many integers $b \in \mathcal{O}_K$ such that $f_{A_i}(X)$, $(i = 1, 2)$, has a linear factor over $\mathbb{Q}$ where A_i is given in Theorem 1.1 (ii).*

We treat the case of $K = \mathbb{Q}$ and $a = m \in \mathbb{Z}$. We get an application of Theorem 1.1 to a related family of quartic Thue equations as follows.

Consider the parametric family of quartic Thue equations

$$F_m(X, Y) := X^4 - mX^3Y - 6X^2Y^2 + mXY^3 + Y^4 = c$$

for $m, c \in \mathbb{Z}$ with $c \neq 0$. Note that $f_m(X) = F_m(X, 1)$. The equation $F_m(X, Y) = c$ has the following solutions:

$$F_m(0, \pm e) = F_m(\pm e, 0) = e^4, \qquad F_m(\mp e, \pm e) = F_m(\pm e, \pm e) = -4e^4.$$

We call such solutions (x, y) to $F_m(x, y) = c$ with $xy(x + y)(x - y) = 0$ the *trivial* solutions.

For $c \in \{\pm 1, \pm 4\}$, Lettl–Pethö [3] and Chen–Voutier [25] gave a complete solution to Thue equation $F_m(X, Y) = c$ independently (cf. Section 4).

In [26], Lettl–Pethö–Voutier showed that for $m \geq 58$, the only primitive solutions $(x, y) \in \mathbb{Z}^2$, i.e., $\gcd(x, y) = 1$, to the Thue inequality

$$|F_m(x, y)| \leq 6m + 7$$

with $|x| \leq y$ are trivial solutions $(0, 1)$, $(\pm 1, 1)$ and $(\pm 1, 2)$. Note that $F_m(\pm 1, 2) = \pm 6m - 7$. Wakabayashi [27] investigated Thue inequalities $|F_{l,m}(x, y)| \leq k$ with two parameters l, m and $F_{1,m} = F_m$.

For $m \in \mathbb{Z}$, we put

$$L_m := \mathrm{Spl}_{\mathbb{Q}} f_m(X).$$

We give the following correspondence between integer solutions to $F_m(X, Y) = c$ and isomorphism classes of the simplest quartic fields L_m.

Theorem 1.4 *Let $m \in \mathbb{Z} \setminus \{0, \pm 3\}$ and $L_m = \mathrm{Spl}_{\mathbb{Q}} f_m(X)$. There exists an integer $n \in \mathbb{Z} \setminus \{\pm m\}$ such that $L_n = L_m$ if and only if there exists nontrivial solution $(x, y) \in \mathbb{Z}^2$, i.e., $xy(x + y)(x - y) \neq 0$, to the quartic Thue equation*

$$F_m(x, y) = c \qquad (*)$$

where c is a divisor of $4(m^2+16)$. *Moreover, integers m, n, and solutions* $(x, y) \in \mathbb{Z}^2$ *to* $(*)$ *can be chosen to satisfy the equation*

$$N = m + \frac{(m^2 + 16)xy(x + y)(x - y)}{F_m(x, y)} \qquad (**)$$

where either $N = n$ *or* $N = -n$, *and the equation* $(**)$ *occurs for only one of* $N = n$ *and* $N = -n$.

The assumption $m \neq 0, \pm 3$ ensures that $f_m(X)$ is irreducible over $\mathbb{Q}$, $\mathrm{Gal}_{\mathbb{Q}} f_m(X) \cong C_4$ and the equality $(**)$ holds for only one of $N = n$ and $N = -n$. This phenomenon comes from the group theoretical reason (see Sect. 2). Indeed, in the case of $m = \pm 3$, $f_{\pm 3}(X) = (X^2 \pm X - 1)(X^2 \mp 4X - 1)$ and the equation $(**)$ occurs for both of $N = 3$ and $N = -3$. Thus nontrivial solutions $(x, y) \in \mathbb{Z}^2$ to $(*)$ which satisfy $(**)$ for $N = -m$ exist (cf. Theorem 1.1). Hence the assumption $m \neq 0, \pm 3$ also ensures that a nontrivial solution (x, y) to $(*)$ corresponds $N \in \mathbb{Z}\backslash\{\pm m\}$ via $(**)$.

If there exists an integer $n \in \mathbb{Z}\backslash\{\pm m\}$ such that $L_n = L_m$, then we may choose a primitive solution $(x, y) \in \mathbb{Z}^2$ to $(*)$ with $(x, y) \equiv (1, 0) \pmod 2$. Then four solutions $\pm(x, y), \pm(y, -x)$ to $(*)$ for $c = d$ and four solutions $\pm(x', y'), \pm(y', -x')$ to $(*)$ for $c = -4d$ are primitive, where $(x', y') = (x - y, x + y)$ and d is an odd divisor of $m^2 + 16$, and only these eight primitive solutions satisfy $(**)$ for the same N as in Theorem 1.4.

Corollary 1.5 *For* $m \in \mathbb{Z}\backslash\{0, \pm 3\}$, *let* $\mathcal{N}$ *be the number of primitive solutions* $(x, y) \in \mathbb{Z}^2$ *with* $xy(x + y)(x - y) \neq 0$ *to* $F_m(x, y) = c$ *where c is a divisor of* $4(m^2 + 16)$. *Then we have*

$$\#\left\{n \in \mathbb{Z} \setminus \{\pm m\} \;\middle|\; L_n = L_m,\ n > 0\right\} = \frac{\mathcal{N}}{8}$$

where $L_m = \mathrm{Spl}_{\mathbb{Q}} f_m(X)$. *In particular, if there does not exist* $n \in \mathbb{Z}\backslash\{\pm m\}$ *with* $L_n = L_m$ *then* $F_m(x, y) = c$ *where c is a divisor of* $4(m^2 + 16)$ *has only trivial solutions* $(x, y) \in \mathbb{Z}^2$ *with* $xy(x + y)(x - y) = 0$.

However we do not know nontrivial solutions to $(*)$ with $m \geq 0$ except for $m = 1$, 2, 3, 4, 22, 103, 956. By Theorem 1.1, we can check that $L_1 = L_{103}$, $L_2 = L_{22}$, $L_4 = L_{956}$. Some numerical examples are given in Sects. 6, 7 and 8.

2 Preliminaries

In order to prove Theorem 1.1, we recall known results of the resolvent polynomials which are fundamental tools in the computational aspects of Galois theory (cf. [28–30]). We intend to explain how to get an answer to the field intersection problem of $f_s(X) = X^4 - sX^3 - 6X^2 + sX + 1$, i.e., for $a, b \in K$ how to determine the intersection of $\mathrm{Spl}_K f_a(X)$ and $\mathrm{Spl}_K f_b(X)$. An answer to the field isomorphism problem (Theorem 1.1) may be obtained as the special case of the field intersection problem.

Let $\overline{K}$ be a fixed algebraic closure of a field K. Let $f(X) := \prod_{i=1}^{m}(X - \alpha_i) \in K[X]$ be a separable polynomial of degree m with some fixed order of the roots $\alpha_1, \ldots, \alpha_m \in \overline{K}$. By resolvent polynomials with suitable invariants, we may determine the Galois group of the polynomial $f(X)$ over K as follows.

Let $R := K[x_1, \ldots, x_m]$ be the polynomial ring over K with m variables $x_1, \ldots, x_m$. For $\Theta \in R$, take a surjective homomorphism $\omega_f : R \to k(\alpha_1, \ldots, \alpha_m)$, $\Theta(x_1, \ldots, x_m) \mapsto \Theta(\alpha_1, \ldots, \alpha_m)$, which is called the specialization map. The kernel of ω_f is the ideal $I_f := \{\Theta \in R \mid \Theta(\alpha_1, \ldots, \alpha_m) = 0\}$ in R.

Let S_m be the symmetric group of degree m. We extend the action of S_m on m letters $\{1, \ldots, m\}$ to that on R by $\pi(\Theta(x_1, \ldots, x_m)) := \Theta(x_{\pi(1)}, \ldots, x_{\pi(m)})$. We define the Galois group of $f(X)$ over K by $\mathrm{Gal}(f/K) := \{\pi \in S_m \mid \pi(I_f) \subseteq I_f\}$.

Then the Galois group of the splitting field $\mathrm{Spl}_K f(X)$ of $f(X)$ over K is isomorphic to $\mathrm{Gal}(f/K)$. If we take another ordering of roots $\alpha_{\pi(1)}, \ldots, \alpha_{\pi(m)}$ of $f(X)$ for some $\pi \in S_m$, the corresponding realization of $\mathrm{Gal}(f/K)$ is conjugate in S_m. Hence, for arbitrary ordering of the roots of $f(X)$, $\mathrm{Gal}(f/K)$ is determined up to conjugacy in S_m.

For $H \le U \le S_m$, an element $\Theta \in R$ is called a U-primitive H-invariant if $H = \mathrm{Stab}_U(\Theta) := \{\pi \in U \mid \pi(\Theta) = \Theta\}$. For a U-primitive H-invariant Θ, the polynomial

$$\mathcal{RP}_{\Theta,U}(X) = \prod_{\overline{\pi} \in U/H} (X - \pi(\Theta)) \in R^U[X]$$

where $\overline{\pi}$ runs through the left cosets of H in U, is called the *formal* U-relative H-invariant resolvent by Θ. The polynomial

$$\mathcal{RP}_{\Theta,U,f}(X) := \omega_f(\mathcal{RP}_{\Theta,U}(X))$$

is called the U-relative H-invariant resolvent of f by Θ. The following theorem is fundamental in the theory of resolvent polynomials (see e.g. [28, p. 95]).

Theorem 2.1 *Let $G = \mathrm{Gal}(f/K)$, $H \le U \le S_m$ be finite groups with $G < U$ and Θ a U-primitive H-invariant. Suppose that $\mathcal{RP}_{\Theta,U,f}(X) = \prod_{i=1}^{l} h_i^{e_i}(X)$ gives the decomposition of $\mathcal{RP}_{\Theta,U,f}(X)$ into a product of powers of distinct irreducible polynomials $h_i(X)$, $(i = 1, \ldots, l)$, in $K[X]$. Then we have a bijection*

$$\begin{array}{ccc} G\backslash U/H & \longrightarrow & \{h_1^{e_1}(X), \ldots, h_l^{e_l}(X)\} \\ G\pi H & \longmapsto & h_\pi(X) = \displaystyle\prod_{\tau H \subseteq G\pi H} (X - \omega_f(\tau(\Theta))) \end{array}$$

where the product runs through the left cosets τH of H in U contained in $G\pi H$, that is, through $\tau = \pi_\sigma \pi$ where π_σ runs a system of representative of the left cosets *of* $G \cap \pi H \pi^{-1}$; *each* $h_\pi(X)$ *is irreducible or a power of an irreducible polynomial with* $\deg(h_\pi(X)) = |G\pi H|/|H| = |G|/|G \cap \pi H \pi^{-1}|$.

Corollary 2.2 *If* $G \le \pi H \pi^{-1}$ *for some* $\pi \in U$ *then* $\mathcal{RP}_{\Theta,U,f}(X)$ *has a linear factor over* K. *Conversely, if* $\mathcal{RP}_{\Theta,U,f}(X)$ *has a non-repeated linear factor over* K *then there exists* $\pi \in U$ *such that* $G \le \pi H \pi^{-1}$.

Note that when $\mathcal{RP}_{\Theta,U,f}(X)$ is not squarefree, there exists a suitable Tschirnhausen transformation $\hat{f}$ of f over K such that $\mathcal{RP}_{\Theta,U,\hat{f}}(X)$ is squarefree (cf. [31], [29, Algorithm 6.3.4]).

We apply Theorem 2.1 to the cyclic quartic case. Let $f^1(X) \in K[X]$ and $f^2(X) \in K[X]$ be separable quartic polynomials over K respectively.

We put

$$f(X) := f^1(X) f^2(X)$$

and

$$G_1 := \mathrm{Gal}(f^1/K), \quad G_2 := \mathrm{Gal}(f^2/K), \quad G := \mathrm{Gal}(f/K).$$

We assume that $G_1, G_2 \le C_4$ and apply Theorem 2.1 to $m = 8$, $f(X) = f^1(X)f^2(X)$, $U = \langle\sigma\rangle \times \langle\tau\rangle$, $H = \langle\sigma\tau\rangle$ or $\langle\sigma\tau^3\rangle$ where $\sigma, \tau \in S_8$ act on $R = K[x_1, \ldots, x_8]$ by

$$\sigma : x_1 \mapsto x_2 \mapsto x_3 \mapsto x_4 \mapsto x_1,$$
$$\tau : x_5 \mapsto x_6 \mapsto x_7 \mapsto x_8 \mapsto x_5.$$

We put $U := \langle\sigma\rangle \times \langle\tau\rangle$. Let Θ_1 (resp. Θ_2) be a U-primitive $\langle\sigma\tau\rangle$-invariant (resp. $\langle\sigma\tau^3\rangle$-invariant). Then we have the U-relative $\langle\sigma\tau\rangle$-invariant (resp. $\langle\sigma\tau^3\rangle$-invariant) resolvent polynomial of $f(X) = f^1(X)f^2(X)$ by Θ_1 (resp. Θ_2) as

$$\mathcal{R}^i_f(X) := \mathcal{RP}_{\Theta_i,U,f}(X), \quad (i = 1, 2).$$

This kind of resolvent polynomial is also called (absolute) *multi-resolvent polynomial* (cf. [32, 33]).

For a squarefree polynomial $\mathcal{R}(X) \in K[X]$ of degree l, we define the *decomposition type* $\mathrm{DT}(\mathcal{R})$ of $\mathcal{R}(X)$ by the partition of l induced by the degrees of the irreducible factors of $\mathcal{R}(X)$ over K. By Theorem 2.1, we get the intersection field $\mathrm{Spl}_K f^1(X) \cap \mathrm{Spl}_K f^2(X)$ via the decomposition types $\mathrm{DT}(\mathcal{R}^1_f)$ and $\mathrm{DT}(\mathcal{R}^2_f)$.

Theorem 2.3 *For* $f(X) = f^1(X)f^2(X) \in K[X]$ *with* $G_1, G_2 \le C_4$, *we assume that* $\#G_1 \ge \#G_2$ *and both* $\mathcal{R}^1_f(X)$ *and* $\mathcal{R}^2_f(X)$ *are squarefree. Then the Galois group* $G = \mathrm{Gal}(f/K)$ *and the intersection field* $\mathrm{Spl}_K f^1(X) \cap \mathrm{Spl}_K f^2(X)$ *are given by the decomposition types* $\mathrm{DT}(\mathcal{R}^1_f)$ *and* $\mathrm{DT}(\mathcal{R}^2_f)$ *as on Table* 1.

We checked the decomposition types $\mathrm{DT}(\mathcal{R}^i_f)$, $(i = 1, 2)$, on Table 1 by GAP [34].

Table 1 The decomposition types

G_1	G_2	G		$\mathrm{DT}(\mathscr{R}_f^1)$	$\mathrm{DT}(\mathscr{R}_f^2)$
		$C_4 \times C_4$	$L_1 \cap L_2 = K$	4	4
	C_4	$C_4 \times C_2$	$[L_1 \cap L_2 : K] = 2$	2, 2	2, 2
		C_4	$L_1 = L_2$	2, 2	1, 1, 1, 1
C_4				1, 1, 1, 1	2, 2
	C_2	$C_4 \times C_2$	$L_1 \cap L_2 = K$	4	4
		C_4	$L_1 \supset L_2$	4	4
	$\{1\}$	C_4	$L_1 \supset L_2 = K$	4	4
	C_2	$C_2 \times C_2$	$L_1 \cap L_2 = K$	2, 2	2, 2
C_2		C_2	$L_1 = L_2$	1, 1, 1, 1	1, 1, 1, 1
	$\{1\}$	C_2	$L_1 \supset L_2$	2, 2	2, 2
$\{1\}$	$\{1\}$	$\{1\}$	$L_1 = L_2 = K$	1, 1, 1, 1	1, 1, 1, 1

Now we get an answer to the field isomorphism problem of

$$f_s(X) = X^4 - sX^3 - 6X^2 + sX + 1$$

via multi-resolvent polynomials $\mathscr{R}^i_{f_{a,b}}(X) := \mathscr{RP}_{\Theta_i,\langle\sigma\rangle\times\langle\tau\rangle,f_{a,b}}$, $(i = 1, 2)$, where

$$f_{a,b}(X) := f_a(X) f_b(X)$$

as the special case of Theorem 2.3. Note that $\mathrm{disc}(f_s(X)) = 4(s^2 + 16)^3$.

Theorem 2.4 *For $a, b \in K$ with $(a^2 + 16)(b^2 + 16) \neq 0$, we assume that both $\mathscr{R}^1_{f_{a,b}}(X)$ and $\mathscr{R}^2_{f_{a,b}}(X)$ are squarefree. Then two splitting fields of $f_a(X)$ and of $f_b(X)$ over K coincide if and only if $\mathscr{R}^1_{f_{a,b}}(X)$ or $\mathscr{R}^2_{f_{a,b}}(X)$ splits completely into four linear factors over K.*

This is an analog of the classical result by Kummer theory. Namely for a field K which contains a primitive fourth root of unity and $a, b \in K$, the splitting fields of $X^4 - a$ and of $X^4 - b$ over K coincide if and only if $X^4 - ab$ or $X^4 - ab^3$ has a linear factor (equivalent to split completely) over K. It is remarkable that Theorem 2.4 does not need the assumption that K contains a primitive fourth root of unity.

3 Proof of Theorem 1.1

We give an explicit answer to the field intersection problem of the simplest quartic polynomials $f_s(X)$ via suitable invariants Θ_1 and Θ_2. As a special case, we obtain Theorem 1.1.

Let $K(z)$ be the rational function field over K and σ a K-automorphism of $K(z)$ of order four which is defined by

$$\sigma : z \mapsto \frac{z-1}{z+1} \mapsto -\frac{1}{z} \mapsto -\frac{z+1}{z-1} \mapsto z.$$

We consider the fixed field $K(z)^{\langle\sigma\rangle}$ and the C_4-extension $K(z)/K(z)^{\langle\sigma\rangle}$. Then we get

$$f_s(X) = \prod_{x\in\mathrm{Orb}_{\langle\sigma\rangle}(z)} (X-x) = (X-z)\left(X-\frac{z-1}{z+1}\right)\left(X+\frac{1}{z}\right)\left(X+\frac{z+1}{z-1}\right)$$
$$= X^4 - sX^3 - 6X^2 + sX + 1$$

where

$$s = \frac{z^4-6z^2+1}{z(z^2-1)} = \frac{(z^2+2z-1)(z^2-2z-1)}{z(z+1)(z-1)}$$

as the generating polynomial of the field extension $K(z)/K(z)^{\langle\sigma\rangle}$. It follows that $K(z)^{\langle\sigma\rangle} = K(s)$ and the Galois group of the polynomial $f_s(X)$ over $K(s)$ is isomorphic to C_4.

We also take another rational function field $K(w)$ over K with indeterminate w, $\tau \in \mathrm{Aut}_K K(w)$ with

$$\tau : w \mapsto \frac{w-1}{w+1} \mapsto -\frac{1}{w} \mapsto -\frac{w+1}{w-1} \mapsto w$$

and $f_t(X) = X^4 - tX^3 - 6X^2 + tX + 1$ where

$$t = \frac{w^4-6w^2+1}{w(w^2-1)} = \frac{(w^2+2w-1)(w^2-2w-1)}{w(w+1)(w-1)}$$

in the same manner as $K(z)$, σ, and $f_s(X)$.

Put $U := \langle\sigma\rangle \times \langle\tau\rangle$. Then the field $K(z,w)$ is $(C_4 \times C_4)$-extension of $K(z,w)^U = K(s,t)$.

In order to apply Theorem 2.4, we should find suitable U-primitive $\langle\sigma\tau\rangle$-invariant Θ_1 and U-primitive $\langle\sigma\tau^3\rangle$-invariant Θ_2.

We may find the following two U-primitive $\langle\sigma\tau\rangle$-invariants which are candidates to Θ_1:

$$\Theta_1 = \sum_{l=0}^{3}(\sigma\tau)^l(zw) = \frac{(w+z)(wz-1)(zw-w-z-1)(zw+w+z-1)}{zw(z^2-1)(w^2-1)}, \quad \text{or}$$

$$\Theta_1 = \prod_{l=0}^{3}(\sigma\tau)^l(z+w)$$
$$= \frac{(z^2w^2-zw^2-z^2w+2zw+w+z+1)(z^2w^2+zw^2+z^2w+2zw-w-z+1)}{zw(z^2-1)(w^2-1)}.$$

However, the multi-resolvent polynomial $\mathscr{R}^1_{f_{a,b}}(X)$ where $f_{a,b}(X) = f_a(X)f_b(X)$ becomes complicated in both the cases above.

Example 3.1 We present two explicit examples of the multi-resolvent polynomials $\mathscr{R}^i_{f_{a,b}}(X) := \mathscr{R}\mathscr{P}_{\Theta_i,\langle\sigma\rangle\times\langle\tau\rangle,f_{a,b}}(X)$ for $i = 1, 2$ where $f_{a,b}(X) := f_a(X)f_b(X)$.

(i) In [20], we gave an answer to the field isomorphism problem of $f_s(X)$ by taking

$$\Theta_1 = \frac{2(z^2+1)(w^2+1)(zw+1)(z-w)}{zw(z^2-1)(w^2-1)}, \quad \Theta_2 = \frac{2(z^2+1)(w^2+1)(zw-1)(z+w)}{zw(z^2-1)(w^2-1)}.$$

Then the corresponding multi-resolvent polynomials are given as

$$\mathscr{R}^1_{f_{a,b}}(X) = X^4 - (a^2+16)(b^2+16)(X^2 - 4(a-b)^2),$$
$$\mathscr{R}^2_{f_{a,b}}(X) = X^4 - (a^2+16)(b^2+16)(X^2 - 4(a+b)^2).$$

(ii) If we take another U-primitive $\langle\sigma\tau\rangle$-invariant Θ_1 and $\langle\sigma\tau^3\rangle$-invariant Θ_2 as

$$\Theta_1 = \frac{2(zw+1)(z-w)}{(z^2+1)(w^2+1)}, \quad \Theta_2 = \frac{2(zw-1)(z+w)}{(z^2+1)(w^2+1)}$$

then we get

$$\mathscr{R}^1_{f_{a,b}}(X) = X^4 - X^2 + \frac{4(a-b)^2}{(a^2+16)(b^2+16)},$$
$$\mathscr{R}^2_{f_{a,b}}(X) = X^4 - X^2 + \frac{4(a+b)^2}{(a^2+16)(b^2+16)}.$$

The multi-resolvent polynomials in Example 3.1 are useful since they are biquadratic, i.e., quadratic polynomial with respect to X^2. However, we do not understand for a fixed $a \in K$ whether there exist infinitely many $b \in K$ such that $\mathscr{R}^i_{f_{a,b}}(X)$, $(i = 1, 2)$, splits completely over K or not. By Theorem 2.4, this question means that for a fixed $a \in K$ whether there exist infinitely many $b \in K$ such that $\mathrm{Spl}_K f_a(X) = \mathrm{Spl}_K f_b(X)$.

It follows from [35, Theorem 1.4] that there exist $\langle\sigma\tau\rangle$-invariant Θ_1 and $\langle\sigma\tau^3\rangle$-invariant Θ_2 such that $K(z, w) = K(z, \Theta_1) = K(z, \Theta_2)$. The following gives such invariants Θ_1 and Θ_2 which is the key lemma of this paper.

Lemma 3.2 *Let* $U = \langle\sigma\rangle \times \langle\tau\rangle$. *We take*

$$\Theta_1 := \frac{zw+1}{-z+w} \quad \text{and} \quad \Theta_2 := \frac{zw-1}{z+w}.$$

Then the following assertions hold:
(i) *the element* Θ_1 *is a* U*-primitive* $\langle\sigma\tau\rangle$*-invariant*;
(ii) *the element* Θ_2 *is a* U*-primitive* $\langle\sigma\tau^3\rangle$*-invariant*;
(iii) *the* U*-orbit of* Θ_i *is given by the same as* $\langle\sigma\rangle$*-orbit of* z;

$$\mathrm{Orb}_U(\Theta_i) = \left\{\Theta_i, \ \frac{\Theta_i - 1}{\Theta_i + 1}, \ -\frac{1}{\Theta_i}, \ -\frac{\Theta_i + 1}{\Theta_i - 1}\right\}, \quad (i = 1, 2).$$

proof We can check the assertions by direct computations. □

The multi-resolvent polynomials $\mathscr{R}^i_{f_{a,b}}(X) := \mathscr{RP}_{\Theta_i, \langle\sigma\rangle\times\langle\tau\rangle, f_a f_b}(X)$, $(i = 1, 2)$, with respect to

$$\Theta_1 = \frac{zw + 1}{-z + w} \quad \text{and} \quad \Theta_2 = \frac{zw - 1}{z + w}$$

as in Lemma 3.2 are given by

$$\mathscr{R}^1_{f_{a,b}}(X) = f_{A_1}(X) = X^4 - \frac{ab + 16}{-a + b}X^3 - 6X^2 + \frac{ab + 16}{-a + b}X + 1, \qquad (1)$$
$$\mathscr{R}^2_{f_{a,b}}(X) = f_{A_2}(X) = X^4 - \frac{ab - 16}{a + b}X^3 - 6X^2 + \frac{ab - 16}{a + b}X + 1$$

where

$$A_1 = \frac{ab + 16}{-a + b} \quad \textit{and} \quad A_2 = \frac{ab - 16}{a + b}.$$

Note that

$$\mathrm{disc}(\mathscr{R}^1_{f_{a,b}}(X)) = \frac{4(a^2 + 16)^3(b^2 + 16)^3}{(a - b)^6}, \quad \mathrm{disc}(\mathscr{R}^2_{f_{a,b}}(X)) = \frac{4(a^2 + 16)^3(b^2 + 16)^3}{(a + b)^6}.$$

By Theorem 2.3, we get the intersection field $\mathrm{Spl}_K f_a(X) \cap \mathrm{Spl}_K f_b(X)$ via Table 1.

Theorem 3.3 *Let* $\mathscr{R}^i_{f_{a,b}}(X)$, $(i = 1, 2)$, *be as in* (1). *For* $a, b \in K$ *with* $a \neq \pm b$ *and* $(a^2 + 16)(b^2 + 16) \neq 0$, *we assume that* $\#\mathrm{Gal}_K f_a(X) \geq \#\mathrm{Gal}_K f_b(X)$. *Then the intersection field* $\mathrm{Spl}_K f_a(X) \cap \mathrm{Spl}_K f_b(X)$ *is given by the decomposition types* $\mathrm{DT}(\mathscr{R}^1_{f_{a,b}})$ *and* $\mathrm{DT}(\mathscr{R}^2_{f_{a,b}})$ *as on Table* 1 *in Theorem* 2.3.

Proof of Theorem 1.1. As the special case of Theorem 3.3, we see the conditions (i) and (ii) are equivalent (cf. also Theorem 2.4).

The condition (iii) is just a restatement of (ii). Indeed, we may check that $z \in K$ is a root of $f_{A_1}(X)$ (resp. $f_{A_2}(X)$) if and only if z satisfies the condition (iii) for $B = b$ (resp. $B = -b$). Note that if z is a root of $f_{A_i}(X)$ then $\frac{z-1}{z+1}, -\frac{1}{z}, -\frac{z+1}{z-1}$ are also roots of $f_{A_i}(X)$ for $i = 1, 2$. By Table 1 as in Theorem 2.3, if $\mathrm{Spl}_K f_a(X) \cong C_4$ (resp. C_2 or $\{1\}$) and $f_{A_i}(z)$ splits completely then $f_{A_j}(X)$ is irreducible (resp. splits completely) over K for $(i, j) = (1, 2)$ and $(2, 1)$. This completes the proof. □

4 Simplest Quartic Fields and Related Thue Equations

In this section, we treat the case of $K = \mathbb{Q}$ and $a = m \in \mathbb{Z}$ for

$$f_a(X) = f_m(X) = X^4 - mX^3 - 6X^2 + mX + 1 \in \mathbb{Z}[X].$$

We first see

Lemma 4.1 (i) *For* $m \in \mathbb{Z}\backslash\{0, \pm 3\}$, $f_m(X)$ *is irreducible over* $\mathbb{Q}$;
(ii) $f_{\pm 3}(X) = (X^2 \pm X - 1)(X^2 \mp 4X - 1)$, $f_0(X) = (X^2 + 2X - 1)(X^2 - 2X - 1)$.

proof It follows from $\mathrm{Gal}_{\mathbb{Q}(s)} f_s(X) \cong C_4$ that $\mathrm{Gal}_{\mathbb{Q}} f_m(X) \cong C_4$, C_2 or $\{1\}$ for $m \in \mathbb{Z}$. Thus if $f_m(X)$ is reducible over $\mathbb{Q}$ then there exist $a, b, c \in \mathbb{Z}$ such that $f_m(X) = (X^2 + aX + c)(X^2 + bX + c)$ with $c = \pm 1$. By comparing the coefficients, we see $c = -1$, because if $c = 1$ then $(m, a, b) = (0, \pm 2\sqrt{2}, \mp 2\sqrt{2})$. Also by comparing the coefficients, if $c = -1$ then $(m, a) = (-b + 4/b, -4/b) \in \mathbb{Z}^2$. Hence $b \in \{\pm 1, \pm 2, \pm 4\}$. In each case, we have $(m, a, b) = (3, -4, 1), (-3, 4, -1)$, $(0, -2, 2)$, $(0, 2, -2)$, $(-3, -1, 4)$, $(3, 1, -4)$. □

For $m \in \mathbb{Z}\backslash\{0, \pm 3\}$, the splitting fields $L_m := \mathrm{Spl}_{\mathbb{Q}} f_m(X)$ of $f_m(X)$ over $\mathbb{Q}$ are totally real cyclic quartic number fields and called the simplest quartic fields.

We consider the related parametric family of quartic Thue equations

$$F_m(X, Y) := X^4 - mX^3Y - 6X^2Y^2 + mXY^3 + Y^4 = c$$

for $m, c \in \mathbb{Z}$ with $c \neq 0$. Note that $f_m(X) = F_m(X, 1)$.

We may assume that $m \geq 0$ because if $(x, y) \in \mathbb{Z}^2$ is a solution to $F_m(x, y) = c$ then (y, x) becomes a solution to $F_{-m}(y, x) = c$. The equation $F_m(x, y) = c$ has the following solutions:

$$F_m(0, \pm e) = F_m(\pm e, 0) = e^4, \qquad F_m(\mp e, \pm e) = F_m(\pm e, \pm e) = -4e^4.$$

We call such solutions $(x, y) \in \mathbb{Z}^2$ to $F_m(x, y) = c$ with $xy(x + y)(x - y) = 0$ the *trivial* solutions.

If $(x, y) \in \mathbb{Z}^2$ is a solution to $F_m(x, y) = c$ then four pairs $\pm(x, y), \pm(y, -x)$ are also solutions because $F_m(X, Y)$ is invariant under the action $X \longmapsto Y \longmapsto -X$ of order four.

For $c \in \{\pm 1, \pm 4\}$, Lettl and Pethö [3] and Chen and Voutier [25] gave a complete solution to Thue equation $F_m(X, Y) = c$ independently. For $c \in \{\pm 1, \pm 4\}$ and $m \geq 0$, all solutions to $F_m(X, Y) = c$ are given by eight trivial solutions

$$F_m(0, \pm 1) = F_m(\pm 1, 0) = 1, \qquad F_m(\mp 1, \pm 1) = F_m(\pm 1, \pm 1) = -4$$

for arbitrary $m \geq 0$ and additionally

$$F_1(\mp 2, \pm 1) = F_1(\pm 1, \pm 2) = -1, \qquad F_1(\pm 3, \pm 1) = F_1(\mp 1, \pm 3) = 4,$$
$$F_4(\mp 3, \pm 2) = F_4(\pm 2, \pm 3) = 1, \qquad F_4(\pm 5, \pm 1) = F_4(\mp 1, \pm 5) = -4.$$

We put

$$(x', y') := (x - y, x + y).$$

Then if $(x, y) \in \mathbb{Z}^2$ is a solution to $F_m(x, y) = c$ then $(x', y') \in \mathbb{Z}^2$ is a solution to $F_m(x', y') = -4c$. Conversely, if (x', y') is a solution to $F_m(x', y') = -4c$ then $x' \equiv y' \pmod 2$ and $(x, y) = (\frac{x'+y'}{2}, \frac{-x'+y'}{2}) \in \mathbb{Z}^2$ is a solution to $F_m(x, y) = c$.

In [26], Lettl–Pethő–Voutier showed that for $m \geq 58$, the only primitive solutions $(x, y) \in \mathbb{Z}^2$, i.e., $\gcd(x, y) = 1$, to the Thue inequality

$$|F_m(x, y)| \leq 6m + 7$$

are trivial solutions $\pm(0, 1)$, $\pm(1, 0)$, $\pm(1, 1)$, $\pm(-1, 1)$ and nontrivial solutions $\pm(2, 1)$, $\pm(-1, 2)$, $\pm(-2, 1)$, $\pm(1, 2)$. We note that $F_m(\pm 2, 1) = F_m(\mp 1, 2) = \mp 6m - 7$.

If $(x, y) \in \mathbb{Z}^2$ is a primitive solution to $(*)$ then four pairs $\pm(x, y)$, $\pm(y, -x)$ are primitive solutions to $(*)$. We also see the following lemma:

Lemma 4.2 *Put* $(x', y') := (x - y, x + y)$. *If* $(x, y) \in \mathbb{Z}^2$ *with* $(x, y) \equiv (0, 1)$ *or* $(1, 0) \pmod 2$ *is a primitive solution to* $F_m(x, y) = c$ *then* c *is an odd integer and* $(x', y') \equiv (1, 1) \pmod 2$ *is a primitive solution to* $F_m(x', y') = -4c$. *Conversely, if* $(x', y') \in \mathbb{Z}^2$ *with* $(x', y') \equiv (1, 1) \pmod 2$ *is a primitive solution to* $F_m(x', y') = d$ *then* $d = -4c$ *for an odd integer* c *and* $(x, y) = (\frac{x'+y'}{2}, \frac{-x'+y'}{2}) \equiv (0, 1)$ *or* $(1, 0) \pmod 2$ *is a primitive solution to* $F_m(x, y) = c$.

proof Assume that $(x, y) \in \mathbb{Z}^2$ with $(x, y) \equiv (0, 1)$ or $(1, 0) \pmod 2$ is a primitive solution to $F_m(x, y) = c$. Then c is odd because $F_m(0, 1) = F_m(1, 0) = 1$. It follows by the definition that $(x', y') \equiv (1, 1) \pmod 2$ and $F_m(x', y') = -4c$. If $\gcd(x', y') = k' > 1$ then k' is odd and k' divides both $x = \frac{x'+y'}{2}$ and $y = \frac{-x'+y'}{2}$. This contradicts to $\gcd(x, y) = 1$. Hence we have $\gcd(x', y') = 1$.

Conversely, we assume that $(x', y') \in \mathbb{Z}^2$ with $(x', y') \equiv (1, 1) \pmod 2$ is a primitive solution to $F_m(x', y') = d$. Then $F_m(1, 1) = -4 \equiv 0 \pmod 4$ and $\not\equiv 0 \pmod 8$. Thus $d = -4c$ for an odd integer $c \in \mathbb{Z}$. We also see $F_m(x, y) = c$. If $\gcd(x, y) = k > 1$ then k divides both $x' = x - y$ and $y' = x + y$. This contradicts to $\gcd(x', y') = 1$. Therefore we have $\gcd(x, y) = 1$. Because if $(x, y) \equiv (1, 1) \pmod 2$ then $F_m(x, y) \equiv 0 \pmod 4$, we also obtain $(x, y) \equiv (0, 1)$ or $(1, 0) \pmod 2$. □

5 Proof of Theorem 1.4: the Correspondence

The aim of this section is to establish the correspondence between isomorphism classes of the simplest quartic fields L_m and nontrivial solutions to quartic Thue equations $(*)$ as follows:

Theorem (Theorem 1.4) *Let $m \in \mathbb{Z}\backslash\{0, \pm 3\}$ and $L_m = \mathrm{Spl}_{\mathbb{Q}} f_m(X)$. Then there exists an integer $n \in \mathbb{Z}\backslash\{\pm m\}$ such that $L_m = L_n$ if and only if there exists nontrivial solution $(x, y) \in \mathbb{Z}^2$, i.e., $xy(x+y)(x-y) \neq 0$, to the quartic Thue equation*

$$F_m(x, y) = x^4 - mx^3y - 6x^2y^2 + mxy^3 + y^4 = c \qquad (*)$$

where c is a divisor of $4(m^2+16)$. Moreover, integers m, n, and solutions $(x, y) \in \mathbb{Z}^2$ to $(*)$ can be chosen to satisfy the equation

$$N = m + \frac{(m^2+16)xy(x+y)(x-y)}{F_m(x, y)} \qquad (**)$$

where either $N = n$ or $N = -n$, and the equation $(**)$ occurs for only one of $N = n$ and $N = -n$.

Proof of Theorem 1.4. We use Theorem 1.1 in the case where $K = \mathbb{Q}$.

For $m \in \mathbb{Z}\backslash\{0, \pm 3\}$, we assume that there exists an integer $n \in \mathbb{Z}\backslash\{\pm m\}$ such that $L_m = L_n$. By Theorem 1.1, there exists $z \in \mathbb{Q}$ such that

$$N = m + \frac{(m^2+16)z(z+1)(z-1)}{f_m(z)}$$

where either $N = n$ or $N = -n$. Write $z = x/y$ with $x, y \in \mathbb{Z}$ and $\gcd(x, y) = 1$ then we have

$$N = m + \frac{(m^2+16)xy(x+y)(x-y)}{F_m(x, y)} \in \mathbb{Z}.$$

By the assumption $n \neq \pm m$, we have $xy(x+y)(x-y) \neq 0$.

We show that $c := F_m(x, y)$ divides $4(m^2+16)$. We make use of a resultant and the Sylvester matrix (cf. [36], [37, Section 1.3], see also [23, Theorem 6.1], [24, Chap. 8, Sect. 5]).

Put $h(z) := (m^2+16)z(z+1)(z-1)$ and $f(z) := F_m(z, 1)$. We take the resultant

$$R_m := \mathrm{Res}_z(h(z), f(z)) = 16(m^2+16)^4$$

of $h(z)$ and $f(z)$ with respect to z. We see that R_m is also given by

$$R_m = \begin{vmatrix} m^2+16 & 0 & -m^2-16 & 0 & 0 & 0 & h(z)z^3 \\ 0 & m^2+16 & 0 & -m^2-16 & 0 & 0 & h(z)z^2 \\ 0 & 0 & m^2+16 & 0 & -m^2-16 & 0 & h(z)z \\ 0 & 0 & 0 & m^2+16 & 0 & -m^2-16 & h(z) \\ 1 & -m & -6 & m & 1 & 0 & f(z)z^2 \\ 0 & 1 & -m & -6 & m & 1 & f(z)z \\ 0 & 0 & 1 & -m & -6 & m & f(z) \end{vmatrix}$$
$$= 4(m^2+16)^3\Big(h(z)p(z)+f(z)q(z)\Big)$$

where

$$p(z) = 5z^3 - 5mz^2 - 29z + 4m, \quad q(z) = -(m^2+16)(5z^2-4).$$

Hence we have

$$h(z)p(z) + f(z)q(z) = 4(m^2+16).$$

Put

$$\begin{aligned} H(x,y) &:= (m^2+16)xy(x+y)(x-y), \\ P(x,y) &:= 5x^3 - 5mx^2y - 29xy^2 + 4my^3, \\ Q(x,y) &:= -(m^2+16)y(5x^2-4y^2). \end{aligned}$$

Then it follows from $z = x/y$ that

$$H(x,y)P(x,y) + F_m(x,y)Q(x,y) = 4(m^2+16)y^7.$$

Because the cubic forms $F_m(X,Y)$ and $H(X,Y)$ are invariants under the action of $\sigma : X \mapsto Y, Y \mapsto -X$, we also get

$$H(x,y)P(y,-x) + F_m(x,y)Q(y,-x) = 4(m^2+16)(-x)^7.$$

Hence we have

$$\begin{aligned} \frac{H(x,y)P(x,y)}{F_m(x,y)} + Q(x,y) &= \frac{4(m^2+16)y^7}{F_m(x,y)} \in \mathbb{Z}, \\ \frac{H(x,y)P(y,-x)}{F_m(x,y)} + Q(y,-x) &= -\frac{4(m^2+16)x^7}{F_m(x,y)} \in \mathbb{Z}. \end{aligned}$$

Since x and y are relatively prime, we conclude that $c = F_m(x,y)$ divides $4(m^2+16)$.

Conversely, if there exists $(x, y) \in \mathbb{Z}^2$ with $xy(x+y)(x-y) \neq 0$ such that $c = F_m(x, y)$ divides $4(m^2+16)$ then we can take $N \in \mathbb{Q}\backslash\{m\}$ by

$$N = m + \frac{(m^2+16)xy(x+y)(x-y)}{F_m(x, y)}.$$

From the assumption $m \in \mathbb{Z}\backslash\{0, \pm 3\}$, we have $\mathrm{Gal}_{\mathbb{Q}} f_m(X) \cong C_4$. Hence it follows from Theorem 1.1 (i) and (iii) that $N \in \mathbb{Q}\backslash\{\pm m\}$ and $L_m = L_N$.

We should show that $N \in \mathbb{Z}$. If $x \not\equiv y \pmod 2$ then $c = F_m(x, y)$ divides $m^2 + 16$ and hence $N \in \mathbb{Z}\backslash\{\pm m\}$, because $F_m(x, y) \equiv F_m(x, y) \equiv 1 \pmod 2$.

If $x \equiv y \pmod 2$ then $c = F_m(x, y)$ divides $(m^2 + 16)xy(x+y)(x-y)$ and hence $N \in \mathbb{Z}\backslash\{\pm m\}$, because $xy(x+y)(x-y) \equiv 0 \pmod 4$. □

6 Primitive Solutions

By the proof of Theorem 1.1 and Theorem 1.4, a nontrivial solution $(x, y) \in \mathbb{Z}^2$ to $(*)$ may be obtained as $z = x/y$ with $\gcd(x, y) = 1$ where $z \in \mathbb{Q}$ is a root of $f_A(X) = X^4 - AX^3 - 6X^2 + AX + 1$ for $A = A_1$ or $A = A_2$ with

$$A_1 = \frac{mn+16}{-m+n} \quad \text{and} \quad A_2 = \frac{mn-16}{m+n}$$

as in Theorem 1.1 (ii). Hence we now consider only primitive solutions $(x, y) \in \mathbb{Z}^2$, i.e., $\gcd(x, y) = 1$, to $(*)$.

Lemma 6.1 *Let $m \in \mathbb{Z}\backslash\{0, \pm 3\}$ and $L_m = \mathrm{Spl}_{\mathbb{Q}} f_m(X)$. We assume that there exists $n \in \mathbb{Z}\backslash\{\pm m\}$ such that $L_n = L_m$.*
(i) *We may choose nontrivial primitive solution $(x, y) \in \mathbb{Z}^2$ with $(x, y) \equiv (0, 1) \pmod 2$ to $F_m(x, y) = d$ where d is an odd divisor of $m^2 + 16$. Then four pairs $\pm(x, y), \pm(y, -x)$ are primitive solutions to $F_m(X, Y) = d$ and four pairs $\pm(x', y')$, $\pm(y', -x')$ are primitive solutions to $F_m(X, Y) = -4d$ where $(x', y') = (x - y, x + y)$.*
(ii) *All primitive solutions to $(*)$ which satisfy $(**)$ for either $N = n$ or $N = -n$ are given by the eight solutions as in* (i), *and such solutions exist for only one of $N = n$ and $N = -n$.*

proof By Theorem 1.4, there exists nontrivial solution $(x_0, y_0) \in \mathbb{Z}^2$ to $F_m(x_0, y_0) = c$ for a divisor c of $4(m^2 + 16)$. By Lemma 4.2, we may choose $(x, y) \in \mathbb{Z}^2$ with $(x, y) \equiv (0, 1)$ or $\equiv (1, 0) \pmod 2$ to $F_m(x, y) = d$ where d is an odd divisor of $m^2 + 16$. Again by Lemma 4.2, eight pairs $\pm(x, y), \pm(y, -x)$ for d and $\pm(x - y, x + y), \pm(x + y, -x + y)$ for $-4d$ are all primitive solutions. Hence we may assume that $(x, y) \equiv (0, 1) \pmod 2$.

These eight solutions correspond to the same N as in Theorem 1.4 because

$$\frac{xy(x+y)(x-y)}{F_m(x,y)}$$

is invariant under the actions of $x \longmapsto y \longmapsto -x$ and $x \longmapsto x-y$, $y \longmapsto x+y$.

For $z = x/y$, the equation (**) holds if and only if $f_A(z) = 0$ where

$$A = \frac{mN+16}{-m+N}$$

(see the proof of Theorem 1.1). We see that if $z = x/y$ is a root of $f_A(X)$ then the other three roots of $f_A(X)$ are given by

$$\frac{z-1}{z+1} = \frac{x-y}{x+y}, \quad -\frac{1}{z} = \frac{y}{-x}, \quad -\frac{z+1}{z-1} = \frac{x+y}{-x+y}.$$

Hence only the primitive solutions to (*) which satisfy (**) for N are the eight solutions above. □

Corollary (Corollary 1.5) *For $m \in \mathbb{Z}\backslash\{0, \pm 3\}$, let $\mathcal{N}$ be the number of primitive solutions $(x, y) \in \mathbb{Z}^2$ with $xy(x+y)(x-y) \neq 0$ to $F_m(x, y) = c$ where c is a divisor of $4(m^2+16)$. Then we have*

$$\#\left\{n \in \mathbb{Z} \setminus \{\pm m\} \;\middle|\; L_n = L_m,\ n > 0\right\} = \frac{\mathcal{N}}{8}$$

where $L_m = \mathrm{Spl}_{\mathbb{Q}} f_m(X)$. In particular, if there does not exist $n \in \mathbb{Z}\backslash\{\pm m\}$ with $L_n = L_m$ then $F_m(x, y) = c$ where c is a divisor of $4(m^2+16)$ has only trivial solutions $(x, y) \in \mathbb{Z}^2$ with $xy(x+y)(x-y) = 0$.

By Theorem 1.1, we obtain

$$L_1 = L_{103}, \quad L_2 = L_{22}, \quad L_4 = L_{956}.$$

Hence for $m \in \{1, 2, 4, 22, 103, 956\}$, we get nontrivial eight primitive solutions to (*) via Theorem 1.4 as in Table 2.

7 Reducible Case

In the reducible case $m \in \{0, \pm 3\}$, $f_m(X)$ splits as $f_0(X) = (X^2+2X-1)(X^2-2X-1)$ and $f_{\pm 3}(X) = (X^2 \pm X - 1)(X^2 \mp 4X - 1)$ over $\mathbb{Q}$, and hence $\mathrm{Spl}_{\mathbb{Q}} f_0(X) = \mathbb{Q}(\sqrt{2})$ and $\mathrm{Spl}_{\mathbb{Q}} f_{\pm 3}(X) = \mathbb{Q}(\sqrt{5})$.

If $m = 0$, the trivial solutions correspond to $N = \pm m = 0$.

If $m = 3$, then the eight trivial solutions $\pm(0, 1), \pm(1, 0)$ for $c = 1$ and $\pm(-1, 1)$, $\pm(1, 1)$ for $c = -4$ give $N = 3$ and nontrivial eight solutions $\pm(2, 1), \pm(-1, 2)$ for $c = -25$ and $\pm(-3, 1), \pm(1, 3)$ for $c = 100$ give $N = -3$ (see Table 3).

8 Computational Result

We do not know nontrivial primitive solutions to (∗) for $m \geq 0$ except in Tables 2 and 3. By the correspondence as in Theorem 1.4, in order to find primitive solutions to (∗) we should get $L_m = L_n$ for some $m \neq \pm n$. In [20, Example 5.4], however, we checked with the aid of computer that for integers $0 \leq m < n \leq 10^5$, $L_m = L_n$ if and only if $(m, n) \in \{(1, 103), (2, 22), (4, 956)\}$. Using Magma [38], we can get the following:

Theorem 8.1 *For $0 \leq m \leq 1000$, all nontrivial primitive solutions $(x, y) \in \mathbb{Z}^2$, i.e., $xy(x + y)(x - y) \neq 0$ and $\gcd(x, y) = 1$, to (∗) are given in Tables 2 and 3. In particular, for $0 \leq m \leq 1000$ with $m \notin \{1, 2, 4, 22, 103, 956\}$ and $n \in \mathbb{Z}$, $L_m = L_n$ implies $m = n$ or $m = -n$.*

Table 2 Nontrivial primitive solutions

m	N	$F_m(x, y) = c$	$m^2 + 16$	$xy(x+y)(x-y)$	(x, y)
1	103	-1	17	-6	$\pm(-2, 1), \pm(1, 2)$
1	103	4	17	24	$\pm(3, 1), \pm(-1, 3)$
2	-22	5	20	-6	$\pm(-2, 1), \pm(1, 2)$
2	-22	-20	20	24	$\pm(3, 1), \pm(-1, 3)$
4	-956	1	32	-30	$\pm(-3, 2), \pm(2, 3)$
4	-956	-4	32	120	$\pm(5, 1), \pm(-1, 5)$
22	-2	$125 = 5^3$	$500 = 2^2 5^3$	-6	$\pm(-2, 1), \pm(1, 2)$
22	-2	$-500 = -2^2 5^3$	$500 = 2^2 5^3$	24	$\pm(3, 1), \pm(-1, 3)$
103	1	$-625 = -5^4$	$10625 = 5^4 17$	6	$\pm(2, 1), \pm(-1, 2)$
103	1	$2500 = 2^2 5^4$	$10625 = 5^4 17$	-24	$\pm(-3, 1), \pm(1, 3)$
956	-4	$28561 = 13^4$	$913952 = 2^5 13^4$	-30	$\pm(-3, 2), \pm(2, 3)$
956	-4	$-114244 = -2^2 13^4$	$913952 = 2^5 13^4$	120	$\pm(5, 1), \pm(-1, 5)$

Table 3 Nontrivial primitive solutions (reducible case)

m	N	$F_m(x, y) = c$	$m^2 + 16$	$xy(x+y)(x-y)$	(x, y)
3	-3	-25	25	6	$\pm(2, 1), \pm(-1, 2)$
3	-3	100	25	-24	$\pm(-3, 1), \pm(1, 3)$

Acknowledgments This work was partially supported by Rikkyo University Special Fund for Research.

References

1. Gras, M.N.: Table numérique du nombre de classes et des unités des extensions cycliques réelles de degré 4 de **Q**, Publ. Math. Fac. Sci. Besançon, fasc 2, 1977/1978
2. Lazarus, A.J.: On the class number and unit index of simplest quartic fields. Nagoya Math. J. **121**, 1–13 (1991)
3. Lettl, G., Pethö, A.: Complete solution of a family of quartic Thue equations. Abh. Math. Sem. Univ. Hamburg **65**, 365–383 (1995)
4. Gaál, I.: Diophantine equations and power integral bases. New computational methods. Birkhäuser Boston Inc, Boston (2002)
5. Duquesne, S.: Elliptic curves associated with simplest quartic fields. J. Théor. Nombres Bordeaux **19**, 81–100 (2007)
6. Hashimoto, K., Hoshi, A.: Families of cyclic polynomials obtained from geometric generalization of Gaussian period relations. Math. Comp. **74**, 1519–1530 (2005)
7. Kim, H.K.: Evaluation of zeta functions at $s = -1$ of the simplest quartic field, In: Proceedings of the 2003 Nagoya Conference Yokoi-Chowla Conjecture and Related Problems, Saga Univ., Saga, 63–73 (2004)
8. Louboutin, S.R.: Efficient computation of root numbers and class numbers of parametrized families of real abelian number fields. Math. Comp. **76**, 455–473 (2007)
9. Rikuna, Y.: On simple families of cyclic polynomials. Proc. Amer. Math. Soc. **130**, 2215–2218 (2002)
10. Hashimoto, K., Miyake, K.: Inverse Galois problem for dihedral groups, number theory and its applications (Kyoto, 1997), 165–181, Dev. Math., 2, Kluwer Acad. Publ. Dordrecht (1999)
11. Miyake, K.: Linear fractional transformations and cyclic polynomials, algebraic number theory (Hapcheon/Saga, 1996). Adv. Stud. Contemp. Math. (Pusan) **1**, 137–142 (1999)
12. Komatsu, T.: Arithmetic of Rikuna's generic cyclic polynomial and generalization of Kummer theory. Manuscripta Math. **114**, 265–279 (2004)
13. Kida, M.: Kummer theory for norm algebraic tori. J. Algebra **293**, 427–447 (2005)
14. Ogawa, H.: Quadratic reduction of multiplicative group and its applications, (Japanese) Algebraic number theory and related topics (Kyoto, 2002). Sūrikaisekikenkyūsho Kōkyūroku No. **1324**, 217–224 (2003)
15. Chapman, R.J.: Automorphism polynomials in cyclic cubic extensions. J. Number Theory **61**, 283–291 (1996)
16. Hoshi, A.: On correspondence between solutions of a parametric family of cubic Thue equations and isomorphic simplest cubic fields. J. Number Theory **131**, 2135–2150 (2011)
17. Hoshi, A., Miyake, K.: A Geometric Framework for the Subfield Problem of Generic Polynomials via Tschirnhausen Transformation, Number Theory and Applications. Hindustan Book Agency, New Delhi, 65–104 (2009)
18. Morton, P.: Characterizing cyclic cubic extensions by automorphism polynomials. J. Number Theory **49**, 183–208 (1994)
19. Hoshi, A., Miyake, K.: Tschirnhausen transformation of a cubic generic polynomial and a 2-dimensional involutive cremona transformation. Proc. Japan Acad. Ser. A **83**, 21–26 (2007)
20. Hoshi, A., Miyake, K.: On the field intersection problem of quartic generic polynomials via formal Tschirnhausen transformation. Comment. Math. Univ. St. Pauli **58**, 51–86 (2009)
21. Hoshi, A., Miyake, K.: On the field intersection problem of generic polynomials: a survey. RIMS Kôkyûroku Bessatsu **B12**, 231–247 (2009)
22. Hoshi, A., Miyake, K.: On the field intersection problem of solvable quintic generic polynomials. Int. J. Number Theory **6**, 1047–1081 (2010)

23. Lang, S.: Elliptic Curves: Diophantine Analysis, Grundlehren der Mathematischen Wissenschaften, 231. Springer, Berlin (1978)
24. Lang, S.: Fundamentals of Diophantine geometry. Springer, New York (1983)
25. Chen, J., Voutier, P.: Complete solution of the Diophantine equation $X^2 + 1 = dY^4$ and a related family of quartic Thue equations. J. Number Theory **62**, 71–99 (1997)
26. Lettl, G., Pethö, A., Voutier, P.: Simple families of Thue inequalities. Trans. Amer. Math. Soc. **351**, 1871–1894 (1999)
27. Wakabayashi, I.: Simple families of Thue inequalities. Ann. Sci. Math. Québec **31**, 211–232 (2007)
28. Adelmann, C.: The Decomposition of Primes in Torsion Point Fields. Lecture Notes in Mathematics. Springer, Berlin (2001)
29. Cohen, H.: A Course in Computational Algebraic Number Theory, Graduate Texts in Mathematics, 138. Springer, Berlin (1993)
30. Cohen, H.: Advanced Topics in Computational Number Theory, Graduate Texts in Mathematics, 193. Springer, New York (2000)
31. Girstmair, K.: On the computation of resolvents and Galois groups. Manuscripta Math. **43**, 289–307 (1983)
32. Rennert, N., Valibouze, A.: Calcul de résolvantes avec les modules de cauchy. Exp. Math. **8**, 351–366 (1999)
33. Rennert, N.: A parallel multi-modular algorithm for computing Lagrange resolvens. J. Symb. Comput. **37**, 547–556 (2004)
34. The GAP Group, GAP – Groups, Algorithms, and Programming, Version 4.4.10. (2007). (http://www.gap-system.org)
35. Ahmad, H., Hajja, M., Kang, M.: Negligibility of projective linear automorphisms. J. Algebra **199**, 344–366 (1998)
36. Poulakis, D., Voskos, E.: On the practical solution of genus zero Diophantine equations. J. Symb. Comput. **30**, 573–582 (2000)
37. Sendra, J.R., Winkler, F., Pérez-Díaz, S.: Rational Algebraic Curves. A computer Algebra Approach. Algorithms and Computation in Mathematics, 22. Springer, Berlin (2008)
38. Bosma, W., Cannon, J., Playoust, C.: The Magma algebra system. I. The user language. J. Symb. Comput. **24**, 235–265 (1997)

On the Implementation of Boolean Gröbner Bases

Shutaro Inoue and Akira Nagai

Abstract We show how we can make Boolean Gröbner base computations feasible on standard computer algebra systems which have a routine to compute Gröbner bases in polynomial rings over the Galois field $\mathbb{GF}_2$. We also show that we can even compute a comprehensive Boolean Gröbner basis using only computations of Gröbner bases in a polynomial ring over $\mathbb{GF}_2$. Our implementation on the computer algebra system Risa/Asir achieves tremendous speedup compared with previous implementations of Boolean Gröbner bases.

1 Introduction

A commutative ring $\mathbf{B}$ with an identity is called a *Boolean ring* if every element of which is idempotent. A residue class ring $\mathbf{B}[X_1, \ldots, X_n]/\langle X_1^2 - X_1, \ldots, X_n^2 - X_n\rangle$ with an ideal $\langle X_1^2 = X_1, \ldots, X_n^2 - X_n\rangle$ also becomes a Boolean ring, which is called a *Boolean polynomial ring* and denoted by $\mathbf{B}(X_1, \ldots, X_n)$. A Gröbner basis in a Boolean polynomial ring (called a *Boolean Gröbner basis*) was first introduced in [1, 2] and further developments were done in [3–7]. The original computation algorithm introduced in [1, 2] used a special monomial reduction which was more complicated than a usual monomial reduction in a polynomial ring over a field. It is also directly applicable for computations of comprehensive Boolean Gröbner bases. This algorithm was first implemented in Prolog as a free software [4] for computations of both Boolean Gröbner bases and comprehensive Boolean Gröbner bases for the case that $\mathbf{B}$ is a Boolean ring $\mathscr{P}_{FC}(S)$ that consists of all finite or co-finite subsets of S. (Here, S is a set of all strings of the computer language.)

It seems a natural and easy way to implement them in a computer algebra system which has a facility to manipulate polynomials, however, it is not very simple

S. Inoue (✉)
Tokyo University of Science, Tokyo, Japan
e-mail: sinoue@rs.kagu.tus.ac.jp

A. Nagai
NTT Information Sharing Platform Laboratories, Tokyo, Japan
e-mail: nagai.akira@lab.ntt.co.jp

© Springer-Verlag Berlin Heidelberg 2014

R. Feng et al. (eds.), *Computer Mathematics*, DOI 10.1007/978-3-662-43799-5_8

to implement computations of the above Boolean ring (including how to represent their data structures) in standard computer algebra systems. In [5], an alternative algorithm is introduced where we can obtain a Boolean Gröbner basis by only computing usual Gröbner bases in a polynomial ring over the Galois field $\mathbb{GF}_2$. Its implementation brought us a much faster program than [4]. Unfortunately, this algorithm is not applicable for computations of comprehensive Boolean Gröbner bases, and no implementations had been done in any computer algebra system.

After a decade of pioneering work on Boolean Gröbner bases, further developments were recently done by [3, 6, 7]. Based on these results, we have implemented a software [8] to compute both Boolean Gröbner bases and comprehensive Boolean Gröbner bases for the Boolean ring $\mathscr{P}_{FC}(S)$ in the computer algebra system Risa/Asir [9]. Our software achieves tremendous speedup compared with the previous one. It enables us to do our recent work [10] of a nontrivial application of Boolean Gröbner bases.

In this short paper, we describe how we can implement computations of both Boolean Gröbner bases and comprehensive Boolean Gröbner bases for the Boolean ring $\mathscr{P}_{FC}(S)$ in the computer algebra system Risa/Asir. We can also easily modify our method for any other computer algebra systems which have a routine to compute Gröbner bases in polynomial rings over the Galois field $\mathbb{GF}_2$.

The reader is referred to [3, 7] for detailed descriptions of the properties of Boolean Gröbner bases which we use in this paper together with important definitions such as a Boolean closed polynomial, a reduced Boolean Gröbner basis, and a stratified Boolean Gröbner basis.

2 Several Key Facts of Boolean Gröbner Bases

Given a finite set F of Boolean polynomials in $\mathbf{B}(X_1, \ldots, X_n)$, let $\mathbf{B}^{'}$ be its smallest Boolean subring that contains all coefficients of polynomials in F. Obviously, any Boolean Gröbner basis G of the ideal $\langle F \rangle$ in $\mathbf{B}^{'}(X_1, \ldots, X_n)$ is also a Boolean Gröbner basis of the ideal $\langle F \rangle$ in $\mathbf{B}(X_1, \ldots, X_n)$. Note that $\mathbf{B}^{'}$ is a finite Boolean ring, so it is isomorphic to a direct product $\mathbb{GF}_2^k$ of the Galois field $\mathbb{GF}_2$ for some natural number k.

In what follows, c_i denotes the ith component of $c \in \mathbb{GF}_2^k$ for each $i = 1, \ldots, k$, f_i denotes the Boolean polynomial of $\mathbb{GF}_2(\bar{X})$ obtained from a Boolean polynomial f of $\mathbb{GF}_2^k(\bar{X})$ by replacing each coefficient c with c_i. $\bar{X}$ is an abbreviation of $X_1, \ldots, X_n$.

Computation of Boolean Gröbner bases in $\mathbb{GF}_2(\bar{X})$ is quite easy.

Theorem 1 *For a finite set $\{f_1, \ldots, f_l\}$ of Boolean polynomials in $\mathbb{GF}_2(\bar{X})$, let G be a (reduced) Gröbner basis of the ideal $\langle f_1, \ldots, f_l, X_1^2 - X_1, \ldots, X_n^2 - X_n \rangle$ in the polynomial ring $\mathbb{GF}_2[\bar{X}]$ over the field $\mathbb{GF}_2$ w.r.t. some term order. Then $G \backslash \{X_1^2 - X_1, \ldots, X_n^2 - X_n\}$ is a (reduced) Boolean Gröbner basis of the ideal $\langle f_1, \ldots, f_l \rangle$ in $\mathbb{GF}_2(\bar{X})$ w.r.t. the same term order.*

The next theorem which is essentially a special instance of Theorem 2.3 of [11] plays an important role in the computation algorithm of Boolean Gröbner bases employed in [5].

Theorem 2 *In a Boolean polynomial ring* $\mathbb{GF}_2^k(\bar{X})$*, let G be a finite set of Boolean closed Boolean polynomials. Then, G is a (reduced) Boolean Gröbner basis of an ideal I in* $\mathbb{GF}_2^k(\bar{X})$ *if and only if* $G_i = \{g_i | g \in G\} \setminus \{0\}$ *is a (reduced) Gröbner bas is of the ideal* $I_i = \{f_i | f \in I\}$ *in* $\mathbb{GF}_2(\bar{X})$ *for each* $i = 1, \ldots, k$.

Example 1 The following left constraint with unknown set variables X and Y is equivalent to the right system of equations of a Boolean polynomial ring $\mathbf{B}(X, Y)$, where $\mathbf{B}$ is a Boolean ring $\mathscr{P}_{FC}(S)$.

$$\begin{cases} X \cup Y \subseteq \{s_1, s_2\} \\ s_1 \in X \\ s_2 \in Y \\ X \cap Y = \emptyset \end{cases} \iff \begin{cases} (1 + \{s_1, s_2\})(XY + X + Y) = 0 \\ \{s_1\}X + \{s_1\} = 0 \\ \{s_2\}Y + \{s_2\} = 0 \\ XY = 0 \end{cases}$$

Let $F = \{(1 + \{s_1, s_2\})(XY + X + Y), \{s_1\}X + \{s_1\}, \{s_2\}Y + \{s_2\}, XY\}$. $\mathbf{B}'$ is a finite subring of $\mathbf{B}$ that consists of $\{0, 1, \{s_1\}, \{s_2\}, \{s_1, s_2\}, 1 + \{s_1\}, 1 + \{s_2\}, 1 + \{s_1, s_2\}\}$. It is isomorphic to $\mathbb{GF}_3$ with the isomorphism ψ given by $\psi(\{s_1\}) = (1, 0, 0)$, $\psi(\{s_2\}) = (0, 1, 0)$ and $\psi(1 + \{s_1, s_2\}) = (0, 0, 1)$. Considering $\mathbf{B}'$ as $\mathbb{GF}_3$ with this isomorphism, $F_1 = \{0, X + 1, 0, XY\}$, $F_2 = \{0, 0, Y + 1, XY\}$ and $F_3 = \{XY + X + Y, 0, 0, XY\}$. Reduced Gröbner bases of them in a Boolean polynomial ring $\mathbb{GF}_2(X, Y)$ w.r.t. a lexicographic term order such that $X > Y$ are $G_1 = \{(1, 0, 0)Y, (1, 0, 0)(X + 1)\}$, $G_2 = \{(0, 1, 0)(Y + 1), (0, 1, 0)X\}$ and $G_3 = \{(0, 0, 1)X, (0, 0, 1)Y\}$ respectively. And we have a reduced Boolean Gröbner basis $G = \{\{s_1\}Y, \{s_1\}(X + 1), \{s_2\}(Y + 1), \{s_2\}X, (1 + \{s_1, s_2\})X, (1 + \{s_1, s_2\})Y\}$ of F. Stratified Boolean Gröbner basis is obtained by simply adding elements that have the same leading monomial. For this G, its stratified Boolean Gröbner basis is $\{X + \{s_1\}, Y + \{s_2\}\}$.

The computation of comprehensive Boolean Gröbner bases is much simpler than the computation of usual comprehensive Gröbner bases in polynomial rings over fields. Given a finite set F of Boolean polynomials in $\mathbf{B}(A_1, \ldots, A_m, X_1, \ldots, X_n)$, let G be a (stratified) Boolean Gröbner basis of the ideal $\langle F \rangle$ in the Boolean polynomial ring $(\mathbf{B}(A_1, \ldots, A_m))(X_1, \ldots, X_n)$ over the coefficient Boolean ring $\mathbf{B}(A_1, \ldots, A_m)$, then G is a (stratified) comprehensive Boolean Gröbner basis of F with parameters $A_1, \ldots, A_m$. We can also apply the above method for them, however, when m is not very small we need a huge natural number k for the isomorphism between $\mathbf{B}(A_1, \ldots, A_m)$ and $\mathbb{GF}_2^k$, namely $k \geq 2^m$. Therefore the above method is not feasible when we have many parameters. The next result recently reported in [3] enables us to apply the above method for the computation of comprehensive Boolean Gröbner bases.

Theorem 3 *Let* $G = \{g_1(\bar{A}, \bar{X}), \ldots, g_k(\bar{A}, \bar{X})\}$ *be a Boolean Gröbner basis of* $\langle F \rangle$ *for* $F = \{f_1(\bar{A}, \bar{X}), \ldots, f_l(\bar{A}, \bar{X})\}$ *in a Boolean polynomial ring* $\mathbf{B}(\bar{A}, \bar{X})$ *w.r.t. a* *block term order $>$ such that $\bar{X} \gg \bar{A}$. Then G is a comprehensive Boolean Gröbner basis of F w.r.t. $>_{\bar{X}}$ (restriction of $>$ on $T(\bar{X})$).*

Example 2 The following left constraint is the same as the previous example except that we have another unknown variable a for an element. Using another set variable A to represent a singleton set $\{a\}$, it is equivalent to the right system of equations of a Boolean polynomial ring $\mathbf{B}(A, X, Y)$.

$$\begin{cases} X \cup Y \subseteq \{s_1, s_2\} \\ s_1 \in X \\ a \in Y \\ X \cap Y = \emptyset \end{cases} \iff \begin{cases} (1 + \{s_1, s_2\})(XY + X + Y) = 0 \\ \{s_1\}X + \{s_1\} = 0 \\ AY + A = 0 \\ XY = 0 \end{cases}$$

Let $F = \{(1 + \{s_1, s_2\})(XY + X + Y), \{s_1\}X + \{s_1\}, AY + A, XY\}$.
The stratified Boolean Gröbner basis G of F w.r.t. a lexicographic term order such that $X > Y > A$ has the following form:

$$\begin{aligned} G = \{&\{s_2\}XY, \{s_2\}YA + \{s_2\}A, (1 + \{s_2\})Y, \\ &\{s_2\}XA, (1 + \{s_2\})X + \{s_1\}, (1 + \{s_2\})A\}. \end{aligned}$$

This is not reduced as a Boolean Gröbner basis in $(\mathbf{B}(A))(X, Y)$. In fact, if we specialize A by $\{s_2\}$, $G(\{s_2\})\backslash\{0\}$ becomes $\{\{s_2\}XY, \{s_2\}Y+\{s_2\}, (1+\{s_2\})Y, \{s_2\}X, (1+\{s_2\})X+\{s_1\}\}$, which is not reduced. But, from the above comprehensive Boolean Gröbner basis we can easily construct the stratified Boolean Gröbner basis of F in $(\mathbf{B}(A))(X, Y)$:

$$\begin{aligned} \{&(\{s_2\}A + \{s_2\})XY, (A + 1 + \{s_2\})X + \{s_1\}A + \{s_1\}, \\ &(A + 1 + \{s_2\})Y + \{s_2\}A, (1 + \{s_2\})A\}. \end{aligned}$$

3 Implementation

We show how we can implement the computation method of Boolean Gröbner bases described in the last section using only the manipulations of polynomial rings over the Galois field $\mathbb{GF}_2$.
In our implementation, an element of $\mathcal{P}_{FC}(S)$ is represented as a polynomial over $\mathbb{GF}_2$. For example, an element $1+\{s_1, s_2\}$ of $\mathcal{P}_{FC}(S)$ is represented as a polynomial $1+s_1+s_2$ of $\mathbb{GF}_2[s_1, s_2]$. Using this representation, a polynomial f of $\mathcal{P}_{FC}(S)[\bar{X}]$ is translated into a polynomial in $\mathbb{GF}_2[s_1, \ldots, s_t, \bar{X}]$, where $s_1, \ldots, s_t$ are all the strings which occurs in some coefficient of f. The smallest Boolean subring of $\mathcal{P}_{FC}(S)$ that contains $s_1, \ldots, s_t$ is isomorphic to the direct product $\mathbb{GF}_2^{t+1}$. Using an isomorphism ψ such that $\psi(\{s_i\})$ is the $(t+1)$-tuple of 0, 1 such that only the ith

component is 1 and the others are all 0 and $\psi(1 + \{s_1, \ldots, s_t\})$ is the $(t+1)$-tuple of 0, 1 such that only the last component is 1, we can consider f as a polynomial of $\mathbb{GF}_2^{t+1}[\bar{X}]$. Under this isomorphism, f_i can be computed by simply specializing s_i with 1 and other s_j with 0 for $i = 1, \ldots, t$, for $i = t+1$ by specializing all variables $s_1, \ldots, s_t$ with 0.

Example 3 $f = (1 + \{s_1, s_2\})XY$ is translated into $XY + s_1 s_2 XY$.
$g = \{s_1, s_2\}X(\{s_1\}Y)$ is translated into $(s_1 + s_2)Xs_1Y = s_1^2 XY + s_1 s_2 XY$.

Note that the second polynomial g could be further simplified to $s_1 XY$, however, we do not employ this simplification since it does not affect the above specializations. By this rather lazy strategy together with the computation technique of Boolean Gröbner bases described in Theorem 1, we can construct most part of Boolean Gröbner bases computations by only using facilities of Risa/Asir. Our codes for the computations of both Boolean Gröbner bases and comprehensive Boolean Gröbner bases consist less than 300 lines.

Example 4 The following are computation examples of Examples 1 and 2 by our software.

```
[1378] G=cbgb([(1+(s1+s2))*(x*y+x+y),s1*x+s1,s2*y+s2,x*y],
                                        [x,y],[],[s1,s2],2,1,1,1)$
[1379] bp_str(G,[x,y],[]);
 [1*y+[s2],1*x+[s1]]

[1380] G=cbgb([(1+(s1+s2))*(x*y+x+y),s1*x+s1,a*y+a,x*y],
                                        [x,y],[a],[s1,s2],2,1,1,1)$
[1382] bp_str(G,[x,y],[a]);
 [([s2]+1)*a,1*a*y+([s2]+1)*y+[s2]*a,
   1*a*x+([s2]+1)*x+[s1]*a+[s1],[s2]*a*y*x+[s2]*y*x]
```

4 Conclusions and Remarks

Our program achieves tremendous speedup compared with the old implementation of [5]. It enables us to do our recent work [10] of a nontrivial application of boolean Gröbner bases. Table 1 contains computation time(in terms of seconds) of seven Boolean Gröbner bases for solving seven Sudoku puzzles which are ranked as extremely difficult. The row Risa/Asir and Klic contain computation times of the same Boolean Gröbner basis in each column by our new program and by the old program of [5] respectively, the symbol ∞ means that the computation did not terminate within 2 h. All computations are done on a PC with 2 GB memory and Core2Duo2 GHZ CPU (Table 1).

Table 1 Timing data

Puzzle	1	2	3	4	5	6	7
Risa/Asir	41.7	43.6	48.1	40.1	44.3	48.9	76.2
Klic	134.1	398.3	1025.3	∞	1242.3	686.5	∞

References

1. Sakai, K., Sato, Y.: Boolean Gröbner bases. ICOT Technical Momorandum 488. http://www.icot.or.jp/ARCHIVE/Museum/TRTM/tm-list-E.html (1988)
2. Sakai, K., Sato, Y., Menju, S.: Boolean Gröbner bases (revised). ICOT Technical Report 613. http://www.icot.or.jp/ARCHIVE/Museum/TRTM/tr-list-E.html (1991)
3. Inoue, S.: On the computation of comprehensive boolean gröbner bases. In: Proceedings of the 11th International Workshop on Computer Algebra in Scientific Computing (CASC 2009), LNCS 5743, pp. 130–141. Springer, Berlin (2009)
4. Sato, Y., et al.: Set constrains solvers (prolog version). http://www.icot.or.jp/ARCHIVE/Museum/FUNDING/funding-96-E.html (1996)
5. Sato, Y. et al.: Set constrains solvers (klic version). http://www.icot.or.jp/ARCHIVE/Museum/FUNDING/funding-98-E.html (1998)
6. Sato, Y., Inoue, S.: On the construction of comprehensive Boolean Gröbner bases. In: Proceedings of the 7th Asian Symposium on Computer Mathematics (ASCM 2005), pp. 145–148 (2005)
7. Sato, Y., Nagai, A., Inoue, I.: On the computation of elimination ideals of boolean polynomial rings. In: Proceedings of the 8th Asian Symposium on Computer Mathematics (ASCM 2007), LNAI 5081, pp. 338–348. Springer, Berlin (2008)
8. Inoue, S.: BGSet - a software to compute Boolean Gröbner bases. http://www.mi.kagu.tus.ac.jp/inoue/BGSet (2009)
9. Noro, M., et al.: A computer algebra system Risa/Asir. http://www.math.kobe-u.ac.jp/Asir/asir.html (2009)
10. Sato, Y., Inoue, S., Suzuki, A., Nabeshima, K.: Boolean Gröbner Bases and Sudoku. Submitted for publication
11. Weispfenning, V.: Gröbner bases in polynomial ideals over commutative regular rings. In Davenport Ed., J.H. (ed.) EUROCAL'87, pp. 336–347. Springer, LNCS 378 (1989)

Comprehensive Gröbner Bases in a Java Computer Algebra System

Heinz Kredel

Abstract We present an implementation of the algorithms for computing comprehensive Gröbner bases in a Java computer algebra system (JAS). Contrary to approaches to implement comprehensive Gröbner bases with minimal requirements to the computer algebra system, we aim to provide all necessary algebraic structures occurring in the algorithm. In the implementation of a condition we aim at the maximal semantic exploitation of the occurring algebraic structures: the set of equations that equal zero are implemented as an ideal (with Gröbner base computation) and the set of inequalities are implemented as a multiplicative set which is simplified to polynomials of minimal degrees using squarefree or irreducible decomposition. The performance of our implementation is compared on well-known examples. With our approach we can also make the transition of a comprehensive Gröbner system to a polynomial ring over a regular coefficient ring and test or compute Gröbner bases.

1 Introduction

In this paper we present an implementation of the algorithms for computing comprehensive Gröbner bases [1–4] in a Java computer algebra system (JAS) [5–8]. This paper contains revised and expanded parts of [9].

JAS uses Java to implement a computer algebra library with special emphasis on object-oriented programming in an algebraic setting. The emphasis of this paper is also on the library design for comprehensive Gröbner bases. Contrary to approaches to implement comprehensive Gröbner bases with minimal requirements to the computer algebra system, like that of Suzuki and Sato [10], we aim to provide and utilize all necessary algebraic structures occurring in the algorithm. For example there are parametric polynomials, colored polynomials, or coefficients in residue class rings.

H. Kredel (✉)
IT-Center, University of Mannheim, 68131 Mannheim, Germany
e-mail: kredel@rz.uni-mannheim.de

© Springer-Verlag Berlin Heidelberg 2014
R. Feng et al. (eds.), *Computer Mathematics*, DOI 10.1007/978-3-662-43799-5_9

In the implementation of a condition we aim at the maximal semantic exploitation of the occurring algebraic structures: the set of equations that equal zero are implemented as an ideal (with Gröbner base computation and ideal membership test) and the set of inequalities are implemented as a multiplicative set which is simplified to polynomials of minimal degrees using squarefree or irreducible decomposition. This approach has partially been taken by [11–13].

With our approach we can moreover make the transition of a comprehensive Gröbner system to an ideal basis in a polynomial ring over a (von Neuman) regular coefficient ring [14, 15]. In this polynomial ring we can then check if the resulting boolean closed ideal basis is a Gröbner base.

1.1 Related Work

Comprehensive Gröbner bases have been introduced by Weispfenning [3] and improved to obtain canonical properties in [4]. Further improvements are by Montes and Manubens [1, 16] and alternative approaches are presented by Sato and Suzuki [2, 10, 17].

A first implementation comprehensive Gröbner bases was by [13] in ALDES/SAC-2 and MAS, which was improved in [11, 12]. Newer implementations are presented in [10, 18–20].

Due to limited space we do not discuss the related mathematical work on Gröbner bases and other computer algebra algorithms, which can be found in standard text books.

1.2 Overview of JAS

JAS provides a well-designed library for algebraic computations implemented with the aid of Java's generic types. The library can be used as any other Java software package or it can be used interactively or interpreted through an jython (Java Python) front end. JAS implements interfaces and classes for basic arithmetic of arbitrary precision integers, rational numbers and multivariate polynomials with such coefficients. Further implemented packages are: algorithms for unique factorization domains including polynomial factorization, algorithms for polynomial and solvable polynomial Gröbner bases, and ideal arithmetic including (commutative) primary ideal decomposition and real root isolation, algorithms for multivariate power series and standard bases. Additionally, there are thread parallel and distributed versions of Buchberger's algorithm.

For an introduction to the JAS type system see [8, 21]. Details on the JAS library can be found, as already mentioned, in [5, 7, 8, 21, 22].

1.3 Outline

Section 2 presents the design of the classes for the implementation of comprehensive Gröbner bases. In the subsections we cover conditions and colored polynomials, parametric reduction and colored systems, Gröbner systems and comprehensive Gröbner bases. In Sect. 3 we present some examples with new performance measurements and the transition to regular coefficient rings. Finally Sect. 4 draws some conclusions. The sections contain revised and expanded parts of [9].

2 Comprehensive Gröbner Bases

Recall some definitions from [3]. Let K be a field, $R = K[U_1, \ldots, U_m]$ a polynomial ring over K in the variables $U_1, \ldots, U_m$. Let $S = R[X_1, \ldots, X_n]$ be a polynomial ring over R in the variables $X_1, \ldots, X_n$ and let $\prec_S$ be a term order of S. S is called a parametric polynomial ring with parameters $U_1, \ldots, U_m$ in the main variables $X_1, \ldots, X_n$. $K[U_1, \ldots, U_m][X_1, \ldots, X_n]$ will be abbreviated by $K[\mathbf{U}][\mathbf{X}]$. For polynomials $f \in S$, the highest term, the leading coefficient, and the leading monomial of f with respect to $\prec_S$ is denoted by $\mathrm{HT}(f)$, $\mathrm{HC}(f)$, and $\mathrm{HM}(f) = \mathrm{HT}(f)\mathrm{HC}(f)$ as usual.

A *specialization* σ of S is a ring homomorphism $\sigma : R \longrightarrow K'$ into some field K'. Let F be a subset of S and let ideal(F) denote the ideal generated by F. A finite subset $G \subset S$ is a *comprehensive Gröbner base* for ideal(F) (with respect to $\prec$), if for all fields K' and all specializations $\sigma : R \longrightarrow K'$ of S, $\sigma(G)$ is a Gröbner base for ideal$(\sigma(F))$ in $K'[X_1, \ldots, X_n]$ (with respect to $\prec$).

Comprehensive Gröbner bases can be computed, for example, via Gröbner systems. A *Gröbner system* $\mathcal{G}$ for an ideal(F), $F \subset S$ is a finite set of pairs (γ, G_γ) where γ is a condition and $G_\gamma \subset S$ is a finite set of polynomials, determined by γ. A comprehensive Gröbner base G for an ideal(F) is then obtained as the union of all G_γ, where each γ also determines F. The meaning of 'condition' and 'determined' is explained next. If in S we have ideal(F) = ideal(G) then G is called a *faithful* comprehensive Gröbner base.

A *condition* γ is a finite set $\{z_i(\mathbf{U}) = 0\} \cup \{n_j(\mathbf{U}) \neq 0\}$ of polynomial equations and inequalities. A *coloring* of the ring R by a condition γ associates a color, namely *green*, *red* and *white*, with each polynomial in R. For $a \in R$, a is colored *green* if $a(\mathbf{U}) = 0$ can be deduced from γ, a is colored *red* if $a(\mathbf{U}) \neq 0$ can be deduced from γ, else a is colored *white*. If a is colored c we write color$(a) = c$. The coloring of R is extended to a coloring of S by the coloring of the coefficients. For $p \in S$ we write $p = p_{green} + p_{red} + p_{white}$ with the restriction $p_{green} \succ p_{red} \succ p_{white}$ for $p_c \neq 0$ (for a color c). Note that we allow p_{white} to contain *green*, *red* and *white* coefficients, but p_{green} and p_{red} may only contain *green* respectively *red* coefficients, if they are not zero. The wording 'deduced' is left unspecified. It may mean simple inspection of the polynomials in γ or the usage of more sophisticated methods, like ideal membership tests.

A polynomial p is said to be *determined* with respect to a condition γ, if $p_{red} \neq 0$ or if $p_{red} = 0$ and $p_{white} = 0$. A set F of polynomials is said to be determined w.r.t. γ, if each $p \in F$ is determined w.r.t. γ. A polynomial p is said to be determined with respect to a set of conditions Γ, if p is determined w.r.t. each $\gamma \in \Gamma$.

More on the mathematical background can be found in [3, 4, 15], see also [2, 10, 18, 19].

2.1 Class Layout

We turn now to the algorithms for the computation of comprehensive Gröbner bases in JAS. Due to space restrictions, we must assume some knowledge of Java, object-oriented programming and JAS [7, 8, 21] in the following. An overview of the object-oriented approach to computer algebra software is given in [23].

The overall layout of the implemented classes is shown in Fig. 1. The computation of comprehensive Gröbner bases in class `ComprehensiveGroebnerBaseSeq` is done via Gröbner systems, class `GroebnerSystem`. Gröbner systems are implemented as lists of colored systems in class `ColoredSystem`. The colored systems consist of a tuple of a condition in class `Condition`, a list of colored polynomials and data structure `OrderedCPairlist` representing the critical pairs to be considered. Class `ColorPolynomial` implements a polynomial colored with respect to a certain condition.

The last class `CReductionSeq` provides methods for parametric reductions relative to conditions and also methods for computing conditions which determine poly-

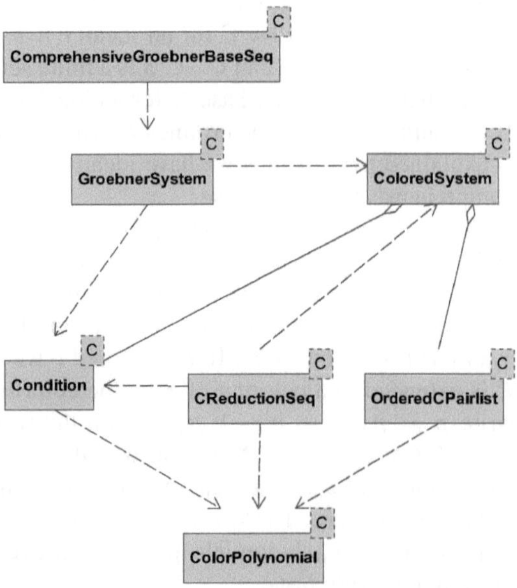

Fig. 1 Overview of involved classes

nomials and sets of polynomials. All classes are parameterized by a type parameter `C` which extends the interface `RingElem⟨C⟩`. The implementation is defined for polynomials with polynomial coefficients over a coefficient ring of type `C`, namely `GenPolynomial⟨GenPolynomial⟨C⟩⟩`.

In the following sections we discuss the functionality of each of the mentioned classes.

2.2 Colored Polynomials and Conditions

Figure 2 shows a class diagram with attributes and methods of the classes `Condition` and `ColorPolynomial`. A condition is defined by a finite set of polynomial equations, polynomials equal to zero $z(\mathbf{U}) = 0$, and a finite set of polynomial inequalities, polynomials not equal to zero $n(\mathbf{U}) \neq 0$. A condition then 'colors' the coefficients of a parametric polynomial in the following way: if a coefficient is contained in the "equals zero" set, it is colored *green*, if a coefficient is contained in the "not equals zero" set, it is colored *red*. In case a coefficient is not contained in

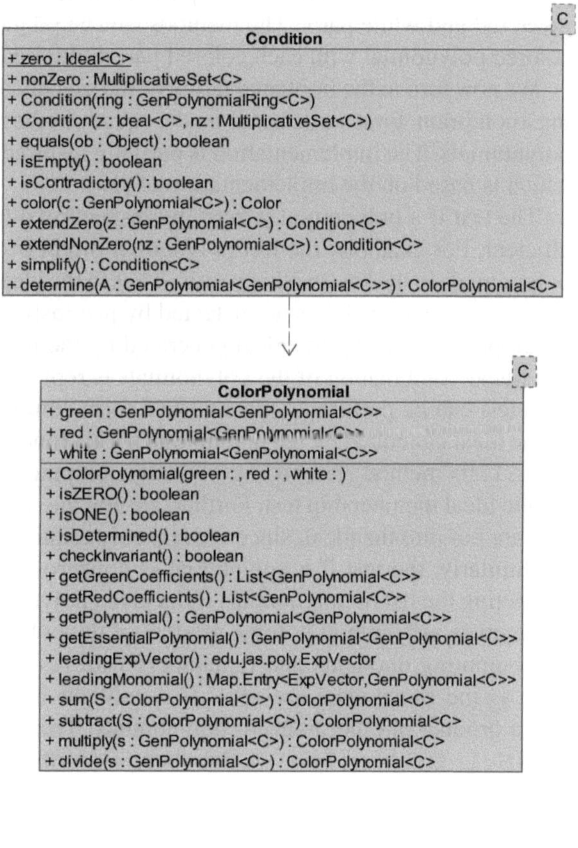

Fig. 2 Conditions and colored polynomials

one of these sets, it is colored *white*. Before we discuss the implementation of these sets, we first explain the rest of the functionality and the colored polynomials.

The class `Condition` provides the method `color()` to deduce if a given (parametric) coefficient is zero or not with respect to this condition. The method `determine()` takes a parametric polynomial as input and returns a colored polynomial with respect to this condition. The methods `extendZero()` and `extendNonZero()` add a (parametric) coefficient to the set of zero, respectively the set of nonzero, polynomial equations.

A colored polynomial `ColorPolynomial` consists of these three colored parts determined by a condition, with the following restriction on the ordering on the terms. A nonzero green part `green` is greater with respect to the term order of the main variables than a nonzero red part `red`, which is greater than a nonzero white part `white`. In case one or more of these parts are zero the restriction holds on the remaining nonzero parts. The method `checkInvariant()` provides a test, if these restrictions are fulfilled. The method `isDetermined()` tests if the red part is nonzero or the white part is also zero. Methods `isZERO()` and `isONE()` ignore the green part in performing the respective test. The `getPolynomial()` method returns the sum of all colored parts. The methods `sum()` and `subtract()` compute a colored polynomial which consists of the sum (difference) of the green parts, a zero red part, and a white part computed from the sum (difference) of the given red and white parts. The methods `multiply()` and `divide()` compute a colored polynomial with each colored part multiplied (divided) by a coefficient.

We now turn to the implementation of the sets of equations and inequalities defining a condition. First we do not store the equations themselves, but only the respective polynomials. The implementation is partially inspired by the implementation in [12] which is based on the implementation of [13].

The test if a polynomial is zero, by inspecting a list of polynomials, is not very efficient. For example, the test polynomial might be a linear combination of some polynomials in the list (in other words, it lies in the ideal generated by the polynomials in the list), a fact which is not detected by just inspecting the list. So we replace the list of polynomials by the ideal generated by the list. Then the test if a polynomial is a linear combination of the polynomials is replaced by an ideal membership test. This test can be performed via a normal form computation modulo a Gröbner base of the ideal generated by the polynomials. This functionality is provided by the class `Ideal`. Its method `contains()` lazily computes a Gröbner base if it is required for the ideal membership test. Further, we add the squarefree part of the polynomials that are put into the ideal, since the test only requires a radical membership test.

Similarly, the test if a polynomial is nonzero can be improved. Instead of just inspecting the list of polynomials if the given polynomial is contained, we can check if the given polynomial is some product of the polynomials in the list. This is done by computing quotients and remainders with respect to polynomials in the list as long as the remainders are zero. If a quotient is constant, the given polynomial was a product of other nonzero polynomials. These algorithms are implemented in class `MultiplicativeSet`. In subclasses further optimizations are implemented, for example, making the polynomials in the set co-prime `MultiplicativeSetCoPrime`, co-prime and squarefree `MultiplicativeSetSquarefree` or irreducible `MultiplicativeSetFactors`. The irreducible factors version relies on the new factorization package, which became usable in 2010. Since it uses Kronecker substitution to reduce multivariate to univariate problems its performance can be improved when Hensel lifting becomes available (see Sect. 3). The default is to use squarefree and co-prime multiplicative sets which are also not too expensive to compute.

The methods `extendZero()` and `extendNonZero()` of `Condition` use the tests just described to avoid adding unnecessary polynomials and to add only maximally reduced polynomials to the respective sets. The methods further try to simplify the condition with method `simplify()` and perform checks for contradictions and return `null` as condition in such a case. Contradictions can show up during the extension operations, as a polynomial in the nonzero list might be contained in the extended ideal generated by the zero polynomials. Similarly a polynomial in the zero polynomial ideal could be a product of polynomials in the extended nonzero polynomials set, again a contradiction. In particular the ideal of zero polynomials might contain one at some extension operation. Such contradictory conditions can then be given special treatment.

2.3 Parametric Reductions and Colored Systems

Class `CReductionSeq` implements parametric reductions with respect to conditions. The class diagram is shown in Fig. 3. The methods `isNormalform()` and `normalform()` test if a polynomial is in reduced form with respect to a list of polynomials or compute such a reduced form relative to a condition and a list of polynomials. All polynomials are colored polynomials and must be colored consistently and be determined. `isNormalform()` checks if a term with a red coefficient is divisible by a red head term of a polynomial in the list.

The computation of the normal form proceeds by inspecting the first non-green term (of the main variables) in the polynomial to be reduced. If it is actually colored green with respect to the condition, then it is put to the green terms of the result polynomial. If it is colored red or white, the term is reduced with respect to a suitable polynomial in the list. If no such polynomial is found, the process ends for top reduction. For non top-reduction the term is put to the result polynomial and the process continues with the next term. Method `SPolynomial()` computes the S-polynomial of two determined polynomials.

The other methods in class `CReductionSeq` implement the computation of sets of conditions and a list of determined polynomial lists. Method `determine()` takes a list of parametric polynomials as input `List⟨GenPolynomial⟨GenPolynomial ⟨C⟩⟩⟩` and returns a list of colored systems `List⟨ColoredSystem⟨C⟩⟩` (explained later). The method first computes a set of conditions for the list of input

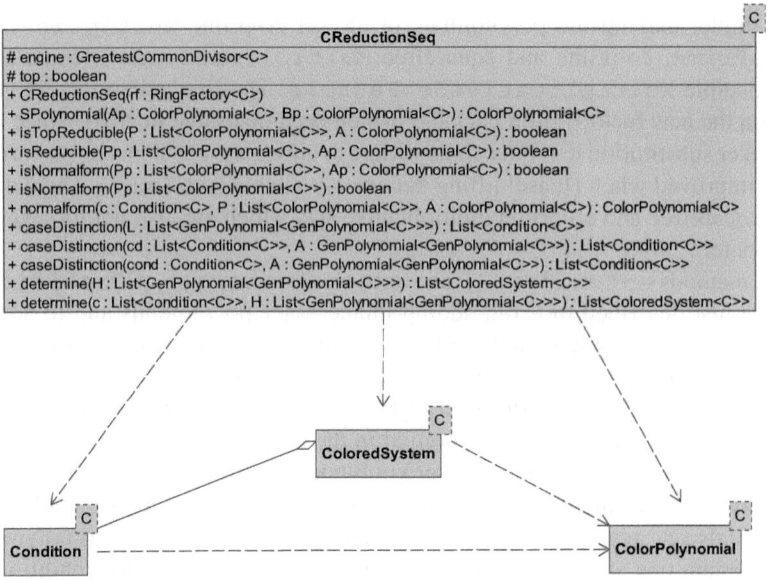

Fig. 3 Parametric reduction

polynomials with method `caseDistinction()` and then determines the polynomials with a method `determine()` which takes a case distinction as input.

A case distinction (a set of conditions) is represented by a list of `Condition` objects. The conditions are constructed in a way, such that every polynomial will have a red head term (or the white part is zero). In the construction of the condition, each (parametric) coefficient of the given polynomial is checked if it is already colored red or green relative to a given condition. If this is not the case, i.e. the coefficient is colored white, the condition is extended two times. First it is extended by adding the coefficient to the set of nonzero polynomials and then it is extended again by adding the coefficient to the set of zero polynomials (as explained in the previous section). If such a newly computed condition is not contradictory and is not already contained in the list of conditions, it is added to the list of conditions.

The two methods `determine()` take a list of parametric polynomials and return a list of `ColoredSystems`, see Fig. 4. A `ColoredSystem` is a container for a `Condition`, which determines a list of `ColorPolynomials` and a `OrderedCPairlist`. Besides the pair-list which is explained below, the `ColoredSystem` class provides methods similar to the `ColorPolynomial` class. Namely, there are methods to check for the validity of the term order invariants or to check if the list of polynomials is correctly determined. Further there are methods to extract lists of the green or red coefficients, the essential parts or the parametric polynomials them-selfs. Other methods just return respective parts of the condition.

To construct a list of `ColoredSystems`, method `determine()` with a list of `Condition` parameter, uses a list variant of method `determine()` of class

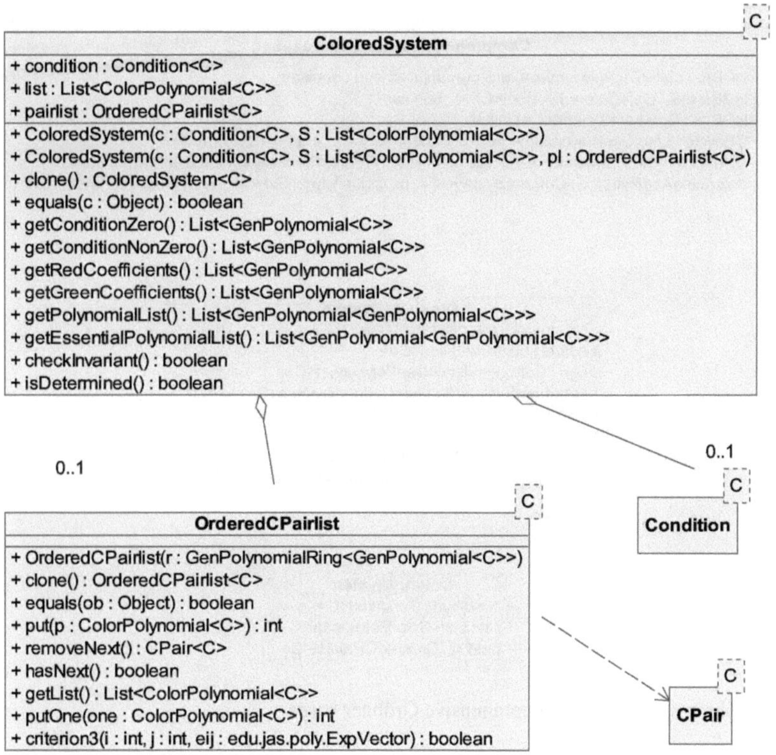

Fig. 4 Colored systems and critical pair lists

`Condition` to compute a list of colored polynomials from the list of the given parametric polynomials. The condition together with the list of determined colored polynomials are then the building parts for the `ColoredSystem` container. The `determine()` method without a `Condition` parameter, first constructs a set of conditions and then constructs a colored system for each condition in the case distinction.

Class `OrderedCPairlist` implements a data structure for the critical pairs to be considered during the course of the Buchberger algorithm. It encapsulates pair selection strategies and book keeping for criteria to avoid critical pairs.

2.4 Gröbner Systems and Comprehensive Gröbner Bases

A `GroebnerSystem` is a container for a list of `ColoredSystems`, see Fig. 5. Like class `ColoredSystem`, it has a method `isDetermined()` to test if all contained colored systems are determined and a method `checkInvariant()` to check all invariants of all contained colored polynomials. The method

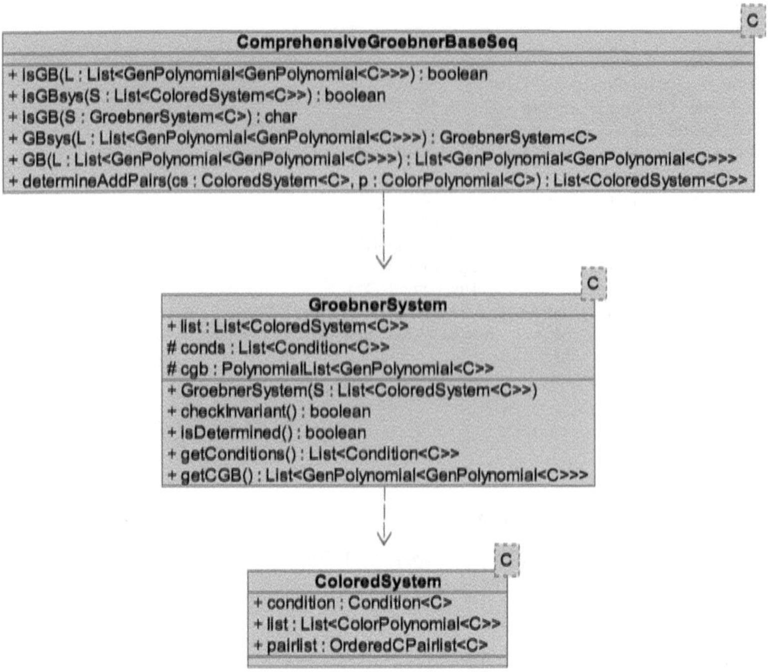

Fig. 5 Gröbner systems and comprehensive Gröbner bases

`getConditions()` extracts a list of all `Conditions` from all `Colored-Systems` and stores them for later access in attribute `conds`. Method `getCGB()` extracts a list of all parametric polynomials as a union of all parametric polynomials from all colored systems.

The computation of comprehensive Gröbner bases via Gröbner systems is implemented in class `ComprehensiveGroebnerBaseSeq`. This class has methods to test if a given list of parametric polynomials is a comprehensive Gröbner base (method `isGB()`) and to test if a given list of colored systems is a Gröbner system (method `isGBsys()`). Both methods are over-loaded to allow also `GroebnerSystems` as parameters and perform the respective checks. Internally there exist two tests, if a list of parametric polynomials is a comprehensive Gröbner base. `isGBcol()` determines the given list of polynomials and calls method `isGBsys()` on the list of `ColoredSystems`. The second test `isGBsubst()` also determines the given list of polynomials but then maps the polynomials to each residue class modulo the zero polynomial ideal contained in the `Condition` and test if it is a Gröbner base over these coefficient rings. To this end we transform the polynomials from `GenPolynomial⟨GenPolynomial⟨C⟩⟩` to `GenPolynomial⟨Residue⟨C⟩⟩` and use method `isGB()` from implementation `GroebnerBasePseudoSeq` for the test, that is, we map

$$K[U_1, \ldots, U_m][X_1, \ldots, X_n] \longrightarrow K[U_1, \ldots, U_m]_{/\mathrm{ideal}(Z_l)}[X_1, \ldots, X_n],$$

where Z_i is the set of polynomials to be treated as zero in condition i. As one last test a random ideal in the coefficient polynomial ring is generated (an ideal generated by random polynomials) and the test is performed modulo this random ideal. Note, that the ideal(Z_i) of the zero conditions might not be a prime ideal. However, the head terms of the polynomials have moreover been colored red by the multiplicative set of nonzero conditions. And, if the condition is not a contradiction, it is guaranteed that they miss all prime ideals which contain the residue class ideal.

To compute a faithful comprehensive Gröbner base, the method `GB()` first computes a `GroebnerSystem` and then extracts the comprehensive Gröbner base with `getCGB()`.

The main work is performed in method `GBsys()`, which takes a list of parametric polynomials as input an returns a `GroebnerSystem` container. In this method, first the method `determine()` of the parametric reduction engine is used to construct a list of determined colored systems. Each `ColoredSystem` is then augmented by a pair list `OrderedCPairlist` containing all critical pairs of the colored polynomials of it. For each `ColoredSystem`, an inner loop iterates over all critical pairs of this system. For each critical pair a parametric S-polynomial and a parametric normalform of it with respect to the list of colored polynomials is computed. If the normal-form polynomial is nonzero, the condition is refined so that it becomes determined with method `determineAddPairs()`. This method also adds new pairs to the critical pair list if required. The method returns a list of `ColoredSystems`, consisting of successors of the actual condition, an updated list of colored polynomials and an updated list of critical pairs. This list of new `ColoredSystems` is then merged with the existing list of `ColoredSystems` and the actual `ColoredSystem` is replaced by a suitable new system. In this way, a depth first search for a `ColoredSystem` with empty pair list is performed. If all critical pairs of the actual `ColoredSystem` are done, it is moved to the result list and the next colored system is taken. By the termination argument for the computation of Gröbner systems only finitely many new colored systems are added at each step and for each colored system only finitely many new critical pairs are generated. So by Königs tree lemma combined with Dickson's lemma the two interleaved loops eventually terminate. Upon termination all critical pairs in a colored system have been processed, so the polynomials form a Gröbner base relative to the given condition. Since the conditions of the colored system cover the empty condition, the list of colored systems form a Gröbner system for the list of given input polynomials.

3 Examples and Gröbner Bases Over Regular Rings

In this section we summarize some old (from [9]) and report on new performance measurements. Then we discuss relations to Gröbner bases for regular rings. We first show the performance on the examples from Raksanyi and Hawes2, see [24]. The examples are contained in the `examples` directory of [21].

Table 1 Gröbner system for Raksanyi and Hawes examples

Example	MAS time	Conditions	JAS time	Conditions
Raksanyi, S, Gr	40	3	520/229/190	5
Raksanyi, L	5,630	22	511/225/175	4
Raksanyi, G	30	3	337/147/99	3
Hawes2, G	>20 min	–	1,119/603/578	5

Time in ms, Term order: G = graded, L = lexicographical, S = Gr = reverse graded, timings in slashes are for subsequent runs

Table 2 Gröbner system for Nabeshima examples

Example	From [19]	Cond	JAS, L	Cond	JAS, G	Cond	JAS, sqf 64	Cond
F_1	31	4	285/97	7	270/99	7	205/162	G, 6
F_2	93	6	2,299/1,664	12	509/165	10	1,070/287	L, 9
F_3	2,203	22	1,186/660	29	1,199/681	29	1,507/363	L, 24
F_4	234	15	1,231/674	34	1,365/751	34	1,084/226	L, 18
F_5	109	6	359/126	11	367/125	8	303/97	L, 8
F_6	359	17	5,280[a]/1,223[a]	9[a]	2,187[a]/861[a]	9[a]	1,312/175	G, 9
F_7	375	7	392/117	6	424/128	6	552/261	G, 6
F_8	133,200	458	2,548/1,788	32	4,883/3,664	32	1,941/416	G, 20

Time in ms, Term order: G = graded, L = lexicographical, cond = number of conditions, timings after slashes are for third run, sqf = squarefree multiplicative sets, 64 = 64 bit JVM
[a] corrected entries from [9]

In Table 1 (parts from [9]) we compared the computation with MAS [12] on the same computer. For JAS we compute the same Gröbner system three times in the same instance of a Java virtual machine. The time for the second and third run is separated by a slash. We see, that for the first run there is considerable time spend in JVM code profiling and just-in-time compilation. In subsequent runs we then see performance improvements. In most cases, the computing times for the third run are less than half to one third of the computing for the first run. We see that for small examples the MAS code runs faster, but for bigger examples the JAS code runs faster.

In Table 2 (with corrected parts from [9]) we present the JAS computation times of the examples from Nabeshima. The timings of Nabeshima are the original timings from the article [19]. These timings are measured on a Pentium M running at 1.73 GHz. The computing times for JAS are in ms on one older AMD 2.1 GHz CPU and a newer AMD 3.0 GHz Dual Core CPU, with JDK 1.6 and 32-bit respectively 64-bit server JVM. So the timings are not directly comparable, but the CPU influence should not be more than a factor of two for the older CPU. Nevertheless we can draw the conclusion, that the CPU or the system software (C, Risa/Asir vs. Java, JAS) is qualitatively in the same speed region and the timing differences seem to be mainly caused by different algorithms. That is, the mathematical optimization to produce a minimal number of conditions, is more important than the relative CPU speed. The new timings and condition count in the last columns are for the algorithms as

Table 3 Gröbner system for Montes examples

Example	From [1]	Cond	JAS	Cond	JAS, sqf 64	Cond	JAS, irr 64	Cond
11.1, L	8,800	6	777/327	23	762/408	16	1,722/554	16
11.2, L	5,200	6	490/143	10	425/175	9	>60 s	–
11.3, L	115,900	7	1,013/516	9	1,244/694	8	1,420/663	8
11.4, L	33,000	7	371,939/355,794	7	64,575/36,653	7	80,865/53,192	7
5.1 simp, L	8,400	4	248/86	3	180/109	3	344/117	3

Time in ms, Term order: L = lexicographical, timings after slashes are for third run, sqf = squarefree, irr = irreducible multiplicative sets, 64 = 64 bit JVM

they have improved over the last year. Again, the times do not reflect the different CPU speeds but the different algorithms which is indicated by the smaller number of conditions (converging to the counts of [19]).

In Table 3 (parts from [9]) we present the JAS computation times of the examples from Montes. The timings of Montes are the original timings from [1]. The last four columns are new timings on the 64-bit CPU and JVM. The colum labeled 'sqf' presents times for the algorithm with squarefree multiplicative sets and column 'irr' with irreducible multiplicative sets. Only for example '11.4' we see an improvement by the faster CPU. The new algorithms have improved in the number of conditions. However, the irreducible factorization does not pay off and increases the computation time considerably without improving the number of conditions. As in the examples above we conclude that the different algorithms are most important for the different computing times. As it is not the primary focus of this paper to compare different algorithmic details the timings show that our object-oriented approach with Java is not slower than other approaches.

As pointed out in [15, 25, 26] there is some strong relation between comprehensive Gröbner bases and Göbner bases over (von Neuman) regular rings [14]. Since we also have Gröbner bases over (finite) regular rings implemented in JAS, we can check, for example, if a comprehensive Gröbner base is indeed also a Gröbner base over a corresponding regular ring.

We take a list of colored polynomials from `GroebnerSystem` (which could be a Gröbner system). From the condition of each colored system we construct a residue class ring modulo the ideal generated from the condition zero polynomials. The (finite) product of these residue class rings are then used as coefficient ring for a polynomial ring with type

```
GenPolynomialRing<Product<Residue<C>>>,
```

in mathematical notation $\left(\prod_{i=1}^{k} K[U_1, \ldots, U_m]_{/\mathrm{ideal}(Z_i)}\right)[X_1, \ldots, X_n]$, where k is the number of colored systems. Note, as mentioned above, the ideal(Z_i) of the zero conditions might not be a prime ideal. However, the head terms of the parametric polynomials have also been colored red by the multiplicative set of nonzero conditions. So, if the condition is not contradictory, it is guaranteed that the head coefficients miss all prime ideals which contain the residue class ideal.

Then the union of the parametric polynomials from the colored polynomials is mapped to this polynomial ring. The described conversion is implemented by the (static) method `toProductRes()` of class `PolyUtilApp`. The boolean closure of such a list of polynomials is constructed by method `booleanClosure()` of class `RReductionSeq`. The computation of Gröbner bases for a regular ring is provided via class `RGroebnerBasePseudoSeq<C>` and method `GB()`. The test for a Gröbner base is implemented by method `isGB()`. So if we start with a boolean closed set derived from a comprehensive Gröbner system, the method `isGB()` of the regular coefficient ring Gröbner base will return `true`. An example is contained in the jython file `examples/raksanyi_cr.py` in [21]. For this example the output of the polynomial ring object looks like this:

```
PolyRing(RR([
    RC(Ideal(PolyRing(QQ(),"a1, a2, a3, a4",PolyRing.lex),
            list=[])),
    RC(Ideal(PolyRing(QQ(),"a1, a2, a3, a4",PolyRing.lex),
            list=[((a4 - a2))])),
    RC(Ideal(PolyRing(QQ(),"a1, a2, a3, a4",PolyRing.lex),
            list=[((a4 - a2)), (a1 * a2 * a3)])),
    RC(Ideal(PolyRing(QQ(),"a1, a2, a3, a4",PolyRing.lex),
            list=[(a1 * a3 * a4)]))
        ]),"x1, x2, x3, x4",PolyRing.lex)
```

The regular ring coefficients `RR` consist of four residue class rings `RC` corresponding to the four conditions of the Gröbner system.

4 Conclusions

We have presented an implementation of the algorithms for computing comprehensive Gröbner bases in a JAS. We provide and utilize all necessary algebraic structures occurring in the comprehensive Gröbner bases algorithm, such as parametric polynomials, colored polynomials, conditions, or colored systems. A condition is implemented as an ideal, with normal Gröbner base computations to decide ideal membership and a multiplicative set which is targeted to produce polynomials of minimal degrees using squarefree or irreducible decomposition.

The computing times for our object-oriented approach using Java are at least as fast as other implementations. Faster CPUs give improvements only for bigger examples and using irreducible multiplicative sets increase computing times considerably without improving the number of conditions. Recent improvements of our implementations have reduced the number of conditions. So we conclude that differences in the computing times are mainly due to different mathematical details (which have not been the primary focus of investigation in this paper). With our explicit algebraic types approach we showed how to transform a comprehensive Gröbner system to a polynomial ring over a (von Neuman) regular coefficient ring and test for Gröbner bases in such polynomial rings. This was possible since the software representations of all rings in JAS snap together like 'LEGO blocks' to build up arbitrary structured rings.

Further, we plan to implement comprehensive Gröbner bases for parametric solvable polynomial rings [27]. There are also many opportunities for utilizing parallelism, see [18, 28, 29] for a start.

Acknowledgments I thank Thomas Becker for discussions on the implementation of a polynomial template library and Raphael Jolly for the discussions on the generic type system suitable for a computer algebra system. JAS itself was improved by requirements from various users, especially Axel Kramer and by valuable feedback from other colleagues, in particular by Dongming Wang, Thomas Sturm, and Wolfgang K. Seiler, to name a few. This paper profited moreover from comments and feedback we received at the conference. Thanks also to Markus Aleksy and Hans-Günther Kruse for encouraging and supporting this work.

References

1. Montes, A.: An new algorithm for discussing Gröbner basis with parameters. J. Symb. Comput. **33**(1–2), 183–208 (2002)
2. Suzuki, A., Sato, Y.: An alternative approach to comprehensive Gröbner bases. J. Symb. Comput. **36**(3–4), 649–667 (2003)
3. Weispfenning, V.: Comprehensive Gröbner bases. J. Symb. Comp. **14**(1), 1–29 (1992)
4. Weispfenning, V.: Canonical comprehensive Gröbner bases. In: ISSAC 2002, pp. 270–276. ACM (2002)
5. Kredel, H.: On the design of a Java computer algebra system. In: Proceedings of PPPJ 2006, pp. 143–152. University of Mannheim (2006)
6. Kredel, H.: Evaluation of a Java computer algebra system. In: Proceedings ASCM 2007, pp. 59–62. National University of Singapore (2007)
7. Kredel, H.: Evaluation of a Java computer algebra system. Lect. Notes Artif. Intell. **5081**, 121–138 (2008)
8. Kredel, H.: On a Java computer algebra system, its performance and applications. Sci. Comput. Program. **70**(2–3), 185–207 (2008)
9. Kredel, H.: Comprehensive Gröbner bases in a Java computer algebra system. In: Proceedings ASCM 2009, pp. 77–90. Kyushu University, Fukuoka, Japan (2009)
10. Suzuki, A., Sato, Y.: A simple algorithm to compute comprehensive Gröbner bases using Gröbner bases. In: Proceedings of ISSAC 2006, pp. 326–331 (2006)
11. Dolzmann, A., Sturm, T.: Redlog: computer algebra meets computer logic. ACM SIGSAM Bull. **31**(2), 2–9 (1997)
12. Kredel, H., Pesch, M.: MAS: The Modula-2 Algebra System. Computer Algebra Handbook, pp. 421–428. Springer, Heidelberg (2003)
13. Schönfeld, E.: Parametrische Gröbnerbasen im Computer Algebra System ALDES / SAC-2. Diplomarbeit. Universität Passau, Passau (1991)
14. Weispfenning, V.: Gröbner bases for polynomial ideals over commutative regular rings. In: Davenport, H. (ed.) Proceedings of ISSAC'87, pp. 336–347. Springer (1987)
15. Weispfenning, V.: Comprehensive Gröbner bases and regular rings. J. Symb. Comput. **41**, 285–296 (2006)
16. Montes, A., Manubens, M.: Improving DISPGB algorithm using the discriminant ideal. J. Symb. Comput. **41**, 1245–1263 (2006)

17. Kapur, D.: An approach for solving systems of parametric polynomial equations. Principles and Practices of Constraint Programming, pp. 217–244. MIT Press, Cambridge (1995)
18. Inoue, S., Sato, Y.: On the parallel computation of comprehensive Gröbner systems. In: Proceedings of PASCO'07, pp. 99–101 (2007)
19. Nabeshima, K.: A speed-up of the algorithm for computing comprehensive Gröbner systems. In: Proceedings of ISSAC 2007, pp. 299–306 (2007)
20. Nabeshima, K.: PGB: A package for computing parametric polynomial systems. In: Proceedings ASCM 2009, pp. 111–122. Kyushu University, Fukuoka, Japan (2009)
21. Kredel, H.: The Java algebra system (JAS). Technical report, http://krum.rz.uni-mannheim.de/jas/ (2000)
22. Kredel, H.: Multivariate greatest common divisors in the Java computer algebra system. In: Proceedings of Automated Deduction in Geometry (ADG), pp. 41–61. East China Normal University, Shanghai (2008)
23. Jolly, R., Kredel, H.: Generic, type-safe and object oriented computer algebra software. In: Proceedings of CASC 2010, pp. 162–177. Springer LNCS 6244 (2010)
24. Gräbe, H.G.: The Symbolicdata project. Technical report, http://www.symbolicdata.org (2000–2006) Accessed March 2010
25. Sato, Y., Suzuki, A.: Gröbner bases in polynomial rings over von Neumann regular rings - their applications. In: Proceedings ASCM 2000, World Scientific Publications, Lecture Notes Series on Computing, vol. 8, pp. 59–62 (2000)
26. Sato, Y., Nagai, A., Inoue, S.: On the computation of elimination ideals of boolean polynomial rings. In: ASCM, pp. 334–348 (2007)
27. Kredel, H.: Solvable Polynomial Rings. Dissertation, Universität Passau (1993)
28. Kredel, H.: Distributed parallel Gröbner bases computation. In: Proceedings of Workshop on Engineering Complex Distributed Systems at CISIS 2009, CD-ROM. Fukuoka Insitute of Technology, Japan (2009)
29. Kredel, H.: Distributed hybrid Gröbner bases computation. In: Proceedings of Workshop on Engineering Complex Distributed Systems at CISIS 2010. CD-ROM. University of Krakow, Poland (2010)

A Practical Method for Floating-Point Gröbner Basis Computation

Tateaki Sasaki

Abstract Computing Gröbner bases with inexact coefficients is eagerly desired in industrial applications, but the computation with floating-point numbers is quite unstable if performed naively. In previous papers, the present author clarified that large term-cancellations occur frequently making the computation unstable, and he proposed a method of removing the harm of exact term-cancellations and a method of estimating the amounts of inexact term-cancellations. However, the estimation is very rough and never satisfactory. The inexact cancellations cause the accuracy loss of the output system, hence it is important to estimate their amounts precisely. In this chapter, we propose a practical method of estimating the amounts of inexact cancellations fairly well and very simply. We have tested our method by several examples, and obtained reasonable results. We also propose a simple device to reduce the amounts of exact term-cancellations.

1 Introduction

By "floating-point Gröbner basis" we mean a Gröbner basis of polynomial ideal computed with floating-point numbers. There are two kinds of floating-point Gröbner bases. The first kind is that the coefficients of the input polynomials are exact (say algebraic numbers or transcendental numbers) but we utilize the floating-point numbers for some reasons. The second kind is that the coefficients of input polynomials are inexact; hence, we inevitably express the coefficients by floating-point numbers. If the numerical errors increase during the computation, we can replay the computation with higher precision for the first kind; however, for the second kind, we must devise to preserve the initial accuracies of the given polynomials as far as possible. In this chapter, we deal with the second kind.

Work supported in part by Japan Society for the Promotion of Science under Grants 19300001.

T. Sasaki (✉)
Institute of Mathematics, University of Tsukuba, Tsukuba, Japan
e-mail: sasaki@math.tsukuba.ac.jp

© Springer-Verlag Berlin Heidelberg 2014
R. Feng et al. (eds.), *Computer Mathematics*, DOI 10.1007/978-3-662-43799-5_10

The first kind of floating-point Gröbner bases were studied by Shirayanagi and Sweedler [16–18]. The second kind of floating-point Gröbner bases were studied by Stetter [19], Kalkbrener [5], Fortuna et al. [3], Traverso and Zanoni [24], Traverso [23], Weispfenning [25], Kondratyev et al. [7], Gonzalez-Vega et al. [4], Stetter [21], Bodrato and Zanoni [1], and so on. In spite of these studies, computation of floating-point Gröbner bases of the second kind has been a serious problem until recent years; the computation was so unstable in most cases if performed naively. This seriousness forced Mourrain to propose the so-called "border bases" [9, 10]. Furthermore, recently, Suzuki [22] and Nagasaka [11] proposed to compute Gröbner bases by reducing large numerical matrices by Gaussian elimination. The border basis is constructed from a set of monomials surrounding the monomials which constitute a basis of the residue class ring of the ideal, hence it is definable only for zero-dimensional ideals. The author is wondering whether the border bases can be computed stably when the input polynomials are inexact. The linear algebra algorithms are currently very time-consuming.

Why the floating-point Gröbner basis computation is so unstable? Shirayanagi [16] pointed out the appearance of *fully-erroneous terms* and Sasaki and Kako [13] pointed out the occurrence of *main-term cancellations*. In the computation with floating-point numbers, subtraction of mathematically the same terms often gives a term with coefficient of no significant bit, and we call such a term fully-erroneous term. In [13], Sasaki and Kako clarified a simple mechanism of term-cancellations in the Gröbner basis computation. If there appears a polynomial of small or large leading term then, in the subsequent computation, cancellations of all the main terms often occur causing large errors. The fully-erroneous terms can be removed simply and efficiently by expressing the input coefficients either by intervals [16] or by "effective floating-point numbers" ("efloats" in short) [13]. Here, the efloats are floating-point numbers proposed by Kako and Sasaki [6] to detect the cancellation errors efficiently but approximately; see Appendix for details. Thus, the biggest problem is how to protect the initial accuracy from large cancellation errors.

Sasaki and Kako classified the term cancellations into two types, *exact* and *inexact*. In [14], Sasaki and Kako pointed out that if we increase the precision then we can remove the harm of the exact cancellations completely, which is the *high-precision method*; we will explain this method in Sect. 2. The systematic inexact cancellation occurs when the input system is ill-conditioned; we will explain this in Sect. 3. The systematic inexact cancellation cannot be removed unless we perform some preconditioning of the input system. Since they cause accuracy loss in the output system, we must know the amount of inexact cancellation. In [14], Sasaki and Kako proposed also a method to estimate the amounts of inexact cancellations occurred on the coefficients. However, the method is very complicated yet it is incomplete theoretically and gives only rough estimations. Therefore, the remaining problem is how to estimate the amounts of inexact cancellations precisely.

In this chapter, we propose a simple and practical method of estimating the amounts of inexact cancellations occurring on the coefficients of intermediate as well as final polynomials fairly precisely. In new method, the high-precision method is combined with a "marking" method which we propose in this chapter. We explain our new method in Sect. 4.1 and implementation details in Sect. 4.2 . We have tested our method by simple examples, and the results are given in Sect. 4.3 . We also discuss a fault of our current marking method in Sect. 4.4 . In Sect. 5, we propose a simple device to reduce the amounts of exact term-cancellations.

2 On Exact Cancellation and High-Precision Method

By F, P, etc., we denote polynomials in $\mathbb{C}[x, y, \ldots, z]$; the coefficients are actually represented by floating-point numbers. By $\|F\|$, we denote the *norm* of F; we employ the infinity norm. The term (monomial) with no coefficient is called the *power product*. By $\mathrm{lt}(F)$, $\mathrm{lc}(F)$, and $\mathrm{rt}(F)$, we denote the *leading term* (monomial), the *leading coefficient*, and the *rest terms*, of F, respectively, with respect to a given order $\succ$: $F = \mathrm{lt}(F) + \mathrm{rt}(F)$, $\mathrm{lt}(F) \succ \mathrm{rt}(F)$. By $\mathrm{Spol}(F, G)$ and $\mathrm{Lred}(F, G)$, we denote the *S-polynomial* of F and G and the *leading-term reduction* of F by G, respectively. We express $\mathrm{Lred}(F, G)$ also as $F \xrightarrow{G}$.

In this chapter, *we restrict the reduction of polynomials only to the leading term reduction*: we compute the Gröbner bases by constructing S-polynomials and performing the leading-term reductions successively. If we need the reduced Gröbner basis then we perform the reduction of nonleading terms after computing an unreduced Gröbner basis.

No matter how the exact cancellation is caused, Sasaki and Kako showed in [14] that the following method protects the initial accuracy from the cancellation errors, so long as the cancellation is exact.

High-precision method Let the accuracy of polynomials of the given initial basis be $\varepsilon_{\mathrm{init}}$ ($\varepsilon_{\mathrm{init}} \ll 1$). Let C_{exct} be the maximum of the exact cancellations occurred on the coefficients of floating-point Gröbner basis computed. Then, initially, set the computational precision $\varepsilon_{\mathrm{cal}}$ to $\varepsilon_{\mathrm{init}}/C_{\mathrm{exct}}$, convert all the coefficients of the input polynomials to efloats or intervals of precision $\varepsilon_{\mathrm{cal}}$, and perform the Gröbner basis computation (Buchberger's procedure). After computing the Gröbner basis, recover the original precision.

Of course, C_{exct} is not known before the computation but we can know it by increasing the precision as $\varepsilon_{\mathrm{cal}} = 10^{-30} \Rightarrow 10^{-60} \Rightarrow 10^{-90} \Rightarrow \cdots$, for example.

We explain why this method protects the initial accuracy even if large main-term cancellations occur. Consider, for example, the computation of $c + d - c$, where c and d are floating-point numbers such that $c \simeq 10^{10}$ and $d \simeq 1$ and that $c+d$ is computed first, hence the exact cancellation of about 10^{10} occurs in the computation of $(c+d) - c$. Suppose that both c and d have relative accuracy 10^{-15} initially; see Fig. 1 at the next page. In the high-precision method, c and d are converted to numbers of 30 decimal figures, for example, among which 15 figures padded at tails are completely meaningless. However, all the 30 figures of c and d are treated by the computer as definite numbers with formal relative errors 10^{-30}. When c and d

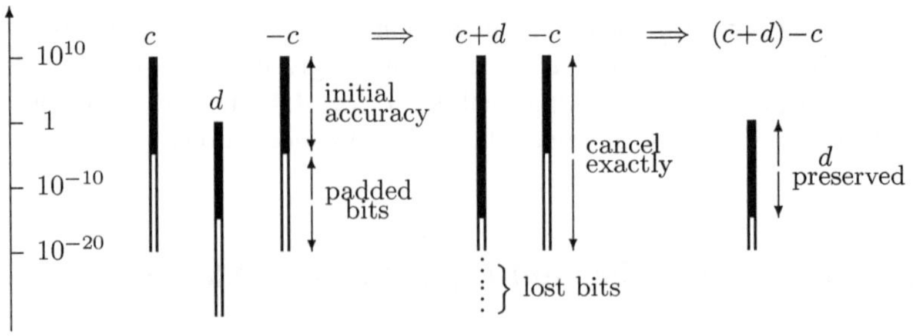

Fig. 1 Computation of $c+d-c$, $|c| \gg |d|$, by high-precision method

are added, $c+d$ becomes a number of formal error about 10^{-20}. After subtracting c from $c+d$, the result becomes a number of magnitude 1 with formal relative error about 10^{-20}, because c and $-c$ cancel exactly. Hence the initial accuracy 10^{-15} of d is preserved in the computation.

3 On Inexact Cancellation and Matrix Formulation

The high-precision method is ineffective for errors caused by inexact cancellation; the mechanism illustrated by Fig. 1 is not applicable to inexact cancellations. We show a simple example in which inexact cancellations occur.

Example 1 (inexact cancellation) Let P_1, P_2, P_3 be as follows.

$$\left\{ \begin{array}{l} P_1 = 57/56\,x^2y + 68/67\,xz^2 - 79/78\,xy + 89/88\,x, \\ P_2 = \qquad\qquad\qquad xyz^3 - \qquad xy^2z + \qquad xyz, \\ P_3 = 56/57\,xy^2 - 67/68\,yz^2 + 78/79\,y^2 \; - 88/89\,y. \end{array} \right\} \tag{3.1}$$

Note that there is the following approximately linear dependent relation among P_1, P_2, P_3:

$$\|56/57\,yzP_1 - 57/56\,xzP_3 - 2P_2\| = 0.000041.$$

We convert the coefficients into double-precision floating-point numbers, which introduces relative errors of 2×10^{-16} into the coefficients. Computing a Gröbner basis w.r.t. the total-degree order with floating-point numbers of 10^{-30}-precision, we obtained the following basis (cf. Example 4 in Sect. 4.3); the underlines show correct figures found by comparing with the computation over $\mathbb{Q}$.

$$
\begin{cases}
P_1 \text{ and } P_3 \text{ are unchanged}, \quad P_2 \to Q_2 \text{ (no cancellation)}, \\
P_6 = y^2z^2 - \underline{2.9954369477}32552644538319700370\, xy^2 + \cdots, \\
P_7 = xz^2 - \underline{0.00176431634237}0426661429391997\, yz^2 \\
\qquad\quad - \underline{0.99472324501868}0541945733244338\, xy \\
\qquad\quad + \underline{0.00176798297372}6193638564792753\, y^2 \\
\qquad\quad + \underline{0.99649064171123}0003668442003074\, x \\
\qquad\quad - \underline{0.177052976567}91266626082187529 2\, y.
\end{cases}
$$

We see that cancellations of about 10^4–10^5 and 10^2 occurred on the coefficients even in high-precision computation. These large errors are due to inexact main-term cancellations. ◇

3.1 Matrix Formulation of Buchberger's Procedure

We can formulate Buchberger's procedure by matrices composed of coefficient vectors of polynomials concerned, just as we can formulate Euclidean algorithm for polynomials by subresultants; see [8, 11]. First, we define $\mathrm{Spol}(F, G)$ and $\mathrm{Lred}(F, G)$ similarly as the pseudo-division of polynomials. Let

$$F = f_1S_1 + f_2S_2 + \cdots + f_mS_m, \qquad G = g_1T_1 + g_2T_2 + \cdots + g_nT_n, \tag{3.2}$$

where $f_1, \ldots, f_m$ and $g_1, \ldots, g_n$ are numbers and $S_1, \ldots, S_m$ and $T_1, \ldots, T_n$ are power-products satisfying $S_1 \succ S_2 \succ \cdots \succ S_m$ and $T_1 \succ T_2 \succ \cdots \succ T_m$. Let $S = M/S_1$ and $T = M/T_1$, where $M = \mathrm{lcm}(S_1, T_1)$. In defining $\mathrm{Spol}(F, G)$, we assume that we have $SS_i = TT_i$ $(i = 2, \ldots, \min\{m, n\})$ (we pad zero-coefficient terms to satisfy these equalities, if necessary). We define $\mathrm{Spol}(F, G)$ to be $\mathrm{Spol}(F, G) = g_1SF - f_1TG$, then we can represent $\mathrm{Spol}(F, G)$ by the following matrix (eliminating f_1 by the second row, we obtain the coefficient vector of $\mathrm{Spol}(F, G)$).

$$\mathrm{Spol}(F, G) \Longleftarrow \begin{pmatrix} f_1 & f_2 & \cdots & f_m \\ g_1 & g_2 & \cdots & g_n \end{pmatrix}. \tag{3.3}$$

Similarly, we define $\mathrm{Lred}(F, G)$ to be $\mathrm{Lred}(F, G) = g_1F - f_1U_1G$, where $U_1 = S_1/T_1$. Hence, if $F \xrightarrow{G}\!\!\!\!\to \tilde{F}$ is k times reductions of F by G then we must multiply g_1^k to G. Let $\tilde{F} = g_1^kF - b_1U_1G - \cdots - b_kU_kG$, where $U_1, \ldots, U_k$ are power-products, then $\tilde{F}$ can be represented by the following matrix (eliminating left elements of the last row, we obtain the coefficient vector of $\tilde{F}$).

$$\tilde{F} \Longleftarrow \begin{pmatrix} g_1 & g_2 & \cdots & g_n & & & \\ & \ddots & \ddots & \cdots & \ddots & & \\ & & & g_1 & g_2 & \cdots & g_n \\ f_1 & f_2 & \cdots & f_m & & & \end{pmatrix}. \tag{3.4}$$

Here, each column corresponds to the same power-product (we pad zero-coefficient terms to satisfy this condition), and the upper k rows are coefficient vectors of $U_1G, \ldots, U_kG$; the g_1 elements may not be on the diagonal positions.

3.2 Characterization of Inexact Cancellations

The matrix formulation tells us what kinds of term cancellations occur in Buchberger's procedure. In Sect. 1, we mentioned that the term cancellations can be classified into two types, exact and inexact. The term cancellations can be classified in another way, *accidental* and *systematic*, as was done in [15]. The accidental term-cancellation happens when the coefficients of two terms coincide accidentally, hence it is not interesting theoretically, although we cannot neglect them practically. On the other hand, in the computation of polynomial remainder sequence, we encounter that many terms cancel each time the remainder is computed except for the first often remainder. Such cancellation is the systematic term-cancellation, and it occurs frequently in Buchberger's procedure, too. The mechanism of systematic exact term-cancellation in Buchberger's procedure was unclear until recently, although a typical mechanism was clarified by Sasaki and Kako [13, 14]. Recently, the present author clarified the mechanism [12]. Since the exact term-cancellations can be made harmless by the high-precision method, we do not consider this type of cancellations any more in this chapter.

The matrix theory tells us that the systematic inexact term-cancellations occur only when there is an approximately linear-dependent relation among coefficient vectors of initial polynomials and their monomial multiples.

Let the initial basis be $\Phi_0 = \{F_1, \ldots, F_r\}$. Any polynomial P appearing in executing Buchberger's procedure can be expressed by $F_1, \ldots, F_r$ as $P = A_1F_1 + \cdots + A_rF_r$, where $A_1, \ldots, A_r$ are polynomials. The tuple $(A_1, \ldots, A_r)$ is called the *syzygy* for P. (The syzygy is not unique and we can compute a syzygy for P by extending Buchberger's procedure slightly). Let $A_i = a_{i,1}U_{i,1} + a_{i,2}U_{i,2} + \cdots$, $(i = 1, \ldots, r)$, where $U_{i,1}, U_{i,2}, \ldots$ are power-products satisfying $U_{i,1} \succ U_{i,2} \succ \cdots$, then P can be represented by the following matrix:

$$P \Longleftarrow \begin{pmatrix} \text{coefficient vector of} U_{1,1}F_1 \\ \quad \text{coefficient vector of} U_{1,2}F_1 \\ \ddots \quad \ddots \quad \ddots \quad \ddots \\ \vdots \quad \vdots \quad \vdots \quad \vdots \quad \vdots \\ \text{coefficient vector of} U_{r,1}F_r \\ \quad \text{coefficient vector of} U_{r,2}F_r \\ \ddots \quad \ddots \quad \ddots \quad \ddots \end{pmatrix}. \tag{3.5}$$

By computing approximately linear-dependent relations among rows of the above matrix, we can know the amounts of inexact cancellations occurred on P.

One may think that we can know the amounts of inexact cancellations by computing syzygies, as was proposed in [13, 14], but it is wrong. We show an example (Example 6 in [13]).

Example 2 (syzygy and cancellation) Let P_1, P_2, P_3 be as follows.

$$\left\{\begin{array}{l} P_1 = x^2\,(2yz+1)/2.0, \\ P_2 = (x\,(3xz-2) - (2yz+1))/3.0, \\ P_3 = (yz\,(3xz-2))/3.0. \end{array}\right\} \tag{3.6}$$

A Gröbner basis w.r.t. the total-degree order is $\{Q_1, Q_2\}$. Let the syzygy for Q_1 be (A_{11}, A_{12}, A_{13}). We have $\|A_{11}\| = 1021.8$, $\|A_{12}\| = 560.0$ and $\|A_{13}\| = 443.9$, if we normalize Q_1 to be 1.0, and $\max\{\|A_1P_1\|, \|A_2P_2\|, \|A_3P_3\|\} \approx 2044$. However, the matrix representation of Q_1 shows that there is no approximately linear dependent relation among its rows. ◇

The matrix representation allows us to define *condition number* for polynomial P. We omit the discussion because the condition number is not useful in our case.

4 A New Method: High-Precision and Marking Method

In this section, we propose a new practical and stable method for floating-point Gröbner basis computation. From now on, by inexact cancellations, we mean the systematic inexact term-cancellations. Before describing the method, we specify the conditions we are given and point out problems we must solve.

- Each coefficient of the input system contains an error, and we assume that we know the amount of error. Let $\varepsilon_{\text{init}}$ be the accuracy of the coefficients of the initial system: $\varepsilon_{\text{init}}$ is an average value of $|\text{error}(c_i)|/|\text{value}(c_i)|$ $(i = 1, 2, \ldots)$, where c_i is the i-th coefficient of the input system. If large inexact cancellations occur on an intermediate polynomial and its initial accuracy is almost lost then the polynomial is meaningless and we must discard it. Similarly, if the accuracy of a coefficient has been lost fully then we must discard the term. Therefore, we are necessary to estimate the amounts of inexact cancellations occurred on the coefficients of not only the output system but also all the polynomials appearing in the computation.
- The high-precision method does not reduce the amounts of exact cancellations, and efloats allow us to detect only the sum of exact and inexact cancellations in each coefficient. In many cases, syzygy for P allows us to estimate the amount of inexact cancellation occurred on P, but the estimation is very bad in some cases. We need some clever device to estimate the amount of inexact cancellation occurred on each coefficient fairly precisely.

The above first point indicates that we are going to compute an "approximate Gröbner basis" if the input coefficients are inexact.

4.1 Description of the New Method

Our new method, which we call *high-precision and marking method*, is based on the following three devices; we introduce "marks" to solve the problem mentioned above.

1. We remove the harm of exact cancellations by the high-precision method, as before; we convert all the coefficients of the input system to high-precision efloats. Let $\varepsilon_{\rm cal}$ be the computational precision, and let $C_{\rm total}$ be the maximum of the total cancellations (exact + inexact) occurred on the coefficients of output system. If

$$\varepsilon_{\rm cal} < \varepsilon_{\rm init}/C_{\rm total}, \tag{4.1}$$

then $\varepsilon_{\rm cal}$ is sufficient for the computation. If $\varepsilon_{\rm cal}$ is found to be insufficient then we increase the precision and replay the Gröbner basis computation.
2. We prepare two initial systems which are marked differently, and compare two Gröbner bases computed from two initial systems. Here, the marking is done as follows. Consider an initial coefficient c_0 of high-precision efloat, with an error r_0. The mark to this coefficient is a tiny number of magnitude $r_0/|c_0|$ and either added to or subtracted from c_0. By this, a bit of c_0 at the head position of r_0 is changed (marked).
3. Let $\Phi'_0 = \{F'_1, \ldots, F'_r\}$ and $\Phi''_0 = \{F''_1, \ldots, F''_r\}$ be two initial systems with high-precision coefficients, where $F'_1, \ldots, F'_r$ are marked in one way while $F''_1, \ldots, F''_r$ are marked in another way. Hence, mutually corresponding coefficients in Φ'_0 and Φ''_0 are the same only in their leading bits above the marks. Suppose $\varepsilon_{\rm cal}$ satisfies (4.1), and consider mutually corresponding coefficients c' and c'' appearing in intermediate bases started from Φ'_0 and Φ''_0, respectively. We divide c' and c'' into two parts, as follows.

$$c' = \bar{c} + d', \quad c'' = \bar{c} + d'', \tag{4.2}$$

where $\bar{c} = (c' + c'')/2$ and d' and d'' are different in their leading bits. We can regard that d' and d'' are errors while $\bar{c}$ is accurate down to the bit just before the leading bits of d' and d''. Then, we can estimate the amount of inexact cancellation occurred on the unmarked coefficient c corresponding to c' and c'' by $|c' - c''|/(\varepsilon_{\rm init}\bar{c})$.

Figure 2 below illustrates the high-precision and marking method. The c'_0 and c''_0 are mutually corresponding coefficients (normalized to 1.0) of initial systems marked differently at $\varepsilon_{\rm init}$: symbols $\circ\circ\circ$ and $***$ show different figures padded to make their precision $\varepsilon_{\rm cal}$. Although padded figures are different, the computations are performed almost the same because the computer treats c'_0 and c''_0 as numbers of accuracy

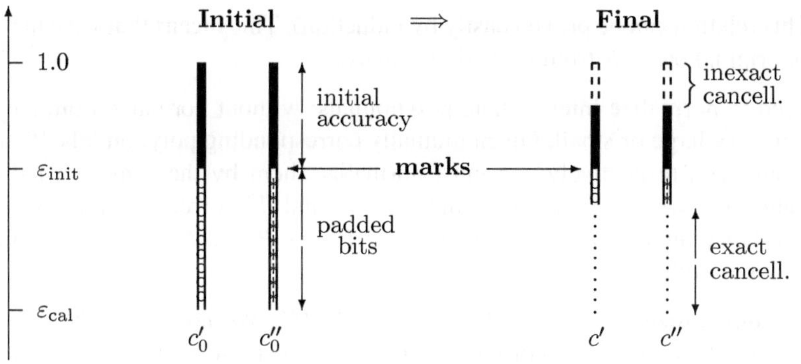

Fig. 2 Illustration of high-precision and marking method

ε_{cal}. The c' and c'' are mutually corresponding coefficients in an intermediate or the final systems marked differently. Tail figures of c' and c'' will be lost by exact cancellations and some leading figures will be lost by inexact cancellations. Since the same inexact cancellation must occur on c' and c'', we can estimate the amount of inexact cancellation by dividing c' and c'', as in (4.2).

4.2 Implementation Details

Although our idea mentioned above is very simple, we need careful considerations on not only how to make marking but also how to normalize intermediate polynomials. For example, if we put mark $\varepsilon_{\text{init}}$ to every coefficient in the system Φ_0' and mark $-\varepsilon_{\text{init}}$ to every coefficient in the system Φ_0'', then systems Φ_0' and Φ_0'' are essentially the same.

We tested several methods of marking and have chosen the following one.

How to mark: We put marks only on nonleading coefficients of the polynomials. We generate random numbers distributed in $(-0.6,\ -0.3)$ and $(0.3,\ 0.6)$. Let c be a nonleading coefficient of the initial system Φ_0, and let r be a random number not used in previous marking. We mark on coefficients c_0' and c_0'' in Φ_0' and Φ_0'', respectively, as follows:

$$c_0' = (1 + r\varepsilon_{\text{init}})\, c_0 \quad \text{and} \quad c_0'' = (1 - r\varepsilon_{\text{init}})\, c_0. \tag{4.3}$$

Merit of our method: Let c', c'', and c be mutually corresponding coefficients in any corresponding intermediate bases (may be final bases) Φ', Φ'' and Φ, respectively. Then, so long as the cancellation errors and rounding errors are ignored, we have

$$(c' + c'')/2 = c. \tag{4.4}$$

(This relation can be proved easily by induction). This means that we can recover the original basis Φ from Φ' and Φ'' easily.

We need to normalize intermediate polynomials: without normalization, they may become very large or small. Given mutually corresponding polynomials P' and P'' in Φ' and Φ'', respectively, we must normalize them by the same normalization constant; otherwise, we cannot compute $\bar{c}$, d', and d'' correctly. Thus, we should not normalize them as $P' \rightarrow P'/\mathrm{lc}(P')$ and $P'' \rightarrow P''/\mathrm{lc}(P'')$. Our normalization method is as follows:

How to normalize: Let $c = [\mathrm{lc}(P') + \mathrm{lc}(P'')]/2$. We normalize P' and P'' by dividing them by c. The normalization is made just after reducing P' and P'' by $\Phi' \setminus \{P'\}$ and $\Phi'' \setminus \{P''\}$, respectively.

Merit of our method: The relation in (4.4) is preserved, so we can make re-marking during the computation.

Remark 1 After many operations, values of d' and d'' in (4.2) will distribute probabilistically, and the distribution is often pretty wide. We can determine the amount of inexact cancellation occurred on P precisely, as follows. Let P be obtained by reducing $\check{P}$ by $\Phi \setminus \check{P}$. Then, express this reduction by a matrix as in (3.5), and eliminate the matrix with pivoting as was done in [15]. In Sect. 4.3 below, we show data obtained by using marks only.

Remark 2 One may think that, if the distribution of marks becomes wide, we had better make re-marking by using the relation $(P' + P'')/2 = P$. However, the re-marking destroys the mechanism of removing harm of systematic exact cancellations. If, however, we know that the computation before re-marking does not cause large systematic exact cancellations in the subsequent computation, then we can make re-marking. Such a case occurs if no polynomial with small/large leading coefficient appears before the re-marking.

4.3 Experiments by Simple Examples

In examples given below, we performed the computation as follows: The input rational coefficients are converted into double-precision floating-point numbers, which introduces about 2×10^{-16} relative errors, then converted to efloats of 30 decimal precision. We set $\varepsilon_{\mathrm{init}}$ to 2×10^{-16}, and put marks to the coefficient as mentioned in Sect. 4.2 . We employ the total-degree order. In the S-polynomial construction, we choose a pair of lowest-order polynomials not chosen yet, and in the reduction, we choose lower-order polynomial earlier as the reducer. We perform only the leading-term reduction, hence the Gröbner basis obtained is unreduced. By #BE$[f, e]$ we denote the big-efloat, where f and e are the value-part and the error-part, respectively. The e was set to $10^{-28}|f|$ initially.

Example 3 (marking method for the system given in Example 2) The unreduced Gröbner basis obtained is $\{Q_1, Q_2\}$. We show Q'_1 and Q''_1:

$$\begin{aligned} Q'_1 &= \#\mathrm{BE}[1.000000000000000297762204178224,\ 6.7\mathrm{e}{-25}]\,x \\ &\quad + \#\mathrm{BE}[1.333333333333337927134496326\,36,\ 8.9\mathrm{e}{-25}]\,y, \\ Q''_1 &= \#\mathrm{BE}[0.999999999999997022377958217\,75,\ 6.7\mathrm{e}{-25}]\,x \\ &\quad + \#\mathrm{BE}[1.333333333333329035591643573\,69,\ 8.9\mathrm{e}{-25}]\,y. \end{aligned}$$

Note that $Q'_1 + Q''_1$ gives Q_1 very accurately.

Remember that the syzygy polynomials for Q_1 are of large norms. The error-parts of coefficients in Q'_1 suggest us that about 6,700~8,900 cancellations (exact + inexact) occurred. Observing the computation steps carefully, we found that no big systematic exact term-cancellation occurs. Hence, we must conclude that small errors in big-efloats accumulate to a big one in this example. The marking method shows that the inexact cancellations occurred on the x- and y-terms are about 15 and 17, respectively. (The matrix representation tells us that there occurs almost no systematic inexact cancellation on Q_1). ◇

Example 4 (marking method for the system given in Example 1) Computing $P_4 :=$ Spol(P_3, P_1), and reducing P_2 as $P_2 \xrightarrow{P_4} \xrightarrow{P_3}\!\!\!\rightarrow Q_2$, we obtain

$$\begin{aligned} Q'_2 &= \#\mathrm{BE}[1.000000000019858335105170719\,54,\ 1.2\mathrm{e}{-23}]\,yz^3 \\ &\quad -\#\mathrm{BE}[1.998309129878761012727503558\,32,\ 1.2\mathrm{e}{-23}]\,xyz \\ &\quad -\#\mathrm{BE}[1.002078216532274410772353569\,79,\ 1.2\mathrm{e}{-23}]\,y^2z \\ &\quad +\#\mathrm{BE}[1.003521717276342545948109561\,65,\ 1.2\mathrm{e}{-23}]\,yz, \\ Q''_2 &= \#\mathrm{BE}[0.999999999980141664894829280\,45,\ 1.2\mathrm{e}{-23}]\,yz^3 \\ &\quad -\#\mathrm{BE}[1.998309129922940053117700770\,80,\ 1.2\mathrm{e}{-23}]\,xyz \\ &\quad -\#\mathrm{BE}[(1.002078216492475240762636648\,7,\ 1.2\mathrm{e}{-23}]\,y^2z \\ &\quad +\#\mathrm{BE}[1.003521717236486404783260874\,68,\ 1.2\mathrm{e}{-23}]\,yz. \end{aligned}$$

We see that large inexact cancellations occurred on Q_2: Q'_2 and Q''_2 show that inexact cancellations of amounts 99300, 55300, 99300, and 99300 occurred on yz^3-, xyz-, y^2z-, and yz-terms, respectively.

The Gröbner basis obtained is $\{P_7, P_3, P_1, Q_2, P_6\}$. We show P'_7 and P''_7.

$$\begin{aligned} P'_7 &= \#\mathrm{BE}[0.999999999969087468358781738\,41,\ 7.0\mathrm{e}{-21}]\,xz^2 \\ &\quad -\#\mathrm{BE}[0.001764316342370377749124736\,99,\ 2.5\mathrm{e}{-23}]\,yz^2 \\ &\quad -\#\mathrm{BE}[0.994723244987876991691976542\,04,\ 7.0\mathrm{e}{-21}]\,xy \\ &\quad +\#\mathrm{BE}[0.001767982973726144589287748\,83,\ 2.5\mathrm{e}{-23}]\,y^2 \\ &\quad +\#\mathrm{BE}[0.996490641680425951644077436\,26,\ 7.0\mathrm{e}{-21}]\,x \\ &\quad -\#\mathrm{BE}[0.001770529765679077225248203\,52,\ 2.5\mathrm{e}{-23}]\,y, \end{aligned}$$

$$
\begin{aligned}
P_7'' &= \#\mathrm{BE}[1.00000000003091253164121826158,\ 7.0\mathrm{e}{-21}]\,xz^2\\
&-\#\mathrm{BE}[0.00176431634237047557373188086,\ 2.5\mathrm{e}{-23}]\,yz^2\\
&-\#\mathrm{BE}[0.99472324504948409219949209434,\ 7.0\mathrm{e}{-21}]\,xy\\
&+\#\mathrm{BE}[0.00176798297372624268783966603,\ 2.5\mathrm{e}{-23}]\,y^2\\
&+\#\mathrm{BE}[0.99649064174203405569280657009,\ 7.0\mathrm{e}{-21}]\,x\\
&-\#\mathrm{BE}[0.00177052976567917609996606021,\ 2.5\mathrm{e}{-23}]\,y.
\end{aligned}
$$

We again see that the sum of coefficients of xz^2-terms is very close to 1.0.

P_7' and P_7'' show that inexact cancellations of amounts 154600, 139, 145800, 139, 154600, and 140 occurred on xz^2-, xy-, y^2- x-, and y-terms, respectively. On the other hand, the error-parts show that the amounts of exact cancellations are not so large. This result is roughly consistent with that in Example 1. ◇

Example 5 (a system which causes very large exact cancellations) Let the input system be as follows:

$$
\left\{
\begin{aligned}
P_1 &= x^3/30 + x^2y + y^2/3,\\
P_2 &= x^2y^2/3 - xy^2 - xy/3,\\
P_3 &= y^3/20 + x^2.
\end{aligned}
\right\}
$$

This is the system investigated in [13] and known to cause large systematic exact cancellations of amounts $O(10^9)$.

The Gröbner basis obtained is $\{Q_2, Q_4, Q_5\}$, where $Q_2 = y^2$ and

$$
\begin{aligned}
Q_4 &= xy\\
&\quad + \#\mathrm{BE}[0.08440225504521958676284694861,\ 3.1\mathrm{e}{-20}]\,y^2,\\
Q_5 &= x^2\\
&\quad + \#\mathrm{BE}[7.14849689746270700639023440400,\ 4.2\mathrm{e}{-19}]\,xy\\
&\quad + \#\mathrm{BE}[0.57371613952464572257619626087,\ 3.4\mathrm{e}{-20}]\,y^2.
\end{aligned}
$$

The error-parts show that big cancellations of $O(10^8)$–$O(10^9)$ have occurred. Q_5' and Q_5'', for example, are as follows:

$$
\begin{aligned}
Q_5' &= \#\mathrm{BE}[1.00000000000000043403304762146,\ 2.4\mathrm{e}{-19}]\,x^2\\
&\quad + \#\mathrm{BE}[7.14849689746270864028809323982,\ 8.4\mathrm{e}{-19}]\,xy\\
&\quad + \#\mathrm{BE}[0.57371613952464591195996112900,\ 6.8\mathrm{e}{-20}]\,y^2,\\
Q_5'' &= \#\mathrm{BE}[0.99999999999999956596695237853,\ 2.4\mathrm{e}{-19}]\,x^2\\
&\quad + \#\mathrm{BE}[7.14849689746270537249237556819,\ 8.4\mathrm{e}{-19}]\,xy\\
&\quad + \#\mathrm{BE}[0.57371613952464553319243139273,\ 6.8\mathrm{e}{-20}]\,y^2.
\end{aligned}
$$

Q_5' and Q_5'' show that the amounts of inexact cancellations occurred on x^2-, xy-, and y^2-terms are 2, 1, and 2, respectively. Therefore, the cancellations occurred are mostly systematic exact ones. This result is consistent with the estimation by matrix representations. ◇

4.4 Discussion

By the experiments given above, we may say that our high-precision marking method is quite effective for detecting the occurrence of systematic inexact term-cancellations. Looking into details of the computation in Example 4, however, we notice that the estimation of inexact cancellations occurred on xz^2-, xy-, and x-terms are considerably larger than those shown in Example 1. So, we consider the computation in Example 4. There occurs no cancellation in $P_4 = \mathrm{Spol}(P_3, P_1)$. When P_2 is reduced by P_4 to $\tilde{P}_2$, large systematic exact cancellations of amount $O(10^5)$ occur: $\tilde{P}_2 := P_2 - zP_4 \approx 0.8144\mathrm{e}{-5}\, xy^2z - 1.632\mathrm{e}{-5}\, xyz$. In the succeeding reduction of $\tilde{P}_2$ by P_3, the marks in $\mathrm{lc}(\tilde{P}_2')$ and $\mathrm{lc}(\tilde{P}_2'')$ "contaminate" all the terms of $\mathrm{rt}(P_3')$ and $\mathrm{rt}(P_3'')$, respectively. On the other hand, in the computation by pure high-precision method, there is no such contamination. In this way, in the marking method, the accuracy loss in the leading coefficient contaminates the subsequent computations. A more elaborated implementation than the current one will be necessary.

5 Reducing the Systematic Exact Cancellations

In [24], Traverso and Zanoni tested many practical examples and reported that the cancellations of 2^{100} or more occur frequently. The most parts of such big cancellations are exact ones. Therefore, it is strongly desirable to reduce the amounts of exact cancellations, because the smaller the $\varepsilon_{\mathrm{cal}}$ is the more time the computation spends.

Since large main-term cancellations are caused by polynomials of small or large leading terms, we define the abnormality of polynomial as follows:

Definition 1 (abnormality) Let a polynomial P be $P = cT + P'$, where $cT = \mathrm{lt}(P)$ with $c = \mathrm{lc}(P)$ and we assume $P' \neq 0$. We define *abnormality* of P as follows:

$$\text{abnormality} = \begin{cases} \|P'\|/|c| & \text{if} \quad |c| \leq \|P'\|, \\ -|c|/\|P'\| & \text{if} \quad |c| > \|P'\|. \end{cases} \tag{5.1}$$

We expect that, by treating polynomials of large |abnormalities| specially, we will be able to reduce the amounts of exact cancellations. We have tested the following three strategies:

Strategy 1: Divide the computation into two stages. In the first stage, the reduction is performed only by polynomials of small |abnormalities|. The reduction by polynomials of large |abnormalities| are done after the termination of the first stage.

Result: The first stage sometimes does not terminate, because many S-polynomials are constructed and not reduced to 0.

Strategy 2: Divide the computation into two stages. In the first stage, construct only S-polynomials by combining polynomials of similar abnormalities. Combining polynomials of non-similar abnormalities is done in the second stage.

Result: One may expect that many abnormal polynomials are reduced to 0 in the first stage by this strategy. Actually, however, many abnormal polynomials survive to the second stage, and large cancellations occur in the second stage.

Strategy 3: Sort existing polynomials in small-to-large order w.r.t. their |abnormalities|. In constructing S-polynomials and in reducing polynomials, take up polynomials of smaller |abnormalities| earlier.

Result: This strategy succeeds considerably, as shown in Example 6.

Example 6 (test of strategy 3 by the system in Example 5) We compute the Gröbner basis w.r.t. total-degree order, with two strategies for choosing polynomials in the S-polynomial construction and the reduction.

$$\left\{\begin{array}{l} P_1 = x^3/10.0 + 3.0x^2y + 1.0y^2 \\ P_2 = 1.0x^2y^2 - 3.0xy^2 - 1.0xy \\ P_3 = y^3/10.0 + 2.0x^2 \end{array}\right\} \Rightarrow \left\{\begin{array}{l} Q_2 = y^2 \\ Q_4 = xy + \cdots \\ Q_5 = x^2 + \cdots \end{array}\right\}$$

We found that, with the Strategy 3 mentioned above, the exact cancellations occurred on Q_5, Q_2, and Q_4 are 1010, 89, and 1010, respectively. Compared with the amounts of cancellations in Example 5, we see that the Strategy 3 is quite effective for reducing the amounts of exact cancellations. ◇

Appendix: Effective Floating-Point Numbers

The efloat number was proposed by Kako and Sasaki in 1997 [6] so as to detect the cancellation errors automatically. (Similar ideas would be proposed by other authors). We represent an efloat as #E[f, e] (big-efloat as #BE[f, e]), where f is a floating-point number representing the value of this efloat, either of fixed precision or of arbitrary precision, and e is a fixed-precision floating-point number representing an approximate error of f. The efloat arithmetic is as follows: (we put a = #E[f_a, e_a] and b = #E[f_b, e_b]).

$$\begin{aligned} a+b &\Longrightarrow \#\mathrm{E}[f_a + f_b,\ \max\{e_a, e_b\}], \\ a-b &\Longrightarrow \#\mathrm{E}[f_a - f_b,\ \max\{e_a, e_b\}], \\ a\times b &\Longrightarrow \#\mathrm{E}[f_a \times f_b,\ \max\{|f_b e_a|, |f_a e_b|\}], \end{aligned}$$

$$a \div b \Longrightarrow \#\mathrm{E}[f_a \div f_b,\ \max\{|e_a/f_b|, |f_a e_b/f_b^2|\}].$$

Thus, the value-part of efloat is nothing but the conventional floating-point value. On the other hand, the error-part of efloat represents the cancellation error approximately; the rounding errors are neglected in determining the error-parts. Suppose the coefficients of the input polynomials contain relative errors of $\varepsilon_{\rm init}$. Then, we set the initial error-part of each coefficient to $(5 \sim 50)\varepsilon_{\rm init} \times |f|$; we set the error-part to $10^{-15} \times |f|$ for double-precision efloat, and to a much larger value for big-efloat. In our algebra system, the efloat #E[f, e] with $|f| < e$ is set to 0 automatically. Therefore, our algebra system detects fully-erroneous terms and remove them automatically, unless the rounding errors accumulate to $(5 \sim 50)\varepsilon_{\rm init}$ or more, which is extremely rare in practical computations. We note that the computation in Example 1 was done by using efloats; if we use the conventional floating-point numbers we will encounter many very small terms which are fully erroneous.

References

1. Bodrato, M., Zanoni, A.: Intervals, syzygies, numerical Gröbner bases: A mixed study. In: Proceedings of CASC2006 (Computer Algebra in Scientific Computing). LNCS, vol. 4194, pp. 64–76. Springer (2006)
2. Faugère, J.-C.: A new efficient algorithm for computing Gröbner bases (F_4). J. Pure Appl. Algebra **139**, 61–88 (1999)
3. Fortuna, E., Gianni, P., Trager, B.: Degree reduction under specialization. J. Pure Appl. Algebra **164**, 153–164 (2001)
4. Gonzalez-Vega, L., Traverso, C., Zanoni, A.: Hilbert stratification and parametric Gröbner bases. In: Proceedings of CASC2005 (Computer Algebra in Scientific Computing). LNCS, vol. 3718, pp. 220–235. Springer (2005)
5. Kalkbrener, M.: On the stability of Gröbner bases under specialization. J. Symb. Comput. **24**, 51–58 (1997)
6. Kako, F., Sasaki, T.: Proposal of "effective" floating-point number. Preprint of University Tsukuba, May 1997 (unpublished)
7. Kondratyev, A., Stetter, H.J., Winkler, F.: Numerical computation of Gröbner bases. In: Proceedings of CASC2004 (Computer Algebra in Scientific Computing), pp. 295–306, St. Petersburg (2004)
8. Lazard, D.: Gröbner bases, Gaussian elimination and resolution of systems of algebraic equations. In: Proceedings of EUROCAL 1983. LNCS, vol. 162, pp. 146–156. Springer (1983)
9. Mourrain, B.: A new criterion for normal form algorithms. In: Proceedings of AAECC 1999 (Applied Algebra, Algebraic Algorithms, and Error Correcting Codes). vol. 1719, pp. 430–443. Springer (1999)
10. Mourrain, B.: Pythagore's dilemma, symbolic-numeric computation, and the border basis method. Symbolic-Numeric Computations (Trends in Mathematics). Birkhäuser, Switzerland (2007)
11. Nagasaka, K.: A Study on Gröbner basis with inexact input. Lecture Notes in Computer Science. In: Proceedings of CASC2009 (Computer Algebra in Scientific Computing), vol. 5743, pp. 248–258, Kobe (2009)
12. Sasaki, T.: A subresultant-like theory for Buchberger's procedure. Japan J. Indust. Appl. Math. **31**, 137–164 (2014)
13. Sasaki, T., Kako, F.: Computing floating-point Gröbner base stably. In: Proceedings of SNC2007 (Symbolic Numeric Computation), pp.180–189. ACM Press (2007)

14. Sasaki, T., Kako, F.: Floating-point Gröbner basis computation with ill-conditionedness estimation. In: Proceedings of ASCM2007 (Asian Symposium on Computer Mathematics), Singapore, 2007; see also LNAI, vol. 5081, pp. 278–292, Deepak Kapur (Ed.), Springer (2008)
15. Sasaki, T., Kako, F.: Term-cancellations in computing floating-point Gröbner bases. In: Gerdt, P., Koepf, W., Mayr, W. Vorozhtsov, V. (Eds.) Proceedings of CASC2010 (Computer Algebra in Scientific Computing). LNCS, vol. 6244, pp. 220–231, Springer (2010)
16. K. Shirayanagi: An algorithm to compute floating-point Gröbner bases. Mathematical Computation with Maple V. Ideas and Applications, pp. 95–106. Birkhäuser (1993)
17. Shirayanagi, K.: Floating point Gröbner bases. Math. Comput. Simul **42**, 509–528 (1996)
18. Shirayanagi, K., Sweedler, M.: Remarks on automatic algorithm stabilization. J. Symb. Comput. **26**, 761–765 (1998)
19. Stetter, H.J.: Stabilization of polynomial systems solving with Gröbner bases. In: Proceedings of ISSAC'97 (Intern'l Symposium on Symbolic and Algebraic Computation), pp. 117–124, ACM Press(1997)
20. Stetter, H.J.: Numerical Polynomial Algebra. SIAM Publication, Philadelphia (2004)
21. Stetter, H.J.: Approximate Gröbner bases—an impossible concept?. In: Proceedings of SNC2005 (Symbolic-Numeric Computation), pp. 235–236, Xi'an, China(2005)
22. Suzuki, A.: Computing Gröbner bases within linear algebra. Lecture Notes in Computer Science. In: Proceedings of CASC2009 (Computer Algebra in Scientific Computing). LNCS, vol. 5743, pp. 310–321. Springer (2009)
23. Traverso, C.: Syzygies, and the stabilization of numerical Buchberger algorithm. In: Proceedings of LMCS2002 (Logic, Mathematics and Computer Science), pp. 244–255. RISC-Linz, Austria (2002)
24. Traverso, C., Zanoni, A.: Numerical stability and stabilization of Gröbner basis computation. In: Proceedings of ISSAC2002 (Intern'l Symposium on Symbolic and Algebraic Computation), pp. 262–269. ACM Press (2002)
25. Weispfenning, V.: Gröbner bases for inexact input data. In: Proceedings of CASC2003 (Computer Algebra in Scientific Computing), pp. 403–411, Passau (2003)

Series-Expansion of Multivariate Algebraic Functions at Singular Points: Nonmonic Case

Tateaki Sasaki and Daiju Inaba

Abstract In a series of papers, we have developed a method of expanding multivariate algebraic functions at their singular points. The method applies the Hensel construction to the defining polynomial of the algebraic function, so we named the resulting series "Hensel series". In [1], we derived a concise representation of Hensel series for the monic defining polynomial, and clarified several characteristic properties of Hensel series theoretically. In this paper, we study the case of nonmonic defining polynomial. We show that, by determining the so-called Newton polynomial suitably, we can construct Hensel series which show reasonable behaviors at zero-points of the leading coefficients and we can derive a representation of Hensel series in the nonmonic case just similarly as in the monic case. Furthermore, we investigate the convergence/divergence behavior and many-valuedness of Hensel series in the nonmonic case.

1 Introduction

Series expansion is a fundamental tool in numerical analysis, such as for solving differential equations [2], for tracing analytic functions numerically [3], and so on. So far, the Taylor expansion, multivariate as well as univariate, has mostly been used in numerical analysis. The Taylor expansion breaks down at singular points, and most numerical algorithms are constructed to avoid the singular points. However, singular points are very important in mathematics; many important mathematical properties are concentrated at singular points. It is strongly desirable to establish

Work supported in part by Japan Society for the Promotion of Science under Grants 19300001.

T. Sasaki (✉)
Institute of Mathematics, University of Tsukuba, Tsukuba, Japan
e-mail: sasaki@math.tsukuba.ac.jp

D. Inaba (✉)
Mathematics Certification Institute of Japan, Katsushika-ku, Japan
e-mail: d.inaba@su-gaku.net

© Springer-Verlag Berlin Heidelberg 2014
R. Feng et al. (eds.), *Computer Mathematics*, DOI 10.1007/978-3-662-43799-5_11

numerical analysis at and near the singular points. For this purpose, we need useful and tractable method of series expansion at singular points.

In a series of papers [1, 4–6, 8, 9], we have developed a method of expanding multivariate algebraic functions into series at singular points, where the algebraic functions are defined as roots of a given multivariate polynomial. Since our method is based on the Hensel construction, we named the series obtained by our method *Hensel series*. For bivariate polynomials, our method computes Puiseux series roots simultaneously and efficiently; for Puiseux series, see [10]. For multivariate polynomials, different kinds of expansions are possible, and the series obtained by our method are very different from multivariate Puiseux series [11–13].

The Hensel series was used so far to the analytic continuation of algebraic functions via singular points [14], solving multivariate algebraic equations in series forms [8, 9], the analytic factorization of polynomials of more than two variables [15, 16], and so on. For other researches of utilizing series expansions at singular points, see [17–21].

Let $F(x, \boldsymbol{u}) \stackrel{\text{def}}{=} F(x, u_1, \ldots, u_\ell)$ be a given multivariate polynomial, and let $f_n(\boldsymbol{u})$ be its leading coefficient w.r.t. x. The algebraic function is affected strongly and delicately by $f_n(\boldsymbol{u})$. If $\boldsymbol{u}$ approaches a zero-point of $f_n(\boldsymbol{u})$ then at least one branch of algebraic function goes to infinity, and if $\boldsymbol{u}$ approaches another zero-point then another branch will go to infinity. One will think that generalization to nonmonic case is straightforward. For example, we can employ a famous transformation which converts any nonmonic polynomial to a monic one. However, with such a transformation, the effect of $f_n(\boldsymbol{u})$ is not fully taken into the Hensel series. We want to treat the nonmonic case so that the effect of leading coefficient is taken into the Hensel series as fully as possible.

In this paper, we propose a reasonable treatment of the leading coefficient. In our method, the so-called Newton polynomial plays an essential role, so we include the leading coefficient into the Newton polynomial. This treatment seems to be peculiar at the first glance. However, the Hensel series are expressed in the roots of Newton polynomial and many analytic properties of Hensel series are determined by the roots. Therefore, by including the leading coefficient into the Newton polynomial, the effect of the leading coefficient is taken into Hensel series largely. Furthermore, with our treatment, the Hensel series are represented by the same formula as that in the monic case, and we can clarify the properties of Hensel series simply. It is, however, emphasized that this paper treats only the case that the Newton polynomial is squarefree; treatment of general case is complicated, as [5, 8, 9] show.

2 Our Approach to Nonmonic Case

Let $F(x, \boldsymbol{u}) \stackrel{\text{def}}{=} F(x, u_1, \ldots, u_\ell) \in \mathbb{C}[x, u_1, \ldots, u_\ell]$ be a given multivariate polynomial. By $\deg(F)$ and $\mathrm{lc}(F)$, we denote the degree and the leading coefficient, respectively, w.r.t. x, of $F(x, \boldsymbol{u})$. We put $\deg(F) = n$ and $\mathrm{lc}(F) = f_n(\boldsymbol{u})$. By tdeg($f$), with $f(\boldsymbol{u}) \in \mathbb{C}[\boldsymbol{u}]$, we denote the *total-degree* of $f(\boldsymbol{u})$ w.r.t. subvariables $u_1, \ldots, u_\ell$; if $T = c\, u_1^{e_1} \cdots u_\ell^{e_\ell}$, $c \in \mathbb{C}$, then tdeg$(T) = e_1 + \cdots + e_\ell$. By ord$(f)$ we denote the order of $f(\boldsymbol{u})$, i.e., the minimum of the total-degrees of terms of $f(\boldsymbol{u})$. Let $\varphi(\boldsymbol{u})$ be an algebraic function defined as a root w.r.t. x, of $F(x, \boldsymbol{u})$, hence we have $F(\varphi(\boldsymbol{u}), \boldsymbol{u}) = 0$.

Let $\boldsymbol{s} \stackrel{\text{def}}{=} (s_1, \ldots, s_\ell) \in \mathbb{C}^\ell$ be an expansion point. Without loss of generality, we assume that $F(x, \boldsymbol{u})$ is irreducible over $\mathbb{C}$ hence squarefree. If $f_n(\boldsymbol{s}) \neq 0$ and $F(x, \boldsymbol{s})$ is squarefree then every root of $F(x, \boldsymbol{u})$ w.r.t. x can be expanded into Taylor series in $u_1 - s_1, \ldots, u_\ell - s_\ell$.

Definition 1 (*Singular point, singular leading coefficient*) We call the expansion point $\boldsymbol{s} \in \mathbb{C}^\ell$ a *singular point of algebraic function*, or a *singular point* in short, if $F(x, \boldsymbol{s})$ is not squarefree. If $f_n(\boldsymbol{s}) = 0$ then we say *the leading coefficient is singular* at $\boldsymbol{s}$.

Without loss of generality, we assume that the origin $\boldsymbol{u} = \boldsymbol{0}$ is a singular point or $f_n(\boldsymbol{u})$ is singular at the origin, and we consider the series expansion at the origin. We show the importance of treatment of leading coefficient by a simple example (below, "H.T." denotes higher order terms).

Example 1 Let $F_1(x, u, v)$ be as follows.

$$F_1(x, u, v) = [(u+v)\,x - (u^2+v^2)] \cdot [uv\,x^2 - (u^4+v^4)] + \text{H.T.}$$

If deg(H.T.) ≤ 2 then the root $\varphi(u, v)$ of $F_1(x, u, v)$ w.r.t. x behaves as

$$\varphi(u, v) \approx \frac{u^2+v^2}{u+v}, \qquad \varphi(u, v) \approx \pm\Big(\frac{u^4+v^4}{uv}\Big)^{1/2},$$

for small $|u|$ and $|v|$, hence $\varphi(u, v)$ diverges on lines $u+v = 0$ or $u = 0$, $v = 0$. However, if H.T. of $F_1(x, u, v)$ contains $(u^4+v^4)\,x^3$, for example, which changes the leading coefficient of F_1 only a little near the origin, then $\varphi(u, v)$ does not diverge on lines $u+v = 0$, $u = 0$, $v = 0$. That is, the algebraic function is affected strongly by the leading coefficient of its defining polynomial. □

The key concept in the Hensel construction at a singular point is the Newton polynomial $F_{\text{New}}(x, \boldsymbol{u})$; see [8, 9]. In treating nonmonic case, we want to define $F_{\text{New}}(x, \boldsymbol{u})$ so that the resulting Hensel series reflect the effects of leading coefficient as fully as possible yet we can treat the Hensel series near the expansion point as simply as possible.

An approach which we adopt in this paper is quite simple, although it seems rather peculiar; we define the Newton polynomial as follows.

Definition 2 (*Newton polynomial* $F_{\text{New}}(x, \boldsymbol{u})$, *the nonmonic case*) For each monomial $c\, x^i t^j u_1^{j_1} \cdots u_\ell^{j_\ell}$ of $F(x, t\boldsymbol{u})$, with $c \in \mathbb{C}$ and $j = j_1 + \cdots + j_\ell$, plot a dot at the point (i, j) in the (e_x, e_t)-plane. Let $\nu = \text{ord}(f_n)$, and let $\mathcal{L}_{\text{New}}$ be a straight line in

(e_x, e_t)-plane, such that it passes the point (n, ν) and another dot plotted and that any dot plotted is not below $\mathcal{L}_{\rm New}$. Construct $F_{\rm New}(x, t\boldsymbol{u})$ by summing all the monomials plotted on $\mathcal{L}_{\rm New}$ and by replacing $\mathrm{lc}(F_{\rm New})$ by $f_n(\boldsymbol{u})$. (Hence, $\mathrm{lc}(F_{\rm New}) = f_n(\boldsymbol{u})$.)

Let the slope of $\mathcal{L}_{\rm New}$ be $-\lambda$ (note the "minus" sign). We have assumed that the origin $\boldsymbol{u} = \boldsymbol{0}$ is a singular point. In drawing figures of $\varphi(\boldsymbol{u})$ near the origin, we should "regularize" $\varphi(\boldsymbol{u})$. The behavior of $\varphi(\boldsymbol{u})$ near the origin is approximately determined by the Newton polynomial: if the slope of $\mathcal{L}_{\rm New}$ is positive (negative) then $\varphi(\boldsymbol{u})$ goes to infinity (zero, resp.) as $(u_1, \dots, u_\ell) \to (0, \dots, 0)$. Therefore, we define "regularized-root" $\bar{\varphi}(\boldsymbol{u})$ as follows (below, $\|\boldsymbol{u}\|$ has the meaning only when we substitute values to all the variables $u_1, \dots, u_\ell$).

$$\bar{\varphi}(\boldsymbol{u}) \stackrel{\rm def}{=} \|\boldsymbol{u}\|^{-\lambda}\varphi(\boldsymbol{u}), \qquad \|\boldsymbol{u}\| \stackrel{\rm def}{=} (|u_1|^2 + \cdots + |u_\ell|^2)^{1/2}. \tag{2.1}$$

Note that $\bar{\varphi}(\boldsymbol{s}) \to$ [finite value] as $\|\boldsymbol{s}\| \to 0$ for most values of $\boldsymbol{s} \in \mathbf{C}^\ell$. We also regularize Hensel series $\phi(\boldsymbol{u})$ to be determined in the next section.

3 Hensel Series in a Compact Representation

First of all, we confine ourselves to discussing the following restricted case.

Assumption N *We assume that $F_{\rm New}(x, \boldsymbol{u})$ is square-free not only exactly but also approximately, i.e., we assume that $F_{\rm New}(x, \boldsymbol{u})$ has no multiple/close roots. Furthermore, we assume that $F_{\rm New}(x, \boldsymbol{u})$ has no numeric root.*

If $F_{\rm New}(x, \boldsymbol{u})$ has a numeric root then the corresponding branch of the algebraic function can be expanded into Taylor series, and we are not interested in such a case. If $F_{\rm New}(x, \boldsymbol{u})$ has multiple roots then we must apply Hensel construction repeatedly, see [9].

Let the roots of $F_{\rm New}(x, \boldsymbol{u})$ be $\alpha_1(\boldsymbol{u}), \dots, \alpha_n(\boldsymbol{u})$, which we often write as $\alpha_1, \dots, \alpha_n$; the roots are usually algebraic functions.

$$F_{\rm New}(x, \boldsymbol{u}) = f_n(\boldsymbol{u})(x - \alpha_1(\boldsymbol{u})) \cdots (x - \alpha_n(\boldsymbol{u})), \qquad \alpha_i \neq \alpha_j \;\; (\forall i \neq j). \tag{3.1}$$

We define $\widetilde{F}(x, \boldsymbol{u}, \eta)$ by introducing an auxiliary variable η, as follows.

$$\left\{ \begin{array}{l} F(x, \boldsymbol{u}) \stackrel{\rm def}{=} F_{\rm New}(x, \boldsymbol{u}) + F_h(x, \boldsymbol{u}), \\ \widetilde{F}(x, \boldsymbol{u}, \eta) \stackrel{\rm def}{=} F_{\rm New}(x, \boldsymbol{u}) + \eta\, F_h(x, \boldsymbol{u}). \end{array} \right. \tag{3.2}$$

Let $F_{\rm New}(x, \boldsymbol{u})$ be factored as $F_{\rm New}(x, \boldsymbol{u}) = G_1^{(0)}(x, \boldsymbol{u}) \cdots G_r^{(0)}(x, \boldsymbol{u})$, where $G_1^{(0)}, \dots, G_r^{(0)}$ are relatively prime by Assumption N. Then, using $G_1^{(0)}, \dots, G_r^{(0)}$ as initial factors, we perform the Hensel construction of $\widetilde{F}(x, \boldsymbol{u}, \eta)$ with modulus η to satisfy

$$\widetilde{F}(x,\boldsymbol{u},\eta) \equiv G_1^{(k)}(x,\boldsymbol{u},\eta)\cdots G_r^{(k)}(x,\boldsymbol{u},\eta) \pmod{\eta^{k+1}}, \qquad k=1,2,\ldots.$$

Choosing the initial factors as $G_1^{(0)} = x-\alpha_1$ and $\tilde{G}^{(0)} = F_{\mathrm{New}}(x,\boldsymbol{u})/(x-\alpha_1)$, we obtain

$$\begin{cases} \widetilde{F}(x,\boldsymbol{u},\eta) \equiv G_1^{(k)}(x,\boldsymbol{u},\eta)\cdot \tilde{G}^{(k)}(x,\boldsymbol{u},\eta) \pmod{\eta^{k+1}}, \\ G_1^{(0)}(x,\boldsymbol{u},1) = x-\alpha_1, \quad G_1^{(k)}(x,\boldsymbol{u},\eta) = x-\phi_1^{(k)}(\boldsymbol{u},\eta). \end{cases} \tag{3.3}$$

The $\phi_1^{(\infty)}(\boldsymbol{u},1)$ is the Hensel series corresponding to α_1.

Once the Newton polynomial is defined, the Hensel construction is straightforward. Because of the page limit, we omit the details of deriving a representation of Hensel series, for which the reader can refer to [1].

Theorem 1 *Let $F_{\mathrm{New}}(x,\boldsymbol{u})$ be squarefree. Then, the Hensel factors $G_1^{(\infty)}(x,\boldsymbol{u},\eta)$ and $\tilde{G}^{(\infty)}(x,\boldsymbol{u},\eta)$ in (3.3) are expressed as follows ($F'_{\mathrm{New}} = dF_{\mathrm{New}}/dx$).*

$$G_1^{(\infty)}(x,\boldsymbol{u},\eta) = x-\alpha_1 + \sum_{k=1}^{\infty} \eta^k \frac{\delta F^{(k)}(\alpha_1,\boldsymbol{u})}{F'_{\mathrm{New}}(\alpha_1,\boldsymbol{u})}, \tag{3.4}$$

$$\begin{aligned} \tilde{G}^{(\infty)}(x,\boldsymbol{u},\eta) = {} & \frac{F_{\mathrm{New}}(x,\boldsymbol{u})}{x-\alpha_1} \\ & + \sum_{j=2}^{n} \frac{F_{\mathrm{New}}(x,\boldsymbol{u})}{(x-\alpha_1)(x-\alpha_j)} \Big(\sum_{k=1}^{\infty} \eta^k \frac{\delta F^{(k)}(\alpha_j,\boldsymbol{u})}{F'_{\mathrm{New}}(\alpha_j,\boldsymbol{u})} \Big), \end{aligned} \tag{3.5}$$

where $\delta F^{(1)} = F_{\mathrm{h}}(x,\boldsymbol{u})$ and the k*th order residual $\delta F^{(k)}$ $(k \geq 2)$ is given by*

$$\delta F^{(k)}(x,\boldsymbol{u}) = -\sum_{j=2}^{n} \frac{F_{\mathrm{New}}(x,\boldsymbol{u})}{(x-\alpha_1)(x-\alpha_j)} \Big(\sum_{k'=1}^{k-1} \frac{\delta F^{(k')}(\alpha_1,\boldsymbol{u})}{F'_{\mathrm{New}}(\alpha_1,\boldsymbol{u})} \frac{\delta F^{(k-k')}(\alpha_j,\boldsymbol{u})}{F'_{\mathrm{New}}(\alpha_j,\boldsymbol{u})} \Big). \tag{3.6}$$

Example 2 (Numerical evaluation of Hensel series) Let $F_2(x,u,v)$ be

$$F_2(x,u,v) = [(u+v)x-1][uvx^2+(u-v)x+1] + (u^3+v^3)x^2 + (u+v).$$

The Newton polynomial for $F_2(x,u,v)$ is as follows.

$$F_{2\mathrm{New}}(x,u,v) = [(u+v)x-1][uvx^2+(u-v)x+1].$$

Let α_1 be the root of $F_{2\mathrm{New}}(x,u,v)$ corresponding to the factor $[(u+v)x-1]$ and let the roots of $F_2(x,u,v)$ be $\varphi_i(u,v)$ $(i=1,2,3)$.

We have computed Hensel series $\phi_1^{(k)}(u,v)$ starting from α_1 up to $k=4$, and compared with algebraic functions $\varphi_i(u,v)$ $(i=1,2,3)$ which we computed exactly by Mathematica. Table 1 shows values of regularized truncated Hensel series $\bar{\phi}_1^{(4)}(u,v)$

Table 1 Numerical evaluation

u	$\bar{\phi}_1^{(4)}(u, 0.1)$	$\bar{\varphi}_i(u, 0.1)$ $(i = 1, 2, 3)$		
−0.20	−7.11517	0.696242	−2.75189	±1.12207 *i*
−0.15	∞	0.635459	−3.33738	±1.33045 *i*
−0.10	∞	0.585786	−3.41421	∞
−0.05	2.46037	0.573904	2.46039	−3.76102
0.00	∞	0.626789	1.59543	∞
0.05	0.590837	0.590885	0.552400	±1.54123 *i*
0.10	0.565685	0.565685		±1.41421 *i*
0.15	0.559016	0.559010	−0.324574	±1.41098 *i*
0.20	0.552161	0.552125	−0.630106	±1.40194 *i*

and $\bar{\varphi}_1(u, v)$ $(i = 1, 2, 3)$ in the real range $-0.20 \leq u \leq 0.20$, where we fixed v to 0.1.

The "convergence domain" of $\bar{\phi}_1^{(4)}(u, v)$ will be shown in Sect. 5 (Fig. 1a, b). We see that, in the "convergence domain", the Hensel series agrees with one branch of the algebraic function fairly well. Note that the values at $u = -0.05$ show "jumping" of Hensel series among branches of algebraic function, which we will discuss in Sect. 6. □

4 Order Estimations Near the Expansion Point

Formula (3.4) shows that the behavior of each Hensel series depends critically on the roots $\alpha_1, \ldots, \alpha_n$. In this section, we order-estimate $\alpha_1, \ldots, \alpha_n$ etc. when $\boldsymbol{u}$ is near the expansion point (= the origin).

Below, we put $F(x, \boldsymbol{u}) = f_n(\boldsymbol{u})x^n + f_{n-1}(\boldsymbol{u})x^{n-1} + \cdots + f_0(\boldsymbol{u})$, and we define $\|\boldsymbol{u}\|$ to be the Euclidean norm of $\boldsymbol{u}$, $\|\boldsymbol{u}\| \stackrel{\text{def}}{=} (|u_1|^2 + \cdots + |u_\ell|^2)^{1/2}$, after substituting suitable numbers for $u_1, \ldots, u_\ell$. Note that we have $\|\boldsymbol{u}\| \ll 1$ in this section, and that if actual values are substituted for $u_1, \ldots, u_\ell$ then we may have $\|\alpha_i(\boldsymbol{u}) - \alpha_j(\boldsymbol{u})\| = 0$ in special cases. The following order estimations are for generic $\boldsymbol{u}$. By $\mathrm{o}(\|\boldsymbol{u}\|^a)$ we mean either $\mathrm{O}(\|\boldsymbol{u}\|^{a+1})$ or less.

4.1 When $\mathrm{ord}(f_n) = 0$

The condition $\mathrm{ord}(f_n) = 0$ means that $f_n(\mathbf{0}) \neq 0$. Since $f_n(\boldsymbol{u})$ has a nonzero constant term, we obtain the same order estimations as in the monic case.

Lemma 1 *When* $\mathrm{ord}(f_n) = 0$*, we have the following order estimations for small* $\|\boldsymbol{u}\|$*, so long as* $\boldsymbol{u}$ *is not close to the zero-points of* $\alpha_i(\boldsymbol{u})-\alpha_j(\boldsymbol{u})$ $(\forall i \neq j)$.

$$
\begin{aligned}
\|\alpha_1\|,\ \|\alpha_1-\alpha_j\| &= \mathrm{O}(\|\boldsymbol{u}\|^{\lambda}) \quad (j=2,\ldots,n), && (4.1)\\
\|F'_{\mathrm{New}}(\alpha_i,\boldsymbol{u})\| &= \mathrm{O}(\|\boldsymbol{u}\|^{(n-1)\lambda}) \quad (i=1,\ldots,n), && (4.2)\\
\|F_h(\alpha_i,\boldsymbol{u})\| &= \mathrm{o}(\|\boldsymbol{u}\|^{n\lambda}) \quad (i=1,\ldots,n). && (4.3)
\end{aligned}
$$

Proof For small $\|\boldsymbol{u}\|$, $\alpha_1,\ldots,\alpha_n$ are determined mostly by the homogeneous parts of $F_{\mathrm{New}}(x,\boldsymbol{u})$ (that is, we can discard the higher order terms of $f_n(\boldsymbol{u})$ when $\|\boldsymbol{u}\|$ is small). Hence, by Assumption N, we obtain (4.1) and (4.2) at once. Since each coefficient of $F_h(x,\boldsymbol{u})$ is of higher order w.r.t. $u_1,\ldots,u_\ell$ than the corresponding coefficient of $F_{\mathrm{New}}(x,\boldsymbol{u})$ by at least 1, we obtain (4.3). □

4.2 When $\mathrm{ord}(f_n) = \nu > 0$ and $\|f_n(u)\| = \mathrm{O}(\|u\|^{\nu})$

Under Assumption N, condition $\mathrm{ord}(f_n) = \nu > 0$ means that $\mathrm{ord}(f_i) > 0$ for $i \geq 1$ so long as $f_i \not\equiv 0$; $\mathrm{ord}(f_0)$ may be 0 or positive. Hence, $f_i(\mathbf{0}) = 0$ for any $i \geq 1$. If $f_i \equiv 0$ or $\mathrm{ord}(f_i)$ is large for many i then the theoretical treatment becomes complicated. In this section, we make the order estimations by imposing the following assumption.

Assumption F1 We assume the following for any i.

$$\|f_i(\boldsymbol{u})\| = \mathrm{O}(\|u\|^{\nu+(n-i)\lambda}) \ \text{ so long as } \nu+(n-i)\lambda \text{ is an integer.} \tag{4.4}$$

Lemma 2 *For small* $\|\boldsymbol{u}\|$ *satisfying* (4.4)*, we have the following order estimations, so long as* $\boldsymbol{u}$ *is not close to the zero-points of* $f_n(\boldsymbol{u})$ *and* $\alpha_i(\boldsymbol{u})-\alpha_j(\boldsymbol{u})$ $(\forall i \neq j)$.

$$
\begin{aligned}
\|\alpha_1\|,\ \|\alpha_1-\alpha_j\| &= \mathrm{O}(\|\boldsymbol{u}\|^{\lambda}) \quad (j=2,\ldots,n), && (4.5)\\
\|F'_{\mathrm{New}}(\alpha_i,\boldsymbol{u})\| &= \mathrm{O}(\|\boldsymbol{u}\|^{\nu+(n-1)\lambda}) \quad (i=1,\ldots,n), && (4.6)\\
\|F_h(\alpha_i,\boldsymbol{u})\| &= \mathrm{o}(\|\boldsymbol{u}\|^{\nu+n\lambda}) \quad (i=1,\ldots,n). && (4.7)
\end{aligned}
$$

Proof Order estimations in (4.5) and (4.6) are direct consequences of Assumption N and Assumption F1. Since each coefficient of $F_h(x,\boldsymbol{u})$ is of higher order w.r.t. $u_1,\ldots,u_n$ than the corresponding coefficient of $F_{\mathrm{New}}(x,\boldsymbol{u})$ by at least 1, we obtain (4.7). □

Proposition 1 *Let* $\delta\phi_1^{(k)}(\boldsymbol{u})$ *be the coefficient of* η^k*-term of Hensel series* $\phi_1^{(\infty)}(\boldsymbol{u},t)$*. Then, except near the zero-points of* $\alpha_i(\boldsymbol{u})-\alpha_j(\boldsymbol{u})$ $(\forall i \neq j)$*, we have*

$$\|\delta\phi_1^{(k)}(\boldsymbol{u})\| = \mathrm{o}(\|\boldsymbol{u}\|^{\lambda+k-1}) \ \text{ for small } \|\boldsymbol{u}\| \text{ satisfying (4.4).} \tag{4.8}$$

Proof Consider (3.4) with Lemma 2. Since $\delta F^{(1)} = F_h(x, \boldsymbol{u})$, we see

$$\|\delta\phi_1^{(1)}(\boldsymbol{u})\| = \frac{\mathrm{O}(\|F_h(\alpha_1, \boldsymbol{u})\|)}{\mathrm{O}(\|F'_{\mathrm{New}}(\alpha_1, \boldsymbol{u})\|)} = \frac{\mathrm{o}(\|\boldsymbol{u}\|^{\nu+n\lambda})}{\mathrm{O}(\|\boldsymbol{u}\|^{\nu+(n-1)\lambda})} = \mathrm{o}(\|\boldsymbol{u}\|^{\lambda}).$$

Hence, (4.8) is valid for $k = 1$. Suppose the proposition is valid up to $k-1$, so

$$\frac{\|\delta F^{(j)}(\alpha_i, \boldsymbol{u})\|}{\|F'_{\mathrm{New}}(\alpha_i, \boldsymbol{u})\|} = \mathrm{o}(\|\boldsymbol{u}\|^{\lambda+j-1}) \quad \text{for small } \|\boldsymbol{u}\|, \qquad j = 1, \ldots, k-1.$$

Then, we can order-estimate $\|\delta F^{(k)}(\alpha_1, \boldsymbol{u})\|$ in (3.6), $k \geq 2$, as follows.

$$\mathrm{O}\Big(\sum_{j=2}^{n} \frac{\|F'_{\mathrm{New}}(\alpha_1, \boldsymbol{u})\|}{\|\alpha_1 - \alpha_j\|} \times \Big[\sum_{k'=1}^{k-1} \frac{\|\delta F^{(k')}(\alpha_1, \boldsymbol{u})\|}{\|F'_{\mathrm{New}}(\alpha_1, \boldsymbol{u})\|} \frac{\|\delta F^{(k-k')}(\alpha_j, \boldsymbol{u})\|}{\|F'_{\mathrm{New}}(\alpha_j, \boldsymbol{u})\|}\Big]\Big)$$
$$= \frac{\mathrm{O}(\|\boldsymbol{u}\|^{\nu+(n-1)\lambda})}{\mathrm{O}(\|\boldsymbol{u}\|^{\lambda})} \cdot \sum_{k'=1}^{k-1} \mathrm{o}(\|\boldsymbol{u}\|^{\lambda+k'-1})\, \mathrm{o}(\|\boldsymbol{u}\|^{\lambda+(k-k')-1}) = \mathrm{o}(\|\boldsymbol{u}\|^{\nu+n\lambda+k-1}).$$

Therefore, $\|\delta\phi_1^{(k)}(\boldsymbol{u})\| = \|\delta F^{(k)}(\alpha_1, \boldsymbol{u})\| / \|F'_{\mathrm{New}}(\alpha_1, \boldsymbol{u})\| = \mathrm{o}(\|\boldsymbol{u}\|^{\lambda+k-1})$. □

The above order estimation in (4.8) is the same as that in monic case.

4.3 *When* ord$(f_n) = \nu > 0$ *and* $\|f_n(\boldsymbol{u})\| = \mathrm{O}(\|\boldsymbol{u}\|^{\mu})$, $\mu > \nu$

Let $\boldsymbol{s}_0 \neq \boldsymbol{0}$ be a zero-point of $f_n(\boldsymbol{u})$. As we have mentioned in Sect. 2, if $\boldsymbol{u}$ approaches $\boldsymbol{s}_0$ then at least one root among $\alpha_1, \ldots, \alpha_n$ goes to infinity. If $f_n(\boldsymbol{u})$ is a nonconstant multivariate polynomial, there are infinitely many zero-points of $f_n(\boldsymbol{u})$. It will be tedious to consider all the cases of how $\boldsymbol{u}$ approaches $\boldsymbol{s}_0$. In this subsection, we consider only a typical and simple case by setting the following assumption.

Assumption F2 We assume that $f_{n-1}(\boldsymbol{u}) \neq 0$. We also assume that

$$\begin{aligned} &\|f_n(\boldsymbol{u})\| = \mathrm{O}(\|\boldsymbol{u}\|^{\mu}), \quad \mu > \nu, \\ &\|f_i(\boldsymbol{u})\| = \mathrm{O}(\|\boldsymbol{u}\|^{\nu+(n-i)\lambda}) \quad \text{so long as } \nu + (n-i)\lambda \text{ is an integer.} \end{aligned} \tag{4.9}$$

Then, one root of $F_{\mathrm{New}}(x, \boldsymbol{u})$, let it be $\alpha_1(\boldsymbol{u})$, is determined approximately by $f_n(\boldsymbol{u})x - f_{n-1}(\boldsymbol{u}) = 0$, and $\|\alpha_1\|$ is much larger than $\|\alpha_2\|, \ldots, \|\alpha_n\|$.

Lemma 3 *For small* $\|\boldsymbol{u}\|$ *satisfying* (4.9), *the following order estimations hold so long as* $\boldsymbol{u}$ *is not close to the zero-points of* $\alpha_i(\boldsymbol{u}) - \alpha_j(\boldsymbol{u})$ $(\forall i \neq j)$.

$$\|\alpha_1\|,\ \|\alpha_1-\alpha_j\| = \mathrm{O}(\|\boldsymbol{u}\|^{\lambda-\mu+\nu})\quad (j=2,\ldots,n), \tag{4.10}$$
$$\|F'_{\mathrm{New}}(\alpha_1,\boldsymbol{u})\| = \mathrm{O}(\|\boldsymbol{u}\|^{\mu+(n-1)(\lambda-\mu+\nu)}), \tag{4.11}$$
$$\|F_h(\alpha_1,\boldsymbol{u})\| = \mathrm{o}(\|\boldsymbol{u}\|^{\nu+\lambda+(n-1)(\lambda-\mu+\nu)}), \tag{4.12}$$
$$\|\alpha_j\|,\ \|\alpha_j-\alpha_{j'}\| = \mathrm{O}(\|\boldsymbol{u}\|^{\lambda})\quad (j,j'=2,\ldots,n;\ j\neq j'), \tag{4.13}$$
$$\|F'_{\mathrm{New}}(\alpha_j,\boldsymbol{u})\| = \mathrm{O}(\|\boldsymbol{u}\|^{\nu+(n-1)\lambda})\quad (j=2,\ldots,n), \tag{4.14}$$
$$\|F_h(\alpha_j,\boldsymbol{u})\| = \mathrm{o}(\|\boldsymbol{u}\|^{\nu+n\lambda})\quad (j=2,\ldots,n). \tag{4.15}$$

Proof We note that $\|\boldsymbol{u}\|^{\mu} \ll \|\boldsymbol{u}\|^{\nu}$ for small $\|\boldsymbol{u}\|$. The magnitude of α_1 is determined mostly by $f_n(\boldsymbol{u})$ and $f_{n-1}(\boldsymbol{u})$, hence we obtain (4.10). $\|F'_{\mathrm{New}}(\alpha_1,\boldsymbol{u})\|$ and $\|F_h(\alpha_1,\boldsymbol{u})\|$ are dominated by $\|f_n(\boldsymbol{u})\alpha_1^{n-1}\|$ and $\|f_{n-1}(\boldsymbol{u})\alpha_1^{n-1}\|$, respectively, so we obtain (4.11) and (4.12).

For (4.13), see the previous subsection. $\|F'_{\mathrm{New}}(\alpha_j,\boldsymbol{u})\|$ and $\|F_h(\alpha_j,\boldsymbol{u})\|$ are dominated by $\|f_{n-1}(\boldsymbol{u})\alpha_j^{n-2}\|$ and $\|f_{n-1}(\boldsymbol{u})\alpha_j^{n-1}\|$, respectively, so we obtain (4.14) and (4.15). □

Proposition 2 *For small* $\|\boldsymbol{u}\|$ *satisfying* (4.9), *the following order estimations hold, so long as* $\boldsymbol{u}$ *is not close to the zero-points of* $\alpha_i(\boldsymbol{u})-\alpha_j(\boldsymbol{u})$ $(\forall i\neq j)$.

$$\|\delta\phi_1^{(1)}(\boldsymbol{u})\| = \mathrm{o}(\|\boldsymbol{u}\|^{\lambda-\mu+\nu}),\qquad \|\delta\phi_1^{(k)}(\boldsymbol{u})\| = \mathrm{o}(\|\boldsymbol{u}\|^{\lambda+k-1})\ \ (k\geq 2). \tag{4.16}$$

Proof Consider (3.4) with Lemma 3. We first note that

$$\frac{\|F_h(\alpha_1,\boldsymbol{u})\|}{\|F'_{\mathrm{New}}(\alpha_1,\boldsymbol{u})\|} = \mathrm{o}(\|\boldsymbol{u}\|^{\lambda-\mu+\nu}).\qquad \frac{\|F_h(\alpha_j,\boldsymbol{u})\|}{\|F'_{\mathrm{New}}(\alpha_j,\boldsymbol{u})\|} = \mathrm{o}(\|\boldsymbol{u}\|^{\lambda})\ \ (j\geq 2). \tag{4.17}$$

We next order-estimate $\delta F^{(2)}(\alpha_1,\boldsymbol{u})/F'_{\mathrm{New}}(\alpha_1,\boldsymbol{u})$ and $\delta F^{(2)}(\alpha_j,\boldsymbol{u})/F'_{\mathrm{New}}(\alpha_j,\boldsymbol{u})$, $j\geq 2$, as follows.

$$\begin{aligned}
\frac{\|\delta F^{(2)}(\alpha_1,\boldsymbol{u})\|}{\|F'_{\mathrm{New}}(\alpha_1,\boldsymbol{u})\|} &= \mathrm{O}\Big(\sum_{j=2}^{n}\frac{1}{\|\alpha_1-\alpha_j\|}\times\frac{\|\delta F^{(1)}(\alpha_1,\boldsymbol{u})\|}{\|F'_{\mathrm{New}}(\alpha_1,\boldsymbol{u})\|}\frac{\|\delta F^{(1)}(\alpha_j,\boldsymbol{u})\|}{\|F'_{\mathrm{New}}(\alpha_j,\boldsymbol{u})\|}\Big)\\
&= \mathrm{o}(\|\boldsymbol{u}\|^{\lambda+1}),\\
\frac{\|\delta F^{(2)}(\alpha_j,\boldsymbol{u})\|}{\|F'_{\mathrm{New}}(\alpha_j,\boldsymbol{u})\|} &= \mathrm{O}\Big(\sum_{j'=2}^{n}\frac{\delta_{j,j'}}{\|\alpha_j-\alpha_1\|}\times\frac{\|\delta F^{(1)}(\alpha_1,\boldsymbol{u})\|}{\|F'_{\mathrm{New}}(\alpha_1,\boldsymbol{u})\|}\frac{\|\delta F^{(1)}(\alpha_{j'},\boldsymbol{u})\|}{\|F'_{\mathrm{New}}(\alpha_{j'},\boldsymbol{u})\|}\Big)\\
&= \mathrm{o}(\|\boldsymbol{u}\|^{\lambda+1}).
\end{aligned}$$

Since $\delta\phi_1^{(k)} = -\delta F^{(k)}(\alpha_1,\boldsymbol{u})/F'_{\mathrm{New}}(\alpha_1,\boldsymbol{u})$, the proposition is valid for $k=1,2$. Suppose that the proposition and the following order estimations are valid for $k'=2,\ldots,k-1$ (we have showed the validity for $k'=2$ above).

$$\frac{\|\delta F^{(k')}(\alpha_i, \boldsymbol{u})\|}{\|F'_{\mathrm{New}}(\alpha_i, \boldsymbol{u})\|} = \mathrm{o}(\|u\|^{\lambda+k'-1}) \qquad (i=1,2,\ldots,n). \tag{4.18}$$

Then, we can order-estimate $\delta F^{(k)}(\alpha_1, \boldsymbol{u})\|/F'_{\mathrm{New}}(\alpha_1, \boldsymbol{u})$, $k \geq 3$, as follows.

$$\begin{aligned}
&\mathrm{o}\Big(\sum_{j=2}^{n} \frac{1}{\|\alpha_1 - \alpha_j\|} \times \Big[\sum_{k'=1}^{k-1} \frac{\|\delta F^{(k')}(\alpha_1, \boldsymbol{u})\|}{\|F'_{\mathrm{New}}(\alpha_1, \boldsymbol{u})\|} \frac{\|\delta F^{(k-k')}(\alpha_j, \boldsymbol{u})\|}{\|F'_{\mathrm{New}}(\alpha_j, \boldsymbol{u})\|}\Big]\Big) \\
&= \frac{\max\{\mathrm{o}(\|\boldsymbol{u}\|^{\lambda-\mu+\nu})\ \mathrm{o}(\|\boldsymbol{u}\|^{\lambda+k-2}),\ \mathrm{o}(\|\boldsymbol{u}\|^{\lambda+1})\ \mathrm{o}(\|\boldsymbol{u}\|^{\lambda+k-3})\}}{\mathrm{O}(\|\boldsymbol{u}\|^{\lambda-\mu+\nu})} \\
&= \mathrm{o}(\|\boldsymbol{u}\|^{\nu+k-1}).
\end{aligned}$$

Similarly, we can order-estimate $\|\delta F^{(k)}(\alpha_j, \boldsymbol{u})\|/\|F'_{\mathrm{New}}(\alpha_j, \boldsymbol{u})\|$, $k \geq 3$. Therefore, the right equality in (4.16) and (4.18) are valid for $k' = k$. □

5 On Convergence Property Near the Expansion Point

The formula (3.4) with (3.6) is very useful for analyzing the properties of Hensel series theoretically. In [1], we have done such an analysis in the case of monic defining polynomials. In this section, we perform a similar analysis in the nonmonic case. We will see that, except near the zero-points of the leading coefficient, the regularized Hensel series in the nonmonic case show quite similar convergence behaviors as those in the monic case.

Theorem 2 *In a neighborhood of the expansion point, except near the zero-points of $f_n(\boldsymbol{u})$, any divergence domain of regularized Hensel series $\bar{\phi}_1^{(\infty)}(\boldsymbol{u}, 1)$ starts from the expansion point and spreads outside radially along the zero-points of $\alpha_1(\boldsymbol{u}) - \alpha_j(\boldsymbol{u})$, $2 \leq j \leq n$.*

Proof $G_1^{(k)}(x, \boldsymbol{u}, t)$ in (3.4) with (3.6) tells us that, in the neighborhood of the expansion point, $\bar{\phi}_1^{(\infty)}(\boldsymbol{u}, 1)$ diverges on the zero-points of $F_{\mathrm{New}}(\alpha_1, \boldsymbol{u}) = f_n(\boldsymbol{u}) \prod_{j=2}^{n}(\alpha_1(\boldsymbol{u}) - \alpha_j(\boldsymbol{u}))$, may diverge on zero-points of $f_n(\boldsymbol{u})$, and diverges on no other point. □

Theorem 3 *In the neighborhood of the expansion point, two branches $\bar{\varphi}_1(\boldsymbol{u})$ and $\bar{\varphi}_j(\boldsymbol{u})$ $(j \geq 2)$ of the regularized algebraic function cross along the zero-points of $\alpha_1(\boldsymbol{u}) - \alpha_j(\boldsymbol{u})$ if $\alpha_1(\boldsymbol{u})$ and $\alpha_j(\boldsymbol{u})$ cross, or they are tangent to each other if $\alpha_1(\boldsymbol{u})$ and $\alpha_j(\boldsymbol{u})$ are so.*

Proof Theorem 2 tells that, in the neighborhood of the expansion point, the divergence domains spread from the expansion point radially along the zero-points of $\alpha_1(\boldsymbol{u}) - \alpha_j(\boldsymbol{u})$, and Proposition 1 says that the Hensel series are well approximated by the initial terms. Furthermore, the branches of algebraic function are also approximated well by the roots of the Newton polynomial. Therefore, near the zero-points of $\alpha_1(\boldsymbol{u})-\alpha_j(\boldsymbol{u})$, $\bar{\varphi}_1(\boldsymbol{u})$ either crosses with $\bar{\varphi}_j(\boldsymbol{u})$ or is tangent to $\bar{\varphi}_j(\boldsymbol{u})$. □

Theorem 4 *Let S_r be the surface of the hypersphere, $\|\boldsymbol{u}\|^2 = r^2$, where r is a small real positive number. Suppose S_r contains zero-points of $f_n(\boldsymbol{u})$ on which $\alpha_1(\boldsymbol{u})$ diverges, and let δS_r be small neighborhood of the zero-points, on S_r. Let $\check{S}_r$ be $S_r - \delta S_r$. Then, we have*

$$\frac{[\text{divergence area of } \bar{\phi}_1(\boldsymbol{u}) \text{ on } \check{S}_r]}{[\text{convergence area of } \bar{\phi}_1(\boldsymbol{u}) \text{ on } \check{S}_r]} \to 0 \quad \text{as } r \to 0. \tag{5.1}$$

Proof The regularized root $\bar{\alpha}_1(\boldsymbol{u})$ does not diverge on $\check{S}_r$, and the higher order regularized terms $\overline{\delta\phi_1^{(k)}}(\boldsymbol{u})$ $(k \geq 1)$ diverge only at the zero-points of $F_{\text{New}}(\alpha_1, \boldsymbol{u})$. Lemma 2 tells us that $\|\delta\phi_1^{(k)}(\boldsymbol{u})\|/\|\alpha_1(\boldsymbol{u})\| = \mathrm{o}(\|\boldsymbol{u}\|^{k-1})$ for small $\|\boldsymbol{u}\|$ except near the zero-points of $F_{\text{New}}(\alpha_1, \boldsymbol{u})$. Hence, the ratio in (5.1) becomes smaller and smaller as $r \to 0$. □

Corollary 1 *The regularized Hensel series $\bar{\varphi}_1^{(\infty)}(\boldsymbol{u}, 1)$ converges in most area of the small neighborhood of the expansion point.*

Example 3 (Convergence domain; for $F_2(x, u, v)$ in Example 2) We compare the regularized truncated Hensel series $\bar{\phi}_1^{(4)}(u, v)$ with $\bar{\varphi}_i(u, v)$ $(i = 1, 2, 3)$ in the real region $-a \leq u, v \leq a$. Since we do not know the correspondence between the Hensel series and the branches, we compute the quantity

$$d \equiv \min\{\, |\bar{\phi}_1^{(4)}(u, v) - \bar{\varphi}_i(u, v)|,\ i = 1, 2, 3\,\}$$

at many points in the region and draw a domain in which we have $d \leq \delta$ ($\delta = 0.01$ or $\delta = 0.001$). Figure 1a, b show the results; gray areas are "convergence domains" determined as mentioned above.

We see from Fig. 1a that the "convergence domain" is pretty wide and long-range. On the other hand, looking around the origin, we notice that Theorem 4 seems to be unsatisfied. So, we investigate the "convergence domain" around the origin by magnifying the neighborhood of the origin by 10 times. Figure 1b shows the result, which is consistent with Theorem 4.

The $\bar{\phi}_1^{(4)}(u, v)$ drawn in Fig. 1a, b corresponds to the initial factor $(u+v)x - 1$. Hence, $\bar{\phi}_1^{(4)}(u, v)$ diverges on the line $u+v = 0$. We see that the divergence domain near the line is very narrow, showing that the Hensel series truncated at $k = 4$ approximates the corresponding algebraic function well even in a small neighborhood of the zero-points of the leading coefficient. On the other hand, we have $F'_{2\text{New}}(\alpha_1, u, v) = u(2u + 3v)/(u + v)$, hence $\bar{\phi}_1^{(4)}(u, v)$ diverges along lines $u = 0$ and $2u+3v = 0$. □

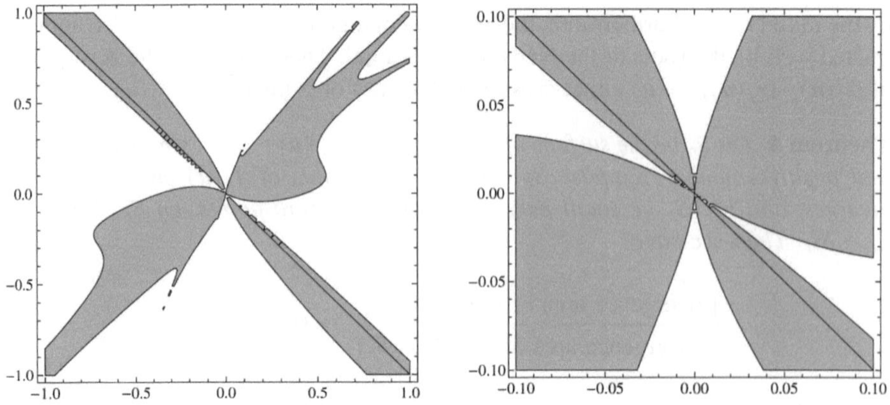

Fig. 1 **a** $a = 1.0$, $\delta = 0.01$. **b** $a = 0.1$, $\delta = 0.001$

6 On Many-Valuedness of Hensel Series

In [4], we observed that a Hensel series may jump from one branch of algebraic function to another when it passes a divergence domain, and we verified this property theoretically in [1] for monic defining polynomials. In this paper, we verify this property for nonmonic defining polynomials.

Formulas in Theorem 1 tells us that if $\alpha_1(\boldsymbol{u})$ is a rational function in $u_1, \ldots, u_\ell$ then so is $\delta F^{(k)}(\alpha_1, \boldsymbol{u})/F'_{\text{New}}(\alpha_1, \boldsymbol{u})$ for any $k > 0$. Hence, we classify the Hensel series as follows.

Definition 3 If $\alpha_1(\boldsymbol{u}) \in \mathbb{C}(\boldsymbol{u})$ then the Hensel series $\phi_1^{(\infty)}(\boldsymbol{u}, 1)$ is called *rational*, otherwise the series is called *algebraic*.

Let $F_{\text{New}}(x, \boldsymbol{u})$ is factored into irreducible factors in $\mathbb{C}[x, \boldsymbol{u}]$ as $F_{\text{New}}(x, \boldsymbol{u}) = F_{\text{New},1}(x, \boldsymbol{u}) \cdots F_{\text{New},r}(x, \boldsymbol{u})$, and α_1 be a root of $F_{\text{New},1}(x, \boldsymbol{u})$. If $\deg(F_{\text{New},1}) = m > 1$ then there are m conjugate roots, let them be $\alpha_{1,1}(= \alpha_1), \ \ldots, \ \alpha_{1,m}$. Let the algebraic Hensel series $\phi_{1,i}^{(\infty)}$ correspond to $\alpha_{1,i}$ $(1 \le i \le m)$.

Lemma 4 (Sasaki–Kako 1993) *The truncated algebraic Hensel series* $\phi_{1,1}^{(k)}, \ldots, \phi_{1,m}^{(k)}$ $(k \ge 1)$ *are mutually conjugate, hence* $\phi_1^{(k)}(\boldsymbol{u}, 1)$ *is* m*-valued.* □

Continuous singular points usually form a line when $\ell = 2$, a surface when $\ell = 3$, or an $(\ell - 1)$-dimensional hypersurface when $\ell \ge 4$, and we call them *singularity lines* for simplicity.

Let P_0 and P_1 be points in $\mathbb{C}^\ell$, which are in a convergence domain of Hensel series $\phi_1^{(\infty)}(\boldsymbol{u}, 1)$, in the neighborhood of the expansion point. Let C_0 be a path which starts P_0, rounds a singularity line one time, and comes back to P_0. Let C_1 be a path which starts P_0 and arrives at P_1, without rounding any singularity line.

Theorem 5 *Tracing the truncated algebraic Hensel series $\phi_1^{(k)}(\boldsymbol{u}, 1)$ $(k \geq 2)$ along the path* C_0*, the series transfers to one of the conjugate Hensel series when it arrives at* P_0.

Proof Obvious from Lemma 4. □

Theorem 6 *Tracing the truncated Hensel series $\bar{\phi}_1^{(k)}(\boldsymbol{u}, 1)$ $(k \geq 2)$ along the path* C_1*, we encounter the same Hensel series at* P_1*, no matter whether the series is rational or algebraic, and no matter whether the path passes a divergence domain or not.*

Proof Since the divergence domain spreads from the expansion point radially and the most area of the small neighborhood of the expansion point is the convergence domain, we can move the path C_1 in $\mathbb{C}^\ell$ so that it does not pass any divergence domain. Then, the value of $\bar{\phi}_1^{(k)}(\boldsymbol{u}, 1)$ changes continuously from P_0 to P_1, proving the theorem. □

Corollary 2 *Let $\bar{\phi}_1^{(\infty)}(\boldsymbol{u}, 1)$ correspond to a branch $\bar{\varphi}_1(\boldsymbol{u})$ of the algebraic function at* P_0*. Tracing $\bar{\phi}_1^{(k)}(\boldsymbol{u}, 1)$ along* C_1 *by passing a divergence domain for $\bar{\phi}_1^{(\infty)}(\boldsymbol{u}, 1)$, it may jump to another branch of the original algebraic function.*

Proof Theorem 3 says that two branches of the algebraic function cross or tangent to each other in each divergence domain near the origin. Therefore, if we trace the Hensel series along C_1, the series either may stay on the same branch of the algebraic function or may jump to another branch, if these branches cross there. □

If $m < n$ then a discrepancy of the many-valuedness seems to arise between Hensel series and the original algebraic function, because the former is m-valued while the latter is n-valued. This superficial discrepancy seems to be resolved by the property mentioned in Corollary 2.

We show the jumping by two simple examples; one is for rational Hensel series and other is for algebraic Hensel series. Note that each branch is continuous except at zero-points of the leading coefficient. Hence, we can recognize the jumping if a Hensel series runs along a branch of the algebraic function in some region and runs along another branch in a different region.

Example 4 (Jumping of rational Hensel series) Let $F_3(x, u, v)$ be

$$F_3(x, u, v) = [(u+v)x - 1]\,(ux - 1)\,(vx + 1) + (u^4+v^4)\,x^2 + u^3v^3.$$

We have $F_{\mathrm{New}}(x, u, v) = [(u+v)x - 1]\,(ux - 1)\,(vx + 1)$, which gives three roots $\alpha_1 = 1/(u+v)$, $\alpha_2 = 1/u$, $\alpha_3 = 1/v$, hence we obtain three rational Hensel series. We trace truncated Hensel series $\phi_3^{(10)}(u, v)$ corresponding to the root α_3 and three branches $\varphi_i(u, v)$ $(i = 1, 2, 3)$ of the algebraic function, along the circle C_3: $(u, v) = 0.1 \times (\cos\theta, \sin\theta)$ $(0 \leq \theta \leq 2\pi)$. (If we make the circle C_3 bigger, the graphs become more complicated.) The Hensel series are real on C_3, while the branches become complex in several narrow regions of C_3.

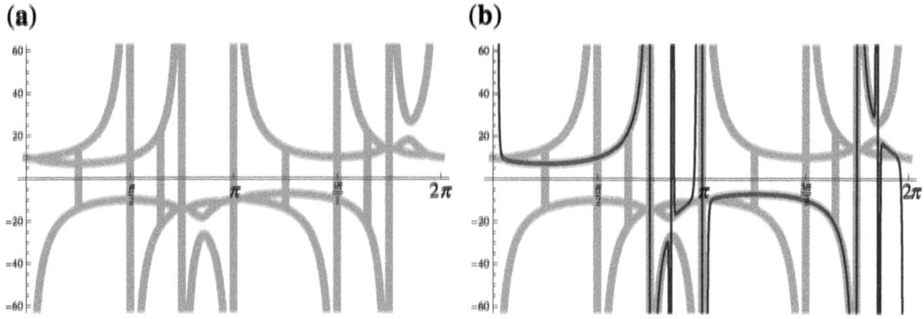

Fig. 2 **a** $\varphi_i(u, v)$ $(i = 1, 2, 3)$. **b** $\phi_3^{(10)}(u, v)$

The $\phi_3^{(10)}(u, v)$ will diverge at $\theta = 0,\ \pi,\ 2\pi$, where the corresponding leading coefficient v becomes 0. Since, we have $F'_{3\mathrm{New}}(-1/v, u, v) = (u+v)(u+2v)/v^2$, the divergence domain will spread along the lines $u + v = 0$ and $u + 2v = 0$; C_3 will cross the divergence domains at about $\theta = 3\pi/4,\ 5\pi/6,\ 7\pi/4,\ 11\pi/6$.

In Fig. 2a at the next page, we show real parts of $\varphi_i(u, v)$ $(i = 1, 2, 3)$ (gray curves) traced along C_3; the value of φ_i becomes complex around $\theta = 0,\ 3\pi/4,\ \pi,\ 7\pi/4,\ 2\pi$. (We have used Mathematica to draw the figures in this paper. Mathematica draws only the principal values of algebraic functions, and four fake vertical lines in Fig. 2a are due to Mathematica.)

In Fig. 2b, we show $\phi_3^{(10)}(u, v) = -1/v + \text{H.T.}$ (black curve) traced along C_3; for comparison, we show $\varphi_i(u, v)$ $(i = 1, 2, 3)$ by gray curves. The graph diverges as we have mentioned above. We see that the Hensel series jumps from one branch to another of the algebraic function, when it passes $\theta = 5\pi/6,\ 11\pi/6$, showing a natural behavior if the behaviors around jumping points are neglected. Note that, no jumping occurs at $\theta = 0,\ \pi,\ 2\pi$ where the leading coefficient vanishes. Note, further that $\phi_3^{(10)}(u, v)$ is real on C_3 and it is in the divergence domain where φ_i becomes complex. Hence, no inconsistency appears between algebraic function and Hensel series. □

Example 5 (Jumping of algebraic Hensel series) Let $F_4(x, u, v)$ be

$$F_4(x, u, v) = [(u+v)\,x - 1]\,[(u^2+v^2)\,x^2 - 1] + (u^5+v^5)\,x^2 + u^4v^4.$$

The Newton polynomial is $F_{\mathrm{New}}(x, u, v) = [(u+v)\,x - 1]\,[(u^2+v^2)\,x^2 - 1]$, which gives three roots $\alpha_1 = 1/(u+v)$ and $\alpha_\pm = \pm 1/\sqrt{u^2+v^2}$. Hence, we obtain one rational Hensel series $\phi_1^{(k)}(u, v)$ and mutually conjugate algebraic Hensel series $\phi_\pm^{(k)}(u, v)$, corresponding to α_1 and $\alpha_\pm$, respectively. The $\phi_1^{(k)}$ and $\phi_\pm^{(k)}(u, v)$ diverge on $u + v = 0$ and $u^2 + v^2 = 0$, respectively, where the leading coefficient of $F_4(x, u, v)$ vanishes. We have

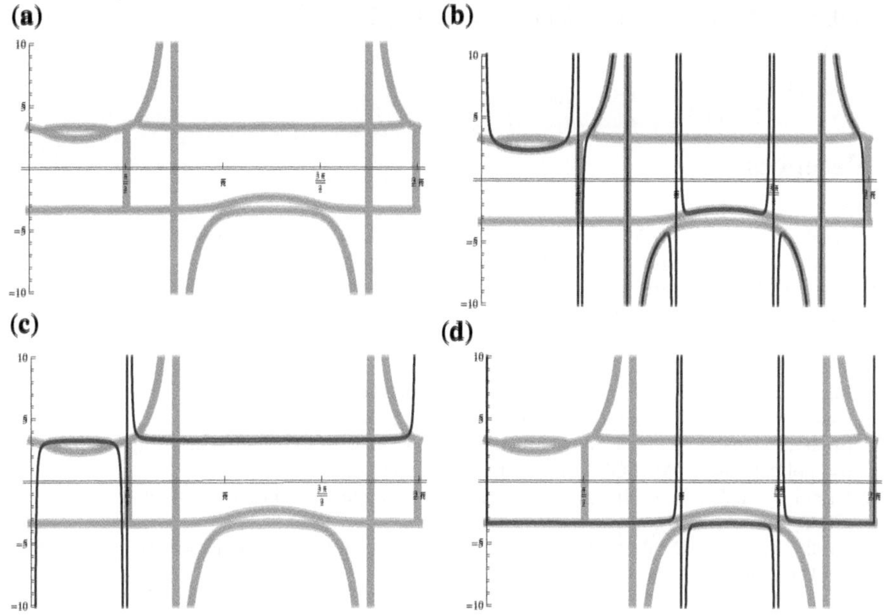

Fig. 3 **a** $\varphi_i(u, v)$ $(i = 1, 2, 3)$. **b** $\phi_1^{(10)}(u, v)$. **c** $\phi_+^{(10)}(u, v)$. **d** $\phi_-^{(10)}(u, v)$

$$F'_{4\mathrm{New}}(\alpha_1, u, v) = \frac{-2uv}{u+v}, \qquad F'_{4\mathrm{New}}(\alpha_\pm, u, v) = 2\,\frac{u+v \mp \sqrt{u^2+v^2}}{\sqrt{u^2+v^2}}.$$

Thus, when observed on the real plane, $\phi_1^{(k)}(u, v)$ diverges on lines $u = 0$ and $v = 0$, while $\phi_+^{(k)}(u, v)$ diverges on half lines $u = 0$ $(v \geq 0)$ and $v = 0$ $(u \geq 0)$, and $\phi_-^{(k)}(u, v)$ diverges on half lines $u = 0$ $(v \leq 0)$ and $v = 0$ $(u \leq 0)$.

We trace $\varphi_i(u, v)$ $(i = 1, 2, 3)$ and $\phi_1^{(10)}(u, v)$, $\phi_\pm^{(10)}(u, v)$ along the circle C_4: $(u, v) = 0.3 \times (\cos\theta, \sin\theta)$ $(0 \leq \theta \leq 2\pi)$.

In Fig. 3a at the next page, we show real parts of $\varphi_i(u, v)$ $(i = 1, 2, 3)$ traced along C_4 (gray curves); among the three gray curves in this figure, the curve which diverges at $\theta = 3\pi/4, 7\pi/4$ corresponds to α_1, and two branches become complex around $\theta = \pi/2, 2\pi$. Three vertical lines are fake due to Mathematica.

In Fig. 3b, we show $\phi_1^{(10)}(u, v)$ (black curves) traced along C_4. For comparison, we show $\varphi_i(u, v)$ $(i = 1, 2, 3)$ by gray curves. Note that, both $\phi_1^{(k)}(u, v)$ and $\phi_\pm^{(k)}(u, v)$ are real on C_4. We see that the Hensel series jumps from one branch to another when it passes $\theta = \pi$, $3\pi/2$, showing a natural behavior if the behaviors around jumping points are neglected. Furthermore, jumping occurs at $\theta = \pi/2, 2\pi$, although it is rather difficult which curve corresponding to the same branch. In Fig. 3c, d, we show $\phi_\pm^{(10)}(u, v)$ (black curves) traced along C_4. We see that jumping occurs only at

$\theta = 0, \pi/2, 2\pi$ on $\phi_+^{(10)}(u, v)$ and at $\theta = \pi, 3\pi/2$ on $\phi_-^{(10)}(u, v)$, as $F'_{4\mathrm{New}}(\alpha_\pm, u, v)$ predicts. □

References

1. Sasaki, T., Inaba, D.: Convergence and many-valuedness of Hensel series near the expansion point. In: Proceedings of the SNC 2009, pp. 159–167. ACM Press (2009).
2. Lambert, J.D.: Numerical Methods for Ordinary Differential Systems: The Initial Value Problem. Wiley, New York (1991)
3. Allgower, E.L., Georg, K.: Numerical Continuation Methods: An Introduction. Springer, New York (1990)
4. Inaba, D., Sasaki, T.: A numerical study of extended Hensel series. In: Verschede, J., Watt, S.T. (eds.) Proceedings of the SNC'07, pp. 103–109. ACM Press (2007).
5. Sasaki, T., Inaba, D.: Hensel construction of $F(x, u_1, \ldots, u_\ell)$, $\ell \geq 2$, at a singular point and its applications. ACM SIGSAM Bulletin **34**(1), 9–17 (2000)
6. Sasaki, T., Inaba, D.: Multivariate Hensel construction in roots. Preprint of University Tsukuba (2008) (unpublished).
7. Sasaki, T., Inaba, D.: A study of Hensel in series in general case. In: Proceedings of the SNC 2011, pp. 34–43, ACM Press (2011).
8. Sasaki, T., Kako, F.: Solving multivariate algebraic equation by Hensel construction. Preprint of University Tsukuba (1993).
9. Sasaki, T., Kako, F.: Solving multivariate algebraic equation by Hensel construction. Jpn. J. Indust. Appl. Math. **16**(2), 257–285 (1999)
10. Walker, R.J.: Algebraic Curves. Springer, New York (1950)
11. Beringer, F., Jung, F.: Multi-variate polynomials and Newton-Puiseux expansions. In: Winkler, W., Langer, U., (eds.) Proceedings of the SNSC 2001, Lecture Notes Computer Science, vol. 2630, pp. 240–254 (2003).
12. McDonald, J.: Fiber polytopes and fractional power series. J. Pure Appl. Algebra **104**, 213–233 (1995)
13. González Pérez, P.D.: Singularités quasi-ordinaires toriques et polyèdre de Newton du discriminant. Canad. J. Math. **52**(2), 348–368 (2000)
14. Shiihara, K., Sasaki, T.: Analytic continuation and Riemann surface determination of algebraic functions by computer. Jpn. J. Indust. Appl. Math. **13**(1), 107–116 (1996)
15. Iwami, M.: Analytic factorization of the multivariate polynomial. In: Ganzha, V.G., Mayr, E.W., Vorozhtsov, E.V. (eds.) Proceedings of the CASC 2003, pp. 213–225. Technishe Universität München Press (2003).
16. Iwami, M.: Extension of expansion base algorithm to multivariate analytic factorization. In: Ganzha, V.G., Mayr, E.W., Vorozhtsov, E.V. (eds.) Proceedings of the CASC 2004, pp. 269–282. Technishe Universität München Press (2004).
17. Della Dora, J., Richard-Jung, F.: About the Newton algorithm for non-linear ordinary differential equation. In: Küchlin, W.W. (ed.) Proceedings of the ISSAC'97, pp. 298–304. ACM Press (1997).
18. Inaba, D.: Factorization of multivariate polynomials by extended Hensel construction. ACM SIGSAM Bulletin **39**(1), 2–14 (2005)
19. McCallum, S.: On testing a bivariate polynomial for analytic reducibility. J. Symb. Comput. **24**, 509–535 (1997)
20. Poteaux, A.: Computing monodromy groups defined by plane algebraic curves. In: Verschede, J., Watt, S.M. (eds.) Proceedings of the SNC'07, pp. 36–45. ACM Press (2007).
21. Sasaki, T., Yamaguchi, S.: An analysis of cancellation error in multivariate Hensel construction with floating-point number arithmetic. In: Gloor, O. (ed) Proceedings of the ISSAC'98, pp. 1–8. ACM Press (1998).

A Sequence of Nearest Polynomials with Given Factors

Hiroshi Sekigawa

Abstract Let K be either $\mathbb{R}$ or $\mathbb{C}$, and p and f_0 be polynomials in $K[x_1, \ldots, x_s]$ such that $p \neq 0$, $\|f_0\| = 1$, where $\|f_0\|$ is the Euclidean norm of f_0, and the coefficient of f_0 with the maximal absolute value is a positive real number. For $j = 1, 2, \ldots$, let $p_{2j-1} = f_{j-1}g_j$ be the nearest polynomial to p such that $f_{j-1}|p_{2j-1}$ and $\deg(p_{2j-1}) \leq \deg(p)$, where deg is the total degree, and $p_{2j} = c_j f_j g_j$ be the nearest polynomial to p such that $c_j \in K$, $g_j|p_{2j}$, $\deg(p_{2j}) \leq \deg(p)$, $\|f_j\| = 1$, and the coefficient of f_j with the maximal absolute value is a positive real number. We investigate the behavior of the sequences $\{p_j\}$, $\{f_j\}$, $\{g_j\}$, and $\{c_j\}$.

1 Introduction

Since the mid 1990s, there have been many studies on the nearest polynomials to given polynomials with given properties. Finding the nearest polynomial with a given zero is a typical problem [3]. This problem can be generalized to finding the nearest polynomial with a given factor.

Let K be either $\mathbb{R}$ or $\mathbb{C}$. We measure the distance between f and g in $K[x_1, \ldots, x_s]$ by using $\|f - g\|$, the Euclidean norm of $f - g$. Let p and f_0 be polynomials in $K[x_1, \ldots, x_s]$ such that $p \neq 0$, $\|f_0\| = 1$, and one of the coefficients of f_0 with the maximal absolute value is a positive real number. For $j = 1, 2, \ldots$, let $p_{2j-1} = f_{j-1}g_j$ be the nearest polynomial to p such that $f_{j-1}|p_{2j-1}$ and $\deg(p_{2j-1}) \leq \deg(p)$, where deg is the total degree, and $p_{2j} = c_j f_j g_j$ be the nearest polynomial to p such that $c_j \in K$, $g_j|p_{2j}$, $\deg(p_{2j}) \leq \deg(p)$, $\|f_j\| = 1$, and one of the coefficients of f_j with the maximal absolute value is a positive real number. Do the sequences $\{p_j\}$, $\{f_j\}$, $\{g_j\}$, and $\{c_j\}$ converge? If they converge to $\tilde{p}$, $\tilde{f}$, $\tilde{g}$, and $\tilde{c}$, respectively, does $\tilde{p} = \tilde{f}\tilde{g}$ or $\tilde{p} = \tilde{c}\tilde{f}\tilde{g}$ hold? The aim of this article is to answer these questions.

H. Sekigawa (✉)
Tokyo University of Science, Tokyo, Japan
e-mail: sekigawa@rs.tus.ac.jp

© Springer-Verlag Berlin Heidelberg 2014

R. Feng et al. (eds.), *Computer Mathematics*, DOI 10.1007/978-3-662-43799-5_12

If f_j and g_{j+1} (or g_j) are not constant, $f_j g_{j+1}$ (or $c_j f_j g_j$) is an approximate factorization of p. Hence, by perturbing f_0, we might obtain a good approximate factorization of p.

2 Partial Answers to the Questions

Let P and F be polynomials in $K[x_1, \dots, x_s]$ with $\deg(P) = n$ and $\deg(F) = d$ $(1 \le d < n)$, V be $\{ H \in K[x_1, \dots, x_s] \mid \deg(H) \le n \}$, and W be $\{ H \in V \mid F \text{ is a factor of } H \}$. V is a finite dimensional K-vector space and W is a K-subspace of V. Let Q be the orthogonal projection of $P \in V$ onto W. Then, Q is the unique nearest polynomial to P such that $F|Q$ and $\deg(Q) \le \deg(P)$.

Since $V \supset \{ H \in K[x_1, \dots, x_s] \mid \deg(H) \le n-d \} \ni X \mapsto FX \in W$ is a linear map, we can represent the product of F as a matrix $M(F)$. Let $v(H)$ be the vector of coefficients of $H \in V$. Then, $P_1 = FG$ is the nearest polynomial to P if and only if $Y = v(G)$ minimizes $\|v(P) - M(F)Y\|$. $Y = v(G)$ is the unique solution of the normal equation

$$M(F)^* M(F) Y = M(F)^* v(P), \tag{1}$$

where $M(F)^*$ is the conjugate transpose of $M(F)$ (Theorem 2.1.2 in [1]). Therefore, we can easily obtain G. Note that $M(F)^* M(F)$ is nonsingular.

We define the sequences $\{ p_j \}$, $\{ f_j \}$, $\{ g_j \}$, and $\{ c_j \}$, as described in Sect. 1. Note that if p and f_0 are real polynomials so are all p_j's, f_j's, and g_j's.

Proposition 1 *When $p_1 \neq 0$, $\|p_j\|$ is monotone nondecreasing and converges, and $\|p_{j+1} - p_j\|$, $\|c_{j+1} f_{j+1} - f_j\|$, $\|g_{j+1} - c_j g_j\| \to 0$ and $|c_j| \to 1$ as $j \to \infty$.*

Remark 1 g_1 and p_1 might be 0. For example, $g_1 = p_1 = 0$ when $s = 1$, $p = x_1^2 + x_1 + 1$, and $f_0 = (x_1 - 1)/\sqrt{2}$. If $g_1 \neq 0$, then p_j's, f_j's, and g_j's are nonzero for every j because $\|p_1\| > 0$ and $\|p_j\|$ is monotone nondecreasing.

If $\|p_{N+1} - p\| = \|p_N - p\|$ holds for some N, then $p_j = p_N$ for every $j \ge N$. This follows from the fact that f_j (resp. g_j) is a factor of p_{2j} and p_{2j+1} (resp. p_{2j-1} and p_{2j}) and that p_{2j+1} (resp. p_{2j}) is the unique nearest polynomial to p having f_j (resp. g_j) as a factor.

To prove Proposition 1, we need the following lemma.

Lemma 1 *For $F, G \in K[x_1, \dots, x_s]$, there exist positive constants m and M depending on only $\deg(FG)$ such that $m \le \frac{\|FG\|}{\|F\| \cdot \|G\|} \le M$ hold.*

Proof First, we consider the case of $s = 1$. Let $FG = a_n x_1^n + \cdots + a_1 x_1 + a_0$ $(a_n \neq 0)$. We can take 2^{-n} as m (Corollary 6.33 in [4]). Since the inequality $|a_j| \le \|F\| \cdot \|G\|$ holds for every $0 \le j \le n$, we can take $\sqrt{n+1}$ as M.

Next we consider the case of $s \ge 2$. Since the Kronecker substitution $\sigma : K[x_1, \dots, x_s] \to K[x]$ that maps x_j to $x^{n^{j-1}}$ $(n = \deg(FG))$ for $1 \le j \le s$ is a ring homomorphism and $\|H\| = \|\sigma(H)\|$ holds for any $H \in K[x_1, \ldots, x_s]$, we can reduce this case to that of $s = 1$. □

Proof of Proposition 1. Let $r_j = p - p_j$. $\|r_j\|$ converges because it is monotone nonincreasing. Thus, $\|p_j\|$ is monotone nondecreasing and converges because $\|p_j\|^2 = \|p\|^2 - \|r_j\|^2$. Note that $\langle p_j, r_j \rangle = 0$, where $\langle \cdot, \cdot \rangle$ is the inner product of $\{ h \in K[x_1, \ldots, x_s] \mid \deg(h) \leq n \}$. Since $\|p_{j+1} - p_j\|^2 = \|r_j\|^2 - \|r_{j+1}\|^2$ holds, $\|p_{j+1} - p_j\| \to 0$. Note that $p_{j+1} - p_j = r_j - r_{j+1}$ and $\langle r_{j+1}, r_{j+1} - r_j \rangle = 0$.

There exist positive constants m and M not depending on j such that

$$\|p_{2j+2} - p_{2j+1}\| = \|g_{j+1}(c_{j+1} f_{j+1} - f_j)\| \geq m \|g_{j+1}\| \cdot \|c_{j+1} f_{j+1} - f_j\|,$$
$$\|p_1\| \leq \|p_{2j+1}\| = \|f_j g_{j+1}\| \leq M \|f_j\| \cdot \|g_{j+1}\| = M \|g_{j+1}\|$$

from Lemma 1. Therefore, $\|c_{j+1} f_{j+1} - f_j\| \leq \frac{M\|p_{2j+2} - p_{2j+1}\|}{m\|p_1\|}$ holds and $\|c_{j+1} f_{j+1} - f_j\|$ converges to 0. Similar arguments for $p_{2j+1} - p_{2j} = f_j(g_{j+1} - c_j g_j)$ implies that $\|g_{j+1} - c_j g_j\|$ converges to 0. The triangle inequality implies that $\|c_{j+1} f_{j+1} - f_j\| \geq |\, \|c_{j+1} f_{j+1}\| - \|f_j\| \,| = ||c_{j+1}| - 1|$. Thus, $\lim_{j \to \infty} |c_j| = 1$. □

The following is the main theorem.

Theorem 1 *We assume that $p_1 \neq 0$. The sequence $\{ p_j \}$ converges or there are infinitely many limit points. Let $\{ p_{j_k} \}$ be a convergent subsequence of $\{ p_j \}$.*

1. *$\{ f_{j_k} \}$ and $\{ c_{j_k} g_{j_k} \}$ are convergent sequences and $\tilde{p} = \tilde{f}\tilde{g}$, where $\tilde{p} = \lim_{k \to \infty} p_{j_k}$, $\tilde{f} = \lim_{k \to \infty} f_{j_k}$, and $\tilde{g} = \lim_{k \to \infty} (c_{j_k} g_{j_k})$.*
2. *$\tilde{p}$ satisfies the conditions (i) $\tilde{p}$ is the unique nearest polynomial to p such that $\tilde{f} | \tilde{p}$ and $\deg(\tilde{p}) \leq \deg(p)$, and (ii) $\tilde{p}$ is the unique nearest polynomial to p such that $\tilde{g} | \tilde{p}$ and $\deg(\tilde{p}) \leq \deg(p)$.*
3. *For the univariate case, if $\tilde{p} \neq p$, then $\deg(\gcd(\tilde{f}, \tilde{g})) \geq 1$ holds.*

Proof $\{ p_j \}$ converges or there are infinitely many limit points from Lemma 2 below.

Let $\tilde{p}_k = p_{j_k}$, $\tilde{f}_k = f_{\lfloor j_k/2 \rfloor}$, and $\tilde{g}_k = g_{\lfloor (j_k+1)/2 \rfloor}$. Then, $\tilde{p}_k = \tilde{c}_k \tilde{f}_k \tilde{g}_k$ holds, where $\tilde{c}_k = 1$ or $c_{\lfloor (j_k+1)/2 \rfloor}$. Let $\{ \tilde{f}_{j_k} \}$ be a convergent subsequence of $\{ \tilde{f}_j \}$. We write $\tilde{p}_{j_k}$, $\tilde{f}_{j_k}$, $\tilde{g}_{j_k}$, and $\tilde{c}_{j_k}$ as $\hat{p}_k$, $\hat{f}_k$, $\hat{g}_k$, and $\hat{c}_k$, respectively. Let $\hat{f}$ be $\lim_{j \to \infty} \hat{f}_j$. Noting that $\hat{p}_j = \hat{c}_j \hat{f}_j \hat{g}_j$, $\hat{f}$ is a factor of $\tilde{p}$, $\hat{c}_j \hat{g}_j$ converges, and $\tilde{p} = \hat{f} \cdot \lim_{j \to \infty} (\hat{c}_j \hat{g}_j)$ $\hat{c}_j \hat{f}_j - \hat{f} = (\hat{c}_j \hat{f}_j - \hat{f}_{j-1}) + (\hat{f}_{j-1} - \hat{f})$ $\tilde{p} - \hat{p}_j$ 0. $\|\hat{f}\| = 1$ and one of the coefficients of $\hat{f}$ with maximal absolute value is a positive real number. Since the number of factors h's of $\tilde{p}$ such that $\|h\| = 1$ and a coefficient of h with maximal absolute value is a positive real number is finite, Lemma 2 implies that $\{ \tilde{f}_j \}$ is a convergent sequence. Thus, $\{\tilde{c}_j \tilde{g}_j \}$ is also a convergent sequence.

The normal equation (1) implies that the nearest polynomial $\hat{p}$ to p such that f is a factor of $\hat{p}$ and $\deg(\hat{p}) \leq \deg(p)$ is continuous with respect to f. If q ($\neq \tilde{p}$) is the nearest polynomial to p such that $\tilde{f}$ is a factor of q and $\deg(q) \leq \deg(p)$, $\|p - q\| < \|p - \tilde{p}\|$ holds. Then, $\|p - \tilde{p}_j\| < \|p - \tilde{p}\|$ holds for sufficiently large j since $\tilde{f}_j \to \tilde{f}$. This contradicts the fact that $\|p - \tilde{p}_j\|$ is monotone nonincreasing and converges to $\|p = \tilde{p}\|$. Similar arguments hold for $\tilde{g}$.

For the univariate case, when $\tilde{p} \neq p$ let $\deg(\tilde{f}) = d$, $\deg(\tilde{g}) = e$, and

$$V' = \{ h \in K[x_1] \mid \deg(h) \leq d + e \},$$
$$W_1 = \{ \tilde{f}h \mid h \in K[x_1], \deg(h) \leq e \}, \; W_2 = \{ \tilde{g}h \mid h \in K[x_1], \deg(h) \leq d \}.$$

W_1 and W_2 are K-subspaces of the K-vector space V'. The dimensions of V', W_1, and W_2 are $d + e + 1$, $e + 1$, and $d + 1$, respectively. For every $v_1 \in W_1$ and $v_2 \in W_2$, $\langle p - \tilde{p}, v_1 \rangle = \langle p - \tilde{p}, v_2 \rangle = 0$ hold. Therefore, $\langle p - \tilde{p}, w \rangle = 0$ for every $w \in W = W_1 + W_2$. Thus, $\dim(W) \leq d + e$ and $\dim(W_1 \cap W_2) \geq 2$ since $p - \tilde{p} \neq 0$. Take a nonzero polynomial $h \in (W_1 \cap W_2) \setminus \{ a\tilde{f}\tilde{g} \mid a \in K \}$. If $\deg(h) = d + e$ we can take $c \in K$ such that $\deg(h - c\tilde{f}\tilde{g}) < d + e$ because $\deg(\tilde{f}\tilde{g}) = d + e$. Since $0 \neq h - c\tilde{f}\tilde{g} \in (W_1 \cap W_2) \setminus \{ a\tilde{f}\tilde{g} \mid a \in K \}$, by replacing h with $h - c\tilde{f}\tilde{g}$, if necessary, we can assume that $h \in W_1 \cap W_2$, $h \neq 0$, and $\deg(h) < d + e$. Therefore, $\deg(\gcd(\tilde{f}, \tilde{g})) \geq 1$ holds. □

Lemma 2 *Let* $\{ a_j \} \subset K^n$ *and* $a \in K^n$. *When* $\|a_j - a\|$ *is monotone nonincreasing and* $\lim_{j \to \infty} \|a_{j+1} - a_j\| = 0$, $\{ a_j \}$ *converges or there are infinitely many limit points.*

Proof Let $\overline{B} = \{ c \in K^n \mid \|c - a\| \leq \|a_1 - a\| \}$. Note that there exists at least one limit point of $\{ a_j \}$ because $\{ a_j \} \subset \overline{B}$ and $\overline{B}$ is compact.

It is sufficient to prove that if the number of limit points of $\{ a_j \}$ is finite, the number is one and $\{ a_j \}$ converges. Let $\tilde{a}_1, \ldots, \tilde{a}_t$ $(t < \infty)$ be all the limit points and $B_j(\delta) = \{ c \in K^n \mid \|c - \tilde{a}_j\| < \delta \}$ $(\delta > 0)$. For any $\epsilon > 0$, we can take δ such that $0 < \delta < \epsilon$ and $B_j(2\delta) \cap B_k(2\delta) = \emptyset$ $(j \neq k)$. If there are infinitely many a_j's in $A(\delta) = B_1(\delta)^c \cap \cdots \cap B_t(\delta)^c$, where $B_k(\delta)^c = K^n \setminus B_k(\delta)$, there exists a limit point in $A(\delta)$ because $A(\delta) \cap \overline{B}$ is compact and $\{ a_j \} \subset \overline{B}$. This contradicts the assumption that $\tilde{a}_1, \ldots, \tilde{a}_t$ are all the limit points. Thus, $\{ a_j \} \cap A(\delta)$ is a finite set. Hence, there exists $N \in \mathbb{N}$ such that $a_k \notin \{ a_j \} \cap A(\delta)$ and $\|a_{k+1} - a_k\| < \delta$ for every $k \geq N$. Let μ be the number such that $a_N \in B_\mu(\delta)$. Then, $a_k \in B_\mu(\delta)$ for every $k \geq N$; that is, $\|a_k - \tilde{a}_\mu\| < \delta < \epsilon$. Therefore, there exists only one limit point $\tilde{a}_\mu$ and $\lim_{k \to \infty} a_k = \tilde{a}_\mu$. □

3 Conclusion

We investigated some properties of the sequences $\{ p_j \}$, $\{ f_j \}$, $\{ g_j \}$, and $\{ c_j \}$. One of the directions of the future research is to study algorithms for finding an approximate factorization of a given polynomial based on the properties of the sequences. The algorithms must be compared with existing methods such as proposed in [2] and in the references there. Another direction is to prove that $\{ p_j \}$ always converges or to find an example $\{ p_j \}$ that does not converge.

Acknowledgments This work was supported by the Japan Society for the Promotion of Science through Grant-in-Aids for Scientific Research (KAKENHI) 21500026 and 24500022.

References

1. Campbell, S.L., Meyer Jr, C.D.: Generalized Inverses of Linear Transformations. Pitman Publishing Ltd, London (1979)
2. Kaltofen, E., May, J.P., Yang, Z., Zhi, L.: Approximate factorization of multivariate polynomials using singular value decomposition. J. Symbolic Comput. **43**(5), 359–376 (2008)
3. Stetter, H.J.: The nearest polynomial with a given zero, and similar problems. ACM SIGSAM Bull. **33**(4), 2–4 (1999)
4. von zur Gathen, J., Gerhard, J.: Modern Computer Algebra, 2nd edn. Cambridge University Press, Cambridge (2003)

Digitization Workflow in the Czech Digital Mathematics Library

Petr Sojka

Abstract Experience in setting up a workflow from scanned images of mathematical writings into a fully fledged mathematical library is described on the example of the project Czech Digital Mathematics Library DML-CZ. An overview of the whole process is given, with detailed description of production steps involving scanned image processing and optical character recognition. Experience gained, lessons learned, and tools prepared during development of DML-CZ are described. DML-CZ now serves more than 30,000 articles (more than 300,000 digitised pages) to the public.

Keywords Digital mathematical library · Mathematical knowledge representation · Digitisation workflow · Optical character recognition · OCR · Retro-digitisation · DML-CZ

"Viva la Workflows! (Carole Goble [1])"

1 Motivation

Digital Library business has moved from data/files centered processing toward process-oriented *workflows*. Workflows enact the machinery of building and running a digital library. Instead of running simple tools and mirroring file repositories more subtle solutions have to be devised: data curatorship changes to workflow curatorship and *services*.

There are communities and systems that start to dominate in some thematic areas: PubMed Central (PMC) is one such system in the medical domain, speeding up research and author's citation indexes in the area. Unfortunately, only domains where global initial funding was available took advantages of the platforms established and tools and workflows developed. In the PMC case, journal publishers are now eager to join the club, and authors enjoy global topical ontology-based search. Researchers

---

P. Sojka (✉)
Faculty of Informatics, Masaryk University, Botanická 68a, 602 00 Brno, Czech Republic
e-mail: sojka@fi.muni.cz

© Springer-Verlag Berlin Heidelberg 2014
R. Feng et al. (eds.), *Computer Mathematics*, DOI 10.1007/978-3-662-43799-5_13

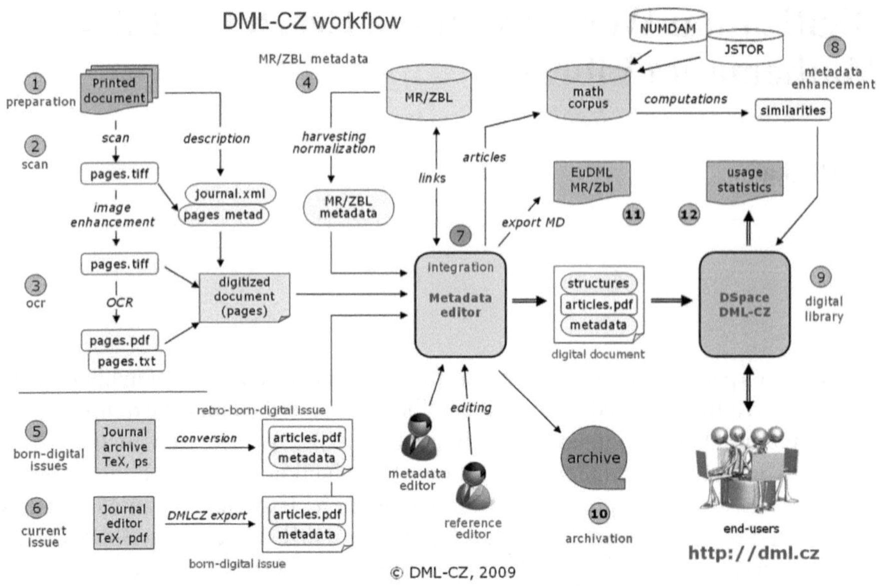

Fig. 1 DML-CZ top-level workflow scheme

send their papers only to journals available in PMC as this leverages their citation indexes. However, the realization of the dream of a World Digital Mathematics Library [2] is yet to come.

We report on the experience gained, lessons learned, and tools prepared during the development of a digitization workflow for The Czech Digital Mathematics Library DML-CZ project. The aim of the project approved for the five years' period 2005–2009 was to digitize the relevant mathematical literature published in the Czech lands. It comprises periodicals, selected monographs, and conference proceedings from the nineteenth century up until currently produced mathematical publications. It has been launched and is readily available on dml.cz, serving almost 30,000 articles on 300,000 pages to the public.

The general workflow of the project, shown in Fig. 1, reflects different types of acquired input data:

full digitisation from print work starts from a paper copy;
full digitisation from bitmap image work starts from an electronic bitmap of pages;
retro-born-digital work starts from an electronic version of the document (usually in POSTSCRIPT or PDF);
born-digital workflow of the journal production is enriched with an automated export of data for the digital library.

Within the project, several general purpose tools have been developed:

1. scripting of transformation pipes of scanned images,
2. DML-CZ OCR workflow allowing recognition of scanned mathematical documents,
3. web-based Metadata Editor [3],
4. tools for classification of mathematical documents and measuring their similarity [4],
5. workflow for born-digital publication production with direct export of metadata for DML [5] and
6. plenty of other smaller tools like: extensions to LUCENE engine allowing indexing of mathematics, batch PDF stamper for digital signing of produced PDF, an optimizer re-compressing image objects in PDF with the new JBIG compression filter supported by Adobe since PDF specification version 1.6 (Adobe Reader 5) or batch article PDF generation with title page by XeLaTeX.

In the following sections we describe part (steps 2, 3, 7 and 8 in Fig. 1) of our digitization workflow with the hope that they can be used by similar projects or even in other domains.

"We are all apprentices in a craft
where no-one ever becomes a master. (Ernest Hemingway)"

2 Scanning and Image Transformations

Processing of scanned images is aimed at final delivery of 600 DPI bi-tonal images suitable for quality OCR and a fine print. This is the quality recommended by the Committee on Electronic Information and Communication (CEIC) and used for example by JSTOR and NUMDAM. Images from the Göttingen Digitisation Centre (GDZ) and images scanned in the Digitisation Centre of the Library of Academy of Sciences, Czech Republic prior to the project DML-CZ have bi-tonal 400 DPI quality. The difference is visible, and leads to a higher OCR error rate. We strongly support the recommendation to scan with a resolution of at least 600 DPI.

We perform our new scans at 600 DPI with 4-bit depth, 'having a depth/space' for geometrical and other transformations done on images before binarization. Scanning is carried out in Digitization Centre of the Library of Academy of Sciences in Jenštejn near Prague on the Zeutschel OS 7000 A2 book scanners.

The primary scans in TIFF format are archived for a possible future reprocessing if needed. We use BOOK RESTORER™ image restoration software by i2S for interactive and batch image processing in an uncompressed TIFF format. Operations performed on images are:

1. *geometrical correction* as narrowing the baselines and widths of the same characters on the same line;
2. *cropping* the page to cut out the speckles at page borders;
3. *blur filter*, 3×3 pixels, to eliminate one or two pixel size variations;
4. *binarization* with manually adjusted parameters for every batch (usually journal volume);

5. *despeckle filter*, with both white and black spotting, 3 × 3 pixels;
6. *publish/export*: processed TIFF s are stored being compressed by the Lempel-Ziv-Welsh method for compressing grayscale and the G4 one for binarized images to speed up further processing (OCR) and to save space.

Both the order of these steps and the parameter adjustments for images of different quality are very important. For the data from GDZ, slightly different operations are needed as the input files are already bi-tonal and some filters are applicable only on grayscale images.

Step 1 employs the algorithms that allow perspective correction of a scanned image. As most of the material to digitize cannot be cut, we scan 2-up page spreads, making the text size nonuniform even when trying to flatten the spread by pane of glass. Book Restorer can also flatten the lighting across the scanned spead. For more details of this step see [6, p. 2].

Step 2 crops the unnecessary border parts of the page shot.

Step 3 aims at better binarization and despeckling by unsharping the shapes in the image.

Step 4 is necessary as most OCR engines work on bi-tonal images. It may be left to the high-quality OCR engine—clever thresholding starts to be a standard part of OCR programs [7], or perform it ourselves adaptively based on OCR feedback [8].

Step 5 is inserted to remove small impurities in the image.

Step 6 is the final step: image is stored as LZW-compressed grayscale or G4-compressed bi-tonal TIFF.

For the lower resolution data from GDZ, slightly different operations are needed as the input files are already bi-tonal (e.g. we did upscaling before unsharping) and because some filters are applicable only on grayscale images.

It is wise to differentiate processing of pages with grayscale images (e.g. photos) so that they are not degraded by image filters suitable for text. To avoid possible difficulties in the later steps it is important from the very beginning to carefully check image quality before proceeding with the remaining steps. At least automated procedures that check the technical metadata (e.g. `tiffinfo`) and image quality (pixel width and height) has to be the part of quality assurance. Metrics of compressibility by the JBIG2 encoder were used to trigger quality checks.

There is a tradeoff between price of image cleanup and quality results within constraints of digitization budget. When acquiring a craftmanship in good image editing software, results very close to (or even better than) the original could be achieved [9]. These handmade touches are usually beyond the budget of most digitization projects, where the highest degree of automation is needed to reduce the cost of digitisation. In DML-CZ, we have prepared [10] set of batches of typical transformation procedures to be used by BOOK RESTORER™ operators to achieve the best price/effort ratio.

Fine-tuning of operations on the pixel level pays back in the following step: the OCR.

"The road to wisdom?
Well, it's plain and simple to express:
Err and err and err again,
but less and less and less. (Piet Hein)"

3 Optical Character Recognition: DML-CZ OCR

To have papers indexed we need to get full text from page bitmaps by the process of optical character recognition. Also, we need to recognize logical page numbers located in every TIFF, to link the page images to article metadata.

Tests with various OCR programs showed that no single one gives acceptable results for mathematical content, with character error rates often above 10 (counting wrong character positions and font types as errors too). For text recognition, FINEREADER by textABBYY® gave the best results, whereas for the structural recognition of mathematics InftyReader [11] had impressive results.

The FINEREADER software development kit (SDK for Windows version 8.1) was used to develop a part of the system for the location and recognition of page numbers, and a batch system DML-CZOCR [12, 13] which takes sequences of TIFF images and produces two-layered one page PDFs (with invisible full-texts behind the images). The processing starts with the recognition of languages used in every paragraph, and then blocks are recognized again with a special setting (language dictionaries used) for every given block of text. With such fine-tuning of parameters, we are able to achieve a one percent character error rate [13].

Among solutions and software evaluated on plain texts, FINEREADER gave the best results, but it has no support for the recognition of mathematical expressions. Texts without recognized maths may be sufficient for basic indexing and search. However, it is not surprising that omitting maths matters when the full texts are used for such tasks as automated text classification and categorization or for computing paper similarity [14]. Therefore we strive to enhance the state-of-the-art possibilities for mathematical OCR.

Neither ABBYY® nor Google responded positively on the near future of math OCR development plans—mathematics is only a small market niche for them. On the other hand, developers of the INFTYREADER system [11] were willing to gradually improve their support for European languages, MATHML and LaTeX export filters and to enrich their recognized database of mathematical symbols.

We found that setting the parameters of the OCR engine (language, word-list consultation) influences the precision significantly. We trained FINEREADER on the type cases used at the printer where journals were typeset.

At the end of extensive experiments, we developed a method of OCR processing consisting of several phases, both in FINEREADER and Infty. Processing using FINEREADER consist of the following:

1. A page or block of text is recognized for the first time using a universal setup (non-language specific). A histogram of character bigrams and trigrams from words with lengths greater than three is created.
2. The computed histogram of the text block is compared [15] to the histograms created from the journal data during the training phase for all languages used (English, French, Russian, German, and Czech). Perl module `Lingua::Ident` is used. Block with bibliography is detected by different algorithms and is treated differently.
3. Page or block of text is processed for the second time with parameters optimized for recognized 'language' in previous step and saved as a two-layer PDF (with text layer used for searching, indexing and similarity computation).

Recognition of mathematical formulae in FINEREADER is not satisfactory, however. The only suitable tool for this domain that we have found and experimented with is INFTY. INFTY's new PDF import capability is very significant to us: it will allow to import our current FINEREADER's two-layer PDFs, use the text part only, throw away badly recognized maths and to detect and recognize maths expressions. A new INFTY version that combines FINEREADER's technology (OCR voting [16]) is in preparation. In the meantime,

1. PDF is passed to INFTYREADER and results are stored in the INFTYREADER Markup Language (IML) and in LaTeX (Human readable LaTeX).
2. IML is postprocessed by a home-grown program in JAVA to fix recognition errors of some of the accented characters that INFTYREADER does not yet have in its glyph database.

Using the process outlined above we have managed to decrease the character error rate from an initial 11.35 % (universal language setup of FineReader) to an average 0.98 % character error rate [17–19]. The whole processing is fully automated after initial font recognition and language detection training. The error rate may be further decreased when INFTYREADER's character database is semiautomatically enriched when processing a new journal.

"When in doubt, use brute force. (Ken Thompson)"

4 Text Postprocessing and Metadata Enhancements

The OCR step is followed by further text processing, and its results are used for editing of metadata and references.

4.1 Metadata Editor

The Metadata Editor (ME) [3, 20] has gradually developed into a fully fledged and efficient web application, https://editor.dml.cz, that allows simultaneous remote editing according to assigned structured access rights. It supports two levels of actions.

On the first one the operator editing the data is provided with page thumbnails so that he can visually check the completeness, scan the quality and configuration of the articles, easily shuffle the pages and cut or merge articles if necessary. On the other level the operator can check the automatically imported metadata, edit and complete them. An integral part of the ME is the module for administration of authority files with authors' names. It enables the most suitable version of the name for the DML-CZ to be selected and to match it with all its other versions.

We consider bibliographical references as important metadata of every paper. Their availability makes it possible to use professional systems like CROSSREF® for cross-publisher citation linking. The work starts from OCR of the text, in which a block of references is found. Citations are tagged by a script based on regular expressions written for the citation style of every journal. The operator then checks, edits, and approves the list of paper citations.

For fixing errors that can be safely detected (such as a Mathematics Subject Classification (MSC) code string that is invalid in the MSC 2000 standard) procedures are formulated and coded in XSchema generated also from a web-based interface (forms). Other sets of constraint checkers run as overnight jobs together with updates of the database and metadata statistics and logs useful for the management of Metadata Editor workflow.

Finally, various detection procedures for possible errors have been suggested, evaluated, and implemented for finding anomalous and suspicious content of metadata fields, with lists of warnings generated, including hyperlinks for easy checking by an operator. An important control concerns the integrity of TeX sequences in metadata to assure seamless typesetting of article cover pages in the later stages: all metadata to be typeset are exported in one big file with unique references to the article, and typeset by XeLaTeX to check the TeX control sequences used in the metadata fields. This ensures that all of the TeX encoded mathematics converts into MathML format smoothly. Similar procedures allow for an efficient and economical increase of metadata completeness and quality.

4.2 Mathematical Document Classification and Categorization

Article full texts have many applications, e.g., for document classification and categorization. Fine document classification allows document filtering to reach higher precision in information retrieval systems such as DML. The most commonly used classification system today is the Mathematics Subject Classification (MSC) scheme (www.ams.org/msc/), We have developed an MSC classifier (guessed MSC) that is able to assign top-level MSC for retro-digitized articles. Our results convincingly demonstrated the feasibility of a machine learning approach to the classification of mathematical papers [4].

Another round of experiments was done with mathematical document similarity computation. We have collected corpus of full texts of more than 40,000 articles (from DML-CZ and NUMDAM) and we have computed paper similarities using *tfidf* [21] and Latent Semantic Analysis (LSA) [22] and Random Projection methods. Methods use a Vector Space Model, first converting articles to vectors and then using the cosine of the angle between the two document vectors to assess their content similarity [23]. The difference between the methods is that while *tfidf* works directly over tokens, LSA first extracts concepts, then projects the vectors into this conceptual space where it only computes similarity.

We are now showing the links to closest document lists in DML-CZ article landing pages to get feedback from authors and readers to evaluate metrics computed in this experiment. Given that we will enrich our full text mathematical corpus significantly (with data from JSTOR, ARXIV and other sources as planned), we hope it will help to tackle plagiarism, too.

> "Automating the creation of useful digital libraries—that is, digital libraries affording searchable text and reusable output—is a complicated process, whether the original library is paper-based or already available in electronic form (Simske and Lin [9])"

5 Summary, Conclusions, and Acknowledgement

We have described several steps of DML-CZ workflow, as introduced and tested developed during the project development. We carried out most of the steps ourselves, to gain expertize and retain control of fine details, allowing us to plug-in new modules arising from leading edge research in the future—there are, currently, many new developments appearing and much research underway in the digitisation area. It is advisable for smaller project to outsource most of the workflow steps.

The most time-consuming and costly step is metadata handling and editing, and image transformation and editing (if it cannot be automated). Bare scanning costs amount to less than 10 % of the total page costs, and even less when pages can be physically cut before being used for batch scanning.

The complexity of the full digitization workflow should not be underestimated, especially when digitizing heterogeneous sources—continuous and flexible workflow adaptation is a must.

We believe that the methods, algorithms, and tools developed do represent important step toward a European (EuDML) or even worldwide framework for a digital mathematics library, evolved, bottom-up, from smaller regional digital library projects.

This research has been partially supported by the grant reg. no. 1ET200190513 of the Academy of Sciences of the Czech Republic, by MŠMT grants MSM0021622419 and 2C06009. The author thanks other DML-CZ colleagues for fruitful discussions that led to the design of the workflow described there and to the paper reviewers for improvement suggestions. Drawing of Fig. 1 by Mirek Bartošek is acknowledged.

References

1. Goble, C.: Curating services and workflows: the good, the bad and the downright ugly, 2008. Keynote presented at ECDL 2008, http://www.ecdl2008.org/keynotes/
2. Jackson, A.: The digital mathematics library. Notices of Am. Math. Soc., 50(4):918–923 (2003). http://www.ams.org/notices/200308/comm-jackson.pdf
3. Bartošek, M., Kovář, P., Šárfy, M.: DML-CZ metadata editor: Content creation system for digital libraries. In: Sojka [24], pp. 139–151. http://www.fi.muni.cz/sojka/dml-2008-program.xhtml
4. Řehůřek, R., Sojka, P.: Automated classification and categorization of mathematical knowledge. In: Serge Autexier, John Campbell, Julio Rubio, Volker Sorge, Masakazu Suzuki, and Freek Wiedijk, editors, Intelligent Computer Mathematics-Proceedings of 7th International Conference on Mathematical Knowledge Management MKM 2008, volume 5144 of Lecture Notes in Computer Science LNCS/LNAI, pages 543–557, Berlin, Heidelberg, July 2008. Springer-Verlag
5. Růžička, M.: Automated processing of TEX-typeset articles for a digital library. In: Sojka [24], pp. 167–176. http://www.fi.muni.cz/sojka/dml-2008-program.xhtml
6. Chevalier, P.: i2S DigiBook Mag, issue no. 2, July 2002. http://ww.i2s-bookscanner.com/pdf/digibook_mag_no2.pdf
7. Smith, R., Newton, C., Cheatle, P.: Adaptive thresholding for OCR: A significant test. Technical Report HPL-1993-22, HP Laboratories Bristol, March 1993
8. Rangoni, Y., Shafait, F., Breuel, T.M.: OCR based thresholding. In: Proceedings of MVA 2009 IAPR Conference on Machine Vision Applications, pp. 3–18, May 2009
9. Simske, S.J., Lin, X.: Creating digital libraries: Content generation and re-mastering. In: Proceedings of First International Workshop on Document Image Analysis for Libraries (DIAL 2004), p. 13, 2004. http://doi.ieeecomputersociety.org/10.1109/DIAL.2004.1263235
10. Pulkrábek, T.: Obrazové transformace při digitalizaci textů (in Czech, Image Transformation during Digitisation). Master's thesis, Faculty of Informatics, 2008. Bachelor's Thesis Masaryk University, Brno, Faculty of Informatics. https://is.muni.cz/th/139908/fi_b/?lang=en
11. Suzuki, M., Tamari, F., Fukuda, R., Uchida, S., Kanahori, T.: INFTY–An integrated OCR system for mathematical documents. In: C. Vanoirbeek, C. Roisin, and E. Munson, editors, Proceedings of ACM Symposium on Document Engineering 2003, pp. 95–104, Grenoble, France, 2003. ACM
12. Sojka, P.: Towards digital mathematical library: optical character recognition of mathematical texts. In: Julius Štuller and Zdenka Linková, editors, Inteligentní modely, algoritmy a nástroje pro vytváření sémantického webu, pp. 110–113, Prague, 2006. Ústav informatiky AV ČR
13. Sojka, P., Panák, R., Mudrák, T.: Optical character recognition of mathematical texts in the DML-CZ project. Technical report, Masaryk University, Brno, September 2006. Presented at CMDE 2006 conference in Aveiro, Portugal
14. Sojka, P., Řehůřek, R.: Classification of multilingual mathematical papers in DML-CZ. In: Petr Sojka and Aleš Horák, editors, Proceedings of Recent Advances in Slavonic Natural Language Processing–RASLAN 2007, pp. 89–96, Karlova Studánka, Czech Republic, December 2007. Masaryk University
15. Dunning, T.: Statistical identification of language. Technical Report MCCS 94–273, New Mexico State University, Computing Research Lab (1994)
16. Marosi, I., Tóth, L.: OCR voting methods for recognizing low contrast printed documents. In: Proceedings of Second International Conference on Document Image Analysis for Libraries (DIAL 2006), pp. 108–115, April 2006
17. Mudrák, T.: Digitalizace matematických textů (in Czech, Digitisation of Mathematical Texts). Master's thesis, Masaryk University, Brno, Faculty of Informatics, April 2006. https://is.muni.cz/th/60738/fi_m/?lang=en
18. Panák, R.: Digitalizácia matematických textov (in Czech, Digitisation of Mathematical Texts). Master's thesis, Masaryk University, Brno, Faculty of Informatics, April 2006. https://is.muni.cz/th/60587/fi_m/?lang=en

19. Sojka, P.: Workflow in the digital mathematics library project: How mathematics is stored and retrieved. In: J. Paralič, J. Dvorský, and M. Krátký, editors, Proceedings of Znalosti 2006, pp. 243–247. VŠB-Technická univerzita Ostrava (2006)
20. DML-CZ. Digitization metadata editor. http://sourceforge.net/projects/dme/, (2009)
21. Salton, Gerard, Buckley, Chris: Term-weighting approaches in automatic text retrieval. Inf. Process. Manag. **24**, 513–523 (1988)
22. Deerwester, Scott C., Dumais, Susan T., Landauer, Thomas K., Furnas, George W., Harshman, Richard A.: Indexing by latent semantic analysis. J. Am. Soc. Inf. Sci. **41**(6), 391–407 (1990)
23. Manning, C.D., Raghavan, P., Schütze, H.: Introduction to information retrieval. Cambridge University Press, Cambridge (2008)
24. Sojka, P. (ed.): Towards a digital mathematics library, Birmingham, UK, July 2008. Masaryk University. http://www.fi.muni.cz/sojka/dml-2008-program.xhtml

The Implementation and Complexity Analysis of the Branch Gröbner Bases Algorithm Over Boolean Polynomial Rings

Yao Sun and Dingkang Wang

Abstract A new branch of Gröbner basis algorithm over boolean ring has been presented in an earlier paper. In this paper, the detailed implementation and a rough complexity analysis is given. The branch Gröbner basis algorithm implements a variation of the F5 algorithm and bases on the ZDD data structure, which is also the data structure of the framework PolyBoRi. This branch Gröbner basis algorithm is mainly used to solve algebraic systems and attack multivariable cryptosystems, and its goal is to lower the complexity in each branch and expect better total complexity. An important proposition ensures the two original criteria of the non-branch F5 algorithm could still reject almost all unnecessary computations in this new branch algorithm. The timings show this branch algorithm performs very well for randomly generated systems as well as a class of stream ciphers which is generated by the linear feedback shift register (LFSR).

1 Introduction

Solving system of polynomial equations is a basic problem in computer algebra, through which many practical problems can be solved easily. Among all the methods for this purpose, Gröbner bases method, the characteristic set method and resultant method are the most famous ones [1, 2].

Since Buchberger proposed the Gröbner bases algorithm in 1965, this algorithm has been improved by many researchers, both from the data structure and the criteria to remove the redundant S-pairs. Now the most famous Gröbner bases algorithms are the F4 and F5 algorithms proposed by Faugère [3, 4]. The F4 algorithm imports

---

Y. Sun (✉)
SKLOIS, Institute of Information Engineering, CAS, 100093 Beijing, China
e-mail: sunyao@iie.ac.cn

D. Wang
KLMM, Academy of Mathematics and Systems Science, CAS,
100190 Beijing, China
e-mail: dwang@mmrc.iss.ac.cn

© Springer-Verlag Berlin Heidelberg 2014
R. Feng et al. (eds.), *Computer Mathematics*, DOI 10.1007/978-3-662-43799-5_14

the matrix technique to make the reduction process more efficient, while the F5 algorithm presents two new criteria to eliminate the useless S-pairs, which can be definitely reduced to 0.

So far, both F4 and F5 algorithms have been implemented. The most efficient implementation of F4 algorithm is presented by Steel, and is available on the computer algebraic system Magma, while the the most efficient version of F5 algorithm is implemented by Faugère himself, which is not open. However, the F4 algorithm is still not perfect, as high efficiency leads to the cost of enormous memories. For example, attacking the cryptographic system HFE80 by using F4 algorithm in Magma will cost nearly 16G memories.

For solving system of boolean polynomial equations, a Gröbner bases algorithm based on the ZDD data structure has been proposed by Brickenstein in 2007 (the PolyBoRi framework) [5]. His algorithm works very well for computing Gröbner basis with the pure lexicographic monomial order. However, since it is expensive to compute the total degree leading monomial for a polynomial in the ZDD form, this algorithm possibly does not perform very well with total degree orders. So in our implementation, we prefer a new graded expression of polynomials such that the leading monomial for total degree order can be calculated extremely fast, so our algorithm is more efficient with graded monomial orders. A characteristic set method for solving system of boolean polynomial equations is presented by Gao, and his method has pretty good performance on the problem of stream cipher systems [6]. Gao's implementation is also based on ZDD data structure and he uses the branch technique to compute the ascending set series.

The success of Gao's algorithm is a motivation for our research on branch Gröbner basis. Our algorithm makes improvements both on saving the usage of memories and limiting the size of matrices. That is, on one hand, we utilize the ZDD data structure to save polynomials and decrease the cost of space, and on the other hand, we make new branches when the matrix grows bigger so that we only need to handle matrix with a reasonable size.

In theory, we employ a modified matrix F5 algorithm. Details can be found in [7]. In this paper, we concentrate on the implementation and complexity analysis of our algorithm. The contents of this paper are organized as follows: the second section involves some preliminaries and the modified algorithm; the third section introduces the ZDD data structure and some sub-algorithms; complexity analysis comes in the fourth section; some examples and timings are given in the fifth section; we end this paper with conclusions.

2 The Algorithm

2.1 Notations

Let F_2 be the finite field with two elements 0 and 1, and $X = x_1, \ldots, x_n$ stands for the set of variables. Let H be the set of field polynomials $\{x_1^2+x_1, \ldots, x_n^2+x_n\}$, then the ring $R_2 = F_2[X]/\langle H\rangle$ is actually a boolean ring, where $\langle H\rangle$ is the ideal generated by H in $F_2[X]$, and in addition, we call the elements in R_2 boolean polynomials.

Let N be the set of nonnegative integer and T be the power set of X, which means $T = \{x_1^{\alpha_1}\cdots x_n^{\alpha_n}|\alpha_i \in \{0,1\}, i = 1,\ldots,n\}$. Assume $\prec$ is an admissible monomial order defined over T, then given $t = x_1^{\alpha_1}\cdots x_n^{\alpha_n} \in T$, we define the degree of t as $\deg(t) = \sum_{i=1}^{n}\alpha_i$. For a polynomial $0 \neq f \in F_2[X]$, we have $f = \sum x_1^{\alpha_1}\cdots x_n^{\alpha_n}$. Define the degree of f as $\deg(f) = \max\{\alpha_1+\cdots+\alpha_n\}$, while the leading monomial of f is $\mathrm{lm}(f) = \max_{\preccurlyeq}\{x_1^{\alpha_1}\cdots x_n^{\alpha_n}\}$.

2.2 Definitions

To introduce our algorithm, some definitions are necessary and the following definitions are imported from [7]:

Definition 1 Let $L = T \times F_2[X] \times N \times F_2[X] \times N$, and a labeled polynomial is a five-tuple vector $\mathscr{G} = (x^\alpha, f, i, g, k) \in L$. We define the signature of $\mathscr{G}$ as $S(\mathscr{G}) = x^\alpha$, the initial as $\mathrm{init}(\mathscr{G}) = f$, the extended signature as $\mathrm{ES}(\mathscr{G}) = S(\mathscr{G})\mathrm{lm}(\mathrm{init}(\mathscr{G})) = x^\alpha \mathrm{lm}(f)$, the index as $\mathscr{G}) = i$, the polynomial as $\mathrm{poly}(\mathscr{G}) = g$ and the number as $\mathrm{num}(\mathscr{G}) = k$.

Definition 2 Let $\mathscr{F}, \mathscr{G} \in L$ be two labeled polynomials. We say $\mathscr{F} \prec_{es} \mathscr{G}$ (or $\mathscr{G} \succ_{es} \mathscr{F}$), if one of the following three cases is satisfied: 1. $\mathrm{ES}(\mathscr{F}) \prec \mathrm{ES}(\mathscr{G})$. 2. $\mathrm{ES}(\mathscr{F}) = \mathrm{ES}(\mathscr{G})$ and $\mathscr{F}) > \mathscr{G})$. 3. $\mathrm{ES}(\mathscr{F}) = \mathrm{ES}(\mathscr{G})$, $\mathscr{F}) = \mathscr{G})$ and $\mathrm{num}(\mathscr{F}) > \mathrm{num}(\mathscr{G})$.

The order of labeled polynomials, which alleviate the influence of the order of the input polynomials or the initial polynomials, is modified from the F5 algorithm. Therefore, the criteria should be revised correspondingly. In the following two definitions, let $\mathscr{F} \in L$ be a labeled polynomial and $B \subset L$ be a set of labeled polynomials.

Definition 3 We say $\mathscr{F}$ is normalized by B, if there do not exist a labeled polynomial $\mathscr{G} \in B$ and a monomial $v \in T$, such that $S(\mathscr{F}) = v\mathrm{lm}(\mathrm{poly}(\mathscr{G}))$, $\mathscr{F} \succ_{es} v\mathrm{lm}(\mathrm{init}(\mathscr{F}))\mathscr{G}$. Particularly, in the boolean ring R_2, the condition $\mathrm{lm}(\mathrm{init}(\mathscr{F})) \nmid S(\mathscr{F})$ should hold as well.

Definition 4 We say $\mathscr{F}$ can be rewritten by B, if there exists a labeled polynomial $\mathscr{G} \in B$ and a monomial $v \in T$, such that $S(\mathscr{F}) = S(v\mathscr{G})$, $(\mathscr{F}) = (\mathscr{G})$ and $\mathrm{num}(\mathscr{F}) < \mathrm{num}(\mathscr{G})$.

2.3 Criteria

Now, we give two new criteria modified from the F5 algorithm without proofs. A partial proof can be found in [7], and the complete one will come in a future paper.

1. **Syzygy criterion:** Given a critical pair: $s(\mathscr{G}_1, \mathscr{G}_2) = (m, u_1, \mathscr{G}_1, u_2, \mathscr{G}_2)$, where $u_1, u_2 \in T$ and $\mathscr{G}_1, \mathscr{G}_2 \in B$. If either $u_1\mathscr{G}_1$ or $u_2\mathscr{G}_2$ is not normalized by B, then the critical pair $s(\mathscr{G}_1, \mathscr{G}_2)$ can be discarded.
2. **Rewritten criterion:** Given a critical pair: $s(\mathscr{G}_1, \mathscr{G}_2) = (m, u_1, \mathscr{G}_1, u_2, \mathscr{G}_2)$, where $u_1, u_2 \in T$ and $\mathscr{G}_1, \mathscr{G}_2 \in B$. If either $u_1\mathscr{G}_1$ or $u_2\mathscr{G}_2$ can be rewritten by B, then the critical pair $s(\mathscr{G}_1, \mathscr{G}_2)$ can be discarded.

Our non-branch Gröbner bases algorithm is nothing else than a general Buchberger algorithm except replacing the polynomials with the labeled polynomials, adding two criteria and using matrix reduction. After revising the two criteria, our algorithm has less influence from the input order of the initial polynomials than the F5 algorithm. Furthermore, the two revised criteria can also eliminate almost all the useless critical pairs as the F5 algorithm does. For semi-regular systems, there does not exist the critical pairs that can be reduced to 0, too. In fact, we proved in [7] that the modifications in the comparison of two labeled polynomials do not affect the function of the criteria. To illustrate how the two new criteria work, we have tested some randomly generated boolean polynomial systems in Sect. 4.

2.4 Trick

Although the two modified criteria can remove almost all the useless critical pairs generated during the computation, the total efficiency is not as good as we wished. One possible reason may be the conflict between the signature and the polynomial.

In fact, the motivation of the signature is to record the origin of the present label polynomial. Take a labeled polynomial, say $\mathscr{G} = (x^\alpha, f, i, g, k) \in L$, for example. The signature tells us that the polynomial g is obtained by reducing the polynomial $x^\alpha f_i$ with 'smaller' labeled polynomials under the order $\prec_{es}$. So the signature actually works as a clue of the computation, and that is exactly why the two criteria work. However, there exists a natural conflict between the signature and the polynomial. That is, during the computation, the label polynomials generated later sometimes have smaller size but with bigger signature. By our algorithm as well as the F5 algorithm, the label polynomial with a bigger signature should be dealt with later. It is possible that some polynomials of smaller size cannot be used immediately and this may lead to more computations. One trick can be used to solve this conflict. The key idea is to clear the signatures of some simple polynomials and to append them to the initial polynomials. We have

Proposition 1 *Adding an initial polynomial with the largest index at any time will not affect the correctness of the two criteria.*

Although adding new polynomials will not affect the correctness of the two criteria, in order to keep the algorithm correct, we must add polynomials that are in the original ideal. Usually, we add new initial polynomials in the following cases:

1. The new generated polynomial has a very low total degree, such as 1 or 2.

2. The leading monomials of some present initial polynomials can be reduced by the new generated polynomial.

However, adding too many polynomials cannot speedup the algorithm. Because adding new initial polynomials is actually to cut off the relationship between this polynomial and the initial polynomials, which will weaken the criteria, since the system of initial polynomials may not be semi-regular any more, and many useless pairs cannot be detected.

2.5 Branch Strategy

In both the F4 and F5 algorithms, the sizes of matrices in the computation grow quickly with the degree of critical pairs. Huge matrices occupy enormous memories and are difficult to deal with. In consideration of the complexity, a natural idea comes to us, and that is, we should prevent the degree of critical pairs growing too high.

In order to control the size of the matrix, we can add polynomials that are not ideal, but this will apparently make the output incorrect. Fortunately, the cases are better in the boolean ring R_2. Since in R_2, any boolean polynomial has only two possible values 0 and 1, which makes the branch algorithm available. We can clone the present system and add a new polynomial with the value 0 and 1 to them respectively such that two polynomial sets are obtained. The original ideal is the intersection of the ideals generated by these two polynomial sets. This fact is important for solving the original system. Furthermore, after adding the new polynomial, each system may have smaller matrices and are easier to be dealt with. This is the motivation of our branch algorithm.

Theoretically, any polynomial can be added. In consideration of the complexity, we usually add polynomials that are simple enough or that can be used to reduce other polynomials. When should we add polynomials? Based on our experiments, we prefer to add polynomials when the degree of critical pairs is high enough. We can set a degree bound $D \in \mathrm{N}$. When the remaining pairs have higher degrees than D, we add a new polynomial, or equivalently we make a new branch.

We have many options to choose the new polynomials and it is difficult to tell which is the best. So we list some alternatives that have good performance in the experiments.

1. The polynomial with degree one.
2. The polynomial which is the highest homogeneous part of some present polynomial.
3. The polynomial with a high reference degree, which is a parameter that comes from the shared ZDD data structure.

After all, we can present our branch Gröbner bases algorithm.

Algorithm 1 : Branch Grönber bases algorithm

Input: An ordered polynomials set $F = (f_1, \ldots, f_m) \subset \mathbb{F}_2[X]$.
Output: The branch Gröbner bases *BranchGB* of the ideal generated by $F \cup H$, where $H = \{x_1^2 + x_1, \ldots, x_n^2 + x_n\} \subset \mathbb{F}_2[X]$ are the field polynomials.

1 Set $\mathcal{F}_i := (1, f_i, i, f_i, i)$, $i = 1, \ldots, m$, $\mathcal{F}_{m+i} := (1, x_i^2 + x_i, m+i, x_i^2 + x_i, m+i)$, $i = 1, \ldots, n$.
  $index := m + n$, $k := m + n$, $BranchSet := \{\{\mathcal{F}_i | i = 1, \ldots, k\}\}$, $BranchGB := \{\}$.
2 While $BranchSet \neq \emptyset$ do
  2.1 Select a $B \in BranchSet$ and $BranchSet := BranchSet \setminus \{B\}$.
  2.2 Generate $CP := \{s(\mathcal{P}, \mathcal{Q}) | \mathcal{P}, \mathcal{Q} \in B\}$.
  2.3 While $CP \neq \emptyset$ do
    2.3.1 $d := \min\{\deg(c) | c \in CP\}$, $D := \{c \in CP | \deg(c) = d\}$ and $CP := CP \setminus D$.
    2.3.2 $D' := \{c \in D | c$ is not satisfied either of the Syzygy or Rewritten criterion $\}$.
    2.3.3 Reduce D' by matrix and collect the new generated labeled polynomials as F^+.
    2.3.4 For $\mathcal{P} \in F^+$ do
      2.3.4.1 If $\mathcal{P}$ is simple enough
          then $index := index + 1$, $\mathcal{P} := (1, \text{poly}(\mathcal{P}), index, \text{poly}(\mathcal{P}), k + 1)$.
          else $\text{num}(\mathcal{P}) := k + 1$.
      2.3.4.2 $k := k + 1$, $CP := CP \cup \{s(\mathcal{P}, \mathcal{Q}) | \mathcal{Q} \in B\}$. $B := B \cup \{\mathcal{P}\}$.
    2.3.5 If the minimal degree of CP is high enough, then
      2.3.5.1 choose a new polynomial $p \in \mathbb{F}_2[X]$.
      2.3.5.2 $index := index + 1$, $k := k + 1$
      2.3.5.3 Set $\mathcal{P}' := (1, p + 1, index, p + 1, k)$, $BranchSet := BranchSet \cup \{B \cup \{\mathcal{P}'\}\}$.
      2.3.5.4 Set $\mathcal{P} := (1, p, index, p, k)$, $CP := CP \cup \{s(\mathcal{P}, \mathcal{Q}) | \mathcal{Q} \in B\}$, $B := B \cup \{\mathcal{P}\}$.
  2.4 $BranchGB := BranchGB \cup \{\{\text{poly}(\mathcal{P}) | \mathcal{P} \in B\}\}$.
3 Return *BranchGB*.

In this algorithm, step 2.3.4.1 is to clear the signature of a polynomial, and step 2.3.5 is to introduce new polynomials so as to add new branches to the algorithm. In order to make the algorithm more efficient, some auxiliary data can be kept in *BranchSet* as well, such as the number *index*, k, the set CP and so on.

3 Complexity Analysis

3.1 The Principle for Making Branch

One of the motivations of our branch Gröbner bases algorithm is to decrease the complexity in each branch so as to lessen the total complexity for solving the boolean polynomial system. Therefore, if we hope our branch algorithm has better efficiency than the non-branch algorithms, we should constrain the branch number within a reasonable bound.

The complexity of Gröbner bases using the $F4$ and $F5$ algorithm is determined by a degree D_{reg}, which is the upper bound of the highest degree of matrices, say D_{p}, constructed during the computation. Then the total complexity of $F4$ or $F5$ algorithm is roughly $O(n^{\omega D_{\text{reg}}})$, where n is the number of variables and ω is the efficiency of matrix elimination that has a bound $2 \leq \omega \leq 3$. Similarly, since we make a new branch when the degree of pairs exceed a degree D, the complexity of each branch is roughly $O(n^{\omega' D})$, where ω' is efficiency of our matrix reduction. Assuming the number of branches is M, the total complexity of our branch algorithm is $O(Mn^{\omega' D})$. If we hope our branch algorithm performs better, the complexity $O(Mn^{\omega' D})$ must smaller than $O(n^{\omega D_{\text{reg}}})$. Roughly speaking, the number M should satisfy the following inequality:

$$Mn^{\omega' D} < n^{\omega D_{\text{reg}}}, \quad \text{or} \quad M < n^{\omega D_{\text{reg}} - \omega' D}.$$

However, it is difficult to predict the number of branches produced in our algorithm. What we can do is to set up a bound such that when the number of branches exceed it, we stop the program. Fortunately, the number of branches for one kind of examples is usually stable, so we can anticipate the general performance for all problems of this kind by induction. The principle above sets up a criterion to check whether it is possible for our algorithm to have good performance.

3.2 *The Estimation of D_{reg}*

Before the system is computed by the $F5$ algorithm, we cannot obtain the practical value of D_{p}, so we have to try to estimate it or give an upper bound for it. Fortunately, the upper bound D_{reg} for F5 algorithm can be obtained from [8]. For a general semi-regular system in the boolean ring, D_{reg} can be achieved from the following series:

$$S_{m,n}(z) = \sum_{d \geq 0} h_{d,m}(n) z^d = (1+z)^n / \prod_{k=1}^{m} (1 + z^{d_k}).$$

where m is the number of the initial polynomials $(f_1, \ldots, f_m)$ and $d_k = \deg(f_k)$. Then D_{reg} is the first d such that $h_{d,m}$ is nonpositive. Therefore, the upper bound can be calculated easily when m, n, and the d_ks are given.

3.3 *Theoretical Analysis of Randomly Generated Systems*

Faugère has claimed that almost all the systems are semi-regular systems when n is not too small. So the randomly generated systems can be supposed to be semi-regular systems naturally, and the estimation of D_{reg} is a powerful tool to analyze the complexity of the F5 algorithm as well as our branch algorithm.

For convenience, we only consider some special cases and the others can be analyzed in a similar way. We assume the initial polynomials are m quadratic polynomials in R_2 and n is the number of variables. So when $m = n$, we can draw a picture of D_{reg}.

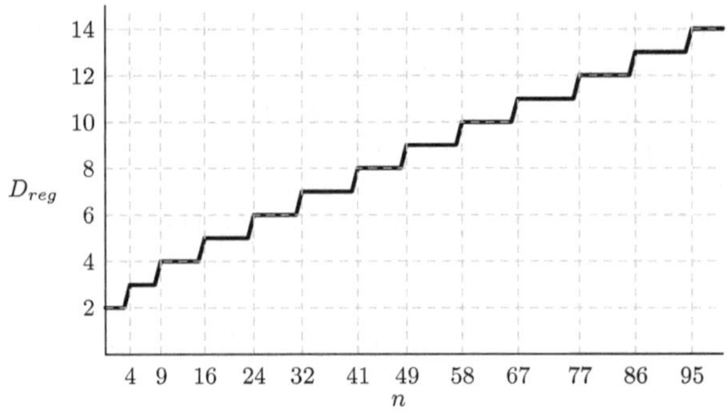

Fig. 1 D_{reg} for $m = n$ quadratic polynomials

Fig. 2 D_{reg} when adding polynomials with degree 1

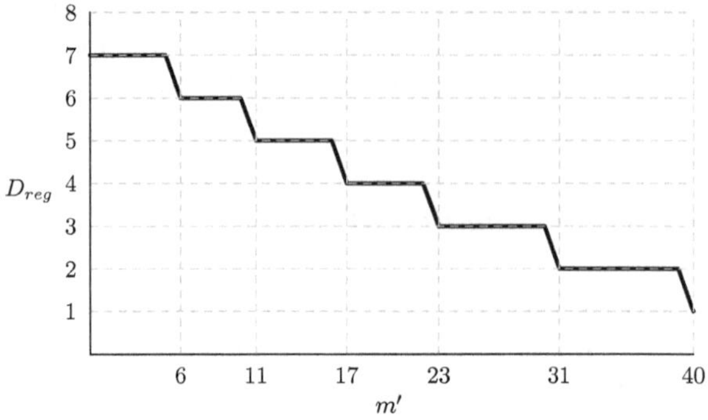

Although D_{reg} is only an upper bound of the practical highest degree D_{p}, the degree D_{reg} is reached almost all the time for randomly generated systems.

The motivation of our branch algorithm is to add new polynomials in order to lower the degree of matrices in the computation. Since randomly generated systems are semi-regular, we can use the series in the last subsection to compute D_{reg} and then find out how many polynomials should be added such that D_{reg} becomes lower. If the number of polynomials is fixed, the number of branches can be obtained easily.

For example, we consider the systems when $m = n = 40$ and in Fig. 1, the corresponding D_{reg} is 7. By the series, we can calculate the specific number of polynomials we should add in order to lower the degree. Since adding different polynomials lead to different results, here we only consider two cases, one is adding polynomials with degree 1 and the other with degree 2. Figure 2 shows how the D_{reg} varies with m', which is the number of new added polynomials. We can see that adding 6 polynomials of degree 1 will lower D_{reg} to 6, while adding 11 polynomials, the D_{reg} becomes 5 and so on. Then we obtain the following table. TH-Num is the theoretic branch number calculated by the inequality in Sect. 3.1 with $\omega = \omega' = 2$. NumPoly is the smallest number of m' to lower D_{reg} to the corresponding degree and EXP-Num is the expected branch number calculated by NumPoly.

Table 1 Adding polynomials with degree 1

D_{reg}	6	5	4	3	2	1
TH-Num	1,600	40^4	40^6	40^8	40^{10}	40^{12}
NumPoly	6	11	17	23	31	40
EXP-Num	64	2,048	2^{17}	2^{23}	2^{31}	2^{40}
TH/EXP(≈)	$2^{4.6}$	$2^{10.3}$	$2^{14.9}$	$2^{19.6}$	$2^{22.2}$	$2^{23.9}$

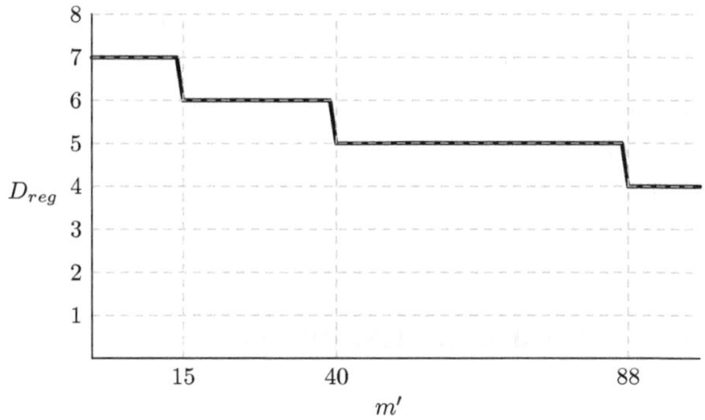

Fig. 3 D_{reg} when adding polynomials with degree 2

In Table 1, all the expected branch numbers are smaller than the theoretical bound, so as discussed in 3.1, our branch algorithm will have better performance than the non-branch algorithm and the practical results in Sect. 4 proves this. Recall that, the proportion in Table 1 shows that adding 40 polynomials will lead to the best efficiency, however, this is not the truth, since the matrices with lower degree are denser than the bigger ones, then the parameter ω' is no longer 2 and combined with other factors, TH-Num will be much smaller than that in the table.

When adding polynomials with degree 2, the cases are not so good as that with degree 1, and Fig. 3 shows the variation of D_{reg}. At least 15 quadratic polynomials should be added to lower D_{reg} to 6, while in Fig. 2, only 6 is enough, so the corresponding expected branch number will be bigger than in Table 1. We can generate a similar table (Table 2).

The expected branch number are all bigger than the theoretic bound, so we cannot expect a better performance of our branch algorithm. However, the practical cases are not as bad as the table shows, since the data in the table are calculated under a hypothesis that the new systems are still semi-regular systems, and so if we add some special polynomials which can reduce the present system significantly, the practical branch number can still be control in a reasonable bound.

Other examples can be analyzed in the same way as well. Based on our experimental results, we can conclude that for almost all randomly generated systems, our branch algorithm has better performance than non-branch algorithms, such as F4 in Magma.

Table 2 Adding polynomials with degree 2

D_{reg}	6	5	4	3	2
TH-Num	1,600	40^4	40^6	40^8	40^{10}
NumPoly	15	40	88	207	740
EXP-Num	2^{15}	2^{40}	2^{88}	2^{207}	2^{704}

Table 3 The criteria

n	6	8	10	12	14	16	18
T-pairs	198	740	2,483	3,296	94,077	144,801	211,385
D-pairs	7	64	727	1,146	46,221	80,308	112,925
0-Polys	0	0	0	0	0	0	0
D/T(%)	3.54	8.65	29.28	34.77	49.13	55.46	53.42

4 Computational Examples

In this section, some timings are imported from [7], but the complexity information is obtained by the method discussed in Sect. 3. The branch strategy we used in the experiments is the first strategy in Sect. 2.5 and all the timings are obtained from a computer (OS Linux, CPU Xion 4*3.0GHz, 16.0GB RAM).

4.1 The Criteria

In this subsection, we see how the two modified criteria work. The initial polynomials are all randomly generated quadratic polynomials with $m = n$.

In Table 3, T-pairs stand for the total number of pairs generated in the computation, while D-pairs are the number of pairs detected by the two criteria and D/T is the proportion of these two numbers. Besides, 0-Polys is the number of polynomials that are reduced to 0 in the algorithm. From the table, all useless pairs are detected and the two criteria play an important role for improving the efficiency during the computation.

4.2 The Randomly Generated Systems

In this section we see how the branch algorithm performs for randomly generated systems. In the following table, three groups of data will be presented. The first group involves the theoretical degree bound D_{reg} and the theoretical upper bound for the branch number M; the second group consists of the practical degree D_{p} computed by the F4 algorithm and the corresponding theoretical upper bound for M; in the

Table 4 The randomly generated systems

n	*EST*		*MGB*			*BGB*		
	TH-Deg	TH-Num	P-Deg	TH-Num	Time	P-Deg	P-Num	Time
18	5	18^4	6	18^6	3.890	3	2,048	0.791
20	5	20^4	6	20^6	14.220	3	8,160	2.992
22	5	22^4	6	22^6	82.790	3	29,046	13.235
24	6	24^6	–	–	–	3	65,400	47.682
26	6	26^6	–	–	–	3	262,018	149.121

last group, the degree D and the practical number of the branches is given. For the second and third groups, the practical timings are given as well.

The initial polynomials are randomly generated quadratic polynomials with $m = n$. The practical degree D_p in the second group are obtained by the F4 algorithm in Magma. The upper bound of M is estimated with $\omega = \omega' = 2$. In Table 4, we use *EST* to represent the estimated data and *MGB* is the experiments data from the F4 algorithm, while *BGB* is that from our branch algorithm. TH- is short for theoretical and P- for practical and "–" means Magma runs out of memory.

Remark that in Table 4, the theoretical degrees for $n = 18, 20, 22$ are lower than the corresponding practical degrees, and the reason is that the F4 algorithm does not have a powerful criteria to remove all the useless pairs, so some useless computations are still done in the F4 algorithm. From the table we can also see that the practical numbers of branches in our algorithm is always kept within the two theoretical bounds, so our branch algorithm will have better performance and the timings prove that.

4.3 The Stream Ciphers

In this section, we use our branch algorithm to attack a class of stream ciphers, which is an important class of encryption algorithm. Here we only consider stream ciphers based on the LFSR, and the filter functions are from [9].

- CanFil 2, $x_1x_2x_3 + x_1x_2x_4 + x_1x_2x_5 + x_1x_4 + x_2x_5 + x_3 + x_4 + x_5$
- CanFil 3, $x_2x_3x_4x_5 + x_1x_2x_3 + x_2x_4 + x_3x_5 + x_4 + x_5$
- CanFil 8, $x_1x_2x_3 + x_2x_3x_6 + x_1x_2 + x_3x_4 + x_5x_6 + x_4 + x_5$
- CanFil 9, $x_2x_4x_5x_7 + x_2x_5x_6x_7 + x_3x_4x_6x_7 + x_1x_2x_4x_7 + x_1x_3x_4x_7 + x_1x_3x_6x_7 + x_1x_4x_5x_7 + x_1x_2x_5x_7 + x_1x_2x_6x_7 + x_1x_4x_6x_7 + x_3x_4x_5x_7 + x_2x_4x_6x_7 + x_3x_5x_6x_7 + x_1x_3x_5x_7 + x_1x_2x_3x_7 + x_3x_4x_5 + x_3x_4x_7 + x_3x_6x_7 + x_5x_6x_7 + x_2x_6x_7 + x_1x_4x_6 + x_1x_5x_7 + x_2x_4x_5 + x_2x_3x_7 + x_1x_2x_7 + x_1x_4x_5 + x_6x_7 + x_4x_6 + x_4x_7 + x_5x_7 + x_2x_5 + x_3x_4 + x_3x_5 + x_1x_4 + x_2x_7 + x_6 + x_5 + x_2 + x_1$
- CanFil 10, $x_1x_2x_3 + x_2x_3x_4 + x_2x_3x_5 + x_6x_7 + x_3 + x_2 + x_1$

The data here consist of two parts: one part involves the practical degree D_p, the theoretical upper bound for M and the practical time costed by the F4 algorithm in Magma; the other part includes the degree D which is given by the user, the practical branch number and the time used by our branch algorithm. Again, $\omega = \omega' = 2$

Table 5 The stream ciphers

Filters	n	81			100			128		
		Time	Deg	Num	Time	Deg	Num	Time	Deg	Num
	MGB	18.730	7	81^8	32.930	7	100^8	–	–	–
CanFil2	*BGB*	0.027	3	9	0.172	3	66	0.357	3	40
	MGB	–	–	–	1.360	7	100^8	–	–	–
CanFil3	*BGB*	0.085	3	11	0.150	3	19	1.210	3	17
	MGB	49.460	7	81^8	12.590	7	100^8	–	–	–
CanFil8	*BGB*	0.046	3	27	0.170	3	190	0.371	3	140
	MGB	–	–	–	–	–	–	–	–	–
CanFil9	*BGB*	0.418	4	40	1.230	4	217	20.901	4	77
	MGB	331.880	7	81^8	–	–	–	–	–	–
CanFil10	*BGB*	0.131	3	99	0.612	3	501	1.296	3	340

and *MGB* and *BGB* represent the F4 algorithm in Magma and our branch algorithm respectively.

Table 5 shows that the practical number of branches is smaller than the estimated upper bound, so it is not strange to see that our branch algorithm is more efficient than the F4 algorithm for such problems.

5 Conclusion

In this paper, we present an implementation of the branch Gröbner bases algorithm and analyze the complexity briefly. The experimental results show that for both randomly generated systems and a class of Stream Ciphers problems, our branch algorithm has better efficiency than the F4 algorithm in Magma. However, there are some problems that are still not solved completely, for example, how to find a general strategy for all problems and how to anticipate the number of branches before the computation, and these can be investigated in the future.

Acknowledgments We thank Professor Xiaoshan Gao for his useful suggestions and Zhenyu Huang for discussing the programming codes.

References

1. Wu, W.T.: Basic principles of mechanical theorem-proving in elementary geometries. J. Sys. Sci. Math. Sci. **4**, 207–235 (1984)
2. Wu, W.T.: Basic principles of mechanical theorem-proving in elementary geometries. J. Autom. Reason. **2**, 221–252 (1986)

3. Faugère, J.C.: A new efficient algorithm for computing Gröbner bases (f4). J. Pure Appl. Algebr. **139**(1), 61–88 (1999)
4. Faugère, J.C.: A new efficient algorithm for computing Grönber bases without reduction to zero (F5). Symbolic and Algebraic Computation, Porc. Conferenz ISSAC 2002, 75–83 (2002)
5. Brickenstein, M., Dreyer, A.: PolyBoRi: A framework for Gröbner basis computations with boolean polynomials. MEGA 2007, Austria (2007)
6. Gao, X.S., Chai, F.J., Yuan, C.M.: A characteristic set method for equation solving in F2 and applications in cryptanalysis of stream ciphers. J. Syst. Sci. Complex. **21**, 191–208 (2008)
7. Sun, Y., Wang, D.K.: Branch Gröbner bases algorithm over boolean ring (Chinese). Preprint (2009)
8. Bardet, M., Faugère, J.C., Salvy, B.: Complexity of Grönber basis computation for Semi-regular overdetermined sequences over F_2 with solutions in F_2. In: Proceedings of the ICPPSS International Conference on Polynomial System Solving Paris, November 24–25-26 2004 in honor of Daniel Lazard (2004)
9. Faugère, J.C., Ars, G.: An Algebraic Cryptanalysis of Nonlinear Filter Generators Using Gröbner Bases. Reserch report 4739, Institut National de Recherche en Informatique et en Automatique, Lorraine (2003)

Towards the Calculation of Casimir Forces for Inhomogeneous Planar Media

C. Xiong, T.W. Kelsey, S.A. Linton and U. Leonhardt

Abstract Casimir forces arise from vacuum fluctuations. They are fully understood only for simple models, and are important in nano- and microtechnologies. We report our experience of computer algebra calculations toward the Casimir force for models involving inhomogeneous dielectrics. We describe a methodology that greatly increases confidence in any results obtained, and use this methodology to demonstrate that the analytic derivation of scalar Green's functions is at the boundatry of current computer algebra technology. We further demonstrate that Lifshitz theory of electromagnetic vacuum energy can not be directly applied to calculate the Casimir stress for models of this type, and produce results that indicate the possibility of alternative regularizations. We discuss the relative strengths and weaknesses of computer algebra systems when applied to this type of problem, and suggest combined numerical and symbolic approaches toward a more general computational framework.

1 Introduction

Casimir forces result from zero-point vacuum fluctuations confined between two dielectric materials [1]. Although these forces were predicted theoretically in the 1940s, empirical evidence confirming the theory has only been obtained in recent years [2–6]. Casimir forces are important in nanotechnology and microtechnology: repulsive Casimir forces can reduce friction in nano- and micromechanical devices, whereas attractive forces can "glue" components together that are designed to be free-moving. Lifshitz theory [7] is a theoretical approach to the calculation of Casimir

---

C. Xiong · T.W. Kelsey (✉) · S.A. Linton
School of Computer Science, University of St Andrews, St Andrews KY16 9SX, UK
e-mail: twk@st-andrews.ac.uk

C. Xiong
e-mail: xc@cs.st-andrews.ac.uk

S.A. Linton
e-mail: sl4@st-andrews.ac.uk

U. Leonhardt
School of Physics and Astronomy, University of St Andrews, St Andrews KY16 9SS, UK
e-mail: ulf@st-andrews.ac.uk

© Springer-Verlag Berlin Heidelberg 2014
R. Feng et al. (eds.), *Computer Mathematics*, DOI 10.1007/978-3-662-43799-5_15

forces, in which the Green's tensor for the electric field is used to derive electromagnetic stress and energy density.

The standard planar model is to have two plates, L and R, of uncharged dielectric materials separated in the x direction by a few micrometers. The materials have permittivities $\varepsilon_L(x, i\xi)$ and $\varepsilon_R(x, i\xi)$, depending on displacement and frequency ξ, which completely describe the media since we assume that there is no magnetic response (we enforce $\mu_L(x, i\xi) = 1 = \mu_R(x, i\xi)$ for the magnetic permeabilities involved). The gap between the plates, C, is either a quantum vacuum or third dielectric with $\varepsilon_C(x, i\xi)$ equal to a constant; we consider such a model to be *homogeneous*. For this model, and for variations of this model that include moving plates [8], the Casimir force can be both calculated analytically and measured empirically [9]. For extensions of this model involving more than two plates, numerical methods can be used to obtain the Casimir forces for specific types of plate [10].

In this paper we consider *inhomogeneous* models where the permittivity of the central region, $\varepsilon_C(x, i\xi)$, varies with x. The primary aim of the paper is to see which, if any, inhomogeneous models allow the analytic derivation of their Casimir forces from calculations performed in the widely-used computer algebra systems Maple (Waterloo Maple Inc., London, Ontario, Canada) and Mathematica (Wolfram Research Inc., Champaign, IL, USA). All computations were run on an Intel Xeon E5430 2.66GHz with 8 cores and 16 Gb memory, using Maple version 13.0 and Mathematica version 7.0 under CentOS 5. In particular we explore the applicability of Lifshitz theory to inhomogeneous models. The Lifshitz regularization process described in Sect. 2 was derived with homogeneous media in mind, we therefore explore the possibility that this could be a confounding factor in attempts to calculate Casimir forces in the inhomogeneous case.

In Sect. 2 we give the standard Lifshitz theoretic approach to deriving Casimir stresses for homogeneous media, and discuss how these may be calculated using Maple and Mathematica. In Sect. 3 we describe our methodology for performing and checking similar calculations for inhomogeneous models, and outline the strengths and weakness of the two computer algebra systems. We present results that suggest that many, but not all, inhomogeneous models can not be dealt with analytically using current computer algebra capabilities. Section 4 contains our analysis of standard Lifshitz theory applied to the model in which the central permittivity decays exponentially ($\varepsilon_C(x, i\xi) = ae^{-bx}$), together with results that suggest that suitable alternative regularizations may be deriveable. In Sect. 5 we discuss computational aspects, such as the limitations of existing computer algebra systems, and the possibility of future numeric-symbolic approaches.

2 The Calculation of Casimir Stress in Planar Media

In this section we describe the mathematical and physical concepts involved in our computations, and present the sequence of calculations involved in determining the Casimir force for planar models. A more detailed exposition of the underlying physics, together with full derivation of the equations involved and descriptions of theoretic approaches other than that of Lifshitz, is given in [9]. The resulting sequence of calculations can, in principle, be done by hand, using symbolic computer algebra, via numeric techniques, or by a combined numeric-symbolic approach. We report on our experiences of the second of these options in Sect. 3.

Stresses on objects in electromagnetic fields are given by Maxwell's stress tensor, in which $\hat{\mathbf{E}}$ and $\hat{\mathbf{H}}$ are respectively the electric and magnetic fields, $\hat{\mathbf{B}}$ is the magnetic induction and $\hat{\mathbf{D}}$ is the electric displacement.

$$\hat{\sigma} = \hat{\mathbf{E}} \otimes \hat{\mathbf{D}} + \hat{\mathbf{B}} \otimes \hat{\mathbf{H}} - \frac{1}{2}(\hat{\mathbf{E}} \cdot \hat{\mathbf{D}} + \hat{\mathbf{B}} \cdot \hat{\mathbf{H}})\mathbb{I}_3 \tag{1}$$

For stationary electromagnetic fields, the divergence of the Maxwell's stress tensor gives the force density $\hat{\mathbf{f}}$,

$$\hat{\mathbf{f}} = \nabla \cdot \hat{\sigma}. \tag{2}$$

The expectation values (also known as correlation functions) for the tensor products in Eq. (1) are related to the retarded Green's function as follows:

$$\langle \hat{\mathbf{E}}(\mathbf{r}, t) \otimes \hat{\mathbf{D}}(\mathbf{r}', t) \rangle = -\frac{\hbar}{\pi c^2} \int_0^\infty d\xi \, \varepsilon(\mathbf{r}, i\xi) \xi^2 \mathbf{G}(\mathbf{r}, \mathbf{r}', i\xi), \tag{3}$$

$$\langle \hat{\mathbf{B}}(\mathbf{r}, t) \otimes \hat{\mathbf{H}}(\mathbf{r}', t) \rangle = \frac{\hbar}{\pi} \int_0^\infty d\xi \frac{1}{\mu(\mathbf{r}, i\xi)} \nabla \times \mathbf{G}(\mathbf{r}, \mathbf{r}', i\xi) \times \overleftarrow{\nabla}' \tag{4}$$

The notation $\times \overleftarrow{\nabla}'$ denotes a curl on $\mathbf{r}'$ in $\mathbf{G}(\mathbf{r}, \mathbf{r}', i\xi)$ from the right. $\mathbf{G}(\mathbf{r}, \mathbf{r}', i\xi)$ is the retarded Green's tensor for the vector potential in a Coulomb gauge, and is defined as the solution of the following inhomogeneous electromagnetic wave equation

$$\nabla \times \frac{1}{\mu} \nabla \times \mathbf{G}(\mathbf{r}, \mathbf{r}', i\xi) + \varepsilon \frac{\xi^2}{c^2} \mathbf{G}(\mathbf{r}, \mathbf{r}', i\xi) = \delta(\mathbf{r} - \mathbf{r}')\mathbb{I}_3. \tag{5}$$

The Green's function should always obey the reciprocity relation:

$$\mathbf{G}(\mathbf{r}, \mathbf{r}', i\xi) = \mathbf{G}(\mathbf{r}', \mathbf{r}, -i\xi); \tag{6}$$

we describe our extensive use of this as a check for correctness of our calculated scalar Green's functions in Sect. 3.

We are considering planar dielectrics, for which the permittivity $\varepsilon(\mathbf{r}, i\xi) = \varepsilon(x, i\xi)$ and magnetic permeability $\mu(\mathbf{r}, i\xi) = \mu(x, i\xi)$, i.e. depend only on the x-coordinate. The Green's function in terms of its Fourier transform in y and z is

$$\mathbf{G}(x, x', u, v, i\xi) = \int_{-\infty}^{\infty} \mathrm{d}y \int_{-\infty}^{\infty} dz \mathbf{G}(\mathbf{r}, \mathbf{r}', i\xi) e^{-iu(y-y')-iv(z-z')}. \quad (7)$$

The Fourier-transformed Green's function $\mathbf{G}(x, x', u, v, i\xi)$ is given by the Fourier-transformed wave equation:

$$\nabla \times \frac{1}{\mu(x, i\xi)} \nabla \times \mathbf{G}(x, x', u, v, i\xi) + \varepsilon(x, i\xi) \frac{\xi^2}{c^2} \mathbf{G}(x, x', u, v, i\xi) = \delta(x - x'). \quad (8)$$

The Casimir force depends only on the xx-component of Maxwell's stress tensor because the force density is also independent of y and z. In the limit $\mathbf{r} \rightarrow \mathbf{r}'$, the result for σ_{xx} is

$$\begin{aligned} \sigma_{xx} &= -\frac{\hbar c}{8\pi^3} \int_{-\infty}^{\infty} \int_{-\infty}^{\infty} \int_{0}^{\infty} u \left(\frac{1}{\mu} (w^2 - \partial_x \partial_{x'}) \tilde{g}_{Es} + \frac{1}{\varepsilon} (w^2 - \partial_x \partial_{x'}) \tilde{g}_{Ms} \right) \mathrm{d}u \, \mathrm{d}v \, \mathrm{d}\kappa \, |_{\mathbf{x}'=\mathbf{x}} \\ &= -\frac{\hbar c}{4\pi^2} \int_0^{\infty} \mathrm{d}u \int_0^{\infty} \mathrm{d}\kappa \, u \left(\frac{1}{\mu} (w^2 - \partial_x \partial_{x'}) \tilde{g}_{Es} + \frac{1}{\varepsilon} (w^2 - \partial_x \partial_{x'}) \tilde{g}_{Ms} \right) |_{v=0, \mathbf{x}'=\mathbf{x}}, \quad (9) \end{aligned}$$

with

$$w = \sqrt{u^2 + v^2 + \varepsilon\mu\kappa^2}, \quad \kappa = \frac{\xi}{c},$$

$$\tilde{g}_{Es} = \tilde{g}_E - \mu\tilde{g}_0, \quad \text{and} \quad \tilde{g}_{Ms} = \tilde{g}_M - \varepsilon\tilde{g}_0.$$

In Lifshitz theory, $\tilde{g}_{Es}$ and $\tilde{g}_{Ms}$ are the regularized electric and magnetic Green's functions (where regularization involves subtraction of the relevant divergent part). $\tilde{g_0}$ is the infinite contribution from the retarded Green's function in a space with homogeneous medium:

$$\tilde{g}_0 = -\frac{1}{2w} e^{(-w|x-x'|)}. \quad (10)$$

In summary, to obtain the Casimir force for a planar dielectric model, the sequence of calculations is:

1. calculate the scalar Green's functions—Eq. (8)—for the permittivity of the specific media under consideration (recalling the modeling assumption $\mu(x, i\xi) = 1$ described in Sect. 1)
2. perform the regularization that removes the infinite parts from the above scalar Green's functions
3. solve the double integral—Eq. (9)—to obtain the stress tensor σ_{xx}

4. the divergence of σ_{xx} is the theoretically predicted Casimir force—Eq. (2).

In the homogeneous case all the calculations can be performed analytically using either Maple or Mathematica, since $\varepsilon(x, i\xi)$ does not vary with x. Eq. (8) reduces to

$$\frac{\mathrm{d}^2}{\mathrm{d}x^2}\tilde{g}(x) - (u^2 + v^2 + \varepsilon\kappa^2)\tilde{g}(x) = \delta(x - x') \tag{11}$$

where $\tilde{g}$ denotes either the electric or magnetic Green's function, and in which none of the left-hand parameters depends on x. The general solution involves trigonometric functions and the Heaviside function; specific solutions are easily obtained from the boundary conditions. The subtractions involved in stage 2 are also straightforward, again since neither μ nor ε varies with x. The double integrals with infinite ranges produce finite results, since the integrand converges to zero with increasing u and κ. Stage 4 is relatively simple. In Sect. 3 we commence our analysis of how the computational details are affected when homogeneity is no longer assured.

3 Specific Inhomogeneous Permittivity Models

For a given model of the permittivity $\varepsilon(x)$ and $\mu = 1$ (no magnetic response), we first need to find the scalar Green functions,

$$\frac{\mathrm{d}^2\tilde{g}_E}{\mathrm{d}x^2} - \left(u^2 + \varepsilon(x)\kappa^2\right)\tilde{g}_E = \delta(x - x'), \tag{12}$$

$$\frac{\mathrm{d}}{\mathrm{d}x}\left(\frac{1}{\varepsilon(x)}\frac{\mathrm{d}}{\mathrm{d}x}\tilde{g}_M\right) - \left(\frac{u^2}{\varepsilon(x)} + \kappa^2\right)\tilde{g}_M = \delta(x - x'). \tag{13}$$

An intrinsic problem is the assessment of the validity of any results. The output from a computer algebra system will consist of symbolic expressions, which can be evaluated as real numbers for supplied values of the parameters involved. Empirical validation is not known to be possible for all models, and is expensive, time-consuming and technically demanding.

Our solution is to constrain our results to be the same when solving from the left and from the right. This is not a guarantee that any results obtained are correct, but it does increase confidence and allows us to detect models for which analytic solution is intractable using current computer algebra capabilities. Our methodology therefore is to perform the calculations using two computer algebra systems (Maple and Mathematica) and to use the reciprocity relations given in Eq. (6). If the results from the two independent systems coincide, and if the results are the same from either the left or from the right, then we have a high level of confidence in their correctness.

For models such as $\varepsilon(x) = 1 + e^{-x}$ and $\varepsilon(x) = x^2$ using Maple we find that the reciprocity relations are violated in the magnetic case. Using Mathematica to

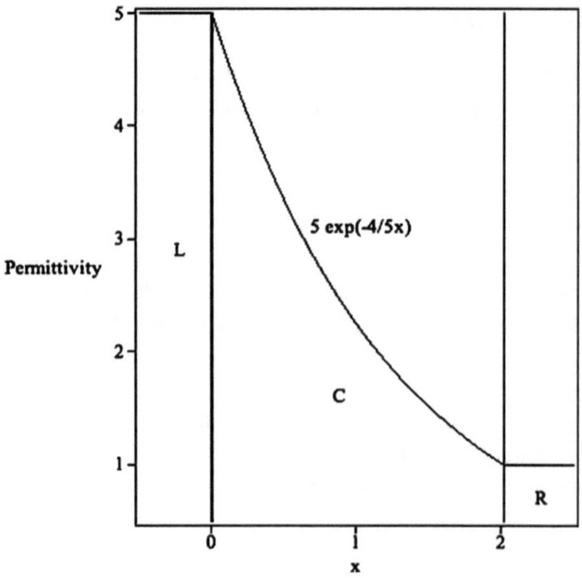

Fig. 1 The inhomogeneous model given by $\varepsilon_L = 5$, $\varepsilon_C = 5\exp(-4/5x)$ and $\varepsilon_R = 1$. The central part runs from $x = 0$ to $x = \log(a)/b \approx 2.0118$

perform the same calculations, we were unable to solve the equations analytically, and were therefore unable to perform the reciprocity check. The difference between the two systems is that recent versions of Maple contain an implementation of the use of Heun's functions [11] to solve ODEs. Heun's functions are the solutions of the Heun form of 2nd order linear ODEs; any such ODE can be converted into Heun form. Equation (12) is a linear 2nd order ODE, so, in Maple, it can be converted to Heun form and solved directly, with perfect agreement from the right and left directions. Equation (13) is nonlinear, however. We can convert it into a modified Heun ODE and obtain a solution in the form of the product of a Heun C function and an exponential correction factor. There is a discrepancy between the left and right solutions, which appears to be introduced when the correction factor is computed. Unfortunately, we are never sure which, if either, of these results is correct, hence further work is needed to correct the Maple implementation. Mathematica, on the other hand, has no Heun ODE or function capabilities, and neither the linear nor nonlinear ODEs could be solved analytically.

We have found only one permittivity model for which the scalar Green's functions can be derived analytically (i.e., without using any numeric calculation options) with the reciprocity relations fully satisfied (Fig. 1 is an illustrative example). This is exponentially decaying permittivity,

$$\varepsilon(x) = ae^{-bx}, \quad \text{for positive constants } a \text{ and } b, \tag{14}$$

bounded on each side by homogeneous dielectrics.

For this case, the general solutions for $\tilde{g}_E$ and $\tilde{g}_M$ were found to be

$$\begin{aligned}
\tilde{g}_E &= C_{E1} I_{\nu 1}(-\lambda) + C_{E2} K_{\nu 1}(\lambda) + \frac{2}{b}(I_{\nu 1}(-\lambda) K_{\nu 1}(\lambda') \\
&\quad - I_{\nu 1}(-\lambda') K_{\nu 1}(\lambda))\text{Heaviside}(\lambda - \lambda'), \\
\tilde{g}_M &= C_{M1} \lambda I_{\nu 2}(-\lambda) + C_{M2} \lambda K_{\nu 2}(\lambda) \qquad (15) \\
&\quad + \frac{b\lambda\lambda'}{2\kappa^2}(I_{\nu 2}(-\lambda) K_{\nu 2}(\lambda') \\
&\quad - I_{\nu 2}(-\lambda') K_{\nu 2}(\lambda))\text{Heaviside}(\lambda - \lambda'), \qquad (16) \\
&\text{where } \lambda = \frac{2\kappa\sqrt{a}}{b} e^{-bx/2}, \quad \nu 1 = 2u^2/b, \quad \nu 2 = \sqrt{1 + \nu 1^2}, \qquad (17)
\end{aligned}$$

and in which the Cs are arbitrary coefficients determined by the continuity of

$$\tilde{g}_E, \quad \tilde{g}_M, \quad \frac{1}{\mu(x, i\kappa)} \partial_x \tilde{g}_E, \quad \text{and} \quad \frac{1}{\varepsilon(x, i\kappa)} \partial_x \tilde{g}_M \qquad (18)$$

at the boundaries, and I and K are the modified Bessel functions.

Two interesting computational aspects were encountered. First, Mathematica does not return solutions of the PDEs in terms of Bessel K function; it instead returns expressions involving Gamma functions which are mathematically equivalent, but which are lengthy and hard for humans to interpret. Secondly, intermediate Maple output suggested the variable changes involving λ, $\nu 1$ and $\nu 2$—Eq. (17). These simplifying re-arrangements both (i) greatly aid the efficiency of the remaining calculations, and (ii) helped us to interpret and check the results. The calculations were therefore easier to perform in Maple than in Mathematica, but, for this model, both systems returned the same results, from the left and from the right, when evaluated as floats. We are therefore confident that our scalar Green's functions are exactly those needed for the Lifshitz regularization process.

4 The Testing of Standard Lifshitz Regularization

Standard Lifshitz theory involves the subtraction of the contribution to the stress that does not arise from material inhomogeneity. This is known to produce accurate (i.e. empirically verifiable) results for the standard model where the gap between the two plates is either empty or filled with a homogeneous dielectric. However, this approach is known to result in an infinite Casimir force in models involving cylinders and spheres [12]. In certain cases an alternative regularization has been found (but not always agreed upon by the expert community), whilst for others the problem of calculating a finite stress using any theoretical approach remains unsolved. For our model (exponentially decaying permittivity for the central medium) we therefore expect to either (i) use standard Lifshitz theory to calculate a finite Casimir force,

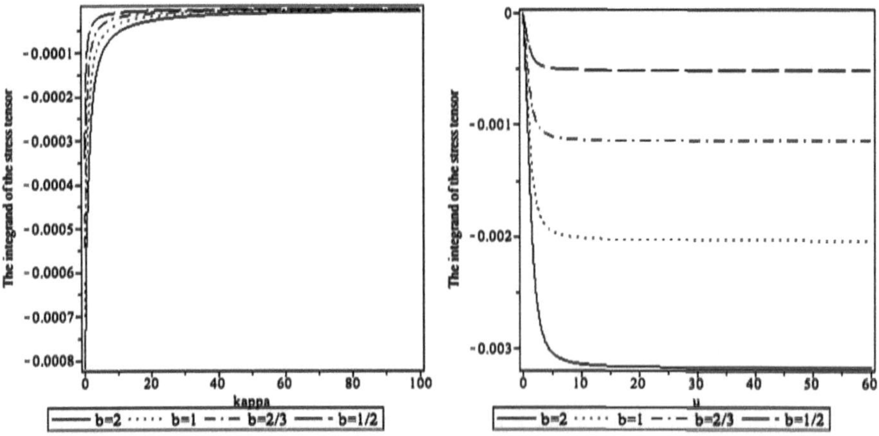

Fig. 2 The integrand of the stress tensor obtained from Green's functions regularized using Lifshitz theory. We use the conventions that the physical constants $\hbar = c = 1$. The central permittivity model is $\varepsilon(x) = 5\exp(-bx)$. As κ increases, the integrand converges to zero (*left plot*). As u increases, the integrand converges to a nonzero constant, the value of which depends on the model parameter b but not on a (*right plot*)

(ii) derive an infinite force, with the structure of the results indicating the possibility of alternative regularization, or (iii) derive an infinite force, with no clues on how to proceed.

Our results, displayed for illustrative parameter choices in Fig. 2, indicate that the for one of the wave parameters (κ) we obtain convergence to a finite integrand, but for the other (u) the integrand diverges. Unfortunately, we can no longer consider κ and u to be the respective x and y wave components, since we have performed a Fourier transformation. Further investigation of the divergence shows that the constant nonzero value depends neither on a or x, but only on b (Fig. 3). Routine simplification, (setting $x = x'$) gives the divergence constant DC for the stress tensor as a function of b:

$$DC \approx -0.0036\, b^2, \tag{19}$$

This divergent behavior leads to the prediction of an infinite Casimir force, which is not a physically realistic outcome. However, the predictability of the amount of divergence suggests that it may be possible to modify Lifshitz theory for this model, so that a plausible finite Casimir force is predicted. Such as regularization has recently been proposed [13], with the resulting Casimir forces calculated using the methodology presented in this paper.

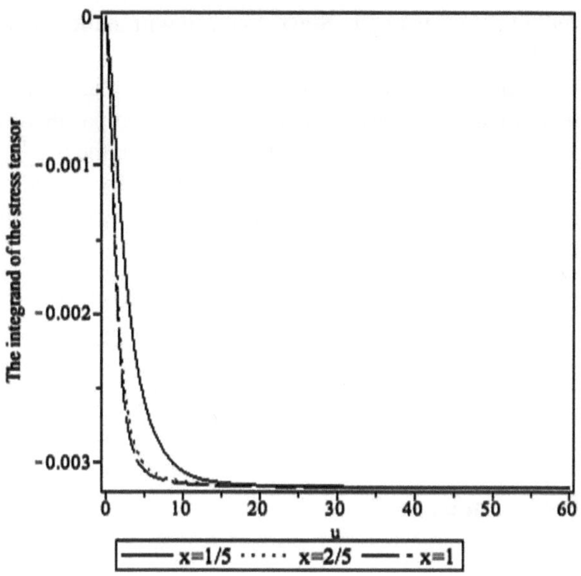

Fig. 3 The integrand of the stress tensor obtained from Green's functions regularized using Lifshitz theory. We use the conventions that the physical constants $\hbar = c = 1$. The central permittivity model is $\varepsilon(x) = 5\exp(-bx)$. Model parameters have been set as $a = 5$ and $b = 4/5$. We observe that for fixed b, the nonzero convergence value is the same for all choices of x in the central region

5 Conclusions

Our findings suggest that the calculation of scalar Green's functions for arbitrary inhomogeneous media is at the boundary of the current capabilities of Maple and Mathematica. Using Maple we can get satisfactory results for exactly one model, and believe we could increase the number of such models if the modified Heun function implementation within Maple were to be improved. Mathematica is less useful for these calculations, as no Heun function implementation is present in the current system. However, we have successfully replicated Maple results using Mathematica, indicating that the lengthy and complex Mathematica expressions produced as intermediate output are completely correct.

We are highly confident that our scalar Green's function calculations are accurate. In addition to the approach described in Sect. 3, we split the Green's functions into bare and scattered parts, allowing us to derive the specific solution of the bare part using initial rather than boundary conditions. The results of these calculations agree with those described in this paper for both systems, and hence also satisfy out reciprocity constraints.

In Sects. 3 and 4 we discussed the first two stages of Casimir force prediction using Lifshitz theory. The third stage, a complicated double integration over infinite ranges, is, in general, not computable analytically in either Maple or Mathematica for inhomogeneous models. Instead, we substitute a realistically high finite value for the infinities and obtain numeric approximations. For example, in the model presented in Figs. 2 and 3 with $u = 150$ the integrand has a magnitude of 10^{-13}, descreasing to zero with increasing u. Stage 4 is well within the capabilities of any decent computer algebra system.

Future avenues of research include (i) the testing of any proposed alternative regularization using our methodology of comparing results from two systems for both the the right and left limits, (ii) the development of a combined numeric-symbolic framework that agrees with the symbolically derived results presented here for the exponential model, and which can be used to calculate Casimir forces for those inhomogeneous model that lie beyond the current analytic capabilities of Maple and Mathematica.

Acknowledgments The authors are supported by EPSRC grant EP/CS23229/1. UL is also supported by a Royal Society Wolfson Research Merit Award. We would like to acknowledge the critical discussions we have had with Tom Philbin.

References

1. Casimir, H.B.G.: On the attraction between two perfectly conducting plates. Proc. K. Ned. Akad. Wet. **51**, 793–795 (1948)
2. Bressi, G., Carugno, G., Onofrio, R., Ruoso, G.: Measurement of the Casimir force between parallel metallic surfaces. Phys. Rev. Lett. **88**(4), 041804 (2002)
3. Decca, R.S., López, D., Fischbach, E., Krause, D.E.: Measurement of the Casimir force between dissimilar metals. Phys. Rev. Lett. **91**(5), 050401 (2003)
4. Hertlein, C., Helden, L., Gambassi, A., Dietrich, S., Bechinger, C.: Direct measurement of critical Casimir forces. Nature **451**(175), 172–175 (2008)
5. Lamoreaux, S.K.: Demonstration of the Casimir force in the 0.6 to 6 μm range. Phys. Rev. Lett. **78**(1), 5–8 (1997)
6. Munday, J.N., Capasso, F., Parsegian, A.V.: Measured long-range repulsive Casimir-Lifshitz forces. Nature **457**(7226), 170–173 (2009)
7. Dzyaloshinskii, I.E., Lifshitz, E.M., Pitaevskii, L.P.: The general theory of van der Waals forces. Adv. Phys. **10**(38), 165–209 (1961)
8. Philbin, T.G., Leonhardt, U.: No quantum friction between uniformly moving plates. New J. Phys. **11**(3), 033–035 (2009)
9. Bordag, M., Klimchitskaya, G.L., Mohideen, U., Mostepanenko, V.M.: Advances in the casimir effect. Oxford University Press, Oxford (2009)
10. Landau, L.D., Lifshitz, E.M., Pitaevskii, L.P.: Statistical physics, part 2. Butterworth-heinemann, Oxford (1980)
11. Heun, K.: Theorie der Riemann'schen Functionen zweiter Ordnung mit vier Verzweigungspunkten. Mathematische Annalen **3**, 161 (1899)
12. Boyer, T.H.: Quantum electromagnetic zero-point energy of a conducting spherical shell and the Casimir model for a charged particle. Phys. Rev. **174**(5), 1764–1776 (1968)
13. Philbin, T.G., Xiong, C., Leonhardt, U.: Casimir stress in an inhomogeneous medium. arXiv:0909.2998, 2009 (Preprint)

Part III
Contributed Papers of ASCM2012

Sparse Polynomial Interpolation by Variable Shift in the Presence of Noise and Outliers in the Evaluations

Brice Boyer, Matthew T. Comer and Erich L. Kaltofen

Abstract We compute approximate sparse polynomial models of the form $\widetilde{f}(x) = \sum_{j=1}^{t} \widetilde{c}_j (x - \widetilde{s})^{e_j}$ to a function $f(x)$, of which an approximation of the evaluation $f(\zeta)$ at any complex argument value ζ can be obtained. We assume that several of the returned function evaluations $f(\zeta)$ are perturbed not just by approximation/noise errors but also by highly perturbed outliers. None of the $\widetilde{c}_j$, $\widetilde{s}$, e_j and the location of the outliers are known beforehand. We use a numerical version of an exact algorithm by [4] together with a numerical version of the Reed–Solomon error correcting coding algorithm. We also compare with a simpler approach based on root finding of derivatives, while restricted to characteristic 0. In this preliminary report, we discuss how some of the problems of numerical instability and ill-conditioning in the arising optimization problems can be overcome. By way of experiments, we show that our techniques can recover approximate sparse shifted polynomial models, provided that there are few terms t, few outliers and that the sparse shift is relatively small.

Keywords Approximate function recovery · Numeric and exact polynomial interpolation · Outlier detection · Sparse representations

1 Introduction

Sparse polynomial interpolation algorithms, where the number of values required depends on the number of nonzero terms in a chosen representation base rather than

---

This material is based on work supported in part by the National Science Foundation under Grants CCF-0830347 and CCF-1115772.

B. Boyer · M.T. Comer (✉) · E.L. Kaltofen (✉)
Department of Mathematics, North Carolina State University, Raleigh, NC 27695-8205, USA
e-mail: mcomer@math.ncsu.edu

B. Boyer
e-mail: bboyer@math.ncsu.edu

E.L. Kaltofen
e-mail: kaltofen@math.ncsu.edu

© Springer-Verlag Berlin Heidelberg 2014
R. Feng et al. (eds.), *Computer Mathematics*, DOI 10.1007/978-3-662-43799-5_16

on the degree of the polynomial, originate from two sources. One is Prony's 1795 algorithm for reconstructing an exponential sum [18] (see also [2]) and another is Blahut's exact sparse polynomial interpolation algorithm in the decoding phase of the Reed–Solomon error correcting code. Both algorithms first determine the term structure via the generator ("error locator polynomial") of the linear recurrent sequence of the values $f(\omega^i)$, $i = 0, 1, 2, \ldots$, of the sparse function f. Blahut's algorithm has led to a rich collection of exact sparse multivariate polynomial interpolation algorithms, among them [1, 12, 13, 16, 20]. Prony's algorithm suffers from numerical instability unless randomization controls, with high probability and for functions of significant sparsity, the conditioning of intermediate Hankel matrices. The probabilistic spectral analysis in the GLL algorithm [5, 7] adapts the analysis of the exact early termination algorithm of [13]. The resulting numerical sparse interpolation algorithms have recently had a high impact on medical signal processing; see the web site http://smartcare.be of Wen-shin Lee and her collaborators. The GLL algorithm can be generalized to multivariate polynomial and rational function recovery via Zippel's variable-by-variable sparse interpolation [14].

Already in the beginning days of symbolic computation, the choice of polynomial basis was recognized: $(x - 2)^{100} + 1$ is a concise representation of a polynomial with 101 terms in power basis representation. The discrete-continuous optimization problem of computing the sparsest shift of an exact univariate polynomial surprisingly has a polynomial-time solution [4, 9, 10]. Our subject is the computation of an approximate interpolant that is sparsified through a shift. One can interpret our algorithm as a numerical version of the exact sparsest shift algorithms. As in least squares fitting, noise can be controlled by oversampling (cf. [8]). The main difficulty is that the shift is unknown. Our numerical algorithm adapts Algorithm `UniSparsestShifts`⟨`one proj, two seq`⟩ in [4] to compute the shift: `UniSparsestShifts`⟨`one proj, two seq`⟩ carries the shift as a symbolic variable z throughout the sparse interpolation algorithm. Since the coefficients of the polynomials in the shift variable z are spoiled by noise, the GCD step becomes an approximate polynomial GCD. A main question answered here is whether the arising nonlinear optimization problems remain well-conditioned. Our answer is a conditional yes: an optimal approximate shift is found among the arguments of all local minima, but the number of local minima is high, preventing the application of standard approximate GCD algorithms. Instead, we perform global optimization, as a fallback, by computing all zeros of the gradient ideal. In addition, our algorithm requires high precision floating point arithmetic.

In [3], we have introduced outlier values to the sparse interpolation problem. There, outlier removal requires high oversampling, as the worst case of k-error linear complexity is $2t(2k + 1)$, where t is the generator degree. However, ours is only an upper bound for sparse interpolation. The situation is different for Algorithm `UniSparsestShifts`⟨`one proj, two seq`⟩. Outliers can be removed at the construction stage of the values containing the shift variable z, by a numeric version of Blahut's decoding algorithm for interpolation with errors. The algorithm, numerical interpolation with outliers, is interesting in its own right. As we will show in Sect. 3, the analysis in [3, 7] does not directly apply, as randomization can only be applied with a limited choice of random evaluation points. We have successfully tested it as a subroutine of our numerical sparsest shift algorithm. Note that a few outliers per interpolation lead to a very small sparse interpolation problem for error location, which can be handled successfully by sparse interpolation with noisy values.

For the sake of comparison with this algorithm, we restrict to characteristic 0 and compare a sparse shift representation to a Taylor expansion expressed at a point that will make the representation sparse. This leads to finding a root common to many derivatives. Combined with a weighted least squares fit for removing outliers and tolerating noise, we manage to compare favorably to the main algorithm.

In Sect. 4, we present the preliminary experimental results that our algorithms can recover sparse models even in the presence of substantial noise and outliers. See Sect. 4.3 for our conclusions.

2 Computing Sparse Shifts

We introduce in this subsection an algorithm to compute a shifted sparse interpolant in a numerical setting. The exact algorithm accepts outliers and uses early termination; we adapt it to a numerical setting, considering noisy and erroneous data. It is based on a numerical version of Blahut's decoding algorithm.

2.1 Main Algorithm with Early Termination

The Early Termination Theorem in [13] is at the heart of computing a sparsest shift. Let

$$g(x) = \sum_{j=1}^{t} c_j x^{e_j}, \quad c_j \neq 0 \quad \text{for all} \quad 1 \leq j \leq t,$$

be a t-sparse polynomial with coefficients in an integral domain D. Furthermore, let

$$\alpha_i(y) = g(y^i) \in \mathsf{D}[y], \quad \text{for} \quad i = 1, 2, \ldots$$

be evaluations of g at powers of an indeterminate y. Prony's/Blahut's theorem states that the sequence of the α_i is linearly generated by $\prod_{j=1}^{t}(\lambda - y^{e_j})$. Therefore, if one considers the Hankel matrices

$$\mathscr{H}_i(y) = \begin{bmatrix} \alpha_1 & \alpha_2 & \ldots & \alpha_i \\ \alpha_2 & \alpha_3 & \ldots & \alpha_{i+1} \\ \vdots & \vdots & \ddots & \vdots \\ \alpha_i & \alpha_{i+1} & \ldots & \alpha_{2i-1} \end{bmatrix} \in \mathsf{D}[y]^{i \times i}, \quad \text{for } i = 1, 2, \ldots$$

one must have $\det(\mathscr{H}_{t+1}) = 0$. Theorem 4 in [13] simply states that $\det(\mathscr{H}_i) \neq 0$ for all $1 \le i \le t$. One can replace the indeterminate y by a randomly sampled coefficient domain element to have $\det(\mathscr{H}_i) \neq 0$ for all $1 \le i \le t$ with high probability (w.h.p.).

We seek an s in any extension of the field of K such that for a given $f(x) \in \mathsf{K}[x]$ the polynomial $f(x+s) = h(x)$ is t-sparse for a minimal t. Now consider $g(x) = f(x+z) \in \mathsf{D}[x]$ with $\mathsf{D} = \mathsf{K}[z]$. We have $\Delta_i(y, z) = \det(\mathscr{H}_i) \in (\mathsf{K}[z])[y]$; note that $\alpha_i(y, z) = g(y^i) = f(y^i + z)$. By the above Theorem 4, the sparsest shift is an s with $\Delta_{t+1}(y, s) = 0$ for the smallest t. Algorithm `UniSparsestShifts` ⟨`one proj, two seq`⟩ computes s as

$$z + s \text{ divides (w.h.p.) } \mathrm{GCD}(\Delta_{t+1}(y_1, z), \Delta_{t+1}(y_2, z)),$$
$$\text{where } y_1, y_2 \text{ are random in } \mathsf{K};$$

note that the first t with a nontrivial GCD is possibly smaller for the projection by $y = y_1$ and $y = y_2$, but with low probability.

For numeric sparse interpolation with a shift, we assume that for $f(x) \in \mathbb{C}[x]$ we can obtain

$$f(\zeta) + noise + outlier\ error, \quad \text{for any} \quad \zeta \in \mathbb{C}.$$

Here only a fraction of the values contain an outlier error, and noise is a random perturbation of $f(\zeta)$ by a relative error of 10^{-10}, say. Our algorithm returns a sparse interpolant $g(x)$ that at all probed values ζ, save for a fraction that are removed as outliers, approximates the returned $f(\zeta)+noise$. Note that probing f at ζ twice may produce a different noise and possibly an outlier.

We now give the outline of our Algorithm `ApproxUniSparseShift` ⟨`one proj, sev seq`⟩. Note that because of the approximate nature of the shifted sparse interpolant, there is a trade-off between backward error and sparsity. Hence we call our algorithm a "sparse shift" algorithm. As in Algorithm `UniSparsest Shifts` ⟨`one proj, two seq`⟩, for L complex values $y = \omega^{[1]}, \omega^{[2]}, \ldots, \omega^{[L]}$ we compute $\widetilde{\delta_i}^{[\ell]}(z) = \widetilde{\Delta_i}(\omega^{[\ell]}, z)$ from $\widetilde{\alpha_i}(\omega^{[\ell]}, z)$, $\ell = 1, 2, \ldots, L$. Here the tilde accent mark $\tilde{}$ indicates that the values have noise in their scalars. As in [7], we choose the $\omega^{[\ell]}$ to be different random roots of unity of prime order. Our algorithm consists of the four following tasks:

Step 1: For $\ell = 1, 2, \ldots, L$, compute the numeric complex polynomials $\widetilde{\alpha_i}(\omega^{[\ell]}, z)$ via a numeric version of the Blahut decoding algorithm; see Sect. 3. Step 1 is assumed to have removed all outliers.

Step 2: Compute the determinants $\widetilde{\delta_i}^{[\ell]}(z)$ of numeric polynomial Hankel matrices $\widetilde{\mathscr{H}_i}^{[\ell]}(z)$ for all ℓ, iterating Steps 3 and 4 on i. We perform the determinant computations with twice the floating point precision as we use for Steps 1, 3 and 4.

Step 3: Determine the sparsity and an approximate shift. Note that the approximate shift $\widetilde{s}$ is an approximate root of the polynomials $\widetilde{\delta_i}^{[1]}(z)$, $\widetilde{\delta_i}^{[2]}(z)$, …, $\widetilde{\delta_i}^{[L]}(z)$. Our method finds the smallest perturbation of the $\widetilde{\delta_i}^{[\ell]}(z)$ that

produces a common root, simultaneously for all ℓ. If that distance is large, we assume that there is no common root and the dimension of the Hankel matrix was too small. It might happen that an accurate shift is diagnosed too early, but then the constructed model produces a worse backward error.
The 2-norm distance to the nearest polynomial system with a common root $\widetilde{s}$ is given by formula (see [11] and the literature cited there):

$$\widetilde{s} = \operatorname*{arginf}_{\zeta \in \mathbb{C}} \sum_{\ell=1}^{L} |\widetilde{\delta}_i^{[\ell]}(\zeta)|^2 \Big/ \Big(\sum_{m=0}^{d} |\zeta^m|^2 \Big),$$

where $d = \max_\ell\{\deg(\widetilde{\delta}_i^{[\ell]}(z))\}$ and in all polynomials, any term coefficients of z^m, where $m \in \{0, 1, \ldots, d\}$, can be deformed.
In our experiments in Sect. 4, we have only considered real shifts $\widetilde{s} \in \mathbb{R}$. The optimization problem is then

$$\widetilde{s}_{\mathrm{real}} = \operatorname*{arginf}_{\xi \in \mathbb{R}} \sum_{\ell=1}^{L} \Big((\Re\widetilde{\delta}_i^{[\ell]})(\xi)^2 + (\Im\widetilde{\delta}_i^{[\ell]})(\xi)^2 \Big) \Big/ \Big(\sum_{m=0}^{d} \xi^{2m} \Big), \tag{1}$$

where $\Re\widetilde{\delta}_i^{[\ell]}$ and $\Im\widetilde{\delta}_i^{[\ell]}$ are the real and imaginary parts of the polynomials $\delta_i^{[\ell]}$, respectively. We find $\widetilde{s}_{\mathrm{real}}$ among the real roots of the numerator of the derivative of the objective function in (1),

$$\frac{\partial \sum_\ell ((\Re\widetilde{\delta}_i^{[\ell]})(z)^2 + (\Im\widetilde{\delta}_i^{[\ell]})(z)^2)}{\partial z} \times \Big(\sum_{m=0}^{d} z^{2m} \Big)$$
$$- \sum_{\ell=1}^{L} ((\Re\widetilde{\delta}_i^{[\ell]})(z)^2 + (\Im\widetilde{\delta}_i^{[\ell]})(z)^2) \times \Big(\sum_{m=0}^{d} (2m) z^{2m-1} \Big), \tag{2}$$

and choose the root that minimizes the objective function in (1).
We have observed that a larger number L of separate $\omega^{[\ell]}$ can improve the accuracy of the optimal shift, at a cost of oversampling. We have also observed that the optimization problem (1) and (2) has numerous local optima, some near the optimal approximate shift, which prevents the use of any local approximate GCD algorithm.

Step 4: With the approximate sparsest shift $\widetilde{s}$, complete the sparse polynomial reconstruction, as in [6] and [3, Sect. 6].
One reuses the evaluations from previous steps, having removed those that were declared outliers in Step 1.

In the remainder of this section, we restrict ourselves to characteristic 0. We now describe a more naïve approach for the same problem. Some early termination can also be achieved here. Unlike the main algorithm, this one cannot recover errors in the exact setting.

2.2 Using Taylor Expansions

Let $f(x) = \sum_{i=1}^{t} c_i(x-s)^{e_i}$ be a t-sparse shifted polynomial of degree d. We can see this expression as a Taylor expansion of f at $x = s$:

$$f(x) = \sum_{i=0}^{\infty} \frac{\left(\partial^i f/\partial x^i\right)(s)}{i!}(x-s)^i .$$

A sparsest shift is then an s that is a root of the maximum number of polynomials in the list $S = \left\{\left(\partial^i f/\partial x^i\right)(x) \mid i \in \{0, \ldots, d-1\}\right\}$.

Remark 1 It is stated in Theorem 1 in [17] that if $t \leq (d+1)/2$ then the shift s is unique and rational. Moreover, the proof gives the stronger statement: for any other shift $\hat{s}$, with a sparsity $\hat{t}$, one has $\hat{t} > d+1-t$.

This statement is not true in characteristic $p \neq 0$: for instance, consider the two shifts -1 and 0 in the polynomial $(x+1)^p = x^p + 1 \mod p$.

Lemma 1 *Let S_{2t} be the list of the last $2t$ elements in S. The root that zeros the maximum number of polynomials in S_{2t} is the sparsest shift.*

Proof We prove this by contradiction. Assume a shift s appears r times in S_{2t} and another shift $\hat{s}$ appears $\hat{r}$ times. We first notice that the number of elements in S for which $\hat{s}$ is a root, and the number of elements for which it is not a root, sum to $d+1$. So we have the inequality $\hat{r} + \hat{t} \leq d+1$.

Suppose now that $\hat{r} \geq r$. The sparsity of f in the s-shifted basis being t, the number of elements in S_{2t} that do not have $\hat{s}$ as a root is $2t - \hat{r} \leq t$, thus $\hat{r} \geq t$. On the other hand, $\hat{t} > d+1-t$. Summing these last two inequalities yields $\hat{r}+\hat{t} > d+1$, which is impossible. □

Early termination can be achieved; indeed, under certain circumstances, one need not compute all $2t$ derivatives. For example, suppose that the degree $(d-1)$ term of f in the sparsest shifted basis is missing and try $\bar{s}$, the root of $\left(\partial^{d-1} f/\partial x^{d-1}\right)(x)$, as a shift; this is the Tschirnhaus transformation (originally introduced for solving cubic equations). If the "back-shifted" polynomial $f(x+\bar{s})$ has fewer than $(d+1)/2$ terms, then by Remark 1, $\bar{s}$ is the unique sparsest shift. We can extend this technique by trying all rational roots in the list S_τ for a small τ.

Now we state the naïve algorithm based on the above, then we modify it for a numeric setting. Consider first the following exact algorithm:

Step 1: Compute the exact interpolant using $D+1$ calls to the black box, where $D \geq d$.
Step 2: Try early termination on S_τ, for a small τ, and return if successful.
Step 3: Compute all remaining derivatives in S_θ, for $\theta = \min(2T, D)$.
Step 4: The sparsest shift is the rational root s that zeros the most derivatives in S_θ.

Step 5: The "back-shifted" polynomial $f(x + s)$ gives the support for the sparse polynomial.

This algorithm can be easily translated to a numerical one, based on least squares fitting:

Step 1: Compute a degree-D weighted least squares fit with $O(D+E)$ calls to the black box.
Step 2: Remove outliers by comparing relative errors, then update the fit.
Step 3: Compute the θ derivatives in S_θ (possibly terminate early and proceed to Step 6).
Step 4: The approximate root s that zeros most derivatives is the sparsest shift.
Step 5: The polynomial $f(x + s)$ gives the support for the sparse polynomial.
Step 6: A Newton iteration can be conducted on the result of Step 5 to increase accuracy.

2.3 Discussion on the Numeric Algorithm

Step 4 is sensitive to noise and requires more sampling from Step 1. The approximate roots are determined to be equal up to a certain tolerance (for instance 10^{-2}). In Steps 5 and 6, the coefficients near 0 may be forced to 0 (which would accelerate convergence in Step 6). Step 6 is conducted on the function $f(m'_1, \ldots, m'_k, s') = \sum_{i=1}^{k} m'_i (x - s')^{h_i}$, with initial condition from Step 5, random samples x_j and noisy evaluations $f(x_j)$; the outliers are removed by checking relative errors. If the random samples x_j are not only taken from data in Step 1, then oversampling will help "denoising" the outputs.

Remark 2 It is unknown to us, in the exact algorithm, how to use a number of calls to the black box in Step 1 depending only on T, in order to compute the derivatives. However, it is reasonable to expose the following: we are only interested in the higher-degree terms of f. Consider the Euclidean division $f(x) = Q(x)x^q + R(x)$; then, with high numeric precision and big random x_i, we can recover an approximation of Q by a least squares fit on samples $f(x_i)/x_i^q \approx Q(x_i)$.

3 Numeric Interpolation with Outliers

Blahut's decoding algorithm for Reed/Solomon codes is based on sparse interpolation. Suppose one has values of

$$f(x) = c_{d-1}x^{d-1} + \cdots + c_0 \in \mathsf{K}[x], \quad \deg(f) \leq d - 1$$

at powers ω^i: $a_i = f(\omega^i)$, $i = 0, 1, 2, \ldots, n - 1$, where $n = d + 2E$. Furthermore suppose for $k \leq E$, where the upper bound E is known, those values are spoiled by

k outlier errors: $b_l = a_l + a'_l$, with $a'_{e_j} \neq 0$ exactly at the indices $0 \leq e_1 < e_2 < \cdots < e_k \leq d+2E-1$. If ω is an nth $=(d+2E)$th primitive root of unity, then the $n \times n$ Fourier (Vandermonde) matrix $V(\omega) = [\omega^{l \cdot m}]_{0 \leq l,m \leq n-1}$ satisfies

$$W = V(\omega)^{-1} = \frac{1}{n} V(\omega^{-1}) \quad \text{where} \quad \omega^{-1} = \omega^{n-1}, \tag{3}$$

hence

$$W\mathbf{b} = W\mathbf{a} + W\mathbf{a}' = \begin{bmatrix} \mathbf{c} \\ 0 \end{bmatrix} + \frac{1}{n} V(\omega^{-1})\mathbf{a}'. \tag{4}$$

The last $2E$ entries in $W\mathbf{b}$ allow sparse interpolation of $g(x) = \sum_{j=1}^{k} a'_{e_j} x^{e_j}$:

$$c'_l = (V(\omega^{-1})\mathbf{b})_l = g(\omega^{-l}) \quad \text{for } d \leq l \leq d+2E-1.$$

Note that all vectors are indexed $0, 1, \ldots, n-1$, e.g.,

$$\mathbf{a} = \begin{bmatrix} a_0 \\ \vdots \\ a_{n-1} \end{bmatrix} \quad \text{and} \quad \mathbf{b} = \begin{bmatrix} b_0 \\ \vdots \\ b_{n-1} \end{bmatrix}.$$

By our convention, primed $'$ quantities contain outlier information. Thus, as in Sect. 2, the sequence $c'_d, c'_{d+1}, \ldots$ is linearly generated by $\Lambda(\lambda) = \prod_{j=1}^{k}(\lambda - \omega^{-e_j})$ (called the "error locator polynomial"), which is a squarefree polynomial by virtue of the primitivity of ω. One may also compute Λ from the reverse sequence $c'_{d+2E-1}, c'_{d+2E-2}, \ldots$, which is linearly generated by the reciprocal polynomial $\prod_{j=1}^{k}(\lambda - \omega^{e_j})$.

Not knowing k, the probabilistic analysis of early termination as in [13] and Sect. 2 does not directly apply, as the choice of ω is restricted to a primitive n'th root of unity. Furthermore, the locations e_j of the outlier errors a'_{e_j} may depend on the evaluation points ω^l. Blahut's decoding algorithm processes all $2E$ values c'_l.

If one has ε_l in each evaluation, namely $\tilde{b}_l = a_l + a'_l + \varepsilon_l$, where $|a_{e_j} - a'_{e_j}|/|a_{e_j}| \gg 0$ (one may assume that $\varepsilon_{e_j} = 0$), then

$$W\tilde{\mathbf{b}} = \begin{bmatrix} \mathbf{c} \\ 0 \end{bmatrix} + \frac{1}{n} V(\omega^{-1})\mathbf{a}' + \frac{1}{n} V(\omega^{-1})\varepsilon,$$

so

$$\tilde{c}'_l = (V(\omega^{-1})\tilde{\mathbf{b}})_l = g(\omega^{-l}) + (V(\omega^{-1})\varepsilon)_l = g(\omega^{-l}) + \bar{\varepsilon}_l \quad \text{for} \quad d \leq l \leq d+2E-1,$$

where $|\bar{\varepsilon}_l| \leq |\varepsilon_1| + \cdots + |\varepsilon_n|$. Again, there is an immediate trade-off between noise and outliers: at what magnitude does noise ε_l become an outlier a'_l? For now we assume that the relative error in noise is small, say 10^{-10}, while the relative error in outliers is big, say 10^5. The recovery of an approximate interpolant $\widetilde{g}(x) = \sum_{j=1}^{k} \widetilde{a}'_{e_j} x^{e_j}$ for the evaluations $\widetilde{c}'_l$ hinges on the condition number of the $k \times k$ Hankel matrix:

$$\widetilde{\mathscr{H}}'_k = \begin{bmatrix} \widetilde{c}'_d & \widetilde{c}'_{d+1} & \cdots & \widetilde{c}'_{d+k-1} \\ \widetilde{c}'_{d+1} & \widetilde{c}'_{d+2} & \cdots & \widetilde{c}'_{d+k} \\ \vdots & \vdots & \ddots & \vdots \\ \widetilde{c}'_{d+k-1} & \widetilde{c}'_{d+k} & \cdots & \widetilde{c}'_{d+2k-2} \end{bmatrix}.$$

If the matrix is well-conditioned, the error locations e_j can be determined from the approximate linear generator $\widetilde{\Lambda}$ as in the GLL algorithm [3, 7]. As is shown there, the conditioning is bounded by $1/|\omega^{e_u} - \omega^{e_v}|$. Large values there are prevented by randomizing ω, as the term exponents e_j are fixed for any evaluation. Using an ω^r instead of ω here, where $\mathrm{GCD}(r, n) = 1$, allows redistributing of the ω^{e_j}, but the e_j may then become different.

A special case is $k = 1$: In that case

$$\widetilde{\mathscr{H}}_1' = [\widetilde{c}'_d] = [g(\omega^{-d}) + \bar{\varepsilon}_d] = [a'_{e_1} \omega^{-de_1} + \bar{\varepsilon}_d],$$

which, by our assumption on a large outlier a'_{e_1} and small noise, is a well-conditioned matrix. This is the case we tested in Sect. 4.

Remark 3 When the relative difference between the magnitudes of the outlier a'_{e_1} and noise $\varepsilon_0, \varepsilon_1, \ldots, \varepsilon_{d+2E-1}$ is not so pronounced, erroneous recovery of the exponent e_1 can occur: we have $(\widetilde{c}'_d, \widetilde{c}'_{d+1}, \ldots, \widetilde{c}'_{d+2E-1}) = (\widetilde{c}'_d, \widetilde{c}'_{d+1})$, so the linear generator $\widetilde{\Lambda}(\lambda) = \lambda - \omega^{-e_1}$ can be approximated by computing

$$\frac{\widetilde{c}'_{d+1}}{\widetilde{c}'_d} = \frac{a'_{e_1}\omega^{-(d+1)e_1} + \bar{\varepsilon}_{d+1}}{a'_{e_1}\omega^{-de_1} + \bar{\varepsilon}_d} = \omega^{-e_1} + \frac{\bar{\varepsilon}_{d+1} - \omega^{-e_1}\bar{\varepsilon}_d}{a'_{e_1}\omega^{-de_1} + \bar{\varepsilon}_d} = \widetilde{\omega}. \tag{5}$$

For this reason, we define a bound $\varepsilon_{\text{noise}} \geq \max_l |\varepsilon_l|$ and assume $n\varepsilon_{\text{noise}} < |a'_{e_1}|$ so that

$$|\widetilde{\omega} - \omega^{-e_1}| = \left|\frac{\bar{\varepsilon}_{d+1} - \omega^{-e_1}\bar{\varepsilon}_d}{a'_{e_1}\omega^{-de_1} + \varepsilon_d}\right| \leq \frac{|\bar{\varepsilon}_{d+1}| + |\bar{\varepsilon}_d|}{|a'_{e_1}| - |\bar{\varepsilon}_d|} \leq \frac{2n\varepsilon_{\text{noise}}}{|a'_{e_1}| - n\varepsilon_{\text{noise}}}. \tag{6}$$

By the distribution of complex roots of unity (of order n) on the unit circle, we have that $|\omega^{s+1} - \omega^s| = |\omega - 1| = 2\sin(\pi/n)$ for any integer s. Thus, $|\widetilde{\omega} - \omega^{-e_1}| < \sin(\pi/n)$ will guarantee $|\widetilde{\omega} - \omega^{-e_1}| < |\widetilde{\omega} - \omega^s|$ for any $s \not\equiv -e_1 \pmod{n}$.

Combining this fact with (6) above, we arrive at the sufficient condition

$$|\widetilde{\omega} - \omega^{-e_1}| \leq \frac{2n\varepsilon_{\text{noise}}}{|a'_{e_1}| - n\varepsilon_{\text{noise}}} < \sin(\pi/n) \quad \Leftrightarrow \quad n\varepsilon_{\text{noise}} < |a'_{e_1}| \cdot \frac{\sin(\pi/n)}{2 + \sin(\pi/n)}. \tag{7}$$

Table 1 shows some experiments of decreasing $\Theta^{[\text{abs}]}_{\text{outlier}}$ for a fixed $\varepsilon^{[\text{abs}]}_{\text{noise}}$. Throughout the experiment, we have $f(x) = 87\,x^{11} - 56\,x^{10} - 62\,x^8 + 97\,x^7 - 73\,x^4 - 4\,x^3 -$

$83x - 10$ and $d - 1 = 11$, evaluating at powers of the order $n = d + 2E = 14$ complex root of unity $\omega = \exp(2\pi i/14)$. We add to each evaluation noise, which is implemented as a complex number with polar modulus uniformly chosen at random in the range $[0, \varepsilon_{\text{noise}}^{[\text{abs}]}]$ and polar argument uniformly chosen at random in the range $[0, 2\pi]$. An absolute outlier value is chosen the same way, but the modulus is in the range $[\Theta_{\text{outlier}}^{[\text{abs}]}, 2\Theta_{\text{outlier}}^{[\text{abs}]}]$; the exponent e_1 is also chosen uniformly at random from $\{0, 1, \ldots, d + 2E - 1 = 13\}$. Each row of the table corresponds to 1,000 realizations of the random variable that generates noise and outliers, re-seeding the random number generator with each run. All computations were performed with 15 floating point digits of precision. In the table, $C_n = \sin(\pi/n)/\big(2 + \sin(\pi/n)\big)$.

The column "% Circle" shows the percentage of runs where $|\widetilde{\omega} - \omega^{-e_1}| < \sin(\pi/n)$; "% Sector" shows the percentage of runs where $|\widetilde{\omega} - \omega^{-e_1}| \geq \sin(\pi/n)$, but $|\widetilde{\omega} - \omega^{-e_1}| < |\widetilde{\omega} - \omega^{s}|$ for any $s \not\equiv -e_1 \pmod{n}$; "% Wrong" shows the percentage of the remainder of the runs. When the ratio $\varepsilon_{\text{noise}}^{[\text{abs}]}/\Theta_{\text{outlier}}^{[\text{abs}]}$ is either sufficiently large or small, one can see from (5) that the value of $\widetilde{\omega}$ is determined mainly by the value of either a'_{e_1} or $\bar{\varepsilon}_{d+1}/\bar{\varepsilon}_d$, respectively; this corresponds with the first and last rows of each section of Table 1, where $\bar{\varepsilon}_{d+1}/\bar{\varepsilon}_d$ is far from ω^{-e_1} in general.

However, between the extreme values of $\widetilde{\omega}$, more interesting behavior can occur. Figure 1 shows two individual algorithm runs of the table rows for $\varepsilon_{\text{noise}}^{[\text{abs}]} = 1$. Each

Table 1 Experiments of varying outlier error in the presence of noise

$\varepsilon_{\text{noise}}^{[\text{abs}]}$	$\Theta_{\text{outlier}}^{[\text{abs}]}$	$\frac{n\varepsilon_{\text{noise}}^{[abs]}}{C_n\Theta_{\text{outlier}}^{[\text{abs}]}}$	$\frac{n\varepsilon_{\text{noise}}^{[abs]}}{\Theta_{\text{outlier}}^{[\text{abs}]}}$	% Circle	% Sector	% Wrong
2.5e−01	8.0e+00	4.4e+00	4.4e−01	99.7	0.3	0.0
2.5e−01	4.0e+00	8.7e+00	8.8e−01	92.4	5.0	2.6
2.5e−01	2.0e+00	1.7e+01	1.8e+00	47.5	27.0	25.5
2.5e−01	1.0e+00	3.5e+01	3.5e+00	17.0	26.1	56.9
2.5e−01	5.0e−01	7.0e+01	7.0e+00	4.2	15.0	80.8
2.5e−01	2.5e−01	1.4e+02	1.4e+01	1.8	9.2	89.0
5.0e−01	1.6e+01	4.4e+00	4.4e−01	99.7	0.3	0.0
5.0e−01	8.0e+00	8.7e+00	8.8e−01	92.4	5.0	2.6
5.0e−01	4.0e+00	1.7e+01	1.8e+00	47.5	27.0	25.5
5.0e−01	2.0e+00	3.5e+01	3.5e+00	17.0	26.1	56.9
5.0e−01	1.0e+00	7.0e+01	7.0e+00	4.2	15.0	80.8
5.0e−01	5.0e−01	1.4e+02	1.4e+01	1.8	9.2	89.0
1.0e+00	3.2e+01	4.4e+00	4.4e−01	99.7	0.3	0.0
1.0e+00	1.6e+01	8.7e+00	8.8e−01	92.4	5.0	2.6
1.0e+00	8.0e+00	1.7e+01	1.8e+00	47.5	27.0	25.5
1.0e+00	4.0e+00	3.5e+01	3.5e+00	17.0	26.1	56.9
1.0e+00	2.0e+00	7.0e+01	7.0e+00	4.2	15.0	80.8
1.0e+00	1.0e+00	1.4e+02	1.4e+01	1.8	9.2	89.0

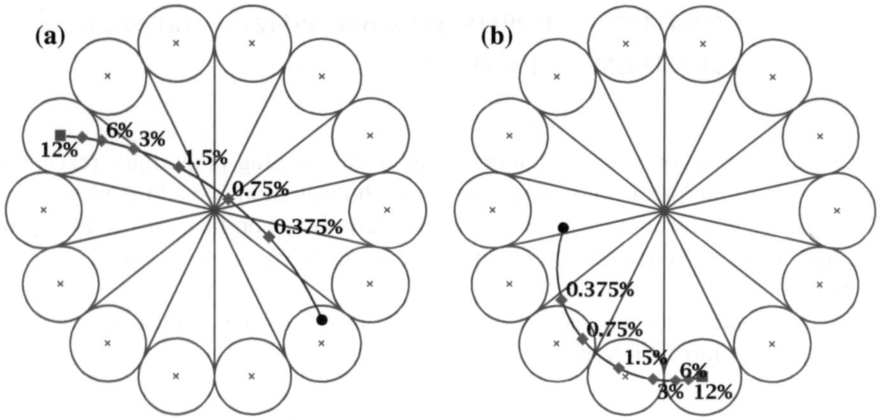

Fig. 1 Examples of varying outlier relative error (*labeled as percentages*). Noise relative error is approximately 0.40 %

power of ω is represented by a "×"; the sphere of radius $\sin(\pi/n)$ is drawn around each power of ω, as well as the corresponding (interior) sector; the solid square denotes ω^{-e_1}, while the solid circle denotes $\bar{\varepsilon}_{d+1}/\bar{\varepsilon}_d$; a complex outlier $a'_{e_1} = \xi$ is fixed, then the function $\widetilde{\omega}(t\xi)$ (for $t \in [2^{-7}, 2^7]$) is plotted as a curve, with several points whose label is the relative error of $t\xi$ compared to ω^{-e_1}. In Fig. 1a, outliers of relative error less than 6 % cause $\widetilde{\omega}$ to approach 0, so that it becomes infeasible to compute a reliable guess for e_1; here, noise constitutes approximately a 0.38 % relative error. By contrast, Fig. 1b shows an example where nearly any outlier relative error greater than 0.375 % would result in $|\widetilde{\omega} - \omega^s| < \sin(\pi/n)$ for one of three values of $s \pmod n$, so that the "nearest ω^s neighbor" criterion is no longer reliable; here, noise constitutes approximately a 0.40 % relative error.

Decoding the interpolant $W\mathbf{b}$ can also be done via the extended Euclidean algorithm for any ω with $\omega^{e_u} \neq \omega^{e_v}$: the Berlekamp–Welch algorithm; see [15]. We will study the numerical properties of variants based on approximate GCD techniques in follow-up work.

4 Implementation and Experiments of `NumericSparsestShift`

4.1 Illustrative Examples for the Main Algorithm

We reversely engineer a noisy black box for

$$\begin{aligned} f_1(x) &= 2\,(x-7)^3 + 3\,(x-7)^6 - 7\,(x-7)^{10} \\ &= -7x^{10} + 490x^9 - 15435x^8 + 288120x^7 - 3529467x^6 \end{aligned}$$

$$+ 29647422x^5 - 172941825x^4 + 691755542x^3 - 1815804312x^2$$
$$+ 2824450258x - 1976974482. \quad (8)$$

Our algorithm computes with a precision of 100 floating-point digits (except in Step 2, where the precision is doubled). To each evaluation, we add random noise causing a relative error of 1×10^{-28}. For each interpolation problem of a given degree i in Step 1, we add one outlier error of relative error 5. We use $L = 3$ different 17th roots of unity $\omega^{[\ell]}$.

Step 1 correctly locates each of the outliers in its $21 = 3 \times 7$ interpolation calls. The relative 2-norm differences

$$\|\delta_4^{[\ell]}(z) - \widetilde{\delta}_4^{[\ell]}(z)\|_2 / \|\delta_4^{[\ell]}(z)\|_2$$

of the coefficient vectors of the 4×4 matrix determinants after Step 2 are 2.126×10^{-27}, 2.681×10^{-27}, 6.596×10^{-27} for $\ell = 1, 2, 3$ all within the added noise (after outlier removal).

The polynomial (2) in Step 3 has 4 real roots, and its minimum objective function value (1) is at $\widetilde{s} = 6.9989162726$ with an objective function value of 2.028×10^{-57}, as opposed to the exact case (without noise) of 2.280×10^{-71} at $s = 7$ (there is one more root with much larger objective value).

The sparse model recovered from $\widetilde{s}$ produces the correct term exponents $e_1 = 3$, $e_2 = 6$, and $e_3 = 10$, and the least squares fit at the non-erroneous $252 = 273 - 21$ prior black box evaluations produces the approximate model for (8),

$$2.009369(x - \widetilde{s})^3 + 2.998102(x - \widetilde{s})^6 - 6.997705(x - \widetilde{s})^{10}, \quad \widetilde{s} = 6.9989162726.$$

The relative 2-norm backward error of the model (with respect to the noisy black box evaluations) is 1.596557×10^{-3}, while that of f_1 itself is 5.774667×10^{-28}. A similar model can be produced with 90 floating-point digits, but not with 80.

When doubling the noise to relative error 2×10^{-28} with 100 floating-point digits, the computed model is

$$2.036489(x - \widetilde{s})^3 + 2.992182(x - \widetilde{s})^6 - 6.991277(x - \widetilde{s})^{10}, \quad \widetilde{s} = 6.9957389337,$$

with relative 2-norm backward error 6.222096×10^{-3}, compared to 1.154933×10^{-27} for f_1. Even with relative noise of 4×10^{-28}, the computed model is

$$2.125876(x - \widetilde{s})^3 + 2.967832(x - \widetilde{s})^6 - 6.972579(x - \widetilde{s})^{10}, \quad \widetilde{s} = 6.9848087178,$$

with relative 2-norm backward error 2.151040×10^{-2}, compared to 2.309866×10^{-27} for f_1. At relative noise of 8×10^{-28}, the algorithm fails to determine a sparse approximant, even when increasing the number of sequences to $L = 10$.

Such failure is deceptive. The lack of sparsity, namely 3 of a maximum of 11 terms, allows for denser models that provide fits. In addition, a shift of 7 produces large evaluations at roots of unity, as indicated in the power basis representation of (8). Making the shift smaller and the degree larger, and considering the polynomial

$$f_2(x) = 2\,(x - 1.55371)^3 + 3\,(x - 1.55371)^6 - 7\,(x - 1.55371)^{15},$$

we can recover from $L = 3$ sequences, with a relative noise in the evaluations of 1×10^{-14}, and again 1 outlier per interpolation, the approximate model

$$1.999718(x - \widetilde{s})^3 + 2.998609(x - \widetilde{s})^6 - 7.000117(x - \widetilde{s})^{15}, \quad \widetilde{s} = 1.5537114392,$$

with relative 2-norm backward error 8.000329×10^{-1}, compared to 8×10^{-1} (to 7 digits) for f_1 itself.

For this particular example, we see a case of the effect mentioned in [3], where the sparse model can fit the noisy evaluations nearly as well (and sometimes better) than the exact black box.

Increasing the noise still, another model with $\widetilde{s} = 1.5537013193$ can be recovered with relative noise of 2×10^{-13}, where now the model and f_1 relative 2-norm backward errors are 5.180450×10^{-2} and 1.108027×10^{-11}, respectively. In this case, a different choice of $L = 3$ different 17th roots of unity was needed in order to compute a sparse model. Both computations used 357 black box evaluations.

4.2 Comparison with the Naïve Algorithm

For the examples given above, the naïve algorithm recovers the sparsest representation with noise such as 1×10^{-10} and precision 20 floating point digits. The precision obtained is close to the level of noise (1×10^{-8} relative error for the shift and 2×10^{-10} maximum relative error on the coefficients in the shifted basis). The number of calls to the black box is below 170.

For a more demanding example such as a degree 55 polynomial with sparsity 8 and a shift between 1 and 2, a level of relative noise 1×10^{-28} is tolerable with precision 200 digits (as in an example above). However, the number of calls was above 600 to get a relative error less than 1×10^{-20} on shift and coefficients. Due to the numerical optimization in Step 3, this is unattainable with the main algorithm, for the moment. The Tschirnhaus early termination was not used yet.

Besides, with more calls to the black box during the Newton iterations, we can further increase the precision on the shift and coefficients, this may however be considered as de-noising.

We can also run experiments on a black box of the type $P + Q_\varepsilon$ where P is a polynomial with a sparse shift representation and Q_ε is a dense polynomial of same degree with coefficients bounded by ε—this may be viewed as perturbation on the coefficients. The algorithms described perform well, however they do not remove outliers if they are introduced as an erroneous term.

4.3 Discussion

Our preliminary experiments lead to the following conclusions: Our correction of 1 outlier per interpolation with Blahut's numerical decoding is highly numerically reliable. The optimization problem in Step 3 requires substantial precision for its real root finding, and is numerically sensitive when the shift is large and there is noise in the evaluations. Our main algorithm works well without noise and outliers, or in high precision with noise when the shift is small and the sparsity is high. We plan to work on a more thorough experimental analysis, including the case of two or more outliers per interpolation. The naïve algorithm gives motivation and potential for improvements to the main one. On the other hand, the number of calls to the black box in the former could be reduced.

References

1. Ben-Or, M., Tiwari, P.: A deterministic algorithm for sparse multivariate polynomial interpolation. In: Proceedings of the Twentieth Annual ACM Symposium on Theory of Computing, pp. 301–309. ACM Press, New York (1988)
2. Brezinski, C.: Computational Aspects of Linear Control. Springer, Heidelberg (2002)
3. Comer, M.T., Kaltofen, E.L., Pernet, C.: Sparse polynomial interpolation and Berlekamp/Massey algorithms that correct outlier errors in input values. In: van der Hoeven, J., van Hoeij, M. (eds.) ISSAC 2012 Proceedings of the 37th International Symposium on Symbolic and Algebraic Computation, pp. 138–145. Association for Computing Machinery, New York (2012). http://www.math.ncsu.edu/~kaltofen/
4. Giesbrecht, M., Kaltofen, E., Lee, W.: Algorithms for computing sparsest shifts of polynomials in power, Chebychev, and Pochhammer bases. J. Symb. Comput. **36**(3–4), 401–424 (2003). (Special issue International Symposium on Symbolic and Algebraic Computation (ISSAC 2002). Guest editors: Giusti, M., Pardo, L.M. http://www.math.ncsu.edu/~kaltofen/
5. Giesbrecht, M., Labahn, G., Lee, W.: Symbolic-numeric sparse interpolation of multivariate polynomials (extended abstract). In: Proceedings of the Ninth Rhine Workshop on Computer Algebra (RWCA'04), pp. 127–139. University of Nijmegen, The Netherlands (2004)
6. Giesbrecht, M., Labahn, G., Lee, W.: Symbolic-numeric sparse interpolation of multivariate polynomials. In: Dumas, J.G. (ed.) ISSAC MMVI Proceedings of the 2006 International Symposium on Symbolic and Algebraic Computation, pp. 116–123. ACM Press, New York (2006). doi: http://doi.acm.org/10.1145/1145768.1145792
7. Giesbrecht, M., Labahn, G., Lee, W.: Symbolic-numeric sparse interpolation of multivariate polynomials. J. Symb. Comput. **44**, 943–959 (2009)
8. Giesbrecht, M., Roche, D.S.: Diversification improves interpolation. In: A. Leykin (ed.) Proceedings of the 2011 International Symposium on Symbolic and Algebraic Computation ISSAC 2011, pp. 123–130. Association for Computing Machinery, New York (2011)

9. Grigoriev, D.Y., Karpinski, M.: A zero-test and an interpolation algorithm for the shifted sparse polynomials. In: Proceedings of the AAECC-10, Lecture Notes in Computer Science, vol. 673, pp. 162–169. Springer, Heidelberg, Germany (1993)
10. Grigoriev, D.Y., Lakshman, Y.N.: Algorithms for computing sparse shifts for multivariate polynomials. Applic. Algebra Engin. Commun. Comput. **11**(1), 43–67 (2000)
11. Hutton, S.E., Kaltofen, E.L., Zhi, L.: Computing the radius of positive semidefiniteness of a multivariate real polynomial via a dual of Seidenberg's method. In: Watt [19], pp. 227–234. http://www.math.ncsu.edu/~kaltofen/
12. Kaltofen, E., Lakshman Y. N., Wiley, J.M.: Modular rational sparse multivariate polynomial interpolation. In: Watanabe, S., Nagata, M. (eds.) Proceedings of the 1990 International Symposium on Symbolic and Algebraic Computation (ISSAC'90), pp. 135–139. ACM Press (1990). http://www.math.ncsu.edu/~kaltofen/
13. Kaltofen, E., Lee, W.: Early termination in sparse interpolation algorithms. J. Symb. Comput. **36**(3–4), 365–400 (2003). (Special issue International Symposium on Symbolic and Algebraic Computation (ISSAC 2002). Guest editors: Giusti, M., Pardo, L.M. http://www.math.ncsu.edu/~kaltofen/
14. Kaltofen, E., Yang, Z., Zhi, L.: On probabilistic analysis of randomization in hybrid symbolic-numeric algorithms. In: Verschelde, J., Watt, S.M. (eds.) SNC'07 Proceedings of the 2007 International Workshop on Symbolic-Numeric Computation, pp. 11–17. ACM Press, New York, (2007). http://www.math.ncsu.edu/~kaltofen/
15. Khonji, M., Pernet, C., Roch, J.L., Roche, T., Stalinsky, T.: Output-sensitive decoding for redundant residue systems. In: Watt [19], pp. 265–272
16. Lakshman, Y.N., Saunders, B.D.: Sparse polynomial interpolation in non-standard bases. SIAM J. Comput. **24**(2), 387–397 (1995)
17. Lakshman, Y.N., Saunders, B.D.: Sparse shifts for univariate polynomials. Applic. Algebra Engin. Commun. Comput. **7**(5), 351–364 (1996)
18. Prony, R.: Essai expérimental et analytique sur les lois de la Dilatabilité de fluides élastiques et sur celles de la Force expansive de la vapeur de l'eau et de la vapeur de l'alcool, à différentes températures. J. de l'École Polytechnique 1, 24–76 (1795). R. Prony is Gaspard(-Clair-François-Marie) Riche, baron de Prony
19. Watt, S.M. (ed.): Proceedings of the 2010 International Symposium on Symbolic and Algebraic Computation ISSAC 2010. Association for Computing Machinery, New York (2010)
20. Zippel, R.: Interpolating polynomials from their values. J. Symb. Comput. **9**(3), 375–403 (1990)

An Incremental Algorithm for Computing Cylindrical Algebraic Decompositions

Changbo Chen and Marc Moreno Maza

Abstract In this paper, we propose an incremental algorithm for computing cylindrical algebraic decompositions. The algorithm consists of two parts: computing a complex cylindrical tree and refining this complex tree into a cylindrical tree in real space. The incrementality comes from the first part of the algorithm, where a complex cylindrical tree is constructed by refining a previous complex cylindrical tree with a polynomial constraint. We have implemented our algorithm in Maple. The experimentation shows that the proposed algorithm outperforms existing ones for many examples taken from the literature.

1 Introduction

Cylindrical algebraic decomposition (CAD) is a fundamental tool in real algebraic geometry. It was invented by Collins in 1973 [1] for solving real quantifier elimination (QE) problems. In the last 40 years, following Collins' original projection-lifting scheme, many enhancements have been performed in order to ameliorate the efficiency of CAD construction, including adjacency and clustering techniques [2], improved projection methods [3–6], partially built CADs [7–9], improved stack construction [10], efficient projection orders [11], making use of equational constraints [12–15], and so on. Moreover, CADs can be computed by several software packages, such as QEPCAD [16, 17], Mathematica [9, 18], Redlog [19], and SyNRAC [20].

In [21], together with Xia and Yang, we presented a different way of computing CADs, based on triangular decomposition of polynomial systems. In that paper, we introduced the concept of cylindrical decomposition of the complex space (CCD),

C. Chen (✉)
Chongqing Key Laboratory of Automated Reasoning and Cognition, Chongqing Institute of Green and Intelligent Technology, Chinese Academy of Sciences, Chongqing, China
e-mail: changbo.chen@hotmail.com

C. Chen · M. Moreno Maza (✉)
ORCCA, University of Western Ontario, London, Canada
e-mail: moreno@csd.uwo.ca

© Springer-Verlag Berlin Heidelberg 2014
R. Feng et al. (eds.), *Computer Mathematics*, DOI 10.1007/978-3-662-43799-5_17

from which a CAD can be easily derived. The concept of CCD is reviewed in Sect. 2. In the rest of the present paper, we use TCAD to denote CAD based on triangular decompositions while PCAD refers to CAD based on Collins' projection-lifting scheme.

The CCD part of TCAD can be seen as an enhanced projection phase of PCAD. However, w.r.t. PCAD (especially when the projection operator is using Collins' [1] or Hong's [5]), the "case discussion" scheme of TCAD avoids unnecessary computations that projection operator performs on unrelated branches. In addition, one observes that the reason why McCallum's [22] (including Brown's [3]) projection operators may fail for some examples is due to the fact that they are missing a "case discussion" scheme. McCallum's operator relies on the assumption that generically all coefficients of a polynomial[1] will not vanish simultaneously above a positive-dimensional component. If this assumption fails, then this operator is replaced by Collins-Hong projection operator [5]. The fact that all coefficients of a polynomial could vanish simultaneously above some component is never a problem in TCAD. For this reason, we view it as an improvement of previous works.

Trying to use sophisticated algebraic elimination techniques to improve CAD constructions is not a new idea. In the papers [23, 24], the authors investigated how to use Gröbner bases to preprocess the input system in order to make the subsequent CAD computations more efficient. The main difference between these two works and the work of [21] is that the former approach is about preprocessing input for CAD while the latter one presents a different way of constructing CADs.

In [21], the focus was on how to apply triangular decomposition techniques to compute CADs. To this end, lots of existing high-level routines were used to facilitate explaining ideas. These high-level routines involve many black boxes, which hide many unnecessary or redundant computations. As a result, the computation time of TCAD is much higher than that of PCAD, although TCAD computes usually less cells as noted in [25].

In the present paper, we abandon those black boxes and compute TCAD from scratch. It turns out that the key solution for avoiding redundant computations is to compute CCD in an *incremental manner*. The same motivation and a similar strategy appeared in [26, 27] in the context of triangular decomposition of algebraic sets. The core operation of such an incremental algorithm is an Intersect operation, which refines an existing cylindrical tree w.r.t. a polynomial. We dedicate Sect. 4 to presenting a complete incremental algorithm for computing TCAD by means of this Intersect operation.

In [28], the author presented an algorithm for computing with semi-algebraic sets represented by cylindrical algebraic formulas. That algorithm also allows computing CAD in an incremental manner. The underlying technique is based on the projection-lifting scheme where one first computes projection factor sets by a global projection operator. In contrast, the incremental algorithm presented here, is conducted by refining different branches of an existing tree via GCD computations.

---

[1] More precisely, a multivariate polynomial regarded as a univariate one with respect to its main variable.

This Intersect operation can systematically take advantage of equational constraints. The problem of making use of equational constraints in CAD has been studied by many researchers [12–15]. In Sect. 6, we provide a detailed discussion on how we solve this problem.

When applied to a polynomial system having finitely many complex solutions, our incremental CCD algorithm specializes into computing a triangular decomposition, say $\mathcal{D}$, such that the zero sets of the output regular chains are disjoint. Moreover, such a decomposition has no critical pairs in the sense of the equiprojectable decomposition algorithm of [29]. This implies that only the "Merge" part of the "Split and Merge" algorithm of [29] is required for turning $\mathcal{D}$ into an equiprojectable decomposition (which is a canonical representation of the input variety, once the variable order is fixed). Consequently, one could hope extending the notion of equiprojectable decomposition (and related algorithms) to positive dimension by means of our incremental CCD algorithm. This perspective can be seen as an indirect application of CAD to triangular decomposition.

As we shall review in Sect. 2, a CCD is encoded by a tree data-structure. Then each path of this tree is a simple system in the sense of [30, 31]. So the work presented here can also be used to compute a Thomas decomposition of a polynomial system [31, 32]. Moreover, the decomposition we compute is not only disjoint, but also cylindrically arranged.

The complexity of our algorithm cannot be better than doubly exponential in the number of variables [33]. So the motivation of our work is to suggest possible ways to improve the practical applicability of CAD. The benchmark in Sect. 7 shows that TCAD outperforms QEPCAD [16, 17] and Mathematica [9] for many well-known examples. The algorithm presented in this paper can support QE. We have realized a preliminary implementation of an algorithm for doing QE via TCAD. We will report on this work in a future paper.

2 Complex Cylindrical Tree

Throughout this paper, we consider a field $\mathbf{k}$ of characteristic zero and denote by $\mathbf{K}$ the algebraic closure of $\mathbf{k}$. Let $\mathbf{k}[\mathbf{x}]$ be the polynomial ring over the field $\mathbf{k}$ with ordered variables $\mathbf{x} = x_1 < \cdots < x_n$. Let $p \in \mathbf{k}[\mathbf{x}]$ be a nonconstant polynomial and $x \in \mathbf{x}$ be a variable. We denote by $\deg(p, x)$ and $\mathrm{lc}(p, x)$ the degree and the leading coefficient of p w.r.t. x. The greatest variable appearing in p is called the *main variable*, denoted by $\mathrm{mvar}(p)$. The leading coefficient, the degree, the reductum of p w.r.t. $\mathrm{mvar}(p)$ are called the *initial*, the *main degree*, the *tail* of p; they are denoted by $\mathrm{init}(p)$, $\mathrm{mdeg}(p)$, $\mathrm{tail}(p)$ respectively. The integer k such that $x_k = \mathrm{mvar}(p)$ is called the *level* of the polynomial p. We denote by $\mathrm{der}(p)$ the derivative of p w.r.t. $\mathrm{mvar}(p)$. The notions presented below were introduced in [21] and they are illustrated at the beginning of Sect. 3.

Separation Let C be a subset of $\mathbf{K}^{n-1}$ and $P \subset \mathbf{k}[x_1, \ldots, x_{n-1}, x_n]$ be a finite set of level n polynomials. We say that P *separates above* C if for each $\alpha \in C$:

- for each $p \in P$, the polynomial $\mathrm{init}(p)$ does not vanish at α,
- the polynomials $p(\alpha, x_n) \in \mathbf{K}[x_n]$, for all $p \in P$, are squarefree and coprime.

Note that this definition allows C to be a semi-algebraic set, see Theorem 3.

Cylindrical Decomposition By induction on n, we define the notion of a *cylindrical decomposition of* $\mathbf{K}^n$ together with that of the *tree associated with a cylindrical decomposition of* $\mathbf{K}^n$. For $n = 1$, a cylindrical decomposition of $\mathbf{K}$ is a finite collection of sets $\mathcal{D} = \{D_1, \ldots, D_{r+1}\}$, where either $r = 0$ and $D_1 = \mathbf{K}$, or $r > 0$ and there exists r nonconstant coprime squarefree polynomials $p_1, \ldots, p_r$ of $\mathbf{k}[x_1]$ such that for $1 \le i \le r$ we have $D_i = \{x_1 \in \mathbf{K} \mid p_i(x_1) = 0\}$, and $D_{r+1} = \{x_1 \in \mathbf{K} \mid p_1(x_1) \ldots p_r(x_1) \neq 0\}$. Note that the D_i's, for all $1 \le i \le r+1$, form a partition of $\mathbf{K}$. The tree associated with $\mathcal{D}$ is a rooted tree whose nodes, other than the root, are $D_1, \ldots, D_r, D_{r+1}$ which all are leaves and children of the root. Now let $n > 1$, and let $\mathcal{D}' = \{D_1, \ldots, D_s\}$ be any cylindrical decomposition of $\mathbf{K}^{n-1}$. For each D_i, let r_i be a non-negative integer and let $\{p_{i,1}, \ldots, p_{i,r_i}\}$ be a set of polynomials which separates above D_i. If $r_i = 0$, set $D_{i,1} = D_i \times \mathbf{K}$. If $r_i > 0$, set

$$D_{i,j} = \{(\alpha, x_n) \in \mathbf{K}^n \mid \alpha \in D_i \text{ and } p_{i,j}(\alpha, x_n) = 0\},$$

for $1 \le j \le r_i$ and set

$$D_{i,r_i+1} = \left\{(\alpha, x_n) \in \mathbf{K}^n \mid \alpha \in D_i \text{ and } \left(\prod_{j=1}^{r_i} p_{i,j}(\alpha, x_n)\right) \neq 0\right\}.$$

The collection $\mathcal{D} = \{D_{i,j} \mid 1 \le i \le s, 1 \le j \le r_i + 1\}$ is called a *cylindrical decomposition* of $\mathbf{K}^n$. The sets $D_{i,j}$ are called the *cells* of $\mathcal{D}$. If T' is the tree associated with $\mathcal{D}'$ then the tree T associated with $\mathcal{D}$ is defined as follows. For each $1 \le i \le s$, the set D_i is a leaf in T' which has all $D_{i,j}$'s for children in T; thus the $D_{i,j}$'s are the leaves of T.

Note that each node N of T is either associated with no constraints, or associated with a polynomial constraint, which itself is either an equation or an inequation. Note also that, if the level of the polynomial defining the constraint at N is ℓ, then ℓ is the length of a path from N to the root. Moreover, the polynomial constraints along a path from the root to a leaf form a polynomial system called a *cylindrical system of* $\mathbf{k}[x_1, \ldots, x_n]$ *induced by* T. Let S be such a cylindrical system. We denote by $Z(S)$ the zero set of S. Therefore, each cell of $\mathcal{D}$ is the zero set of a cylindrical system induced by T.

Let Γ be a subtree of T such that the root of Γ is that of T. Then, we call Γ a *cylindrical tree of* $\mathbf{k}[x_1, \ldots, x_n]$ *induced by* T. This cylindrical tree Γ is said *partial* if it admits a nonleaf node N such that the zero set of the constraint of N is not equal to the union of the zero sets of the constraints of the children of N. If Γ is not partial, then it is called *complete*.

In the algorithms of Sect. 4, the cylindrical tree is an essential data structure. Section 3 discusses the main properties and operations on this data structure.

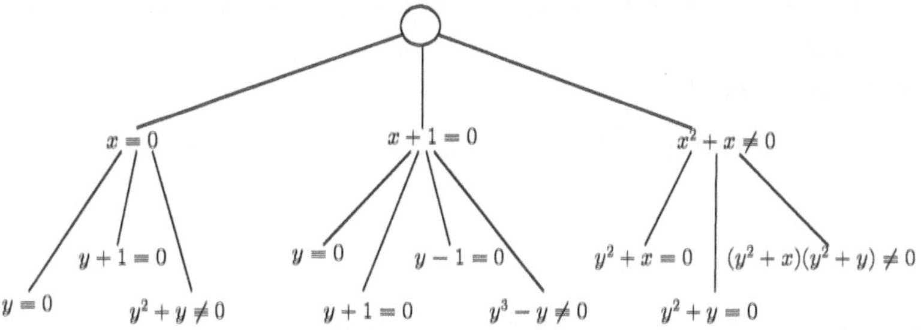

Fig. 1 An $F := \{y^2 + x, y^2 + y\}$ invariant complex cylindrical tree

Let $F = \{f_1, \ldots, f_s\}$ be a finite set of polynomials of $\mathbf{k}[x_1 < \cdots < x_n]$. A cylindrical decomposition $\mathcal{D}$ of $\mathbf{K}^n$ is called *F-invariant* if for any given cell D of $\mathcal{D}$ and any given polynomial $f \in F$, either f vanishes at all points of D or f vanishes at no points of D.

Example 1 Let $F := \{y^2 + x, y^2 + y\}$. An F-invariant cylindrical decomposition of $\mathbb{C}^2$ is illustrated by Fig. 1 .

We observe that every cylindrical system induced by a cylindrical tree is a *simple system*, as defined by Wang in [31]. This notion was first introduced by Thomas in 1937 [30]. Simple systems have many nice properties. For example, if $[A, B]$ is a simple system, then the pair $[A, \prod_{p \in B} p]$ is a squarefree regular system, as defined by Wang in [31, 34].

Let Γ be a cylindrical system of $\mathbf{k}[\mathbf{x}]$ and let p be a polynomial of $\mathbf{k}[\mathbf{x}]$. We say that p is *invertible modulo* Γ if for any $\alpha \in Z(\Gamma)$, we have $p(\alpha) \neq 0$. We say that p is *zero modulo* Γ if for any $\alpha \in Z(\Gamma)$, we have $p(\alpha) = 0$. We say that p is *sign invariant above* Γ if p is either zero or invertible modulo Γ. Let q be another polynomial of $\mathbf{k}[\mathbf{x}]$. We say that $p = q$ modulo Γ if $Z(\Gamma) \cap Z(p) = Z(\Gamma) \cap Z(q)$.

Greatest Common Divisor (GCD)

Let p and f be two level n polynomials in $\mathbf{k}[\mathbf{x}]$. Let Γ be a cylindrical system of $\mathbf{k}[x_1, \ldots, x_{n-1}]$. For any $u \in \mathbf{K}^{n-1}$ of $Z(\Gamma)$, assume at least one of $\mathrm{lc}(p, x_n)(u)$ and $\mathrm{lc}(f, x_n)(u)$ is not zero. A polynomial $g \in \mathbf{k}[\mathbf{x}]$ is called a *GCD of p and f modulo* Γ if for any $u \in \mathbf{K}^{n-1}$ of $Z(\Gamma)$,

- $g(u)$ is a GCD of $p(u)$ and $f(u)$ in $\mathbf{K}[x_n]$, and
- we have $\mathrm{lc}(g, x_n)(u) \neq 0$.

Let $d_p = \deg(p, x_n)$, $d_f = \deg(f, x_n)$. Recall that we assume $d_p, d_f \geq 1$. Let $\lambda = \min(d_p, d_f)$. Let Γ be a cylindrical system of $\mathbf{k}[x_1, \ldots, x_{n-1}]$. Let $S_0, \ldots, S_{\lambda-1}$ be the subresultant polynomials [35, 36] of p and f w.r.t. x_n. Let $s_i = \mathrm{coeff}(S_i, x_n^i)$ be the principle subresultant coefficient of S_i, for $0 \leq i \leq \lambda - 1$. If $d_p \geq d_f$, we define $S_\lambda = f$, $S_{\lambda+1} = p$, $s_\lambda = \mathrm{init}(f)$, and $s_{\lambda+1} = \mathrm{init}(p)$. If $d_p < d_f$, we define $S_\lambda = p$, $S_{\lambda+1} = f$, $s_\lambda = \mathrm{init}(p)$, and $s_{\lambda+1} = \mathrm{init}(f)$.

Theorem 1 *Let j be an integer, with $1 \leq j \leq \lambda + 1$, such that s_j is invertible modulo Γ and such that for any $0 \leq i < j$, we have $s_i = 0$ modulo Γ. Then S_j is a* GCD *of p and f modulo Γ.*

Proof It can be easily proved by the specialization property of subresultant chains. In particular, it is a direct corollary of Theorem 5 in [37].

3 Data Structure for Cylindrical Decomposition

In this section, we describe the data structures that are used by the algorithms presented in this paper for computing cylindrical decompositions. To understand the motivation of our algorithm design, let us consider a simple example with $n = 2$ variables. Let a, b be two coprime squarefree nonconstant univariate polynomials in $k[x_1]$. Observe that $L := k[x_1]/\langle a\, b\rangle$ is a direct product of fields. Let also c, d be two bivariate polynomials of $k[x_1, x_2]$, such that $\deg(c, x_2) > 0$, $\deg(d, x_2) > 0$, and $\mathrm{lc}(c, x_2) = \mathrm{lc}(d, x_2) = 1$ hold and such that c, d are coprime and squarefree univariate as polynomials of $L[x_2]$. Therefore, the following four polynomial systems are simple systems

$$\begin{cases} a(x_1)b(x_1) = 0 \\ c(x_1, x_2) = 0 \end{cases}, \quad \begin{cases} a(x_1)b(x_1) = 0 \\ d(x_1, x_2) = 0 \end{cases}, \quad \begin{cases} a(x_1)b(x_1) = 0 \\ c(x_1, x_2)d(x_1, x_2) \neq 0 \end{cases}, \quad \{a(x_1)b(x_1) \neq 0$$

that we denote respectively by S_1, S_2, S_3, S_4. It is easy to check that the zero sets $Z(S_1)$, $Z(S_2)$, $Z(S_3)$, $Z(S_4)$ are the cells of a cylindrical decomposition $\mathcal{D}$ of $\mathbf{K}^2$.

Let $f \in k[x_1]$ be another univariate polynomial. Assume that one has to *refine* $\mathcal{D}$ into a cylindrical decomposition of $\mathbf{K}^2$ which is required to be $\{f\}$-invariant. That is, one has to test whether f is invertible or zero modulo each of the systems S_1, S_2, S_3, S_4, and further decompose when appropriate. Assume that the polynomial a divides f whereas b, f are coprime. Assume also that the system S_1 is processed first in time. By computing $\gcd(f, ab)$, which yields a, one splits S_1 into the following two subsystems that we denote by $S_{1,1}$ and $S_{1,2}$.

$$\begin{cases} a(x_1) = 0 \\ c(x_1, x_2) = 0 \end{cases}, \quad \text{and} \quad \begin{cases} b(x_1) = 0 \\ c(x_1, x_2) = 0. \end{cases}$$

Assume that S_2 is processed next. By computing $\gcd(f, ab)$ (again) one splits S_2 into the following two subsystems that we denote by $S_{2,1}$ and $S_{2,2}$.

$$\begin{cases} a(x_1) = 0 \\ d(x_1, x_2) = 0 \end{cases}, \quad \text{and} \quad \begin{cases} b(x_1) = 0 \\ d(x_1, x_2) = 0. \end{cases}$$

Consequently, in the course of the creation of $S_{1,1}$, $S_{1,2}$, $S_{2,1}$, and $S_{2,2}$, the same polynomial GCD and the same field extensions (namely $k[x_1]/\langle a\rangle$ and $k[x_1]/\langle b\rangle$) were computed twice. This duplication of calculation and data is a common phenomenon and a performance bottleneck in most algorithms for decomposing polynomial systems.

Mathematically, each constructible set should not be represented more than once in a computer program. To implement this idea, all constructible sets manipulated during the execution of a given computer program should be seen as part of the same *universe*, say $\mathbf{K}^n$. Moreover, the subroutines of this program should have the same view on the universe, which is then a *shared data-structure*, such that whenever a subroutine modifies the universe all subroutines have immediate access to the modified universe. Satisfying these requirements is a well-known challenge in computer science, an instance of which is the question of *memory consistency* for shared-memory parallel computer architectures, such as multicores. With our above example, even if we do not intend to run computations concurrently, we are concerned with the practical efficiency and ease-of-use of the mechanisms that maintain *up-to-date* all views on the universe.

Recall that a cylindrical decomposition can be identified as a tree where each node is a constructible set of $\mathbf{K}^n$ given by either an equation constraint, or an inequation constraint, or no constraints at all. In this latter case, the corresponding constructible set is the whole space. All algorithms in Sect. 4 work on a given cylindrical decomposition $\mathcal{D}$ encoded by a tree T (as defined in Sect. 2). That is, the tree T is regarded as the universe.

We assume that there is a procedure for updating the tree T, which, given a "node-to-be-replaced" N and its "replacing nodes" $N_1, \ldots, N_e$, is called split $(N; N_1, \ldots, N_e)$ and works as follows:

1. for $i = 1, \ldots, e$, for each child C of N deeply copy (thus creating new nodes) the subtree rooted at C and make that copy of C a child of N_i,
2. update the parent of N such that $N_1, \ldots, N_e$ are new children of the the parent of N,
3. remove the entire subtree rooted at N from the universe, including N.

We assume that all updates are performed sequentially (thus using mutual exclusion mechanism in case of concurrent execution of the algorithms of Sect. 4) such that no data races can occur.

We also assume that each node N (whether it is a node in the present or has been removed from the universe) has a unique key, called key(N), and a data field, called value(N), storing various information including:

- a time stamp PAST or PRESENT,
- if PAST, the list of its replacing nodes (as specified with the split procedure) and the list of its children at the time it was replaced,
- if PRESENT, the list of its children and a pointer to the parent.

All nodes are stored in a *dictionary* H which can be accessed by all subroutines. Modifying the universe means updating H using the split procedure. Since all our algorithms stated in Sect. 4 are sequential, no synchronization issue has to be considered. The mechanism described above allows us to achieve our goals.

4 Constructing a Cylindrical Tree Incrementally

In this section, we present an incremental algorithm for computing a cylindrical tree, as defined in Sect. 2. We start by commenting on the style of the pseudo-code. Secondly, we present the specifications of the algorithm and related subroutines. Thirdly, we state all the algorithms in pseudo-code style. Finally, proof sketches of the algorithms are provided at the end of this section.

Following the principles introduced in Sect. 3, our procedures operate on a "universe" (which is a cylindrical tree T) that they modify when needed. These modifications are of two types:

- splitting a node,
- attaching information to a node.

In addition to the attributes described in Sect. 3, a node has attributes corresponding to the results of operations like Squarefree, Gcd, Intersect. In other words, our procedures do not return values; instead they store their results in the nodes of the universe. This technique greatly simplifies pseudo-code.

Since attributes of nodes are intensively used in our pseudo-code, we use the standard "dot" notation of object oriented programming languages. In addition, since a node can have many attributes, we make the following convention. Suppose that a node V is split into two nodes V_1 and V_2. Some attributes are likely to have different values in V_1 (resp. V_2) and V. But most of them will often have the same values in both nodes. Therefore, after setting up the values of the attributes that differ, we simply write V_1.others := V.others to define the attributes of V_1 whose values are unchanged w.r.t. V.

Several procedures iterate through all the paths of the universe T. By path, we mean a path (in the sense of graph theory) from the root of T to a leaf of T. The current path is often denoted by Γ or C. Recall from Sect. 2 that a path in T corresponds to a simple system, say S. Computing modulo S may split S and thus modify the universe *automatically*, that is, in a transparent manner in the pseudo-code. However, splitting S also changes the current path. For clarity, we explicitly invoke a function called UpdatePath, which updates its first argument (namely the current path) from the universe.

In order to iterate through all the paths of the universe T, we use a function NextPathToDo. This command is a *generator* or an *iterator* in the sense of the theory of programming languages. That is, it views T as a stream of paths and returns the next path-to-be-visited, if any. Thanks to the fact that the universe is always up-to-date, the function NextPathToDo is able to return the next path-to-be-visited in the current state of the universe.

A frequently used operation on the universe and its paths is ExtractProjection, see for instance Algorithm 6. When applied to the universe T and an integer k (for $0 \leq k < n$, where n is the length of a path from the root of T to a leaf of T) ExtractProjection returns a "handle" on the universe "truncated" at level k, that is, the universe where all nodes of level higher than k are ignored (thus viewing the level k nodes as leaves). When applied to path, ExtractProjection has a similar output.

We often say that a function (see for instance Algorithm 5) returns a refined cylindrical decomposition. This is another way of saying that the universe is updated to a new state corresponding to a cylindrical decomposition refining (in the sense of a partition of a set refining another partition of the same set) the cylindrical decomposition of the previous state.

After these preliminary remarks on the pseudo-code, we present the specifications of the algorithm and related subroutines.

The top level algorithm for computing a cylindrical tree is described by Algorithm 4. It takes a set F of nonconstant polynomials in $\mathbf{k}[x_1 < \cdots < x_n]$ as input and returns an F-invariant cylindrical decomposition of $\mathbf{K}^n$. This algorithm relies on a core operation, called Intersect, which computes a cylindrical decomposition in an incremental manner.

The Intersect operation is described by Algorithm 5. It takes a cylindrical tree T and a polynomial p of $\mathbf{k}[x_1 < \cdots < x_n]$ as input. It refines tree T such that p is sign invariant above each path of the refined tree T. This operation is achieved by refining each path of T with IntersectPath.

The IntersectPath operation is described by Algorithm 6. It takes a polynomial p, a cylindrical tree T, and a path Γ of T in $\mathbf{k}[x_1 < \cdots < x_n]$ as input. It refines Γ and updates the tree T accordingly such that p is sign invariant above each path derived from Γ in the updated tree T. This operation finds the node N in Γ whose level is the same as that of p. Let Γ_N be the subpath of Γ from N to the root of T. The IntersectPath operation then calls the routine IntersectMain so as to refine Γ_N into a tree T_N such that p becomes sign invariant w.r.t. T_N.

The routine IntersectMain is described by Algorithm 7. It takes a cylindrical tree T, a path Γ of T, and a polynomial of the same level as the leaves of T in $\mathbf{k}[x_1 < \cdots < x_n]$ as input. It refines Γ and updates the tree T accordingly such that p becomes sign invariant above each path derived from Γ in the updated tree.

The routine IntersectMain works in the following way. It first splits Γ such that above the projection C_{n-1} of each new branch C of Γ in $\mathbf{K}^{n-1}$, the number of distinct roots of p w.r.t. x_n is invariant. This is achieved by the operation Squarefree, described by Algorithm 8. The squarefree part of p above a branch C is denoted by sp. If p has no roots or is identically zero above C_{n-1}, the sign of p above C is determined immediately. Otherwise, a case discussion is made according to the structure of the leaf node V of C. If V has no constraints associated to it, then V is simply split into two new nodes $sp = 0$ and $sp \neq 0$. Assume now that V has a constraint, which can be either of the form $f = 0$ or of the form $f \neq 0$, where f is a level n polynomial squarefree modulo C_{n-1}. This case is handled by computing the GCD g of sp and f modulo C_{n-1}. The node V then splits based on the GCD g and the cofactors of sp and f.

The GCD is computed by the operation Gcd, described by Algorithm 9 and 10. The cofactors are computed by Algorithm 11. The Squarefree and Gcd operations rely on the operation MakeLeadingCoefficientInvertible, described by Algorithm 12. This latter operation takes as input a polynomial p of $\mathbf{k}[x_1 < \cdots < x_n]$, a cylindrical tree T of $\mathbf{k}[x_1 < \cdots < x_{n-1}]$, and a path Γ of T. Then, it refines Γ and updates T

accordingly such that above each path C of T derived from Γ, the polynomial p is either zero or its leading coefficient is invertible.

All the algorithms also rely on the following three operations which perform manipulations and traversal of the tree data structure. For these three operations, only specifications are provided below while their algorithms are explained in Sect. 3.

Theorem 2 *For a set of polynomials in* $\mathbf{k}[x_1, \ldots, x_n]$*, Algorithm 4 computes an* F*-invariant cylindrical decomposition of* $\mathbf{K}^n$*.*

Algorithm 1: UpdatePath(Γ, T)

- Input: A cylindrical tree T. A path Γ in some past state of T.
- Output: A subtree ST in present state of T. ST is derived from Γ according to the historical data of T.

Algorithm 2: ExtractProjection(T, k)

- Input: A cylindrical tree T of $\mathbf{k}[x_1 < \cdots < x_n]$. An integer k, $0 \leq k \leq n$.
- Output: A cylindrical tree T_k in $\mathbf{k}[x_1 < \cdots < x_k]$ such that T_k is the projection of T in $\mathbf{k}[x_1 < \cdots < x_k]$.

Algorithm 3: NextPathToDo$_n$(T)

- Input: A cylindrical tree T in $\mathbf{k}[x_1 < \cdots < x_n]$.
- Output: For a fixed traversal order of a tree, return the first "ToDo" path Γ of T.

Algorithm 4: CylindricalDecompose(F)

Input: F is a set of nonconstant polynomials in $\mathbf{k}[x_1 < \cdots < x_n]$.
Output: An F-invariant cylindrical decomposition of $\mathbf{K}^n$.

```
begin
    create a tree T with only one vertex V_0: the root of T;
    for i from 1 to n do
        create a vertex V_i; V_i.signs := ∅; V_i.formula := "any x_i";
        V_{i-1}.child := V_i;
    for p ∈ F do
        Intersect_n(p, T);
    return T;
end
```

Algorithm 5: $\mathsf{Intersect}_n(p, T)$

Input: A cylindrical tree T of $\mathbf{k}[x_1 < \cdots < x_n]$. A nonconstant polynomial p of $\mathbf{k}[x_1 < \cdots < x_n]$.

Output: A refined cylindrical decomposition such that p is sign invariant above each path of T.

while $\Gamma := \mathsf{NextPathToDo}_n(T) \neq \emptyset$ **do**
  $\mathsf{IntersectPath}_n(p, \Gamma, T)$;

Algorithm 6: $\mathsf{IntersectPath}_n(p, \Gamma, T)$

Input: A cylindrical tree T of $\mathbf{k}[x_1 < \cdots < x_n]$. A path Γ of T. A polynomial p of $\mathbf{k}[x_1 < \cdots < x_n]$.

Output: A refined cylindrical decomposition T such that p is sign invariant above each path derived from Γ.

begin
  if $p \in \mathbf{k}$ **then**
    return;
  else
    $k := \text{level}(p)$;
    if $k = n$ **then**
      $\mathsf{IntersectMain}_n(p, \Gamma, T)$;
    else
      $T_k := \mathsf{ExtractProjection}(T, k)$;$\Gamma_k := \mathsf{ExtractProjection}(\Gamma, k)$;
      $\mathsf{IntersectMain}_k(p, \Gamma_k, T_k)$;
      $\mathsf{UpdatePath}(\Gamma, T)$;
      for *each leaf V of Γ* **do**
        Let L_k be the ancestor of V of level k; $V.signs[p] := L_k.signs[p]$;
end

Proof Firstly, we prove the termination. The basic mutual calling graph of its subroutines are:

$$\mathsf{IntersectMain}_n \to \mathsf{Squarefree_n} \to \mathsf{IntersectMain}_{n-1} \to \cdots,$$

and

$$\mathsf{IntersectMain}_n \to \mathsf{Gcd_n} \to \mathsf{IntersectMain}_{n-1} \to \cdots$$

So the termination is easily proved by induction. The correctness follows from the specification of its subroutines and Theorem 1.

Example 2 In this example, we illustrate the operation $\mathsf{IntersectPath}$. Let $F := \{y^2 + x, y^2 + y\}$. The incremental algorithm first computes an $y^2 + x$ sign invariant complex cylindrical tree, which is described by the following tree T.

Algorithm 7: IntersectMain$_n(p, \Gamma, T)$

Input: A cylindrical tree T of $\mathbf{k}[x_1 < \cdots < x_n]$. A path Γ of T. A polynomial p of level n in $\mathbf{k}[x_1 < \cdots < x_n]$.

Output: A refined cylindrical decomposition T such that p is sign invariant above each path derived from Γ.

```
begin
    T_{n-1} := ExtractProjection(T, n − 1); Γ_{n-1} := ExtractProjection(Γ, n − 1);
    Squarefree_n(p, Γ_{n-1}, T_{n-1});
    UpdatePath(Γ, T);
    while C := NextPathToDo_n(Γ) ≠ ∅ do
        V := C.leaf; C_{n-1} := ExtractProjection(C, n − 1);
        sp := C_{n-1}.leaf.Squarefree[p];
        if sp = 0 then
            V.signs[p] := 0;

        else if sp = 1 then
            V.signs[p] := 1;
        else if V.formula is "any x_n" then
            split V into two new vertices V_1 and V_2;
            V_1.formula := sp = 0; V_1.signs := V.signs; V_1.signs[p] := 0;
            V_2.formula := sp ≠ 0; V_2.signs := V.signs; V_2.signs[p] := 1;
            V_1.others := V.others; V_2.others := V.others;
            C_{n-1}.leaf.children := V_1, V_2;
        else
            //V.formula is of the form f = 0 or f ≠ 0
            Gcd_n(sp, f, C_{n-1}, T_{n-1});
            UpdatePath(C, T);
            for each leaf V of C do
                let L be the parent of V;
                cp, g, cf := CoFactor(sp, L.Gcd[sp, f], f);
                if V.formula is of the form f = 0 then
                    if g = 1 then
                        V.signs[p] := 1;

                    else if cf = 1 then
                        V.signs[p] := 0;
                    else
                        split V into two new vertices V_1 and V_2;
                        V_1.formula := g = 0; V_1.signs := V.signs; V_1.signs[p] := 0;
                        V_2.formula := cf = 0; V_2.signs := V.signs; V_2.signs[p] := 1;
                        V_1.others := V.others; V_2.others := V.others;
                        L.children := V_1, V_2;

                else
                    if cp = 1 then
                        V.signs[p] := 1;
                    else
                        split V into two new vertices V_1 and V_2;
                        V_1.formula := cp = 0; V_1.signs := V.signs; V_1.signs[p] := 0;
                        V_2.formula := (f * cp) ≠ 0;
                        V_2.signs := V.signs; V_2.signs[p] := 1;
                        V_1.others := V.others; V_2.others := V.others;
                        L.children := V_1, V_2;
end
```

Algorithm 8: $\mathsf{Squarefree}_n(p, \Gamma, T)$

Input: A cylindrical tree T of $\mathbf{k}[x_1 < \cdots < x_{n-1}]$. A path Γ of T. A polynomial p of level n.

Output: A refined cylindrical tree T of $\mathbf{k}[x_1 < \cdots < x_{n-1}]$. Above each path C of T derived from Γ, there is a dictionary $C.leaf.Squarefree$. Let $p^* := C.leaf.Squarefree[p]$. We have:

- $p = p^*$ modulo C.
- If p^* is of level n, then both $\mathrm{init}(p^*)$ and $\mathrm{discrim}(p^*)$ are invertible modulo C.
- If p^* is of level less than n, then p^* is either 0 or 1.

```
begin
    if n = 1 then
        let r be the root of T; r.Squarefree[p] := SquarefreePart(p);
        return
    MakeLeadingCoefficientInvertible_n(p, p, Γ, T);
    while C := NextPathToDo_{n-1}(Γ) ≠ ∅ do
        f := C.leaf.InvertLc[p];
        if level(f) < n or deg(f, x_n) = 1 then
            C.leaf.Squarefree[p] := f

        else
            Gcd_n(f, der(f), C, T);
            for each leaf L of C do
                g := L.Gcd[f, der(f)];
                if g = 1 then
                    L.Squarefree[p] := f

                else
                    L.Squarefree[p] := pquo(f, g)
end
```

Algorithm 9: $\mathsf{Gcd}_n(p, f, \Gamma, T)$

Input: A cylindrical tree T of $\mathbf{k}[x_1 < \cdots < x_{n-1}]$. A polynomial $p \in \mathbf{k}[x_1 < \cdots < x_n]$ of level n. A path Γ of T. A polynomial f of level n such that $\mathrm{init}(f)$ is invertible modulo Γ.

Output: A refined cylindrical tree T. Above each path C of T derived from Γ, there is a dictionary $C.leaf.Gcd$ such that $C.leaf.Gcd[p, f]$ is a GCD of p and f modulo C.

```
begin
    let S be the subresultant chain of p and f;
    if mdeg(p) ≥ mdeg(f) then
        d := mdeg(f)
    else
        d := mdeg(p) + 1
    return Gcd_n(p, f, S, d, 0, Γ, T);
end
```

Algorithm 10: $\mathsf{Gcd}_n(p, f, S, d, i, \Gamma, T)$

Input:

- A polynomial $p \in \mathbf{k}[x_1 < \cdots < x_n]$ of level n.
- A polynomial f of level n such that $\mathrm{lc}(f)$ is invertible modulo Γ.
- The subresultant chain S of p and f w.r.t. x_n.
- A non-negative integer d (as defined in the pseudo-code of Algorithm 9) and such that the principle subresultant coefficient s_d is invertible modulo Γ.
- A non-negative integer i such that $0 \le i \le d$ and the principle subresultant coefficient s_j is zero modulo Γ, for all $0 \le j < i$.
- A path Γ of T.
- A cylindrical tree T of $\mathbf{k}[x_1 < \cdots < x_{n-1}]$.

Output: A refined cylindrical tree T. Above each path C of T derived from Γ, there is a dictionary $C.leaf.Gcd$ such that $C.leaf.Gcd[p, f]$ is a GCD of p and f modulo C.

```
begin
    if i = d then
        Γ.leaf.Gcd[p, f] := S_i;
        return;
    IntersectPath_{n-1}(s_i, Γ, T);
    while C := NextPathToDo_{n-1}(Γ) ≠ ∅ do
        if C.leaf.signs[s_i] = 1 then
            if i = 0 then
                C.leaf.Gcd[p, f] := 1

            else
                C.leaf.Gcd[p, f] := S_i

        else
            Gcd_n(p, f, S, d, i + 1, C, T)
end
```

$$T := \begin{cases} x = 0 \begin{cases} y = 0 : y^2 + x = 0 \\ y \neq 0 : y^2 + x \neq 0 \end{cases} \\ \\ x \neq 0 \begin{cases} y^2 + x = 0 : y^2 + x = 0 \\ y^2 + x \neq 0 : y^2 + x \neq 0 \end{cases} \end{cases}$$

Let Γ be the path $\{x = 0, y \neq 0\}$ of T. Calling $\mathsf{IntersectPath}(y^2 + y, \Gamma, T)$ will update T into the following tree.

$$\begin{cases} x = 0 \begin{cases} y = 0 : y^2 + x = 0 \\ y = -1 : y^2 + x \neq 0 \wedge y^2 + y = 0 \\ \text{otherwise} : y^2 + x \neq 0 \wedge y^2 + y \neq 0 \end{cases} \\ \\ x \neq 0 \begin{cases} y^2 + x = 0 : y^2 + x = 0 \\ y^2 + x \neq 0 : y^2 + x \neq 0 \end{cases} \end{cases}$$

Algorithm 11: CoFactor(p, g, f)

Input: Two polynomials p and f of level n in $\mathbf{k}[x_1 < \cdots < x_n]$. A polynomial g which is either 1 or of level n in $\mathbf{k}[x_1 < \cdots < x_n]$.

Output: As described by the algorithm.

```
begin
    if g = 1 then
        cp := p; gg := 1; cf := f;

    else if mdeg(g) = mdeg(f) then
        gg := f;
        if mdeg(g) = mdeg(p) then
            cf := 1; cp := 1;
        else
            cf := 1; cp := pquo(p, gg)
    else if mdeg(g) = mdeg(p) then
        gg := p; cf := pquo(f, gg); cp := 1;
    else
        cp := pquo(p, g); cf := pquo(f, g); gg := g;
    return cp, gg, cf;
end
```

Algorithm 12: MakeLeadingCoefficientInvertible$_n(p, \bar{p}, \Gamma, T)$

Input: A polynomial p of $\mathbf{k}[x_1 < \cdots < x_n]$. A polynomial $\bar{p}$ of $\mathbf{k}[x_1 < \cdots < x_n]$ such that $p = \bar{p}$ modulo Γ. A cylindrical tree T of $\mathbf{k}[x_1 < \cdots < x_{n-1}]$. A path Γ of T.

Output: A refined cylindrical tree T of $\mathbf{k}[x_1 < \cdots < x_{n-1}]$. Above each path C of T derived from Γ, there is a dictionary $C.leaf.InvertLc$. Let p^* be the polynomial $C.leaf.InvertLc[p]$. Then, we have:

- $p = p^*$ modulo C.
- If p^* is of level n, then $\mathrm{init}(p^*)$ is invertible modulo the path C.
- If p^* is of level less than n, then p^* is either 0 or 1.

```
begin
    IntersectPath_{n-1}(lc(p̄, x_n), Γ, T);
    while C := NextPathToDo_{n-1}(Γ) ≠ ∅ do
        if C.leaf.signs[lc(p̄, x_n)] = 1 then
            if level(p̄) < n then
                C.leaf.InvertLc[p] := 1

            else
                C.leaf.InvertLc[p] := p̄

        else
            if level(p̄) < n then
                C.leaf.InvertLc[p] := 0

            else
                MakeLeadingCoefficientInvertible_n(p, tail(p̄), C, T)

end
```

5 Building a CAD Tree from a Complex Cylindrical Tree

In this section, we review briefly how to compute a CAD of $\mathbb{R}^n$ from a cylindrical decomposition of $\mathbb{C}^n$. The reader may refer to [21] for more details. Recall that $n \geq 1$ holds. We denote by π_{n-1} the standard projection from $\mathbb{R}^n$ to $\mathbb{R}^{n-1}$ that maps $(x_1, \ldots, x_{n-1}, x_n)$ onto $(x_1, \ldots, x_{n-1})$.

Stack Over a Connected Semi-Algebraic Set Let S be a connected semi-algebraic subset of $\mathbb{R}^{n-1}$. The *cylinder* over S in $\mathbb{R}^n$ is defined as $Z_{\mathbb{R}}(S) := S \times \mathbb{R}$. Let $\theta_1 < \cdots < \theta_s$ be continuous semi-algebraic functions defined on S. The intersection of the graph of θ_i with $Z_{\mathbb{R}}(S)$ is called the θ_i*-section* of $Z_{\mathbb{R}}(S)$. The set of points between two consecutive sections of $Z_{\mathbb{R}}(S)$ is a connected semi-algebraic subset of $\mathbb{R}^n$, called a *sector* of $Z_{\mathbb{R}}(S)$. All the sections and sectors of $Z_{\mathbb{R}}(S)$ form a disjoint decomposition of $Z_{\mathbb{R}}(S)$, called a *stack* over S.

Cylindrical Algebraic Decomposition A finite partition $\mathcal{D}$ of $\mathbb{R}^n$ is called a *cylindrical algebraic decomposition* (CAD) of $\mathbb{R}^n$ if one of the following properties holds.

- Either $n = 1$ and $\mathcal{D}$ is a stack over $\mathbb{R}^0$.
- Or the set of $\{\pi_{n-1}(D) \mid D \in \mathcal{D}\}$ is a CAD of $\mathbb{R}^{n-1}$ and each $D \in \mathcal{D}$ is a section or sector of the stack over $\pi_{n-1}(D)$.

When this holds, the elements of $\mathcal{D}$ are called *cells*.

Sign Invariance and Delineability Let p be a polynomial of $\mathbb{R}[x_1, \ldots, x_n]$, and let S be a subset of $\mathbb{R}^n$. The polynomial p is called *sign invariant* on S if the sign of $p(\alpha)$ does not change when α ranges over S. Let $F \subset \mathbb{R}[x_1, \ldots, x_n]$ be a finite polynomial set. We say S is F-invariant if each $p \in F$ is invariant on S. A cylindrical algebraic decomposition $\mathcal{D}$ is F-invariant if F is invariant on each cell $D \in \mathcal{D}$. Let p be a polynomial of $\mathbb{R}[x_1, \ldots, x_n]$, and let S be a connected semi-algebraic set of $\mathbb{R}^{n-1}$. We say that p is *delineable* on S if the real zeros of p define continuous semi-algebraic functions $\theta_1, \ldots, \theta_s$ such that, for all $\alpha \in S$ we have $\theta_1(\alpha) < \cdots < \theta_s(\alpha)$. In other words, p is delineable on S if its real zeros naturally determine a stack over S. We recall the following theorem introduced in [21].

Theorem 3 *Let $P = \{p_1, \ldots, p_r\}$ be a finite set of polynomials in $\mathbb{R}[x_1 < \cdots < x_n]$ of level n. Let S be a connected semi-algebraic subset of $\mathbb{R}^{n-1}$. If P* separates *above S, then each p_i is delineable on S. Moreover, the product of the $p_1, \ldots, p_r$ is also delineable on S.*

Let F be a finite set of polynomials in $\mathbb{Q}[x_1 < \cdots < x_n]$. Let CT be an F-invariant complete cylindrical tree of $\mathbb{C}^n$. Applying Theorem 3 to polynomials in CT, we can derive an F-invariant cylindrical algebraic decomposition of $\mathbb{R}^n$ by induction on n. A procedure MakeSemiAlgebraic, was introduced in [21] to derive a CAD from a CT via real root isolation of zero-dimensional regular chains.

Example 3 Let $F := \{y^2 + x\}$. An F-invariant cylindrical algebraic decomposition is described by the following tree.

$$T := \begin{cases} x < 0 \begin{cases} y < -\sqrt{|x|} & : y^2 + x > 0 \\ y = -\sqrt{|x|} & : y^2 + x = 0 \\ y > -\sqrt{|x|} \wedge y < \sqrt{|x|} & : y^2 + x < 0 \\ y = \sqrt{|x|} & : y^2 + x = 0 \\ y > \sqrt{|x|} & : y^2 + x > 0 \end{cases} \\ x = 0 \begin{cases} y < 0 : y^2 + x > 0 \\ y = 0 : y^2 + x = 0 \\ y > 0 : y^2 + x > 0 \end{cases} \\ x > 0 \text{ for any } y : \ y^2 + x > 0 \end{cases}$$

6 Making Use of Equational Constraints and Other Optimizations

In this section, we discuss several possible optimizations to algorithms presented in Sect. 4.

Firstly, we discuss how to compute a CAD dedicated to a semi-algebraic system, which provides a systematic solution for making use of equational constraints when computing CADs. The motivation for making use of equational constraints comes from quantifier elimination. Let

$$PF :\equiv (Q_{k+1}x_{k+1} \dots Q_n x_n) FF(x_1, \dots, x_n),$$

be a prenex formula, where FF is a DNF formula. To perform QE by CAD, the first computation step is to collect all the polynomials appearing in FF as a polynomial set F and compute an F-invariant CAD of $\mathbb{R}^n$. This process of computing an F-invariant CAD exhausts all possible sign combinations of F, including those which do not appear in FF, and thus often computes much more than needed for solving the input QE problem. Different techniques in the literature have been proposed for taking advantage of the structure of the input problem. These methods include partial CAD [7] for lazy lifting, simplified projection operator for handling pure strict inequalities [8, 9], smaller projection sets for making use of equational constraints [12–15].

To make the discussion clear, we first quote a paragraph of [12]. "The idea is as follows: if an input formula includes the constraint $f = 0$, then decompose $\mathbb{R}^r$ into regions in which f has invariant sign, and then refine the decomposition so that the other polynomials have invariant sign in those cells in which $f = 0$. The signs of the other polynomials in cells in which $f \neq 0$ are, after all, irrelevant. Additionally, the method of equational constraints seeks to deduce and use constraints that are not explicit in the input formula, but rather arise as consequences of two or more explicit constraints (e.g., if $f = 0$ and $g = 0$ are explicit constraints, then $\mathrm{res}(f, g) = 0$ is also a constraint.)"

This idea, of course, is attractive. Much progress on it has also been made. However, the reason why it is a generally hard problem for CAD is that the framework of PCAD does not have much flexibility to allow propagation of equational constraints. In the world of PCAD, one always tries to obtain a generic projection operator and then applies the same projection operator recursively. To obtain a generic projection operator for handling equational constraints is hard because many problems inherently require different projection operators during projection. Therefore, case discussion is important.

In fact, case discussion is very common in algorithms for computing triangular decompositions. For such algorithms, equational constraints are natural input of these algorithms. The two key ideas "splitting only above $f = 0$" and "if $f = 0$ and $g = 0$ are explicit constraints, then $\text{res}(f, g) = 0$ is also a constraint" have already been systematically taken care of in the Intersect operation of the authors' paper for computing triangular decompositions [26].

Next we explain how to modify algorithms presented in Sect. 4 to automatically implement these ideas.

Suppose now that the input of Algorithm CylindricalDecompose is a system of equations or inequations, this algorithm will then compute a partial cylindrical tree such that its zero set is exactly the zero set of input system. This can be simply achieved by passing an equation or inequation to the function Intersect. W.l.o.g., let us assume that an equation $p = 0$ is passed as an argument of Intersect. Then for this function and all its called subroutines, we will cut the computation branches above which p is known to be nonzero and never proceed with computation branches above which p cannot be zero. For example, we will not create a new vertex at step $15, 32, 42$ in Algorithm IntersectMain. We will delete the vertex V at step 11, 26, 37 since p is nonzero on V.

The first important optimization in IntersectMain which can be implemented is to avoid Squarefree computation at step 3 if $\Gamma.leaf$ is an equational constraint. This idea is quite close to "splitting only above $f = 0$". Another important optimization can be done at step 19 of IntersectMain. Assume that $V.formula$ is an equational constraint $f = 0$, then when Gcd is called, in step 5 of Algorithm 10, we can do as follows. If $i = 0$, then s_i is the resultant of p and f. Thus, we should pass $s_i = 0$ to the IntersectPath operation in order to avoid useless computations on the branch $s_i \neq 0$. This addresses the idea "if $f = 0$ and $g = 0$ are explicit constraints, then $\text{res}(f, g) = 0$ is also a constraint". Moreover, these optimizations are systematically performed during the whole computation.

Next, we briefly mention several other important optimizations. Let V be a leaf of a path Γ of a cylindrical tree. Assume that $V.formula$ is of the form $f \neq 0$ or of the form $f = 0$. We can safely replace f by its primitive part since $\text{lc}(f)$ is invertible modulo Γ_{n-1}. Replacing f by its irreducible factors over $\mathbb{Q}$ is often a more efficient choice. Last but not least, recall that a path Γ in the cylindrical tree is a simple system. Writing Γ as two parts $\Gamma := [T, H]$, where T is a set of equations and H is a set of inequations. We know that T is a regular chain and Γ is a squarefree regular system. Thus, the Zariski closure of Γ is the variety of the saturated ideal of T. We can call the pseudo division operation $\text{prem}(p, T)$ or $\text{prem}(f, T)$ to test whether p or f is zero modulo Γ. And sometimes replacing p by $\text{prem}(p, T)$ and f by $\text{prem}(f, T)$ also ease the computations.

Example 4 Let $F := \{y^2 + x = 0, y^2 + y = 0\}$ be a system of equations. Taking F as input, Algorithm CylindricalDecompose generates a partial cylindrical tree T of $\mathbb{C}^2$ such that the zero set of F is exactly the union of the zero sets of the paths in T, see (Fig. 2).

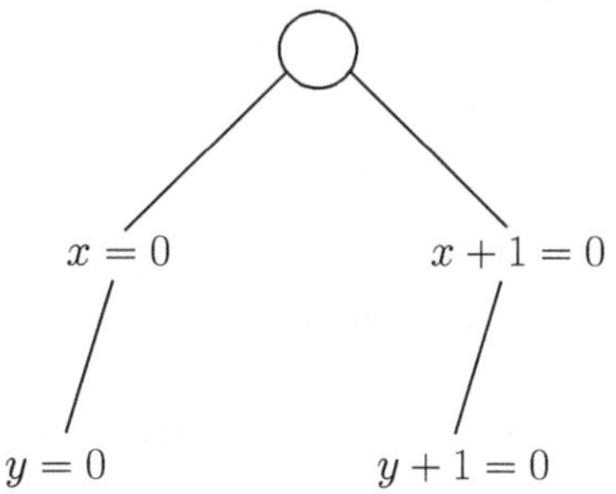

Fig. 2 A partial cylindrical tree T adapted to F

7 Benchmarks

In this section, we report on the experimental results of a preliminary implementation in the `RegularChains` library of MAPLE of the algorithms of Sects. 4 and 5.

The examples in Tables 1 and 2 are from papers on polynomial system solving, such as [38, 39] and the references therein. All the tests were launched on a machine with Intel Core 2 Quad CPU (2.40 GHz) and 8.0 Gb total memory. The time-out is set as 1 h. In the tables, the symbol >1h means time-out.

The MAPLE functions are launched in MAPLE 15 with the latest `RegularChains` library. The memory usage is limited to 60 % of total memory. The software QEPCAD is launched with the option $+N500000000 + L200000$, where the first option specifies the memory to be preallocated (about 23 % of total memory for our machine) and the second option specifies the number of prime numbers to be used.

In Table 1, we report on timings for computing cylindrical decomposition of the complex space with different algorithms and options. Each input system is a set of polynomials. The notation tcd-rec denotes an implementation of the original recursive algorithm in [21], while the notation tcd-inc denotes the incremental algorithm presented in Sect. 4. Both tcd-rec and tcd-inc take a set of polynomials as input. The notation tcd-eqs refers to an optimized version of tcd-inc which makes use of equational constraints, as explained in Sect. 6. With the implementation tcd-eqs, every input polynomial set is regarded as a set of equations (equating each input polynomial to zero). As we can see in Table 1, the incremental algorithm presented in this

Table 1 Timings for computing cylindrical decomposition of the complex space

System	tcd-rec	tcd-inc	tcd-eqs	System	tcd-rec	tcd-inc	tcd-eqs
AlkashiSinus	3373.966	14.568	4.168	MontesS10	>1h	>1h	2.952
Alonso	9.636	1.404	0.700	MontesS12	>1h	>1h	7.528
Arnborg-Lazard-rev	2759.940	2419.543	16.233	MontesS15	>1h	>1h	77.048
Barry	39.346	1.808	0.556	MontesS16	>1h	>1h	8.228
Blood coagulation-2	235.310	9.472	0.808	MontesS4	556.390	102.122	0.488
Bronstein-Wang	255.427	35.990	1.120	MontesS5	1449.810	119.059	1.004
cdc2-cyclin	>1h	68.920	65.976	MontesS7	>1h	>1h	1.060
Circles	276.389	2.280	0.520	MontesS9	269.636	4.212	0.980
genLinSyst-3-2	916.245	19.537	1.384	nql-5-4	>1h	1.056	0.528
genLinSyst-3-3	>1h	160.406	12.408	r-5	68.364	3.232	0.876
Gerdt	>1h	>1h	1.188	r-6	1456.883	46.458	1.200
GonzalezGonzalez	141.072	53.451	0.732	Raksanyi	1471.351	118.227	1.000
hereman-2	>1h	40.042	0.908	Rose	>1h	51.855	1.072
lhlp5	31.069	3.984	0.648	Wang93	>1h	>1h	18.877
Maclane	>1h	>1h	6.420	YangBaxterRosso	54.895	1.560	0.844

paper is much more efficient than the original recursive algorithm. The timings of tcd-eqs show that the optimizations presented in Sect. 6 for making use of equational constraints are very effective.

In Table 2, we report on timings for computing CAD with three different computer algebra packages: QEPCAD, the CylindricalDecomposition command of Mathematica, and the algorithm presented in Sect. 4. Each system is a set of polynomials. Two categories of experimentation are conducted. The first category is concerned with the timings for computing a full CAD of a set of polynomials. For Mathematica, we cannot find any options of CylindricalDecomposition for computing a full CAD of a set of polynomials. Therefore for this category, only the timings of QEPCAD and TCAD are reported. The second category is concerned with the timings for computing a CAD of a variety. For this category, the timings for QEPCAD, Mathematica, and TCAD are all reported.

The notation qepcad denotes computations that QEPCAD performs by (1) treating each input system as a set of nonstrict inequalities and, (2) treating all variables as free variables and, (3) executing with the "full-cad" option. The notation tcad corresponds to computations that TCAD performs by (1) treating each input system as a set of nonstrict inequalities and, (2) computing a sign invariant full CAD of polynomials in the input system and, (3) selecting the cells which satisfy those nonstrict inequalities. In this way, both qepcad and TCAD compute a full CAD of a set of polynomials.

The notation qepcad-eqs denotes the computations that QEPCAD performs by (1) treating each input system as a set of equations and, (2) treating all variables as free variables and, (3) executing with the default option. The notation mathematica-eqs represents computations where the CylindricalDecomposition command of Mathematica treats each input system as a set of equations. The notation

Table 2 Timings for computing CAD

System	qepcad	qepcad-eqs	mathematica-eqs	tcad	tcad-eqs
Alonso	7.516	5.284	0.74	61.591	5.776
Arnborg-Lazard-rev	>1h	>1h	0.952	>1h	17.325
Barry	Fail	216.425	0.032	8.580	1.004
Blood coagulation-2	>1h	>1h	>1h	985.709	7.260
Bronstein-Wang	>1h	>1h	26.726	333.892	2.564
cdc2-cyclin	>1h	>1h	0.208	574.127	503.863
Circles	21.633	5.996	41.211	>1h	40.902
GonzalezGonzalez	10.528	10.412	0.012	214.213	1.136
lhlp2	960.756	5.076	0.016	3.124	0.952
lhlp5	10.300	10.068	0.016	35.338	1.084
MontesS4	>1h	>1h	0.004	2682.391	0.888
MontesS5	Fail	Fail	>1h	>1h	9.400
nql-5-4	93.073	5.420	1303.07	113.675	1.004
r-5	>1h	1802.676	0.016	1282.928	1.208
r-6	>1h	>1h	0.024	>1h	1.500
Rose	Fail	>1h	>1h	606.361	3.136
AlkashiSinus	>1h	>1h	2.232	>1h	58.775
genLinSyst-3-2	Fail	Fail	217.062	3013.764	6.588
MontesS10	>1h	>1h	>1h	>1h	22.797
MontesS12	>1h	>1h	>1h	>1h	330.996
MontesS15	>1h	>1h	0.004	>1h	395.964
MontesS7	>1h	>1h	245.807	>1h	2.452
MontesS9	Fail	Fail	>1h	110.902	4.944
Wang93	Fail	Fail	>1h	>1h	152.673

tcad-eqs corresponds to computations where TCAD treats each input system as a set of equations.

From Table 2, we make the following observations. When full CADs are computed, within one hour time limit, QEPCAD only succeeds on 6 out of 24 examples while TCAD succeeds on 14 out of 24 examples. When CADs of varieties are computed, for all the 10 out of 24 examples that QEPCAD can solve within one hour time limit, both Mathematica and TCAD succeed with usually less time. For the rest 14 examples, TCAD solves all of them while Mathematica only succeeds on 7 of them.

8 Conclusion

In this paper, we present an incremental algorithm for computing CADs. A key part of the algorithm is an Intersect operation for refining a given complex cylindrical tree. If this operation is supplied with an equational constraint, it only computes a partial cylindrical tree, which provides an automatic solution for propagating equational constraints. We have implemented our algorithm in MAPLE. The experimentation shows that the new algorithm is much more efficient than our previous recursive algorithm. We also compared our implementation with the software packages QEPCAD and Mathematica. For many examples, our implementation outperforms the other two. This incremental algorithm can support quantifier elimination. We will present this work in a future paper.

Acknowledgments The authors would like to thank the readers who helped to improve the earlier versions of this paper. This research was supported by the Academic Development Fund ADF-Major-27145 of The University of Western Ontario.

References

1. Collins, G.E.: Quantifier elimination for real closed fields by cylindrical algebraic decomposition. Springer Lect. Notes Comput. Sci. **33**, 515–532 (1975)
2. Arnon, D.S., Collins, G.E., McCallum, S.: Cylindrical algebraic decomposition II: an adjacency algorithm for the plane. SIAM J. Computing **13**(4), 878–889 (1984)
3. Brown, C.W.: Improved projection for cylindrical algebraic decomposition. J. Symb. Comput. **32**(5), 447–465 (2001)
4. Caviness, B., Johnson, J. (eds.): Quantifier Elimination and Cylindical Algebraic Decomposition, Texts and Mongraphs in Symbolic Computation. Springer, Berlin (1998)
5. Hong, H.: An improvement of the projection operator in cylindrical algebraic decomposition. In: ISSAC'90, pp. 261–264. ACM (1990)
6. McCallum, S.: An improved projection operation for cylindrical algebraic decomposition of 3-dimensional space. J. Symb. Comput. **5**(1–2), 141–161 (1988)
7. Collins, G.E., Hong, H.: Partial cylindrical algebraic decomposition. J. Symb. Comput. **12**(3), 299–328 (1991)
8. McCallum, S.: Solving polynomial strict inequalities using cylindrical algebraic decomposition. The Computer Journal **36**(5), 432–438 (1993)
9. Strzeboński, A.: Solving systems of strict polynomial inequalities. J. Symb. Comput. **29**(3), 471–480 (2000)
10. Collins, G.E., Johnson, J.R., Krandick, W.: Interval arithmetic in cylindrical algebraic decomposition. J. Symb. Comput. **34**(2), 145–157 (2002)
11. Dolzmann, A., Seidl, A., Sturm, T.: Efficient projection orders for CAD. In: Proceedings of ISSAC'04, pp. 111–118. ACM (2004)
12. Brown, C.W., McCallum, S.: On using bi-equational constraints in CAD construction. In: ISSAC'05, pp. 76–83 (2005)
13. Collins, G.E.: Quantifier elimination by cylindrical algebraic decomposition—twenty years of progress. In: Caviness, B., Johnson, J., (eds.) Quantifier Elimination and Cylindrical Algebraic Decomposition, pp. 8–23. Springer, Berlin (1998)
14. McCallum, S.: On propagation of equational constraints in CAD-based quantifier elimination. In: Proceedings of ISSAC'01, pp. 223–231 (2001)
15. McCallum, S., Brown, C.W.: On delineability of varieties in CAD-based quantifier elimination with two equational constraints. In: Proceedings of ISSAC'09, pp. 71–78 (2009)
16. Brown, C.W.: Qepcad b: a program for computing with semi-algebraic sets using CADs. SIGSAM Bull. **37**(4), 97–108 (2003)
17. Hong, H., et al.: QEPCAD B, www.usna.edu/Users/cs/qepcad/
18. Strzeboński, A.: Cylindrical algebraic decomposition using validated numerics. J. Symb. Comput. **41**(9), 1021–1038 (2006)

19. Dolzmann, A., Sturm, T.: Redlog computer algebra meets computer logic. ACM SIGSAM Bull. **31**, 2–9 (1996)
20. Iwane, H., Yanami, H., Anai, H., Yokoyama, K.: An effective implementation of a symbolic-numeric cylindrical algebraic decomposition for quantifier elimination. In: Proceedings of SNC'2009, pp. 55–64 (2009)
21. Chen, C., Moreno Maza, M., Xia, B., Yang, L.: Computing cylindrical algebraic decomposition via triangular decomposition. In: ISSAC'09, pp. 95–102 (2009)
22. McCallum, S.: An improved projection operator for cylindrical algebraic decomposition. In: Caviness, B., Johnson, J., (eds.) Quantifier Elimination and Cylindical Algebraic Decomposition, Texts and Mongraphs in Symbolic Computation. Springer (1998)
23. Buchberger, B., Hong, H.: Speeding-up quantifier elimination by Gröbner bases. Technical Report 91–06, RISC (Research Institute for Symbolic Computation), Johannes Kepler University, Linz, Austria, Feb 1991
24. Wilson, D.J., Bradford, R.J., Davenport, J.H.: Speeding up cylindrical algebraic decomposition by Gröbner bases. In: AISC/MKM/Calculemus, pp. 280–294 (2012)
25. Chen, C.: Solving Polynomial Systems via Triangular Decomposition. PhD thesis, University of Western Ontario (2011)
26. Chen, C., Moreno Maza, M.: Algorithms for computing triangular decompositions of polynomial systems. In: Proceedings of ISSAC'11, pp. 83–90 (2011)
27. Moreno Maza, M.: On triangular decompositions of algebraic varieties. Technical Report TR 4/99, NAG Ltd, Oxford, UK, 1999. http://www.csd.uwo.ca/moreno
28. Strzeboński, A.: Computation with Semialgebraic Sets Represented by Cylindrical Algebraic Formulas. In: Proceedings of ISSAC'2010, pp. 61–68, (2010)
29. Dahan, X., Moreno Maza, M., Schost, É., Wu, W., Xie, Y.: Lifting techniques for triangular decompositions. In: ISSAC'05, pp. 108–115. ACM Press (2005)
30. Thomas, J.M.: Differential System. American Mathematical Society, New York (1937)
31. Wang, D.M.: Decomposing polynomial systems into simple systems. J. Symb. Comp. **25**(3), 295–314 (1998)
32. Bächler, T., Gerdt, V., Lange-Hegermann, M., Robertz, D.: Thomas decomposition of algebraic and differential systems. In: Proceedings of CASC'10, pp. 31–54 (2010)
33. Brown, C.W., Davenport, J.H.: The complexity of quantifier elimination and cylindrical algebraic decomposition. In: Proceedings of ISSAC'07, pp. 54–60
34. Wang, D.M.: Computing triangular systems and regular systems. J. Sym. Comp. **30**(2), 221–236 (2000)
35. Ducos, L.: Optimizations of the subresultant algorithm. J. Pure Appl. Algebra **145**, 149–163 (2000)
36. Mishra, B.: Algorithmic Algebra. Springer, New York (1993)
37. Chen, C., Moreno Maza, M.: Algorithms for computing triangular decomposition of polynomial systems. J. Symb. Comput. **47**(6), 610–642 (2012)
38. Boulier, F., Chen, C., Lemaire, F., Moreno Maza, M.: Real root isolation of regular chains. In: Proceedings of ASCM'09, pp. 15–29 (2009)
39. Chen, C., Golubitsky, O., Lemaire, F., Moreno Maza, M., Pan, W.: Comprehensive triangular decomposition. In: Proceedings of CASC'07, vol. 4770 of Lecture Notes in Computer Science, pp. 73–101. Springer (2007)

Finding the Symbolic Solution of a Geometric Problem Through Numerical Computations

Liangyu Chen, Tuo Leng, Liyong Shen, Min Wu, Zhengfeng Yang and Zhenbing Zeng

Abstract In this paper we prove that if L is the maximal perimeter of triangles inscribed in an ellipse with a, b as semi-axes, then

$$(a^2 - b^2)^2 \cdot L^4 - 8(2a^2 - b^2)(2b^2 - a^2)(a^2 + b^2) \cdot L^2 - 432a^4b^4 = 0$$

by accomplishing the following tasks through numeric computations: (1) compute the determinants of matrices of order from 25 to 34 whose entries are polynomials of degree up to 44, (2) construct a series of rectangles $R_1, R_2, \ldots, R_N$ so that if L, a, b satisfies the relation $f(L, a, b) = 0$ then

$$C_1 := \{(b, L) | f(L, 1, b) = 0, 0 \leq b \leq 1\} \subset R_1 \cup R_2 \cup \cdots \cup R_N,$$

This work is supported by the Project No. 20110076110010 from the Ministry of Education of the People's Republic of China.

L. Chen (✉) · M. Wu · Z. Yang
Shanghai Key Laboratory of Trustworthy Computing, East China Normal University, Shanghai 200062, China
e-mail: lychen@sei.ecnu.edu.cn

M. Wu
e-mail: mwu@sei.ecnu.edu.cn

Z. Yang
e-mail: zfyang@sei.ecnu.edu.cn

T. Leng
School of Computer Engineering and Science, Shanghai University, Shanghai 200432, China
e-mail: tleng@shu.edu.cn

L. Shen
School of Mathematical Science, University of Chinese Academy of Sciences, Beijing 100049, China
e-mail: lyshen@ucas.ac.cn

Z. Zeng (✉)
Department of Mathematics, Shanghai University, Shanghai 200432, China
e-mail: zbzeng@shu.edu.cn

© Springer-Verlag Berlin Heidelberg 2014
R. Feng et al. (eds.), *Computer Mathematics*, DOI 10.1007/978-3-662-43799-5_18

and, (3) present a mechanical procedure to decide the validity of

$$R \cap C(F) = \emptyset,$$

where R is a closed rectangle region and $C(F)$ is an algebraic curve defined by $F(x, y) = 0$.

Keywords Symbolic solution · Optimization · Resultant · Lower and upper bounds · Algebraic curves

1 The Problem and Results

In the Sixth ASCM held in October 2003 in Beijing, one of the authors of this paper was asked by a participant of the conference whether or not computer can help to solve the following geometric optimization problem: Given an ellipse with a, b as major-axis and minor-axis, with respectively, let L be the maximal perimeter of triangles inscribed into the ellipse. Find an explicit formula which connects L, a, b [1, 3]. People believe that the solution is a polynomial $f(L, a, b) = 0$ according to the following observation: If the ellipse is expressed by

$$E_{(a,b)} : \quad \frac{x^2}{a^2} + \frac{y^2}{b^2} = 1$$

in the plane, and the following three points:

$$A = \left(\frac{(1-t_1^2)a}{1+t_1^2}, \frac{2t_1 b}{1+t_1^2} \right), \quad B = \left(\frac{(1-t_2^2)a}{1+t_2^2}, \frac{2t_2 b}{1+t_2^2} \right),$$
$$C = \left(\frac{(1-t_3^2)a}{1+t_3^2}, \frac{2t_3 b}{1+t_3^2} \right)$$

form the triangle with maximal perimeter among all triangles inscribed into the ellipse $E_{(a,b)}$, then the triangle PQR formed by the tangent lines T_A, T_B, T_C of the ellipse at points A, B, C, with

$$P = T_B \cap T_C, \quad Q = T_C \cap T_A, \quad R = T_C \cap T_A$$

satisfies that

$$\angle CAQ = \angle BAR, \quad \angle ABR = \angle CBP, \quad \angle BCP = \angle ACQ,$$

as shown in the Fig. 1. For otherwise, if $\angle CAQ \neq \angle BAR$, then the following ellipse

$$K := \{(x, y) \mid PB + PC = AB + AC, P = (x, y)\}$$

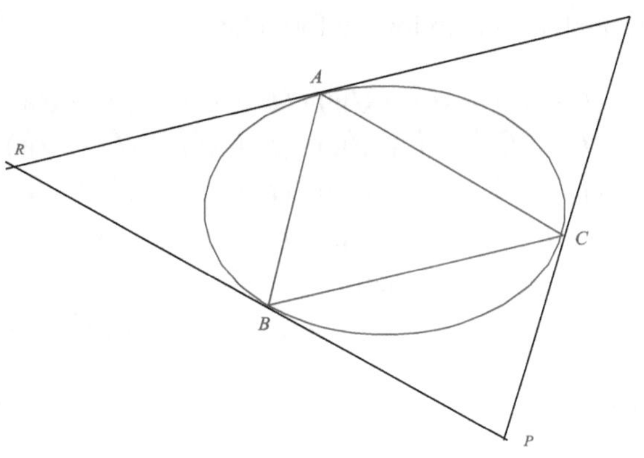

Fig. 1 A necessary condition for $f(L, a, b) = 0$ is $\angle CAQ = \angle BAR$, $\angle ABR = \angle CBP$ and $\angle BCP = \angle ACQ$

intersects the given ellipse $E_{(a,b)}$ transversally at point A, which implies that there is a point $A' \in E_{(a,b)}$ lies in the outside of K, and therefore, $A'B + A'C > AB + AC$, so ABC is not the maximal one. This fact leads to the following three equations:

$$f_1 = \varphi(a, b, t_2, t_3, t_1) = 0, \ f_2 = \varphi(a, b, t_3, t_1, t_2) = 0,$$
$$f_3 = \varphi(a, b, t_1, t_2, t_3) = 0,$$

where

$$\varphi(a, b, t_1, t_2, t_3) := b^2(t_2 + t_3)t_1^4 - 2(b^2(1 + t_2t_3) - 2a^2)t_1^3 + 2(b^2(1 + t_2t_3) - 2a^2t_2t_3)t_1 - b^2(t_2 + t_3).$$

Note that (f_1, f_2, f_3) is not algebraic independent, so the original question can be reduced to solve the following optimization problem:

$$\max \ L = \sum_{1 \le i < j \le 3} \sqrt{a^2\left(\frac{1 - t_i^2}{1 + t_i^2} - \frac{1 - t_j^2}{1 + t_j^2}\right)^2 + b^2\left(\frac{2t_i}{1 + t_i^2} - \frac{2t_j}{1 + t_j^2}\right)^2} \tag{1}$$
$$\text{s.t.} \ \ f_1 = 0, \ f_2 = 0.$$

This problem can be transformed to the following one:

$$\max \ L$$
$$\text{s.t.} \ \ f_0(L, a, b, t_1, t_2, t_3) = 0, \ f_1(a, b, t_1, t_2, t_3) = 0, \ f_2(a, b, t_1, t_2, t_3) = 0,$$
$$-\infty < t_1, t_2, t_3 < +\infty,$$

by applying the following formula:

$$\begin{aligned}
&(L-\sqrt{u}-\sqrt{v}-\sqrt{w})\cdot(L-\sqrt{u}-\sqrt{v}+\sqrt{w})\cdot(L-\sqrt{u}+\sqrt{v}-\sqrt{w})\\
&\cdot(L-\sqrt{u}+\sqrt{v}+\sqrt{w})\cdot(L+\sqrt{u}-\sqrt{v}-\sqrt{w})\cdot(L+\sqrt{u}-\sqrt{v}+\sqrt{w})\\
&\cdot(L+\sqrt{u}+\sqrt{v}-\sqrt{w})\cdot(L+\sqrt{u}+\sqrt{v}+\sqrt{w})\\
&\quad = L^8-4(u+v+w)\cdot L^6+2(3u^2+3v^2+3w^2+2uv+2vw+2wu)\cdot L^4\\
&\quad\ -4((u+v+w)(u^2+v^2+w^2-2uv-2vw-2wu)+16uvw)\cdot L^2\\
&\quad\ +(u^2+v^2+w^2-2uv-2vw-2wu)^2=\psi(L,u,v,w)
\end{aligned}$$

and

$$\begin{aligned}
&u=u(t_1,t_2)=a^2\left(\frac{1-t_1^2}{1+t_1^2}-\frac{1-t_2^2}{1+t_2^2}\right)^2+b^2\left(\frac{2t_1}{1+t_1^2}-\frac{2t_2}{1+t_2^2}\right)^2,\\
&v=u(t_2,t_3),\quad w=u(t_3,t_1).
\end{aligned}$$

to change the objective function $L=\sqrt{u}+\sqrt{v}+\sqrt{w}$ of (1) to a constraint condition $f_0=\psi(L,u,v,w)$. And therefore, for any given a, b, the maximal perimeter L and the parameters (t_1,t_2,t_3) corresponding to the maximal triangle ABC satisfy the following equation system:

$$\begin{cases}
f_{-1}(L,a,b,t_1,t_2,t_3)=0,\\
f_0(L,a,b,t_1,t_2,t_3)=0,\\
f_1(a,b,t_1,t_2,t_3)=0,\\
f_2(a,b,t_1,t_2,t_3)=0,
\end{cases}\tag{2}$$

where

$$f_{-1}(L,a,b,t_1,t_2,t_3)=\begin{vmatrix}
1 & \frac{\partial f_0}{\partial L} & \frac{\partial f_1}{\partial L} & \frac{\partial f_2}{\partial L}\\
0 & \frac{\partial f_0}{\partial t_1} & \frac{\partial f_1}{\partial t_1} & \frac{\partial f_2}{\partial t_1}\\
0 & \frac{\partial f_0}{\partial t_2} & \frac{\partial f_1}{\partial t_2} & \frac{\partial f_2}{\partial t_2}\\
0 & \frac{\partial f_0}{\partial t_3} & \frac{\partial f_1}{\partial t_3} & \frac{\partial f_2}{\partial t_3}
\end{vmatrix}=0,$$

and

$$f_0=\psi(L,u,v,w),\ f_1=\varphi(a,b,t_2,t_3,t_1),\ f_2=\varphi(a,b,t_3,t_1,t_2)$$

are defined as before. So the polynomial $R(L,a,b)=0$ obtained by eliminating t_1,t_2,t_3 from the equation system (2) gives a polynomial relation connecting the maximal perimeter and the semi-axes a, b, that is, $f(L, a, b) | R(L, a, b)$. If

$$R(L, a, b) = c \cdot \prod_{i=1}^{k} r_i(L, a, b)^{d_i}$$

is the irreducible factorization of $R(L, a, b)$, then we define

$$\delta_i := \begin{cases} 0, \text{ if } r_i(L, a, b) \neq f(L, a, b) \text{ for all } a, b, \\ 1, \text{ else.} \end{cases}$$

for $i = 1, 2, \ldots, k$, and call

$$R_0(L, a, b) := \prod_{i=1}^{k} r_i(L, a, b)^{\delta_i}$$

the minimal polynomial of $f(L, a, b)$.

In this paper we devote to compute the elimination and verification problem in this specific example. The paper is organized as follows: In Sect. 2, we investigate the elimination problem, and show that, via interpolations, the result $R(L, a, b)$ is a product of 6 irreducible polynomials $r_i(L, a, b), i = 0, 1, \ldots, 5$. In Sect. 3 we compute the lower bounds $\underline{l}_j$ and upper bounds $\overline{l}_j$ of the maximal perimeter L for $a = 1$ and $b_j = j/256, 0 \leq j \leq 256$ so that the curve

$$C_1 := \{(b, L) | f(L, 1, b) = 0\}$$

can be covered by the union of the following rectangles:

$$R_j := \{(x, y) | b_j \leq x \leq b_{j+1}, \underline{l}_j \leq y \leq \overline{l}_{j+1}\}, \quad j = 0, 1, 2, \ldots, 255.$$

In Sect. 4 we present a mechanical method for checking that a given closed rectangle region R has no intersection with the curve generated by a given polynomial $\Gamma(x, y)$. Applying this procedure to the rectangles $R_j (j = 0, 1, 2, \ldots, 255)$ created in the Sect. 3 and the factors $r_i(L, a, b), i = 1, 2, \ldots, 5$ we prove that the minimal polynomial of $f(L, a, b)$ is $r_0(L, a, b)$.

2 Elimination via Interpolating Resultant

For two polynomials

$$\begin{aligned} P(x, y, \ldots) &= a_m(y, \ldots)x^m + a_{m-1}x^{m-1} + \cdots + a_1(y, \ldots)x + a_0(y, \ldots), \\ Q(x, y, \ldots) &= b_n(y, \ldots)x^n + b_{n-1}x^{n-1} + \cdots + b_1(y, \ldots)x + b_0(y, \ldots), \end{aligned}$$

the Sylvester resultant

$$\text{Resultant}(P, Q, x) := \begin{vmatrix} a_m & a_{m-1} & \cdots & \cdots & a_1 & a_0 & & \\ & a_m & a_{m-1} & \cdots & \cdots & a_1 & a_0 & \\ & & \ddots & & & & & \\ & & & a_m & a_{m-1} & \cdots & \cdots a_1 & a_0 \\ b_n & b_{n-1} & \cdots & b_1 & b_0 & & & \\ & b_n & b_{n-1} & \cdots & b_1 & b_0 & & \\ & & \ddots & & & & & \\ & & & \ddots & & & & \\ & & & & b_n & b_{n-1} & \cdots b_1 & b_0 \end{vmatrix} \begin{matrix} \left.\vphantom{\begin{matrix}a\\a\\a\\a\end{matrix}}\right\} n \text{ rows} \\ \left.\vphantom{\begin{matrix}a\\a\\a\\a\\a\end{matrix}}\right\} m\text{rows} \end{matrix}$$

satisfies that if both $a_m(y, \ldots), b_n(y, \ldots)$ are not zero, then $\text{Resultant}(P, Q, x)$ $(y, \ldots) = 0$ if and only if there exists x_0 such that

$$P(x_0, y, \ldots) = 0, \quad Q(x_0, y, \ldots) = 0.$$

This property can be used to transform the following polynomial equation system $(P = 0, Q = 0)$ into the following three systems:

$$\begin{aligned} &\text{(i)}: \ (P = 0, Q = 0, a_m = 0, b_n = 0), \\ &\text{(ii)}: \ (P = 0, Q = 0, \text{Resultant}(P, Q, x) = 0, a_m \neq 0), \\ &\text{(iii)}: \ (P = 0, Q = 0, \text{Resultant}(P, Q, x) = 0, b_n \neq 0). \end{aligned}$$

Note that in each of the three new systems, at least one polynomial doesn't contain variable x, so this procedure can be regarded as elimination. It is also clear that if P, Q can be factorized, for example,

$$P = c_1 \cdot P_1^{d_1} \cdot P_2^{d_2} \cdot \cdots \cdot P_k^{d_k}, \quad Q = c_2 \cdot Q_1^{d'_1} \cdot Q_2^{d'_2} \cdot \cdots \cdot Q_l^{d'_l},$$

are squarefree factorizations, then

$$\text{f\_Resultant}(P, Q, x) := \prod_{1 \le i \le k, 1 \le j \le l} \text{Resultant}(P_i, Q_j, x)$$

can be used to simplify the computation. To the purpose of calculating the intersecting points in the real plane or space, we can omit the factors of P, Q which are positive or negative definite, namely, if $P = P_0 \cdot P_1$ and $P_0 > 0$ for all points $(x, y, \ldots) \in R^n$, then

$$(P = 0, Q = 0) \Leftrightarrow (P_1 = 0, Q = 0).$$

For the equation system (2) induced from the geometric optimization problem, we have

$$\begin{aligned}
&\text{degs}(f_{-1}) = [8, 10, 10, 12, 12, 9], \quad \text{length}(f_{-1}) = 10818,\\
&\text{degs}(f_0) = [8, 6, 6, 8, 8, 8], \quad \text{length}(f_0) = 1867,\\
&\text{degs}(f_1) = [0, 2, 2, 4, 1, 1], \quad \text{length}(g_{11}) = 10,\\
&\text{degs}(f_2) = [0, 2, 2, 1, 4, 1], \quad \text{length}(g_{21}) = 10,
\end{aligned}$$

here

$$\text{degs}(F) = [\deg(F, L), \deg(F, a), \deg(F, b), \deg(F, t_1), \deg(F, t_2), \deg(F, t_3)].$$

We can eliminate t_3, t_2, t_1 through computing the resultants successively as following. In the first step, eliminate t_3 by:

$$\begin{aligned}
g_0 &:= \text{Resultant}(f_0, f_1, t_3),\\
g_1 &:= \text{Resultant}(f_2, f_1, t_3),\\
g_2 &:= \text{Resultant}(f_{-1}, f_1, t_3),
\end{aligned} \tag{3}$$

factorize the results and remove those explicit nonzero factors (like $t_1 - t_2$ and g shown as below),

$$\begin{aligned}
g_0 &= g_{01} \cdot g_{02},\\
g_1 &= -(t_1 - t_2) \cdot g_{11}(a, b, t_1, t_2),\\
g_2 &= -8g \cdot g_{21}(L, a, b, t_1, t_2),
\end{aligned} \tag{4}$$

where $g := b^2(1 - t_2^2)^2 + 4a^2t_2^2$, and

$$\begin{aligned}
&\text{degs}(g_{01}) = [4, 10, 10, 16, 8, 0], \quad \text{length}(g_{01}) = 582,\\
&\text{degs}(g_{02}) = [4, 12, 12, 24, 8, 0], \quad \text{length}(g_{02}) = 1548,\\
&\text{degs}(g_{11}) = [0, 4, 4, 4, 4, 0], \quad \text{length}(g_{11}) = 25,\\
&\text{degs}(g_{21}) = [8, 26, 26, 44, 21, 0], \quad \text{length}(g_{21}) = 18705.
\end{aligned}$$

In the second step, eliminate t_1 by

$$\begin{aligned}
&h_{01} := \text{Resultant}(g_{01}, g_{11}, t_1),\ h_{02} := \text{Resultant}(g_{02}, g_{11}, t_1),\\
&h_1 := \text{i_Resultant}(g_{21}, g_{11}, t_1),
\end{aligned} \tag{5}$$

where i_Resultant(P, Q, x) means that interpolation is required in the computing of $res(P, Q, x)$. In this step we obtained

$$h_{01} = 1099511627776a^{16}b^{16}(a - b)^4(a + b)^4 \cdot g^{16} \cdot r_0(L, a, b)^2 \cdot h_{011},$$

$$h_{02} = 72057594037927936a^{16}b^{16}(a-b)^{12}(a+b)^{12} \cdot g^{16} \cdot h_{021} \cdot h_{022}^2,$$
$$(h_{021} = h_{011}),$$
$$h_1 = -1424967069597696b^4(b^4 - a^2b^2 + a^4)^2 t_2^4 \cdot g^{42} \cdot r_0(L,a,b)^2 \cdot h_{11} \cdot h_{12}^2 \cdot h_{13},$$

with

$$r_0 = (a-b)^2(a+b)^2 \cdot L^4 - 8(-b^2+2a^2)(-2b^2+a^2)(a^2+b^2) \cdot L^2 - 432a^4b^4, \quad (6)$$

$$h_{11} = 16a^8t_2^4 + (t_2^2-1)^4b^8 + 8a^2b^2t_2^2(t_2^2-1)(2a^4-3a^2b^2+2b^4) > 0,$$
$$h_{12} = 4a^2t_2^2 - b^2t_2^4 - 2b^2t_2^2 + 3b^2,$$

and

$$\text{degs}(h_{011}) = [8, 24, 24, 0, 32, 0], \quad \text{length}(h_{011}) = 734,$$
$$\text{degs}(h_{022}) = [4, 14, 14, 0, 16, 0], \quad \text{length}(h_{022}) = 128,$$
$$\text{degs}(h_{13}) = [12, 34, 34, 0, 36, 0], \quad \text{length}(h_{13}) = 1663.$$

In the third step, eliminate t_2 by doing the following the squarefree resultant:

$$R = r_0 \cdot \text{f\_Resultant}(h_{011} \cdot h_{022}, h_{12}, t_2) \cdot \text{i\_Resultant}(h_{011}, h_{13}, t_2) \cdot \text{i\_Resultant}(h_{022}, h_{13}, t_2). \quad (7)$$

The final eliminating result is

$$R = r_0 \cdot r_1 \cdot \cdots \cdot r_5,$$

where r_0 is as in (6),

$$r_1 = (a-b)^2(a+b)^2L^4 - 8(-2b^2+a^2)(3b^4-3b^2a^2+2a^4)L^2 - 48a^4b^4, \quad (8)$$

and r_2, r_3, r_4, r_5 are polynomials of L, a, b with

$$\text{degs}(r_2) = [8, 22, 24, 0, 0, 0], \quad \text{length}(r_2) = 47,$$
$$\text{degs}(r_3) = [8, 26, 26, 0, 0, 0], \quad \text{length}(r_3) = 51,$$
$$\text{degs}(r_4) = [68, 152, 148, 0, 0, 0], \quad \text{length}(r_4) = 1811,$$
$$\text{degs}(r_5) = [120, 262, 258, 0, 0, 0], \quad \text{length}(r_5) = 5655.$$

Note that in (5) and (7), the resultants can be reduced to compute the determinants of matrices with multivariate polynomials. In (5), we need to compute a determinant of a matrix of order 25 whose entries are polynomials of L, a, b, t_1, t_2 with degree up to 44. In (7), we need to compute the determinants of two matrices of order 26 and 34, whose entries are polynomial of L, a, b, t_2 with degree up to 18. See [2] for more details about the implementation of the interpolation for resultant.

The discussion on the elimination in this section can be summarized as the following result:

Theorem 1 *Let a, b be the semi-axes of an ellipse E and L the largest perimeter of triangles inscribed in E. Then L, a, b satisfies*

$$R(L, a, b) = r_0 \cdot r_1 \ldots r_5 = 0,$$

where r_0, r_1 are given in (6), (8), *with respectively, and*

$$\begin{aligned}
r_2 &= 589824a^8b^{20} + 1624320a^{12}b^{16} + 9L^8a^{20} + \cdots,\\
r_3 &= -10616832a^{14}b^{18} - 589824a^{14}b^{16}L^2 + 5308416a^{12}b^{20} + \cdots,\\
r_4 &= 27660124955458874714672332 8L^{28}a^{42}b^{130}\\
&\quad +186719612622653422385492494605352 96L^{34}a^{90}b^{76}\\
&\quad +14529033298121441055294357504L^2a^{92}b^{106} + \cdots,\\
r_5 &= 3657249285818900176798249985754695770636506479458930488102742917 12a^{160}b^{176}\\
&\quad +5964063145985408875447966286048504266704258730070236358246400 00a^{152}b^{184}\\
&\quad +5769967847054124692787721013381031916316990806053184654540 8L^{88}a^{124}b^{124}\\
&\quad +\cdots.
\end{aligned}$$

In the following two sections, we show that $r_1, r_2, \ldots, r_5$ are extraneous factors so $r_0(L, a, b)$ is the minimal polynomial of $f(L, a, b)$.

3 Lower and Upper Bounds of the Maximal Perimeters

In this section, we will construct a set of rectangles, $R_1, R_2, \ldots, R_N$, so that the point set $\{(L, a, b) | f(L, a, b) = 0\}$ can be covered by the union of these rectangles. It is clear that

$$f(L, a, b) = 0 \Leftrightarrow f(L/a, 1, b/a) = 0,$$

thus we can limit our discussion to the ellipses with $a = 1, 0 < b \leq 1$. The following properties are obvious:

Lemma 1 *If $0 < b_1 < b_2 \leq 1$ and l_1, l_2 satisfy that $f(l_1, 1, b_1) = 0, f(l_2, 1, b_2) = 0$, then $l_1 < l_2$.*

Lemma 2 *If $0 < b_1 < b_2 \leq 1$ and l_1, l_2 satisfy that $f(l_1, 1, b_1) = 0, f(l_2, 1, b_2) = 0$, then for any b, l with $b_1 < b < b_2$, $f(l, 1, b) = 0$ and $\underline{l}_1, \overline{l}_2$ with $\underline{l}_1 \leq l_1, l_2 \leq \overline{l}_2$, we have*

$$(b, l) \subset [b_1, b_2] \times [\underline{l}_1, \overline{l}_2] := \{(x, y) | b_1 \leq x \leq b_2, \underline{l}_1 \leq y \leq \overline{l}_2\}$$

For a fixed number b with $0 < b \leq 1$, we may find a upper bound $\bar{l}$ for the largest perimeter $l(b)$ of triangles inscribed in the ellipse

$$E_b: \quad x^2 + \frac{y^2}{b^2} = 1$$

through the following procedure. For any integer $n > 0$ and real number $b, 0 < b \leq 1$, define

$$l_p(n,b) := \max_{0 \leq i,j,k \leq n} \left\{ A_i B_j + B_j C_k + C_k A_i \mid A_i = \left(\frac{1-\left(\frac{i}{n}\right)^2}{1+\left(\frac{i}{n}\right)^2}, \frac{2b \cdot \left(\frac{i}{n}\right)}{1+\left(\frac{i}{n}\right)^2} \right), \right.$$
$$\left. B_j = \left(-\frac{1-\left(\frac{j}{n}\right)^2}{1+\left(\frac{j}{n}\right)^2}, -\frac{2b \cdot \left(\frac{j}{n}\right)}{1+\left(\frac{j}{n}\right)^2} \right), C_k = \left(\frac{1-\left(\frac{k}{n}\right)^2}{1+\left(\frac{k}{n}\right)^2}, -\frac{2b \cdot \left(\frac{k}{n}\right)}{1+\left(\frac{k}{n}\right)^2} \right) \right\}.$$

Then $l_p(n,b) \leq l(b)$ for all $n \in N$ and $\lim_{n\to\infty} l_p(n,b) = l(b)$. Therefore, if take $n = 500$ and $b_j = j/256$, $j = 0, 1, 2, \ldots, 256$, for example, we can get a sequence $\underline{l}_j := l_p(500, b_j)$, $j = 0, 1, 2, \ldots, 256$ with $\underline{l}_j \leq l(b_j)$. Following data is a part of the sequence of $l_p(500, b)$ for $b = j/256$, $j = 0, 1, 2, \ldots, 256$.

```
[[0,4],
[1/32,  4.000976], [1/16,  4.003910], [3/32, 4.008808],  [1/8, 4.015686],
[5/32,  4.024563], [3/16,  4.035465], [7/32, 4.048423],  [1/4, 4.063475],
[9/32,  4.080663], [5/16,  4.100035], [11/32, 4.121645], [3/8, 4.145549],
[13/32, 4.171811], [7/16,  4.200496], [15/32, 4.231676], [1/2, 4.265420],
[17/32, 4.301804], [9/16,  4.340898], [19/32, 4.382775], [5/8, 4.427501],
[21/32, 4.475137], [11/16, 4.525736], [23/32, 4.579338], [3/4, 4.635974],
[25/32, 4.695657], [13/16, 4.758382], [27/32, 4.824128], [7/8, 4.892855],
[29/32, 4.964504], [15/16, 5.038999], [31/32, 5.116250], [1,   5.196152]]
```

We can construct upper bounds for $l(b)$ from $l_p(n,b)$ according to following estimation:

Lemma 3 *For any integer $n > 0$ and b with $0 < b \leq 1$,*

$$l(b) \leq l_p(n,b) + \frac{6}{n}.$$

Proof It is easy to see that if ABC is a triangle inscribed in the ellipse $x^2 + y^2/b^2 = 1$ and two of A, B, C are on the same quadrant of xOy, the perimeter of ABC is not the largest one. Therefore assume that $A_i A_{i+1}, B_j B_{j+1}, C_k C_{k+1}$ $(0 \leq i, j, k < n)$ are three arcs on the ellipse with

$$A_i = \left(\frac{1-\left(\frac{i}{n}\right)^2}{1+\left(\frac{i}{n}\right)^2}, \frac{2b \cdot \left(\frac{i}{n}\right)}{1+\left(\frac{i}{n}\right)^2} \right), \quad A_{i+1} = \left(\frac{1-\left(\frac{i+1}{n}\right)^2}{1+\left(\frac{i+1}{n}\right)^2}, \frac{2b \cdot \left(\frac{i+1}{n}\right)}{1+\left(\frac{i+1}{n}\right)^2} \right),$$

$$B_j = \left(-\frac{1-\left(\frac{j}{n}\right)^2}{1+\left(\frac{j}{n}\right)^2}, -\frac{2b\cdot\left(\frac{j}{n}\right)}{1+\left(\frac{j}{n}\right)^2}\right), \quad B_{j+1} = \left(-\frac{1-\left(\frac{j+1}{n}\right)^2}{1+\left(\frac{j+1}{n}\right)^2}, -\frac{2b\cdot\left(\frac{j+1}{n}\right)}{1+\left(\frac{j+1}{n}\right)^2}\right),$$

$$C_k = \left(\frac{1-\left(\frac{k}{n}\right)^2}{1+\left(\frac{k}{n}\right)^2}, -\frac{2b\cdot\left(\frac{k}{n}\right)}{1+\left(\frac{k}{n}\right)^2}\right), \quad C_{k+1} = \left(\frac{1-\left(\frac{k+1}{n}\right)^2}{1+\left(\frac{k+1}{n}\right)^2}, -\frac{2b\cdot\left(\frac{k+1}{n}\right)}{1+\left(\frac{k+1}{n}\right)^2}\right),$$

with $A \in \mathrm{arc}(A_iA_{i+1})$, $B \in \mathrm{arc}(B_jB_{j+1})$, $C \in \mathrm{arc}(C_kC_{k+1})$, as shown in the Fig. 2. Then

$$\begin{aligned} AB + BC + CA &\le (AA_i + A_iB) + (BB_j + B_jC) + (CC_k + C_kA) \\ &\le AA_i + (A_iB_j + BB_j) + BB_j + (B_jC_k + CC_k) \\ &\quad + CC_k + (C_kA_i + AA_i) \\ &= A_iB_j + B_jC_k + C_kA_i + 2(AA_i + BB_j + CC_k) \\ &\le l_p(n, b) + 2(AA_i + BB_j + CC_k), \end{aligned}$$

and

$$\begin{aligned} AB + BC + CA &\le A_{i+1}B_{j+1} + B_{j+1}C_{k+1} + C_{k+1}A_{i+1} \\ &\quad + 2(AA_{i+1} + BB_{j+1} + CC_{k+1}) \\ &\le l_p(n, b) + 2(AA_{i+1} + BB_{j+1} + CC_{k+1}), \end{aligned}$$

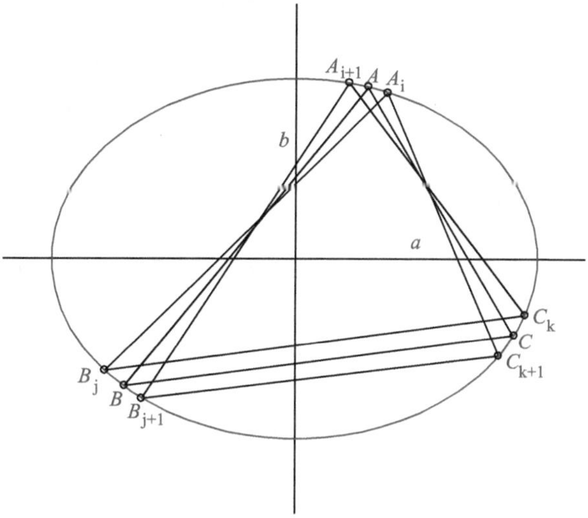

Fig. 2 Perimeter$(ABC) \le$ perimeter$(A_iB_jC_k) + 2(AA_i + BB_j + CC_k)$

which imply that

$$AB + BC + CA \leq l_p(n, b) + (A_iA + AA_{i+1} + B_jB + BB_{j+1} + C_kC + CC_{k+1}).$$

Since $A \in \mathrm{arc}(A_iA_{i+1})$, there exists x, $0 \leq x \leq 1$ such that

$$A = \left(\frac{1 - \left(\frac{i+x}{n}\right)^2}{1 + \left(\frac{i+x}{n}\right)^2}, \frac{2b \cdot \left(\frac{i+x}{n}\right)}{1 + \left(\frac{i+x}{n}\right)^2}\right),$$

Now we consider about $A_iA + AA_{i+1}$. Note that

$$\tan\frac{\theta_1}{2} = \frac{i}{n}, \quad \tan\frac{\theta}{2} = \frac{i+x}{n}, \quad \tan\frac{\theta_2}{2} = \frac{i+1}{n},$$

then $A_i = (\cos\theta_1,\ b\sin\theta_1)$, $A_{i+1} = (\cos\theta_2,\ b\sin\theta_2)$, $A = (\cos\theta,\ b\sin\theta)$.

Suppose the perpendicular line of X axis through the points A_i, A, A_{i+1} intersect the unit circle at M_i, M, M_{i+1}. Notice that $|A_iA| \leq |M_iM|$, $|AA_{i+1}| \leq |MM_{i+1}|$, so

$$\begin{aligned}|A_iA| + |AA_{i+1}| &\leq |M_iM| + |MM_{i+1}| \\ &\leq |\widehat{M_iM_{i+1}}| = |\theta_2 - \theta_1|\end{aligned}$$

By the inequality of common trigonometric function $|\alpha - \beta| \leq |\tan\alpha - \tan\beta|$ (while α, $\beta \in (0, \pi/2)$), hence

$$\begin{aligned}|A_iA| + |AA_{i+1}| \leq |\theta_2 - \theta_1| &= 2\left|\frac{\theta_2}{2} - \frac{\theta_1}{2}\right| \\ &\leq 2\left|\tan\frac{\theta_2}{2} - \tan\frac{\theta_1}{2}\right| \\ &= 2\left|\frac{i+1}{n} - \frac{i}{n}\right| = \frac{2}{n}\end{aligned}$$

Similarly

$$B_jB + BB_{j+1} \leq \frac{2}{n}, \quad C_kC + CC_{k+1} \leq \frac{2}{n}$$

holds for any $B \in \mathrm{arc}(B_jB_{j+1})$, $C \in \mathrm{arc}(C_kC_{k+1})$. Then we get

$$AB + BC + CA \leq l_p(n, b) + \frac{6}{n}$$

holds for any $A \in \mathrm{arc}(A_iA_{i+1})$, $B \in \mathrm{arc}(B_jB_{j+1})$, $C \in \mathrm{arc}(C_kC_{k+1})$ and $0 \leq i, j, k < n$. Lemma 3 is proved. □

Thus, from the obtained lower bound sequence $\underline{l}_j$, $j = 0, 1, 2, \ldots, 256$ we can construct the following sequence of upper bounds $\overline{l}_j$:

$$l(b_j) \le \overline{l}_j = l_p\left(500, \frac{j}{256}\right) + \frac{3}{250}, \; j = 0, 1, 2, \ldots, 256.$$

According to Lemma 2, we have the following inequalities

$$\underline{l}_j = l_p\left(500, \frac{j}{256}\right) \le l(b) \le \overline{l}_{j+1} = l_p\left(500, \frac{j+1}{256}\right) + \frac{3}{250}$$

for all b with $j/256 \le b \le (j+1)/256$, $j = 0, 1, 2, \ldots, 255$. Define R_j, $j = 0, 1, 2, \ldots, 255$ as the following rectangles

$$R_j := \left[\frac{j}{256}, \frac{j+1}{256}\right] \times \left[l_p\left(500, \frac{j}{256}\right), l_p\left(500, \frac{j+1}{256}\right) + \frac{3}{250}\right].$$

Then, we have the following result:

Theorem 2 *Let C_1 be the curve defined by*

$$C_1 := \{(b, L) \mid f(L, 1, b) = 0\}.$$

Then C_1 is covered by the union of $R_0, R_1, \ldots, R_{255}$.

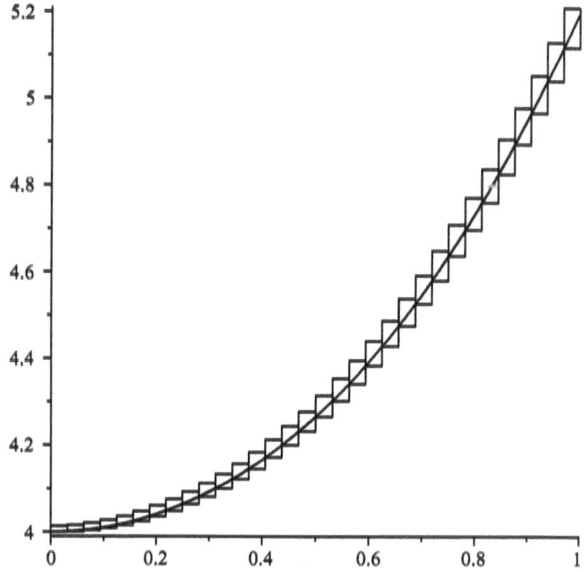

Fig. 3 The 32 rectangles that cover $f(L, 1, b) = 0$ $(0 < b \leqslant 1)$

In Fig. 3 we draw the 32 rectangles that cover $f(L, 1, b) = 0$ from the lower bounds $l_p(500, j/32)$, $j = 0, 1, 2, \ldots, 32$ and the upper bounds $l_p(500, j/32) + 3/250$, $j = 0, 1, \ldots, 32$.

4 Mechanical Verification for the Disjointness of Algebraic Curves to Rectangle Regions

In this section, we will present a general method for proving that a given algebraic curve has no common point with a given rectangle region. The method was established in [4] for mechanization proof to a geometric inequality involving sum of radical functions of three variables, but there is no English introduction to the method until now. Let $F(x_1, x_2, \ldots, x_n)$ be a polynomial with integer coefficients and $R := [a_1, b_1] \times [a_2, b_2] \times \cdots \times [a_n, b_n]$ a rectangle region in the n dimensional Euclidean space. Without loss of generality we may assume that F is expanded into a sum of monomials and $0 \le a_i < b_i$ for $i = 1, 2, \ldots, n$. We claim that if $F(x_1, x_2, \ldots, x_n) = 0$ has no common point with R, then the following procedure terminates after a finite recursive execution.

Procedure: Rectangular-Partition

Input: A 2-tuple (F, R) where F is an expanded polynomial

$$F(x_1, x_2, \ldots, x_n) = \sum_{I=i_1 i_2 \cdots i_n} c_I \cdot {x_1}^{i_1} {x_2}^{i_2} \cdots {x_n}^{i_n},$$

and

$$R := [a_1, b_1] \times [a_2, b_2] \times \cdots \times [a_n, b_n]$$

is a rectangle in R^n with $0 \le a_i < b_i$ for $i = 1, 2, \ldots, n$.

Compute the upper bound of partial derivatives: Let

$$D_i := \frac{\partial F}{\partial x_i}, \quad i = 1, 2, \ldots, n.$$

Apply the sub-procedure Value-Bound to get

$$B_i = B_i(F, R) := \text{Value-Bound}(D_i, [a_1, b_1] \times [a_2, b_2] \times \cdots \times [a_n, b_n])$$

for $i = 1, 2, \ldots, n$.

Partition and Recursive: Let

$$X^0 := \left(\frac{a_1+b_1}{2}, \frac{a_2+b_2}{2}, \ldots, \frac{a_n+b_n}{2}\right)$$

and $F_0 := F(X^0)$. If $F_0 \le 0$ then return -1, else if

$$F_0 > \sum_{i=1}^{n} B_i \cdot (b_i - a_i),$$

then return 1, else construct 2^n sub-rectangles $R_{j_1 j_2 \cdots j_n} (j_k \in \{0, 1\} \text{for} \, k = 1, 2, \ldots, n)$ as follows:

$$R_{j_1 j_2 \cdots j_n} := I_{j_1} \times I_{j_2} \times \cdots \times I_{j_n},$$

in which

$$I_{j_k} = \begin{cases} \left[a_k, \dfrac{a_k+b_k}{2}\right], & \text{if } \; j_k = 0, \\ \left[\dfrac{a_k+b_k}{2}, b_k\right], & \text{if } \; j_k = 1, \end{cases} \qquad (k = 1, 2, \ldots, n),$$

and do the Procedure Rectangular-Partition for the following 2^n pairs of polynomial and rectangles:

$$\text{2-Tuple}(j_1, j_2, \ldots, j_n) : (F, R_{j_1 j_2 \cdots j_n}), \; j_k = 0, 1, k = 1, 2, \ldots, n.$$

Sub-Procedure: Value-Bound of Polynomials

Input: A 2-tuple (F, R) where F is an expanded polynomial

$$F(x_1, x_2, \ldots, x_n) = \sum_{I = i_1 i_2 \cdots i_n} c_I \cdot {x_1}^{i_1} {x_2}^{i_2} \cdots {x_n}^{i_n},$$

and

$$R := [a_1, b_1] \times [a_2, b_2] \times \cdots \times [a_n, b_n]$$

is a rectangle in R^n with $0 \le a_i < b_i$ for $i = 1, 2, \ldots, n$.

Output: Let

$$F^+ = \sum_{C_I > 0} c_I \cdot {x_1}^{i_1} {x_2}^{i_2} \cdots {x_n}^{i_n}, \quad F^- = \sum_{C_I < 0} (-c_I) \cdot {x_1}^{i_1} {x_2}^{i_2} \cdots {x_n}^{i_n},$$

compute

$$d_1 := F^+(a_1, a_2, \ldots, a_n) - F^-(b_1, b_2, \ldots, b_n),$$

$$d_2 := F^+(b_1, b_2, \ldots, b_n) - F^-(a_1, a_2, \ldots, a_n),$$

and return $\max\{|d_1|, |d_2|\}$. It is clear that

$$d_1 \leq F(x_1, x_2, \ldots, x_n) \leq d_2$$

for all $(x_1, x_2, \ldots, x_n) \in [a_1, b_1] \times [a_2, b_2] \times \cdots \times [a_n, b_n]$, and therefore the subprocedure gives an upper bound for $|F(x_1, x_2, \ldots, x_n)|$ over the given rectangle.

We have the following result.

Theorem 3 *Let $F(x_1, x_2, \ldots, x_n)$ be a polynomial and $R := [a_1, b_1] \times [a_2, b_2] \times \cdots \times [a_n, b_n]$ a rectangle in the n-dimensional Euclidean space. Then $F > 0$ for all $(x_1, x_2 \ldots, x_n) \in R$ if and only if the Procedure Rectangle-Partition terminates after finitely many recursive callings.*

Proof If the Procedure Rectangle-Partition does not terminate in any finitely many steps of recursive callings, then there exists a sequence of rectangles

$$R, R_{J_1} = R_{J_1}(R), R_{J_1,J_2} = R_{J_2}(R_{J_1}), \ldots, R_{J_1,J_2,\ldots,J_k} = R_{J_k}(R_{J_1,J_2,\ldots,J_{k-1}}), \ldots$$

so that

$$R_{J_1} \supset R_{J_1,J_2} \supset \cdots \supset R_{J_1,J_2,\ldots,J_k} \supset \cdots$$

and the barycenters $X_{J_1,J_2,\ldots,J_k} := X^0(R_{J_1,J_2,\ldots,J_k})$ of the rectangles satisfy

$$0 \leq F(X_{J_1,J_2,\ldots,J_k}) < \sum_{i=1}^{n} B_i(F, R_{J_1,J_2,\ldots,J_k}) \cdot \frac{b_i - a_i}{2^k}$$

for all $k \geq 1$. Note that

$$B(F, R_{J_1}) \geq B(F, R_{J_1,J_2}) \geq \cdots \geq B(F, R_{J_1,J_2,\ldots,J_k}) \geq \cdots \geq 0,$$

we have

$$0 \leq F(X_{J_1,J_2,\ldots,J_k}) \leq \frac{1}{2^k} \cdot \sum_{i=1}^{n} B_i(b_i - a_i),$$

for all $k \geq 1$, and therefore, $lim_{k \to +\infty} F(X_{J_1,J_2,\ldots,J_k}) = 0$, which implies that $lim_{k \to +\infty} X_{J_1,J_2,\ldots,J_k} = X_0 \in R$ exists and $F(X_0) = 0$. This contradicts to $F(X) > 0$ for all $X \in R$.

Now we prove that if the Procedure Rectangle-Partition terminates after finitely many recursive callings, then $F(X) > 0$ for all $X \in R$. In this case, R is partitioned into a union of finite sub-rectangles $R_1, R_2, \ldots, R_k$ so that on each rectangle R_j,

$$F(X^0(R_j)) > \sum_{i=1}^{n} B_i(F, R_j) \cdot \frac{||x_i(R_j)||}{2}, \tag{9}$$

where $||x_i(R_j)||$ is the diameter of the projection of R_j on x_i-axis, that is,

$$||x_i(R_j)|| = b_i^j - a_i^j, i = 1, 2, \ldots, n$$

provides $R_j = [a_1^j, b_1^j] \times [a_2^j, b_2^j] \times \cdots \times [a_n^j, b_n^j]$. Let $X^0(R_j) = (x_1^0, x_2^0, \ldots, x_n^0)$ be the bary-center of R_j. Then from (9) we prove that for each R_j and any point $(x_1, x_2, \ldots, x_n) \in R_j$ there exists $(\xi_1, \xi_2, \ldots, \xi_n) \in R_j$ such that

$$F(x_1, x_2, \ldots, x_n) = F(x_1^0, x_2^0, \ldots, x_n^0) + \sum_{i=1}^{n} \frac{\partial F}{\partial x_i}(\xi_1, \xi_2, \ldots, \xi_n) \cdot (\xi_i - x_i^0),$$

which implies that

$$\begin{aligned}
|F(x_1, x_2, \ldots, x_n)| &\geq F(x_1^0, x_2^0, \ldots, x_n^0) - |\sum_{i=1}^{n} \frac{\partial F}{\partial x_i}(\xi_1, \xi_2, \ldots, \xi_n) \cdot (\xi_i - x_i^0)| \\
&> \sum_{i=1}^{n} B_i(F, R_j) \cdot \frac{b_i^j - a_i^j}{2} - \sum_{i=1}^{n} |\frac{\partial F}{\partial x_i}(\xi_1, \xi_2, \ldots, \xi_n)| \cdot |\xi_i - x_i^0| \\
&\geq 0,
\end{aligned}$$

in point of view $(\xi_1, \xi_2, \ldots, \xi_n) \in [a_1^j, b_1^j] \times [a_2^j, b_2^j] \times \cdots \times [a_n^j, b_n^j]$ and

$$\begin{aligned}
B_i(F, R_j) &= |D_i^+(b_1^j, b_2^j, \ldots, b_n^j) - D_i^-(a_1^j, a_2^j, \ldots, a_n^j)| \\
&\geq \max\{|\frac{\partial F}{\partial x_i}(x_1, x_2, \ldots, x_n)|, (x_1, x_2, \ldots, x_n) \in R_j\}.
\end{aligned}$$

This proves the Theorem 3. □

According to the following lemma, the Procedure Rectangular-Partition can terminate before recursive callings in the following special case.

Lemma 4 *Let $F(x_1, x_2, \ldots, x_n)$ be a polynomial, $R = [a_1, b_1] \times [a_2, b_2] \times \cdots \times [a_n, b_n]$ a rectangle in the n-dimensional Euclidean space, and ∂R the union of the $n - 1$ dimensional facets of R, i.e.,*

$$\partial R := R|_{x_1=a_1} \cup R|_{x_1=b_1} \cup R|_{x_2=a_2} \cup R|_{x_2=b_2} \cup \cdots \cup R|_{x_n=a_n} \cup R|_{x_n=b_n},$$

where

$$R|_{x_1=a_1} = \{a_1\} \times [a_2, b_2] \times \cdots \times [a_n, b_n],$$
$$R|_{x_1=b_1} = \{b_1\} \times [a_2, b_2] \times \cdots \times [a_n, b_n], \ldots.$$

Then $F(X) > 0$ for all $X \in R$, provided that

1. *$F(X) > 0$ for all $X \in \partial R$,*
2. *there exists no point $X_1 \in R \setminus \partial R$ such that*

$$\frac{\partial F}{\partial x_i}(X_1) = 0 \ \ for \ \ i = 1, 2, \ldots, n.$$

Proof If there exists a point $X_0 \in \text{interior}(R)$ such that $F(X_0) < 0$, then there exists $X_1 \in \text{interior}(R)$ such that $F(X_1) = \min\{F(X), X \in R\} \leq F(X_0) < 0$ and

$$\frac{\partial F}{\partial x_j}(X_1) = 0, \text{for}\, i = 1, 2, \ldots, n,$$

which contradicts to the Assumption 2. This proves Lemma 4. □

For a plane algebraic curve $F(x, y) = 0$ and a rectangle $R := [a_1, a_1] \times [a_2, b_2]$ in the plane, we have

$$\partial R = (\{a_1\} \times [b_1, b_2]) \cup ([a_1, a_2] \times \{b_1\}) \times (\{a_2\} \times [b_1, b_2]) \cup ([a_1, a_2] \times \{b_2\}),$$

and therefore, $F|_{\partial R} > 0$ if and only if $F|_{x_1=a_1}, F|_{x=a_2}$ have no real root in $[b_1, b_2]$ and $F|_{y=b_1}, F|_{y=b_2}$ have no real root in $[a_1, a_2]$. Meanwhile, the number (N) of point set

$$\{(x_1, y_1) \mid (x_1, y_1) \in [a_1, a_2] \times [b_1, b_2], \frac{\partial F}{\partial x}(x_1, y_1) = \frac{\partial F}{\partial y}(x_1, y_1) = 0\}$$

can also be determined by calculating the real root number (N_1) of the univariate equation

$$\text{Resultant}(\frac{\partial F}{\partial x}, \frac{\partial F}{\partial y}, x) = 0$$

in $[b_1, b_2]$ and the real root number (N_2) of the equation

$$\text{Resultant}(\frac{\partial F}{\partial x}, \frac{\partial F}{\partial y}, y) = 0$$

in $[a_1, a_2]$. Namely, if $N_1 = 0$ or $N_2 = 0$, then $N = 0$.

Applying this to the polynomials $r_i(L, 1, b), i = 1, 2, \ldots, 5$ obtained in Theorem 1 and the 256 rectangles R_j, $j = 0, 1, \ldots, 255$ constructed in last section, we get the following result.

Theorem 4 *If L is the largest perimeter of the triangles inscribed in the ellipse with major semi-axis a and minor semi-axis b, then*

$$(a-b)^2(a+b)^2 \cdot L^4 - 8(-b^2+2a^2)(-2b^2+a^2)(a^2+b^2) \cdot L^2 - 432a^4b^4 = 0.$$

Proof For each pair of $F(b, L) = r_i(L, 1, b), i = 1, 2, \ldots, 5$ and $R_j = [b_j, b_{j+1}] \times [\underline{l}_j, \bar{l}_j]$, where

$$b_j = \frac{j}{256}, \ (j = 0, 1, \ldots, 256),$$

and

$$\underline{l}_j = l_p(500, \frac{j}{256}), \quad \bar{l}_j = l_p(500, \frac{j+1}{256}) + \frac{3}{250}, \ (j = 0, 1, \ldots, 255),$$

we can check that

1. the polynomials $F|_{b=b_j}, F|_{b=b_{j+1}}$ have no real root in the interval $[\underline{l}_j, \bar{l}_j]$;
2. the polynomials $F|_{L=\underline{l}_j}, F|_{L=\bar{l}_j}$ have no real root in the interval $[b_j, b_{j+1}]$;
3. the equation system
$$\{\frac{\partial F}{\partial b} = 0, \frac{\partial F}{\partial L} = 0\}$$
has no real root in the rectangle region R_j.

Therefore, each $r_i(L, 1, b) = 0 (i = 1, 2, \ldots, 5)$ has no intersection point with $f(L, 1, b) = 0$. This implies that $r_0 = 0$ is the minimal polynomial of $f(L, a, b)$ and hence Theorem 4 is true. □

References

1. Chen, D.: Mathematical Thinking and Method (in Chinese). Press of Southeast University, Nanjing (2001)
2. Liangyu Chen, Z.Z.: Parallel computation of determinants of matrices with multivariate polynomial entries. Science in China ser. F, (to be published). (2013). doi:10.1007/s11,432-012-4711-7
3. Xiao, W.: Mathematical Proofs (in Chinese). Dalian University of Technology Press, Dalian (2008)
4. Zhenbing Zeng, J.Z.: A mechanical proof to a geometric inequality of zirakzadeh through rectangular partition of polyhedra. J. Sys. Sci. Math. Sci. **30**(11), 1430–1458 (2010)

A Symbolic Approach to Compute a Null-Space Basis in the Projection Method

Mark Giesbrecht and Nam Pham

Abstract We present a hybrid symbolic-numeric approach for the so-called projection method for solving the parameterized differential-algebraic constraint equations associated with multibody mechanical systems. A primary problem in this approach is computing a null-space basis of a matrix of multivariate rational functions, the Jacobian of the symbolic constraint matrix. A purely symbolic approach is untenable in terms of the sheer size of the output, whereas a purely numerical approach does not offer the flexibility of leaving some or all parameters unspecified. Instead we propose a hybrid approach, which does a symbolic preconditioning, followed by representing the null-space basis by straight-line C code, i.e., a black-box null-space basis. We do this in a numerically sensitive way, and show that our black box is numerically robust at almost all parameter settings. This is verified by experimental results on inputs from typical multibody models.

1 Introduction

In recent years, a large amount of work has been devoted to developing symbolic-computation software for modelling and simulating physical systems; for example, MapleSim [15]. The primary motivation has been potential gains over a purely numerical approach that such software could provide engineers in the model-based development process. In this paper, we aim to use and optimize a symbolic approach in solving a constrained mechanical system using the so-called *projection method* [3].

M. Giesbrecht (✉) · N. Pham (✉)
Cheriton School of Computer Science,
University of Waterloo, N2L 3G1, Waterloo, ON, Canada
e-mail: mwg@uwaterloo.ca

N. Pham
e-mail: npham@uwaterloo.ca

© Springer-Verlag Berlin Heidelberg 2014
R. Feng et al. (eds.), *Computer Mathematics*, DOI 10.1007/978-3-662-43799-5_19

Consider a constrained mechanical system represented by a set of m nonlinear algebraic equations:

$$\Phi(q,t) = 0, \tag{1}$$

where q is a set of n generalized coordinates, $q = (x_1, \ldots, x_n)$, and each constraint equation is a function of the coordinates q and the time t. The Jacobian is the $m \times n$ matrix of first partial derivatives of the constraint equations (1):

$$\Phi_q = \left[\frac{\partial(\Phi_1, \Phi_2, \cdots, \Phi_m)}{\partial(x_1, x_2, \cdots, x_n)}\right] = \begin{bmatrix} \frac{\partial\Phi_1}{\partial x_1} & \frac{\partial\Phi_1}{\partial x_2} & \cdots & \frac{\partial\Phi_1}{\partial x_n} \\ \frac{\partial\Phi_2}{\partial x_1} & \frac{\partial\Phi_2}{\partial x_2} & \cdots & \frac{\partial\Phi_2}{\partial x_n} \\ \cdots & \cdots & \ddots & \cdots \\ \frac{\partial\Phi_m}{\partial x_1} & \frac{\partial\Phi_m}{\partial x_2} & \cdots & \frac{\partial\Phi_m}{\partial x_n} \end{bmatrix}. \tag{2}$$

The equations describing the system dynamics in augmented form can then be obtained as

$$M\ddot{x} + \Phi_q^{\mathrm{T}}\lambda = F, \tag{3}$$

where M is an $n \times n$ symmetric generalized mass matrix, Φ_q is the $m \times n$ Jacobian of the constraint matrix Φ with respect to the generalized coordinates, and λ is the $(m \times 1)$ Lagrange multiplier. We will assume that the Jacobian matrix Φ_q above has rank r, which may not be full. This in itself is atypical—full rank is assumed in the standard situation [8]—but is important in the case of automatically generated equations where redundancies may lead to rank deficiencies.

We could solve the $n+m$ differential and algebraic equations formed by (1) and (3) for the coordinate q and the Lagrange multiplier λ by using the appropriate numerical methods. However, solving those nonlinear DAEs is computationally expensive, making it unsuitable for real-time simulation. Hence, it is desirable to express the system equations in a set of purely ordinary differential equations (ODEs).

Reference [2] proposed a new method, called the *projection method*, to eliminate algebraic equations from (3). Specifically, in this method, we find a null-space basis D for Φ_q, an $n \times r$ matrix such that

$$\Phi_q D = 0 \text{ or } D^{\mathrm{T}}\Phi_q^{\mathrm{T}} = 0. \tag{4}$$

By multiplying both sides of (3) by D^T, the resulting equation can be written in an embedded form:

$$D^{\mathrm{T}}M\ddot{x} = D^{\mathrm{T}}F. \tag{5}$$

The tangent velocity u in D is defined as

$$\dot{x} = Du. \tag{6}$$

The determination of the null-space D and its tangent velocity u allows us to analyze the constrained system as two separate problems. First, by substituting (6) into (5), we can derive the dynamic equations of the constrained system in the subspace defined by D:

$$D^{\mathrm{T}} M D \dot{u} = D^{\mathrm{T}}(F - M\dot{D}u). \tag{7}$$

Equations (1) and (7) form a system of $n+r$ purely differential equations in x and u, which can be easily solved using a standard differential equation solver. Given x_0 and v_0, we can compute the required initial tangent speeds for simulation. Secondly, the constraint reactions can be determined at any instant of the simulation by substituting (6) into (3):

$$\lambda = (\Phi_q M^{-1} \Phi_q^{\mathrm{T}})^{-1} \Phi_q (M^{-1} F - \dot{D}u). \tag{8}$$

Thus, the projection method requires an effective algorithm capable of finding a null-space basis D of a large and complicated matrix Φ_q. In the traditional setting, the matrices M and Φ_q are purely numerical, real valued matrices. However, in many applications, these equations depend upon unknown parameters, which would have to be instantiated before the projection method could be applied. A more interesting symbolic approach has been pursued by MapleSim [11, 15], where unknown parameters are simply left as parameters and the equations are handled symbolically, i.e., over the function field defined by the parameters. This has the advantage of completely reflecting knowledge (or lack thereof) of the underlying system, but of course can lead to massive expression swell. The approach we will take here is a hybrid one: from the symbolic system we will generate straight-line code to evaluate solutions to the symbolic system at *any* specific settings of the parameters. This can potentially be used in further manipulations, to reconstruct a symbolic solution (say via sparse interpolation), or as a plugin to a numerical solver. The code produced will (heuristically) be numerically robust in that, assuming reasonable properties of the symbolic system, the straight-line code will have good relative error at most specific parameters.

Numerically, the matrix D is often calculated via the Singular Value Decomposition (SVD) of the Jacobian matrix Φ_q [1]. However, from the symbolic modelling point of view, any numeric processes should be avoided since they require a repeated evaluation of the model. Moreover, a symbolic computation of the SVD (as a symbolic eigenvalue computation) will be both very large, and, when it is used for numerical evaluation, will not necessarily lead to a numerically robust algorithm.

On the other hand, classical symbolic methods (such as fraction-free Gaussian elimination; see, e.g., [4]) to compute the null-space basis will generally work only for small problem instances, or when the number of parameters is very small. The constraint matrix Φ_q in a multi-body system is often condensed with many multivariate functions. For any medium-sized model, symbolic manipulation easily leads to intermediate swell problems that exceed the capabilities of general purpose symbolic software like Maple and Mathematica [13]. To handle the intermediate swell problems, [16] proposed to use fraction-free factoring and implicit reduced involutive form to control the generation of large expression during computation [16, 17].

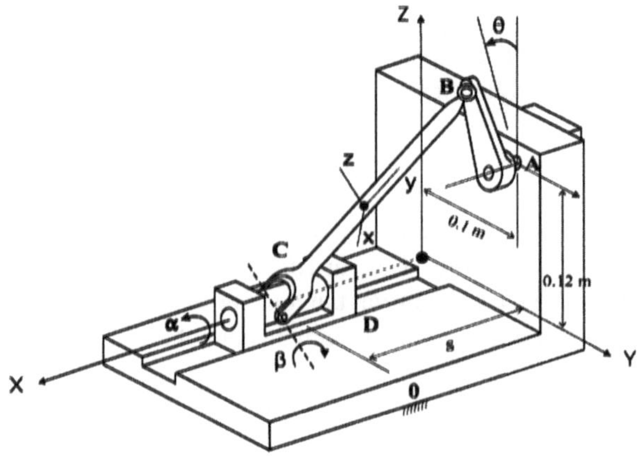

Fig. 1 Slider crank mechanism [7]

In this paper, we make preliminary steps towards a different approach. We aim to develop a hybrid symbolic-numeric computation technique to derive the orthogonal complement matrix D directly. First, we replace the SVD computation by Gaussian elimination with complete row and column pivoting. We then show how to efficiently find row and column permutations of Φ_q such that an LU decomposition of Φ_q proceeds without further pivoting (equivalently, we are determining an *a priori* pivoting strategy we hope is effective almost everywhere). From the permuted matrix, we can then generate straight-line C code to evaluate a clearly defined null-space basis at any settings of the parameters. The method has similarity to the so-called *static pivoting* approach of [10], where a good pivoting strategy is established in advance for the purposes of parallelization.

While we could, at this point, reconstruct a symbolic representation of the null-space (say with sparse interpolation—[5]), it is more likely that the black box is used directly in later numerical solvers. Our method has an important advantage that once the symbolic preconditioning is done, the generated code evaluates a single, unique basis for the null-space.

1.1 Example: Slider Crank Mechanism

In this section, we will introduce the simple slider crank mechanism, which will be used in the following sections to demonstrate the basic ideas of our approach.

As illustrated in Fig. 1 above, a slider-crank consists of three bodies: a crankshaft rotating around the fixed axis, a slider and a connecting rod. The system has one degree of freedom. It is modelled using four generalized coordinates $q = [\alpha, \beta, \theta, s]^T$, which are coupled by three algebraic constraints. Numerical values for system parameters are borrowed from [7].

The Jacobian of the constraint matrix is a 3×4 matrix. Each entry is a function of the coordinates. For example, Φ_q has second column

$$\Phi_q[*, 2] = \begin{bmatrix} -3/10\cos(\beta) \\ -3/10\sin(\beta)\cos(\alpha)\cos(\theta) - 3/10\sin(\beta)\sin(\alpha)\sin(\theta) \\ 3/10\sin(\beta)\cos(\alpha)\sin(\theta) - 3/10\sin(\beta)\sin(\alpha)\cos(\theta) \end{bmatrix}.$$

It is convenient to convert all trigonometric functions (such as sine and cosine) into rational functions by using the tangent half-angle formula $\sin(\alpha) = \frac{2z}{1+z^2}$ and $\cos(\alpha) = \frac{1-z^2}{1+z^2}$ where $z = \tan(\frac{\alpha}{2})$, whence

$$\Phi_q[*; 2] = \begin{bmatrix} \frac{-3}{10} \cdot \frac{1-z_3^2}{1+z_3^2} \\ \frac{-3}{5} \cdot \frac{(1-z_2^2) z_3 (1-z_1^2)}{(1+z_2^2)(1+z_3^2)(1+z_1^2)} - \frac{12}{5} \cdot \frac{z_2 z_1 z_3}{(1+z_2^2)(1+z_3^2)(1+z_1^2)} \\ \frac{6}{5} \cdot \frac{z_2 z_3 (1-z_1^2)}{(1+z_2^2)(1+z_3^2)(1+z_1^2)} - \frac{6}{5} \cdot \frac{(1-z_2^2) z_3 z_1}{(1+z_2^2)(1+z_3^2)(1+z_1^2)} \end{bmatrix}.$$

2 Generating a Straight-Line Program to Compute a Null-Space Basis

Throughout the rest of this paper we will assume that we are working with a matrix A whose entries are rational functions with real (floating point) coefficients. That is, trigonometric and other functions have been replaced by independent indeterminates. This will be represented algebraically by the function field $\mathsf{F} = \mathbb{R}(z_1, z_2, \cdots, z_\ell)$.

Our goal in this section is to demonstrate how to take a matrix $A \in \mathsf{F}^{m \times n}$ and generate efficient code which, at "almost all" settings of the parameters $z_1, \ldots, z_\ell$ from $\mathbb{R}$, produces an evaluation of a specific basis of the null-space of A at those values. The two primary difficulties are that (i) there is no unique null-space basis for A, and (ii) we need to ensure that the output is numerically "good" for any setting of the parameters. A typical symbolic computation of the null-space will require us to make arbitrary decisions on pivots. Under any specialization of the values of $z_1, \ldots, z_\ell$ we must be sure to make the same choices, and hence evaluate the same null-space. We will assume for the remainder of this paper that A has rank r and that entries in A are rational functions whose numerators and denominators have degree less than d.

Our approach is to do a randomized symbolic preconditioning, after which a canonical basis of the null-space is completely determined, even under "almost all" specializations. The idea is similar to the static pivoting strategy of [10], though for entirely different purposes. Moreover, we heuristically proceed in a manner that our choice of pivots is numerically good at "most" numerical specializations.

2.1 Algebraically Effective Static Pivoting

We start by building a preconditioner for A so that all leading $k \times k$ minors are nonsingular, for $1 \le k \le r$. Such a matrix will have a unique, *strict* LU-decomposition, with no need for pivoting choices (i.e., always use the diagonal element as a pivot). An efficient algorithm is provided to do the preconditioning, which involves no expression swell. This is equivalent to recording the pivots in a Gaussian elimination with complete pivoting and applying them every time (but for the efficiency of the generated code we will actually reorder the input matrix). Note that if we only computed the null-space *after* specializing the variables $(z_1, \ldots, z_\ell) \leftarrow (\alpha_1, \ldots, \alpha_\ell) \in \mathbb{R}^\ell$, it would be difficult to determine which basis to choose. In particular, it is important that we are evaluating a *specific* (symbolic) null-space basis at different points $(\alpha_1, \ldots, \alpha_\ell)$, and not simply evaluating a different null-space basis depending upon the parameters (for example, we could not interpolate a symbolic representation from the latter approach).

We start by observing some standard properties of the symbolic matrix A over the function field $\mathsf{F} = \mathbb{R}(z_1, \ldots, z_\ell)$.

Theorem 1 *Let $A \in \mathsf{F}^{m\times n}$ have $m \le n$ and rank r, and be such that all the leading $k \times k$ minors of A are nonzero, for $1 \le k \le r$. Then there exists a* unique *lower triangular matrix $L \in \mathsf{F}^{m\times m}$, a* unique *upper triangular matrix $U \in \mathsf{F}^{n\times n}$, each with only ones on the diagonal, such that $LAU = D$, where $D = \mathrm{diag}(d_1, \ldots, d_r, 0, \ldots, 0) \in \mathsf{F}^{m\times n}$ is a (truncated) diagonal matrix. A basis for the null-space of A is formed by the last $n - r$ columns of U.*

Proof This follows from standard Gaussian elimination with complete (row and column) pivoting. See, for example [12], Sect. 3.10. □

The null-space basis given by the last $n - r$ columns of U in the above theorem is uniquely defined, even up to order; we will refer to it as the *canonical null-space* of A: $w_1, \ldots, w_{n-r} \in \mathsf{F}^{n\times 1}$.

Our first goal then is to find and fix permutation matrices P and Q such that the conditions of the theorem are met by PAQ. Our approach is a very simple randomization technique.

2.1.1 Algebraic Static Pivot Selection

(1) Choose "random" values $\alpha_1, \ldots, \alpha_\ell$ of parameters $z_1, \ldots, z_\ell$ from a finite subset $\mathscr{S} \subseteq \mathbb{C}$;

(2) Return P, Q such that $P \cdot A(\alpha_1, \ldots, \alpha_\ell) \cdot Q$ has an LU-decomposition (without pivoting), using Gaussian elimination with complete pivoting. This is equivalent to recording the choice of pivots used to do the Gaussian elimination with complete pivoting.

Since $P \cdot A(\alpha_1, \ldots, \alpha_\ell) \cdot Q$ has an LU-decomposition, all its leading minors are nonsingular, and hence this must be the case that PAQ is as well. The only thing that could go wrong is that the rank of $A(\alpha_1, \ldots, \alpha_\ell)$ drops below $\operatorname{rank} A$ at this evaluation point. The next theorem shows this happens rarely.

Theorem 2 *Suppose* $A \in \mathbb{R}(z_1, \ldots, z_\ell)^{m \times n}$, *with* $m \leq n$, *has rank* r *and the degrees of all numerators and denominators of entries are less than* d. *Let* $\delta = \mathrm{nm}^2 d \cdot \xi$, *for some user-chosen parameter* ξ, *and* $\mathscr{S}$ *a subset of* $\mathbb{R}$ *with at least* δ *numbers. If* $\alpha_1, \ldots, \alpha_\ell$ *are chosen uniformly and randomly from* $\mathscr{S}$, *then* $\operatorname{rank} A(\alpha_1, \ldots, \alpha_\ell) = \operatorname{rank} A$ *with probability at least* $1 - 1/\xi$.

Proof The product of all denominators in A has degree at most nmd, so we may assume that $A \cdot h \in \mathbb{R}[z_1, \ldots, z_\ell]^{m \times n}$ for some polynomial h of degree at most nmd, and $A \cdot h$ has total degree at most $nm(d+1)$. There exists a nonsingular $r \times r$ minor $\Delta \in \mathbb{R}[z_1, \ldots, z_\ell]$ of $A \cdot h$ which has degree at most $rnm(d+1)$. Thus, if we choose $\alpha_1, \ldots, \alpha_\ell$ such that $h(\alpha_1, \ldots, \alpha_\ell) \neq 0$ and $\Delta(\alpha_1, \ldots, \alpha_\ell) \neq 0$, $\operatorname{rank} A(\alpha_1, \ldots, \alpha_\ell) = \operatorname{rank} A$. It is easy to see that $\deg h + \deg \Delta \leq 3\mathrm{nm}^2 d$, so by the Schwarz–Zippel Lemma [14, 18], the probability that both $(h \cdot \Delta)(\alpha_1, \ldots, \alpha_\ell) \neq 0$ is at least $1 - (nmd + rnm(d+1))/\#S \geq 1 - 1/\xi$. □

A better probability estimate might be obtained by looking more carefully at the degrees of the entries of A and of the common denominator of its entries.

2.2 Numerically Effective Static Pivoting

Since we are concerned also about the numerical properties of our algorithm, it is not reasonable to assume that the choice of points will be effective at all settings of the parameters. To remedy this, we will be more numerically judicious in our choice of static pivots.

First, we recall the following easy theorem about the pivots, assuming that they are all on the diagonal.

Fact 3 Let $A \in \mathbb{R}(z_1, \ldots, z_\ell)^{m \times n}$ be such that all leading $i \times i$ minors are nonsingular, for $1 \leq i \leq m$. Suppose $A = LU$, where $L \in \mathbb{R}(z_1, \ldots, z_\ell)^{m \times m}$ is lower triangular with ones on the diagonal, and $U \in \mathbb{R}(z_1, \ldots, z_\ell)^{m \times n}$ is upper triangular. Then

$$U_{ii} = \det A \begin{bmatrix} 1 \ldots i \\ 1 \ldots i \end{bmatrix} \Big/ \det A \begin{bmatrix} 1 \ldots i-1 \\ 1 \ldots i-1 \end{bmatrix}.$$

In other words the pivots are quotients of minors of A. Whereas from a symbolic point of view, any nonzero pivot is sufficient for effective Gaussian elimination, numerically we want to choose the largest pivot (see [6], Sect. 3.4.8). Our challenge is to estimate the size of the minors/pivots *a priori*. Thus, we will attempt to estimate the "size" of the symbolic pivots via random evaluations. For this, we offer two heuristics: evaluating at all but one variable for each variable in turn (to get an idea of the degree of the pivot), and evaluating at all variables to get an idea of coefficient size. All evaluations are on the unit circle which both supports our analysis and heuristically at least provides some numerical robustness.

2.2.1 Estimating the Degrees of the Chosen Pivots

We first show how to estimate the degree of an unknown, but evaluable, rational function through partial evaluation. This will applied to the pivots in our application. First, the degree of a *univariate* rational function is defined as the degree of the numerator minus the degree of the denominator. Next, the *maximum degree* and *maximum specialized degree* of a multivariate rational function at an evaluation point, respectively, are as defined as follows.

$$\operatorname{maxdeg} f(z_1, \ldots, z_\ell) = \max_i \deg_{z_i} f,$$

$$\operatorname{maxdeg}_{\alpha_1,\ldots,\alpha_\ell} f(z_1, \ldots, z_\ell) = \max \begin{cases} \deg_{z_1} f(z_1, \alpha_2, \ldots, \alpha_\ell), \\ \deg_{z_2} f(\alpha_1, z_2, \ldots, \alpha_\ell), \\ \vdots \\ \deg_{z_\ell} f(\alpha_1, \alpha_2, \ldots, z_\ell). \end{cases}$$

We will regard the maximum degree as a good estimate of the overall size of f, and estimate it by showing that the maximum degree is equal to the maximum specialized degree at most specializations.

Theorem 3 *Let $f \in \mathbb{R}(z_1, \ldots, z_\ell)$ with* total *(of numerator or denominator) degree at most d. Choose a prime $p > 2\mu d\ell$, where $\mu > 1$ is a "confidence" parameter. Let $\omega \in \mathbb{C}$ be a primitive pth root of unity. Then, for uniformly and randomly chosen $e_j \in \{0, \ldots, p-1\}$ we have*

$$\deg_{z_i} f = \deg_{z_i} f(\omega^{e_1}, \ldots, \omega^{e_{i-1}}, z_i, \omega^{e_{i+1}}, \ldots, \omega^{e_\ell}),$$

for all $1 \le i \le \ell$, *with probability at least* $1 - 1/\mu$.

Proof For each i, let $g_i(z_1, \ldots, z_{i-1}, z_{i+1}, \ldots, z_\ell)$ and $h_i(z_1, \ldots, z_{i-1}, z_{i+1}, \ldots, z_\ell)$ be the leading terms of the numerator and denominator f, respectively, as a polynomials in z_i. Then the Schwartz–Zippel Lemma [14, 18] implies that the probability that $g_i(\omega^{e_1}, \ldots, \omega^{e_{i-1}}, \omega^{e_1}, \ldots, \omega^{e_{l-1}}) \neq 0$ and $h_i(\omega^{e_1}, \ldots, \omega^{e_{i-1}}, \omega^{e_1}, \ldots, \omega^{e_{l-1}}) \neq 0$ is at least $1 - d/(\mu d\ell)$. The probability that this is true for at each variable simultaneously is then at least $(1 - 1/(\mu\ell))^\ell > 1 - 1/\mu$. □

2.2.2 Estimating Coefficient Size

For an estimate of the coefficient size of an unknown (but evaluable) multivariate polynomial, the following is useful. First, for

$$f(z_1, \ldots, z_\ell) = \sum_i f_i z_1^{e_{i1}} z_2^{e_{i2}} \cdots z_\ell^{e_{i\ell}},$$

let $\|f\|_2 = \sqrt{\sum_i |f_i|^2}$, the coefficient 2-norm. The next theorem says the expected value of f at an appropriately random point on the unit circle closely approximates $\|f\|_2$ (see also Lemma 3.1 of [9]).

Theorem 4 *Let $f \in \mathbb{R}[z_1, \ldots, z_\ell]$ with $\deg_{z_i} f = d_i$ for $1 \le i \le \ell$. For each i fix a prime $p_i > d_i$, with $p_i \ne p_j$ for $i \ne j$, and let $\omega_i \in \mathbb{C}$ be a primitive p_ith root of unity. Then*

$$\|f\|_2^2 = E\Big[\ |f(\omega_1^{e_1}, \omega_2^{e_2}, \ldots, \omega_\ell^{e_\ell})|^2 : \text{for randomly chosen } 0 \le e_i < p_i\Big].$$

Proof We start by doing a Kronecker-like substitution to convert this to a univariate problem. Let $Q_1, \ldots, Q_\ell$ be the multipliers from the Chinese remainder theorem for $p_1, \ldots, p_\ell$, i.e., $Q_i \equiv 1 \mod p_i$ and $Q_i \equiv 0 \mod p_j$ for $j \ne i$. Let $P = p_1 p_2 \cdots p_\ell$; clearly $Q_i < P$ for $1 \le i \le \ell$. Now let $F(y) = f(y^{Q_1}, y^{Q_2}, \ldots, y^{Q_\ell})$. It is easy to see that there is a 1–1 mapping between nonzero terms in f to nonzero terms in F. Now let ζ be a Pth root of unity, so ζ^{P/p_i} is a p_ith primitive root of unity. Then

$$\|f\|_2^2 = \|F\|_2^2 = \frac{1}{P} \sum_{0 \leqq i < P} |F(\zeta^i)|^2 = \frac{1}{P} \sum_{0 \le i_1 < p_1} \cdots \sum_{0 \le i_\ell < p_\ell} |f(\omega_1^{i_1}, \cdots \omega_\ell^{i_\ell})|^2,$$

where the univariate equality follows from the fact that the Fourier matrix on powers of ζ is orthogonal. □

This theorem is somewhat weaker than we would like: the expected value being indicative of the size does not prove that a random value is indicative. But nonetheless, heuristically we can take this as a good estimate, especially if we make more than one random choice.

2.2.3 Numerical Static Pivot Selection

Suppose $A \in \mathbb{R}(z_1, \ldots, z_\ell)^{m \times n}$, where numerators and denominators have degree at most d_i in variable z_i, and total degree at most d. As above we may assume that we can clear denominators through a common denominator $h \in \mathbb{R}[z_1, \ldots, z_\ell]$ of degree at most nmd_i in each variable z_i, or nmd in total degree, though we will certainly not compute this. Thus, a crude bound on the degree of all pivots is $nm^2(d_i+1)$ in each variable, or $nm^2(d+1)$ in total. Choose a prime $p_0 > 200\,nm^2(d+1)$ and $\omega_0 \in \mathbb{C}$ a p_0th root of unity. Similarly, choose distinct primes $p_i > 2\,nm^2(d_i+1)$, and let $\omega_i \in \mathbb{C}$ be an p_ith root of unity, for $1 \le i \le \ell$.

(1) Choose random evaluation points:

(1.1) For $1 \le j \le \ell$ choose a random $c_j \in \{0, \ldots, p_0-1\}$, and let $\alpha_j^{(0)} = \omega_0^{c_j}$;
(1.2) For $1 \le i \le 4$ and $1 \le j \le \ell$ choose a random $e_j^{(i)} \in \{0, \ldots, p_j-1\}$, and let $\alpha_j^{(i)} = \omega_j^{e_j^{(i)}}$.

(2) Perform Gaussian elimination "simultaneously" with full pivoting on each of the following $(\ell+4)$ matrices:

(2.1) $A(z_1, \alpha_2^{(0)}, \ldots, \alpha_\ell^{(0)})$, $A(\alpha_1^{(0)}, z_2, \ldots, \alpha_\ell^{(0)})$, ..., $A(\alpha_1^{(0)}, \alpha_2^{(0)}, \ldots, z_\ell)$;
(2.2) $A(\alpha_1^{(i)}, \alpha_2^{(i)}, \ldots, \alpha_\ell^{(i)})$, for $i \in \{1, \ldots, 4\}$.

(3) At each stage choose *the same pivot* for all the eliminations, using a pivot that has the highest maxdeg. In case of tie, we choose the one with the largest average evaluation in (2.2). If any choices are zero, or "too small", restart the entire process with different random choices.
(4) Record all the pivot choices and construct permutation matrices $P \in \mathbb{Z}^{m\times m}$, $Q \in \mathbb{Z}^{n\times n}$ such that PAQ has the same elimination with no pivoting.

The idea behind this heuristic strategy is that in each case we estimate the largest pivot as a rational function, first by its degree on each variable, then by its evaluation random points. In particular, by Theorem 3 we expect to get the degree correct with probability at least 99/100. In a tie for highest degree, we expect to find the "largest" pivot by Theorem 4.

In practice we might consider many practical refinements to this approach, especially taking into account the source of the parameters and their meaning in the problem space. For example, some parameters may be known to be large, or near zero. Some parameters may be more important or more numerically sensitive, and hence crucial to get correct. We can also increase the probability of success by considering more choices of random evaluations for our simultaneous Gaussian elimination. We hope to explore these further with application-driven examples, better probabilistic analyses, and better heuristics.

2.3 Example with Numerical Static Pivoting

Continuing the slider crank example from the previous section, we note that the maximum degree in many variable is 2, so we should use 3 distinct primes greater than 216 and proceed to evaluate at random roots of unity of those orders to get evaluations in (1.1) and (1.2).

For the sake of a simple example, we will instead choose $p_0 = 3$, $p_1 = 3$, $p_2 = 5$, and $p_3 = 7$, so $\omega_0 = \exp(2\pi i/3)$, $\omega_1 = \exp(2\pi i/3)$, $\omega_2 = \exp(2\pi i/5)$, and $\omega_3 = \exp(2\pi i/7)$. We then choose $c_1 = 2$, $c_2 = 1$, and $c_3 = 3$ to create three univariate matrices $\Phi_q(z_1, \omega_2^1, \omega_3^3)$, $\Phi_q(\omega_1^2, z_2, \omega_3^3))$, and $\Phi_q(\omega_1^2, \omega_2^1, z_3)$. These matrices are messy to write down, but we note their "degree matrices" as follows:

$$\begin{bmatrix} 0\,2\,0\,2 \\ 2\,2\,2\,0 \\ 2\,2\,2\,0 \end{bmatrix}, \begin{bmatrix} 0\,2\,0\,2 \\ 2\,2\,2\,0 \\ 2\,2\,2\,0 \end{bmatrix}, \begin{bmatrix} 0\,2\,0\,2 \\ 2\,1\,2\,0 \\ 2\,1\,2\,0 \end{bmatrix}.$$

We then perform Gaussian elimination with complete row-column pivoting simultaneously on the three univariate matrices and four random evaluations of $\Phi_q(z_1, z_2, z_3)$:

$$\Phi_q(\omega_1^2, \omega_2^2, \omega_3^2) = \begin{bmatrix} 0.0 & 7.7405\text{e-}12 - 1.4447\text{e-}1i & 0.0 & 1.0 \\ -5.1923\text{e-}1 + 3.7140\text{e-}10i & 1.2421 - 8.6191\text{e-}10i & 3.9562\text{e-}1 - 8.7185\text{e-}2 & 0.0 \\ 3.5456\text{e-}10 + 5.3896\text{e-}1i & -8.5540\text{e-}10 - 1.19671i & -1.4832\text{e-}1 - 4.6630\text{e-}1i & 0.0 \end{bmatrix},$$

$$\Phi_q(\omega_1^1, \omega_2^3, \omega_3^6) = \begin{bmatrix} 0.0 & 4.8246\text{e-}11 - 1.3143i & 0.0 & 1.0 \\ 4.7239 + 1.7945\text{e-}9i & 5.0294 + 2.4527\text{e-}9i & -4.8475 + 8.7185\text{e-}2i & 0.0 \\ -1.7148\text{e-}9 + 4.9033\text{i} & -2.9437 + 4.8454\text{i} & -1.4832\text{e-}1 - 4.9760\text{i} & 0.0 \end{bmatrix},$$

$$\Phi_q(\omega_1^2, \omega_2^1, \omega_3^4) = \begin{bmatrix} 0.0 & 5.2880\text{e-}10 - 0.3761i & 0.0 & 1.0 \\ 4.4241 - 3.2567\text{e-}9i & 5.6790 - 4.7758\text{e-}9i & -4.1005 + 0.3693i & 0.0 \\ -3.3540\text{e-}9 + 4.4400i & -4.8105\text{e-}9 - 5.6586i & 0.3883 - 4.1323i & 0.0 \end{bmatrix},$$

$$\Phi_q(\omega_1^1, \omega_2^3, \omega_3^2) = \begin{bmatrix} 0.0 & 7.7405\text{e-}12 - 0.1444i & 0.0 & 1.0 \\ 0.5192 + 2.6403\text{e-}10i & 1.2421 + 4.1261\text{e-}10i & -0.6428 + 0.0871i & 0.0 \\ -1.9690\text{e-}10 + 0.5389i & -3.8619\text{e-}10 + 1.1967i & -0.1483 - 0.6116i & 0.0 \end{bmatrix}.$$

From this, we ultimately choose permutation matrices P, Q to maximize overall pivot size, in term of their *maxdeg* and then their coefficient sizes, as follows:

$$P = \begin{bmatrix} 0\,1\,0 \\ 0\,0\,1 \\ 1\,0\,0 \end{bmatrix}, \quad Q = \begin{bmatrix} 1\,0\,0\,0 \\ 0\,1\,0\,0 \\ 0\,0\,0\,1 \\ 0\,0\,1\,0 \end{bmatrix}.$$

2.4 Generating Straight-Line Code

Once we have identified permutation matrices P and Q such that PAQ has all leading minors nonsingular, and hence has a unique LU-decomposition as above, the computation of the canonical null-space $w_1, \ldots, w_{n-r} \in \mathsf{F}^{n\times 1}$ is completely determined. In particular, we can then generate straight-line code (i.e., without any conditional statements) which take parameters $\alpha_1, \ldots, \alpha_\ell \in \mathbb{R}$, and will produce $w_i(\alpha_1, \ldots, \alpha_\ell) \in \mathbb{R}^n$ for $1 \le i \le n - r$.

Theorem 5 *Let $A \in m \times n$, with $m \leq n$ and rank r. Suppose that we have determined permutation matrices $P \in \mathbb{Z}^{m \times m}$ and $Q \in \mathbb{Z}^{n \times n}$ such that all leading $k \times k$ leading minors of PAQ are nonsingular, so that the canonical null-space of PAQ is $w_1, \ldots, w_{n-r} \in \mathsf{F}^n$ Then we can generate straight-line (C) code which, given $\alpha_1, \ldots, \alpha_\ell \in \mathbb{R}$, produces*

$$w_1(\alpha_1, \ldots, \alpha_\ell) \in \mathbb{R}^n, \ldots, w_{n-r}(\alpha_1, \ldots, \alpha_\ell) \in \mathbb{R}^n.$$

Proof Assume the input in $\alpha_1, \ldots, \alpha_\ell \in \mathsf{R}$. The code to be generated is simply the code for LU-decomposition without pivoting on the matrix $(PAQ)(\alpha_1, \ldots, \alpha_\ell)$. The null-space basis is precisely the last $n - r$ columns of the computed U, evaluated at $\alpha_1, \ldots, \alpha_\ell$. □

Of course, the generated code is not foolproof. Aside from the (controllably small) probability that the rank of A is incorrectly computed through our randomized specialization (as in Theorem 2), there is also a probability that the instantiation of the parameters lies on a zero or pole of one of the pivots. This will be exceedingly rare, and will immediately cause a floating point exception, which is easy to catch. As in the proof of Theorem 2, any such point must be the root of a specific polynomial (defined by the input system) of degree $O(r^2nmd)$. Hence, the straight-line program will be defined almost everywhere.

Of perhaps greater concern is the numerical robustness of the evaluations provided by the straight-line program. For this our static pivot selection attempted to find pivots which were largest "most" of the time, by estimating the "size" of the pivot functions. Our method is only an effective heuristic, though we are working on stronger probabilistic guarantees and heuristic, as well as numerically robust "fall-back" procedures (i.e., provide black boxes for two distinct bases, one of which is guaranteed to be valid and numerically robust everywhere the input is). Another possibility, explored by [10] is to do some number of steps of iterative refinement on the solution.

3 Implementation and Experimental Results

In this section, we will present some experiment results to illustrate the efficiency of our approach in terms of time and memory management, and to compare its performance with the purely symbolic null-space solver provided by Maple. Note that this comparison is, in some sense, quite unfair: the symbolic solution is a complete solution over the function field, whereas we produce a *black box* which can evaluate that solution. Nonetheless, since the ultimate application is usually to evaluate the solution, we believe this is useful. Moreover, as we will see below, the complete symbolic solution is unattainable for even moderately sized problems.

All experiments are run on Intel Xeon E7330 4 Quad Core CPUS (16 CPUs in total) with 128 GB of RAM. Since the goal of our experiment is to evaluate the usefulness of our technique in real applications, we choose a test set of 4 typical

Table 1 Multibody models from MapleSim

Models	Number of coordinates	Number of constraints	Number of parameters
3D Rigid Slider Crank	4	3	3
Planar Seven Body Mechanism	7	6	7
Quadski Turning	19	11	16
Hydraulic Stewart Platform	24	18	41

Table 2 Running time (in seconds)

Models	Maple's nullspace	Our nullspace
3D Rigid Slider Crank	0.046	0.016
Planar Seven Body Mechanism	0.078	0.031
Quadski Turning	Timeout (>200 s)	0.56
Hydraulic Stewart Platform	Timeout (>200 s)	1.64

multibody models obtained from MapleSim. Refer to Table 1 for the description of each model.

3.1 Time Efficiency

We measured the running time of our algorithm and Maple's NullSpace function to give an idea of the overall performance. Since our algorithm is heuristic in choosing the pivots, its execution time was measured to the average of 5 rounds for each of the models. We also set a time limit of 200 seconds on every call. Table 2 presents the comparison of running time between our algorithm and Maple's NullSpace solver.

We also compare the running time of our algorithm with different numbers of parameters on a same model. Figure 2 shows its performance on the Hydraulic Stewart Platform model. Starting with 41 parameters from the original model, we reduce the number of parameters by substituting numerical values into each of them. When the number of symbolic parameters is reduced to zero, our algorithm is basically run on a purely numerical matrix. It's also worth noting that when the same experiment is performed using Maple's NullSpace, it fails to return an answer in a reasonable time for more than 3 parameters.

Of course, we are not really comparing similar quantities; on one hand Maple is generating a complete solution, whereas we are simply generating C code which

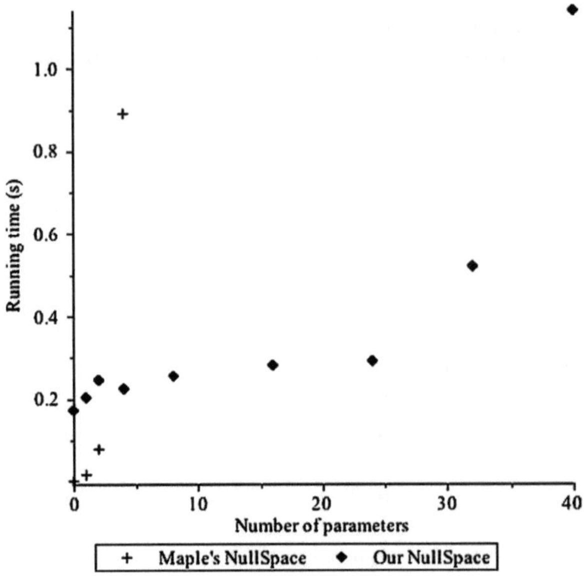

Fig. 2 Running time on hydraulic stewart platformwith different numbers of parameters

can evaluate one specialized solution (in fact the Maple code is remarkably fast in small instances). Nonetheless, the results clearly show that our generated C code is substantially faster than the NullSpace solver in Maple. Much more importantly, we are able to find a null-space basis far beyond when the computing the symbolic solution fails.

3.2 Memory Use

A primary difficulty with the symbolic computation is the need to produce a complete (and very large) presentation of the null-space, and also the need to work with these large entities during the Gaussian elimination. As the size of the intermediate result and the output can grow exponentially, it may well exhaust memory even with a relatively small input, and in particular one which would be quickly evaluable at any specialization of the parameters. We now examine the expected size of the null-space basis that our algorithm generates in term of the length of directed acyclic graph (DAG), which is Maple's internal representation for straight-line code:

Table 3 shows a weakness of our algorithm. Without doing any simplification, the size of the expression in our null-space basis will expand dramatically. However, there is at least some potential for identifying common subexpressions, computing them once, and then re-using these evaluations. While opportunities may be few in a truly generic matrix, in the potentially structured (or at least someone redundant)

Table 3 DAG size of the null-space of Φ_q

Models	Φ_q dimensions	DAG size of D
Planar Slider Crank	3×4	5671
Planar Seven Body Mechanism	6×7	75045
Quadski Turning	11×19	41706824
Hydraulic Stewart Platform	18×24	11849101

matrices from kinematic equations, there may well be many opportunities for such simplifications. Such simplification is, unfortunately, quite nontrivial in general, and a topic of current research.

3.3 Code Generation

A strength of our approach is that the null-space is ultimately computed by straight-line C code. This lends itself to highly optimized compilation. As the null-space is used to computed to the mass matrix $\tilde{M}$ and the force $\tilde{D}$ during simulation and control, it is desirable to port its evaluation to C for numerical computation. The intermediate "symbolic" representation of the straight-line code also has the potential for some optimization. The Maple CodeGeneration package with the *"optimize"* option in Maple can identify at least some common subexpressions while generating the code [13]. For example, in the case of the slider crank mechanism, the null-space basis of Φ_q could be simplified and converted to the following C code for further computation:

```
CodeGeneration[C](NullSpaceD) = t1 = pow(x[2], 0.2e1);
t2 = 0.1e1 + t1; t3 = pow(x[0], 0.2e1); t4 = x[0] * t2;
t5 = 0.5e1 * t4 + 0.3e1 * (-0.1e1 + t3) * t1 + 0.3e1 * t3 - 0.3e1;
t6 = -0.1e1 + t1; t5 =  0.1e1 / t5; t7 = 0.1e1 / t6;
t8 = -0.3e1 + (0.5e1 + 0.3e1 * x[0]) * x[0]; t2 = 0.1e1 / t2;
t8 = 0.1e1 / t8;
cg17[0][0] = (-0.20e2 * t3 + 0.20e2 * t1 + 0.12e2 * t4
                  + 0.10e2 * t1 * t3-0.10e2) * x[2] * t5* t7;
cg17[1][0] = 1; cg17[2][0] = (0.15e2 + 0.15e2 * t3) * x[2] * t8 * t2;
cg17[3][0] = -0.3e1 / 0.10e2 * t6 * t2;
```

A more thorough examination of the preconditioned LU-decomposition we employ, or use of some other equivalent method, will hopefully expose more optimization opportunities.

4 Conclusions and Future Work

In this paper, a new hybrid algorithm to compute the null-space basis of a multivariate matrix has been presented, and employed in the symbolic version of the projection method for solving a constrained mechanical system. The novelty of this algorithm lies in the combination of numeric and symbolic computation to generate fast "black box" code, which can be employed later for numerical or symbolic evaluation. Specifically, instead of computing the null-space directly using LU-decomposition, it is shown that by choosing the ordering of row and column interchanges "randomly", using numerical values from a constrained set, we can emit code to evaluate a symbolic null-space quickly. This is similar to so-called static pivoting schemes. This avoids many of the problems of intermediate expression swell encountered in purely symbolic approaches. Preliminary experiments have been presented to show that this approach is significantly faster than computing null-space symbolically, supporting the use of symbolic computation in engineering problems such as the projection method for multibody systems.

So far we have only done a pilot experiment with small to medium-sized multibody models. It is necessary to test our algorithm with larger models with tens or hundreds of parameters and constraints. There are also several avenues of algorithmic development. First, in the current implementation, we rely on Maple's CodeGeneration package to simplify the straight-line code at the the final step. A more careful analysis of the actual LU-decomposition or equivalent method should lead to identification of common subexpressions. Also, the ultimate "output" of our techniques is C code for a numerical evaluator at specialized values of the parameters. Currently, the numerical robustness of this code is relatively heuristic, estimating the static pivots through randomized evaluation. While we believe this is a good approach, a more rigorous treatment is warranted. Following [10], we will also investigate *post hoc* iterative refinement in cases when numerical stability is questionable.

Acknowledgments The authors thank Dr. Jürgen Gerhard, Maplesoft Inc., for the motivating problem and his many helpful remarks. This work was supported in part by the Natural Sciences and Engineering Research Council of Canada (NSERC) and MITACS Canada.

References

1. Aghili, F., Piedbœuf, J.-C.: Simulation of motion of constrained multibody systems based on projection operator. Multibody Sys. Dyn. **9**(3), 283–309 (2003)
2. Arczewskia, K., Blajer, W.: A unified approach to the modelling of holonomic and nonholonomic mechanical systems. Math. Model. Syst. **2**(3), 157–174 (1996)
3. Blajer, W.: A projection method approach to constrained dynamic analysis. J. Appl. Mech. **59**(3), 643 (1992)
4. Geddes, K.O., Czapor, S.R., Labahn, G.: Algorithms For Computer Algebra. Kluwer Academic Publishers, Alphen aan den Rijn (1992)
5. Giesbrecht, M., Labahn, G., Lee, W-s: Symbolic-numeric sparse interpolation of multivariate polynomials. J. Symbolic Comput. **44**, 943–959 (2009). doi:10.1016/j.jsc.2008.11.003

6. Golub, G., Van Loan, C.: Matrix Computations. Johns Hopkins University Press, Baltimore (1996)
7. Haug, E.J.: Computer Aided Kinematics and Dynamics of Mechanical Systems. volume 1 of Allyn and Bacon series in engineering. Prentice Hall College Div, Upper Saddle River (1989)
8. Jalón, J.G., Bayo, E.: Kinematic and Dynamic Simulation of Multibody Systems: The Real-Time challenge. Springer, New York (1994)
9. Kaltofen, E., Yang, Z., Zhi, L.: On probabilistic analysis of randomization in hybrid symbolic-numeric algorithms. In: Proceedings of SNC'07, pp. 11–17 (2007)
10. Li, X.S., Demmel, J.W.: Making sparse gaussian elimination scalable by static pivoting. In: Proceedings of Supercomputing '98, pp. 1–17 (1998)
11. McPhee, J., Schmitke, C., Redmond, S.: Dynamic modelling of mechatronic multibody systems with symbolic computing and linear graph theory. Math. Comput. Model. Dyn. Syst. **101**(1), 1–23 (2004)
12. Meyer, C.D.: Matrix Analysis and Applied Linear Algebra. SIAM (2000)
13. Moore, B., Piedbœuf, J.-C., Bernardin, L.: Maple as an automatic code generator? Maple Summer Workshop, (2002)
14. Schwartz, J.T.: Fast probabilistic algorithms for verification of polynomial identities. J. Assoc. Comput. Mach. **27**, 701–717 (1980)
15. Waterloo Maple Inc., MapleSim User's Guide. (2011). http://www.maplesoft.com/view.aspx?SF=122742/387839/MapleSimUserGuid.pdf
16. Zhou, W., Jeffrey, D.J., Reid, G.J., Schmitke, C., McPhee, J.: Implicit reduced involutive forms and their application to engineering multibody systems. In IWMM/GIAE, pp. 31–43 (2004)
17. Zhou, W., Carette, J., Jeffrey, D.J., Monagan, M.B.: Hierarchical representations with signatures for large expression management. Proceedings of Artificial Intelligence and Symbolic Computation, Lecture Notes in Computer Science 4120, (2006)
18. Zippel, R.: Probabilistic algorithms for sparse polynomials. In: Proceedings of EUROSAM 79, pp. 216–226, Marseille (1979)

A Simple Quantifier-Free Formula of Positive Semidefinite Cyclic Ternary Quartic Forms

Jingjun Han

Abstract The problem of quantifier elimination of positive semidefinite cyclic ternary quartic forms is studied in this paper. We solve the problem by function `RealTriangularize` in Maple15, the theory of *complete discrimination systems* and the so-called *Criterions on Equality of Symmetric Inequalities method*. The equivalent simple quantifier-free formula is proposed, which is difficult to obtain automatically by previous methods or quantifier elimination tools.

1 Introduction

The elementary theory of real closed fields (RCF) is expressed in a formal language with atomic formulas of the forms $A = B$ and $A > B$, where A and B are multivariate polynomials with integer coefficients. The problem of quantifier elimination (QE) for RCF can be expressed as: for a given formula of RCF, find an equivalent formula containing the same free (unquantified) variables and no quantifiers.

QE is what many researchers have contributed to, including Tarski, who gave a first quantifier elimination method for RCF in the 1930s, although its publishing delayed for nearly 20 years [25], and Collins, who introduced a so-called cylindrical algebraic decomposition (CAD) algorithm for QE problems in the 1970s [12], which has turned into one of the main tools for QE problems, along with its improved variations. Over the years, new algorithms and important improvements on CAD have appeared, including, for instance, [1, 2, 4–7, 10, 13, 19, 20, 22–24, 29]. Most of the works, including Tarskis algorithm, were collected in a book [9].

Many researchers have studied a special quantifier elimination problem (see, for example [3, 10, 21, 28]),

$$(\forall x \in \mathbb{R})[x^4 + px^2 + qx + r \geq 0],$$

J. Han (✉)
Beijing International Center for Mathematical Research and School of Mathematical Sciences, Peking University, Beijing, China
e-mail: hanjingjunfdfz@gmail.com

© Springer-Verlag Berlin Heidelberg 2014
R. Feng et al. (eds.), *Computer Mathematics*, DOI 10.1007/978-3-662-43799-5_20

which is called *quartic problem* in the book just mentioned. There are also many researchers that have studied some special QE problems in other ways. In 1987, Choi etc. obtained the necessary and sufficient condition for the positive semidefiniteness of a symmetric form of degree 3 with n variables [11]. González-Vega etc. proposed a theory on root classification of polynomials in [14] which is based on the Sturm-Habicht sequence and the theory of subresultants. For QE problems in the form $(\forall x)(f(x) \geq 0)$ where the degree of $f(x)$ is a positive even integer, González-Vega proposed a combinatorial algorithm [15] based on the work in [14]. In 1996, Yang etc. proposed the theory of *complete discrimination systems* for polynomials to discuss the root classification problem of one variable polynomial with real parameters [31, 32]. Yang's theory is equivalent to González-Vega's. In 1999, Harris gave a necessary and sufficient condition for the positive semidefiniteness of a symmetric form of degree 4 and 5 with 3 variables [17]. In 2003, Timofte considered the necessary and sufficient condition for the positive semidefiniteness for symmetric forms of degree d with n variables in $\mathbb{R}^n (d \leq 5)$ [26, 27]. By applying Timofte's result and the theory of *complete discrimination systems*, Yao etc. obtained a quantifier elimination of the positive semidefiniteness for symmetric forms of degree d with n variables in $\mathbb{R}^n (d \leq 5)$ [33]. However, the above results are for symmetric forms. Therefore, the author discussed the positive semidefiniteness for more general forms with n variables, including symmetric forms and cyclic forms [16].

In computer-aided geometric design, an interesting problem is to get the quantifier-free formula of *ternary quartic forms problem* [30],

$$(\forall x, y, z \in \mathbb{R})[\sum_{i+j+k=4} a_{ijk} x^i y^j z^k \geq 0].$$

Let $\sum_{\text{cyc}} x^a y^b z^c = x^a y^b z^c + y^a z^b x^c + z^a x^b y^c$, namely the cyclic sum of $x^a y^b z^c$. In this paper, we consider a quantifier-free formula of positive semidefinite cyclic ternary quartic forms, namely the quantifier-free formula of

$$(\forall x, y, z \in \mathbb{R})[F(x, y, z) = \sum_{\text{cyc}} x^4 + k \sum_{\text{cyc}} x^2 y^2 + l \sum_{\text{cyc}} x^2 yz + m \sum_{\text{cyc}} x^3 y + n \sum_{\text{cyc}} xy^3 \geq 0],$$

which is similar to yet also more complex than *quartic problem* and is a special case of *ternary quartic forms problem*. It is difficult to get an answer directly by previous methods or QE tools. Recall Hilbert's 1888 theorem that says, every positive semidefinite ternary quartic (homogeneous polynomial of degree 4 in 3 variables) is a sum of three squares of quadratic forms [18]. Hilbert's proof is non-constructive in the sense that it gives no information about the production of an equivalent quantifier-free formula. Notice that $F(x, y, z) \geq 0$ for $x, y, z \in \mathbb{R}$ is equivalent to the following inequality (see, for example [16]),

$$(\forall x, y, z \in \mathbb{R})[f(x, y, z) = \sigma_1^4 + B\sigma_1^2\sigma_2 + C\sigma_2^2 + D\sigma_1\sigma_3 + E\sigma_1 \sum_{\text{cyc}} x^2 y \geq 0],$$

where $\sigma_1 = x + y + z$, $\sigma_2 = xy + yz + zx$, $\sigma_3 = xyz$ and B, C, D, E satisfying

$$k = 2B + C + E + 6, l = 2C + D + E + 12 + 5B, n = B + 4, m = B + E + 4.$$

The author [16] obtained the following necessary and sufficient condition of $f(x, y, z) \geq 0$,

$$(\forall m \in \mathbb{R})[f(m, 1, km + 1 - k) \geq 0],$$

where k is a real root of the equation

$$Ek^3 - Dk^2 - 3Ek + Dk + E = 0.$$

However, it is still difficult to get a quantifier-free formula by previous methods or QE tools.

The author developed several other methods to solve cyclic and symmetric inequalities including the so-called *Criterions on Equality of Symmetric Inequalities method* [16]. These methods can solve a class of QE problems. This paper is firmly rooted in the author's book [16], especially the technique dealing with the cyclic and symmetric inequalities. In order to be self contained, we will prove some results later in this paper. In order to obtain a simple quantifier-free formula, function `RealTriangularize` [8] of RegularChains package in Maple15 is used to prove inequalities and illustrate semi-algebraic systems without real solution. We also need the theory of *complete discrimination systems* for root classification.

The rest of the paper is organized as follows. Section 2 introduces some basic concepts and results about *complete discrimination systems* for polynomials. Section 3 presents our solution to the positive semidefinite cyclic ternary quartic form.

2 Preliminaries

Given a polynomial

$$f(x) = a_0x^n + a_1x^{n-1} + \cdots + a_n,$$

we write the derivative of $f(x)$ as

$$f'(x) = 0 \cdot x^n + na_0x^{n-1} + (n-1)a_1x^{n-2} + \cdots + a_{n-1}.$$

Definition 1 [31, 32] (*discriminant matrix*) The Sylvester matrix of $f(x)$ and $f'(x)$

$$\begin{vmatrix} a_0 & a_1 & a_2 & \dots & a_n & & & \\ 0 & na_0 & (n-1)a_1 & \dots & a_{n-1} & & & \\ & a_0 & a_1 & \dots & a_{n-1} & a_n & & \\ & 0 & na_0 & \dots & 2a_{n-1} & a_n & & \\ & \vdots & \vdots & \ddots & \vdots & \vdots & & \\ & & & & a_0 & a_1 & \dots & a_n \\ & & & & 0 & na_0 & \dots & a_{n-1} \end{vmatrix}$$

is called the *discrimination matrix* of $f(x)$, and denoted by $Discr(f)$.

Definition 2 [31, 32] (*discriminant sequence*) Denoted by D_k the determinant of the submatrix of $Discr(f)$ formed by the first $2k$ rows and the first $2k$ columns. For $k = 1, \dots, n$, we call the n-tuple

$$\{D_1(f), D_2(f), \dots, D_n(f)\}$$

the *discriminant sequence* of polynomial $f(x)$.

Definition 3 [31, 32] (*sign list*). We call list

$$[\text{sign}(D_1(f)), \text{sign}(D_2(f)), \dots, \text{sign}(D_n(f))]$$

the *sign list* of the *discriminant sequence* $\{D_1(f), D_2(f), \dots, D_n(f)\}$.

Definition 4 [31, 32] (*revised sign list*). Given a *sign list*

$$[s_1, s_2, \dots, s_n],$$

we construct a new list

$$[\varepsilon_1, \varepsilon_2, \dots, \varepsilon_n]$$

as follows (which is called the *revised sign list*): if $[s_i, s_{i+1}, \dots, s_{i+j}]$ is a section of the given list satisfying

$$s_i \neq 0, s_{i+1} = s_{i+2} = \cdots = s_{i+j-1} = 0, s_{i+j} \neq 0, \tag{1}$$

then we replace the subsection

$$[s_{i+1}, s_{i+2}, \dots, s_{i+j-1}]$$

by

$$[-s_i, -s_i, s_i, s_i, -s_i, -s_i, s_i, s_i, \dots],$$

i.e., let

$$\varepsilon_{i+r} = (-1)^{[\frac{r+1}{2}]} \cdot s_i$$

for $r = 1, 2, \ldots, j-1$, where $[a]$ is the floor function i.e., $[a]$ is the greatest integer b such that $b \le a$. Otherwise, $[s_i, s_2, \ldots, s_{i+j}]$ does not satisfy (1), let $\varepsilon_k = s_k$ i.e., no change for other terms.

Lemma 1 [31, 32] *Given a polynomial with real coefficients,* $f(x) = a_0x^n + a_1x^{n-1} + \cdots + a_n$. *If the number of the sign changes of the* revised sign list *of*

$$\{D_1(f), D_2(f), \ldots, D_n(f)\}$$

is v, *then the number of the pairs of distinct conjugate imaginary root of* $f(x)$ *equals* v. *Furthermore, if the number of non-vanishing members of the* revised sign list *is* l, *then the number of the distinct real roots of* $f(x)$ *equals* $l - 2v$.

Theoretically, we can get a quantifier-free formula of the positive semidefinite cyclic ternary quartic form by *complete discrimination systems for polynomials*. But it is impossible because of the complexity.

3 Main Result

Lemma 2 [16] *Let* $x, y, z \in \mathbb{C}$, $x + y + z = 1$ *and* $xy + yz + zx, xyz \in \mathbb{R}$. *The necessary and sufficient condition of* $x, y, z \in \mathbb{R}$ *is* $xyz \in [r_1, r_2]$, *where*

$$r_1 = \frac{1}{27}(1 - 3t^2 - 2t^3), r_2 = \frac{1}{27}(1 - 3t^2 + 2t^3)$$

and $t = \sqrt{1 - 3(xy + yz + zx)} \ge 0$.

Proof We consider the polynomial

$$f(X) = X^3 - (x + y + z)X^2 + (xy + yz + zx)X - xyz,$$

it is obvious that x, y, z are three roots of $f(X) = 0$. By Lemma 1, the equation $f(X) = 0$ has three real roots if and only if

$$D_3(f) \ge 0 \wedge D_2(f) \ge 0,$$

where

$$D_2(f) = (x + y + z)^2 - 3(xy + yz + zx) = 1 - 3(xy + yz + zx),$$
$$D_3(f) = (x - y)^2(y - z)^2(z - x)^2 = \frac{1}{27}(4D_2(f)^3 - (3D_2(f) - 1 + 27xyz)^2).$$

Therefore, using the substitution $t = \sqrt{D_2(f)}$ and $r = xyz$, we have

$$x, y, z \in \mathbb{R} \iff (x-y)^2(y-z)^2(z-x)^2 \geq 0 \wedge (x+y+z)^2 \geq 3xy+3yz+3zx,$$
$$\iff 4t^6 - (3t^2 - 1 + 27r)^2 \geq 0 \wedge t \geq 0$$
$$\iff \frac{1}{27}(1 - 3t^2 - 2t^3) \leq r \leq \frac{1}{27}(1 - 3t^2 + 2t^3) \wedge t \geq 0.$$

That completes the proof. □

Remark 1 The author also got this result by the *Criterions on Equality of Symmetric Inequalities method* [16]. This Lemma implies that if $x, y, z \in \mathbb{R}$ and $x+y+z=1$, then $\sqrt{1-3(xy+yz+zx)} = t \geq 0$, and the range of xyz is $[r_1, r_2]$.

We now try to reduce the number of quantifiers of the positive semidefinite cyclic ternary quartic form which is mentioned in the Introduction,

$$(\forall x, y, z \in \mathbb{R})[F(x,y,z) = \sum_{cyc} x^4 + k\sum_{cyc} x^2y^2 + l\sum_{cyc} x^2yz + m\sum_{cyc} x^3y + n\sum_{cyc} xy^3 \geq 0].$$

Lemma 3 [16] *The inequality $F(x, y, z) \geq 0$ holds for any $x, y, z \in \mathbb{R}$ if and only if*

$$2\sum_{cyc} x^4 + 2k\sum_{cyc} x^2y^2 + 2l\sum_{cyc} x^2yz + (n+m)\sum_{cyc} x^3y + (m+n)\sum_{cyc} xy^3$$
$$\geq |(m-n)(x+y+z)(x-y)(y-z)(z-x)|$$

$x, y, z \in \mathbb{R}$.

Proof It is easy to show that for all $x, y, z \in \mathbb{R}$, $F(x, y, z) \geq 0$ is equivalent to: for all $x, y, z \in \mathbb{R}$,

$$2\sum_{cyc} x^4 + 2k\sum_{cyc} x^2y^2 + 2l\sum_{cyc} x^2yz + (n+m)\sum_{cyc} x^3y + (m+n)\sum_{cyc} xy^3$$
$$\geq (m-n)(x+y+z)(x-y)(y-z)(z-x).$$

On the other hand, if $F(x, y, z) \geq 0$ holds for any $x, y, z \in \mathbb{R}$, then $F(x, z, y) \geq 0$ also holds for any $x, y, z \in \mathbb{R}$. This inequality is equivalent to

$$2\sum_{cyc} x^4 + 2k\sum_{cyc} x^2y^2 + 2l\sum_{cyc} x^2yz + (n+m)\sum_{cyc} x^3y + (m+n)\sum_{cyc} xy^3$$
$$\geq (n-m)(x+y+z)(x-y)(y-z)(z-x)$$

for all $x, y, z \in \mathbb{R}$.

Thus, $F(x, y, z) \geq 0$ for any $x, y, z \in \mathbb{R}$ is equivalent to

$$2\sum_{\text{cyc}} x^4 + 2k\sum_{\text{cyc}} x^2y^2 + 2l\sum_{\text{cyc}} x^2yz + (n+m)\sum_{\text{cyc}} x^3y + (m+n)\sum_{\text{cyc}} xy^3$$
$$\geq |(m-n)(x+y+z)(x-y)(y-z)(z-x)|$$

for all $x, y, z \in \mathbb{R}$. □

Theorem 1 *The positive semidefinite cyclic ternary quartic form*

$$\forall x, y, z \in \mathbb{R} \qquad F(x, y, z) \geq 0$$

holds if and only if the following inequality holds.

$$(\forall t \in \mathbb{R})[g(t) := 3(2+k-m-n)t^4 + 3(4+m+n-l)t^2 + k + 1 + m + n + l$$
$$- \sqrt{27(m-n)^2 + (4k+m+n-8-2l)^2}t^3 \geq 0].$$

Proof Since $\sqrt{27(m-n)^2 + (4k+m+n-8-2l)^2}t^3 \leq 0$ when $t \leq 0$, $(\forall t \geq 0)[g(t) \geq 0]$ implies $(\forall t \in \mathbb{R})[g(t) \geq 0]$, thus they are equivalent. We only need to prove $(\forall x, y, z \in \mathbb{R})[F(x, y, z) \geq 0]$ is equivalent to $(\forall t \geq 0)[g(t) \geq 0]$.

According to Lemma 3, the positive semidefinite cyclic ternary quartic form is equivalent to

$$2\sum_{\text{cyc}} x^4 + 2k\sum_{\text{cyc}} x^2y^2 + 2l\sum_{\text{cyc}} x^2yz + (n+m)\sum_{\text{cyc}} x^3y + (m+n)\sum_{\text{cyc}} xy^3$$
$$\geq |(m-n)(x+y+z)(x-y)(y-z)(z-x)|$$

for all $x, y, z \in \mathbb{R}$.

Substituting $x+y+z, xy+yz+zx, xyz$ with p, q, r, we have

$$\sum_{\text{cyc}} x^4 = p^4 - 4p^2q + 2q^2 + 4pr, \quad \sum_{\text{cyc}} x^2y^2 = q^2 - 2pr,$$
$$\sum_{\text{cyc}} x^2yz = pr, \quad \sum_{\text{cyc}} x^3y + xy^3 = q(p^2 - 2q) - pr,$$

$$|(x-y)(y-z)(z-x)| = \sqrt{(x-y)^2(y-z)^2(z-x)^2}$$
$$= \sqrt{\frac{4(p^2-3q)^3 - (2p^3 - 9pq + 27r)^2}{27}}.$$

The last inequality above becomes

$$\begin{aligned}G(x,y,z)=&2p^4+np^2q-8p^2q+mp^2q+2kq^2-2nq^2-2mq^2\\&+4q^2+2lpr+8pr-npr-mpr-4kpr\\&-|m-n||p|\sqrt{\frac{4(p^2-3q)^3-(2p^3-9pq+27r)^2}{27}}\geq 0.\end{aligned}$$

We first prove the sufficiency.

If $p=0$, then the inequality $G(x,y,z)\geq 0$ becomes

$$2(2+k-m-n)q^2\geq 0.$$

We can deduce $(2+k-m-n)\geq 0$ from $g(t)\geq 0$ for all $t\geq 0$. Since $(2+k-m-n)$ is the *leading coefficient* of $g(t)$.

If $p\neq 0$, since the inequality $F(x,y,z)\geq 0$ is homogenous and the degree of F is an even integer, we can assume that $p=1$. Notice that

$$(x+y+z)^2\geq 3(xy+yz+zx),$$

thus we have $q\leq\frac{1}{3}$. Using the substitution $t=\sqrt{1-3q}\geq 0$, the inequality $G(x,y,z)\geq 0$ is equivalent to

$$\begin{aligned}&2(2+k-m-n)t^4+(16-4k+m+n)t^2-2+2k+m+n\\&\quad+9(8-4k+2l-m-n)r\geq\sqrt{3}|m-n|\sqrt{4t^6-(3t^2-1+27r)^2},\end{aligned}\tag{2}$$

where $t\geq 0$, $r\in[r_1,r_2]$ (r_1 and r_2 are the same as those in the Lemma 2). Since $\frac{2}{3}g(t)\geq 0$ is equivalent to

$$\begin{aligned}&2(2+k-m-n)t^4+(16-4k+m+n)t^2-2+2k+m+n\\&+9(8-4k+2l-m-n)r\geq\frac{2\sqrt{27(m-n)^2+(8-4k+2l-m-n)^2}t^3}{3}\\&+\frac{(8-4k+2l-m-n)(3t^2-1+27r)}{3},\end{aligned}$$

thus, in order to prove $G(x,y,z)\geq 0$, it is sufficient to prove that

$$\begin{aligned}&\sqrt{3}|m-n|\sqrt{4t^6-(3t^2-1+27r)^2}\\&\quad\leq\frac{2\sqrt{27(m-n)^2+(8-4k+2l-m-n)^2}t^3}{3}\\&\quad+\frac{(8-4k+2l-m-n)(3t^2-1+27r)}{3}.\end{aligned}\tag{3}$$

After we square both sides and collect terms, the inequality (3) is equivalent to

$$H^2(r) \geq 0,$$

where

$$H(r) = \frac{2(8-4k+2l-m-n)}{3}t^3 + (3t^2-1+27r)\sqrt{3(m-n)^2+\frac{(8-4k+2l-m-n)^2}{9}}.$$

The inequality (3) is obviously true.

So the sufficiency is proved. Now we prove the necessity, which is equivalent of proving that when the inequality (2) holds for all $x, y, z \in \mathbb{R}$, then $(\forall t \geq 0)$ $[g(t) \geq 0]$. For any $t \geq 0$, if there exist $x, y, z \in \mathbb{R}$ such that $H(r) = 0, x+y+z = 1$ and $1-3(xy+yz+zx) = t^2$, then the equality of inequality (3) could be attained. Choosing such $x, y, z \in \mathbb{R}$, inequality (2) becomes

$$\begin{aligned}
&2(2+k-m-n)t^4 + (16-4k+m+n)t^2 - 2 + 2k + m + n \\
&+ 9(8-4k+2l-m-n)r \geq \frac{2\sqrt{27(m-n)^2+(8-4k+2l-m-n)^2}t^3}{3} \\
&+ \frac{(8-4k+2l-m-n)(3t^2-1+27r)}{3},
\end{aligned}$$

which is equivalent to $(\forall t \geq 0)[g(t) \geq 0]$. Thus, it suffices to show that there exist such $x, y, z \in \mathbb{R}$. Notice that

$$\begin{aligned}
H(r_1)H(r_2) =& \left(\frac{2(8-4k+2l-m-n)}{3}t^3 - 2t^3\sqrt{3(m-n)^2+\frac{(8-4k+2l-m-n)^2}{9}}\right) \\
& \left(\frac{2(8-4k+2l-m-n)}{3}t^3 + 2t^3\sqrt{3(m-n)^2+\frac{(8-4k+2l-m-n)^2}{9}}\right) \\
=& -12t^6(m-n)^2 \leq 0,
\end{aligned}$$

where

$$r_1 = \frac{1}{27}(1-3t^2-2t^3), r_2 = \frac{1}{27}(1-3t^2+2t^3).$$

Therefore, for any given $t = \sqrt{1-3(xy+yz+zx)} \geq 0$, there exists $r_0 \in [r_1, r_2]$, such that $H(r_0) = 0$. By Lemma 2, such $x, y, z \in \mathbb{R}$ exist and we prove the necessity. □

We will apply function `RealTriangularize` of RegularChains package in Maple15 to prove the following lemma.

Lemma 4 *Let $a_0 > 0$, $a_4 > 0$, $a_1 \neq 0$, $a_1, a_2 \in \mathbb{R}$, we consider the following polynomial*

$$f(x) = a_0x^4 + a_1x^3 + a_2x^2 + a_4.$$

The discriminant sequence *of $f(x)$ is*

$$D_f = [D_1(f), D_2(f), D_3(f), D_4(f)],$$

where

$$\begin{aligned}
D_1(f) &= {a_0}^2,\\
D_2(f) &= -8a_0^3a_2 + 3a_1^2a_0^2,\\
D_3(f) &= -4a_0^3a_2^3 + 16a_0^4a_2a_4 + a_0^2a_1^2a_2^2 - 6a_0^3a_1^2a_4,\\
D_4(f) &= -27a_0^2a_1^4a_4^2 + 16a_0^3a_2^4a_4 - 128a_0^4a_2^2a_4^2\\
&\quad - 4a_0^2a_1^2a_2^3a_4 + 144a_0^3a_2a_1^2a_4^2 + 256a_0^5a_4^3.
\end{aligned}$$

For all $x \in \mathbb{R}$, $f(x) \geq 0$ holds if and only if one of the following cases holds,

$$\begin{aligned}
&(1)\ D_4(f) > 0 \wedge (D_2(f) < 0 \vee D_3(f) < 0),\\
&(2)\ D_4(f) = 0 \wedge D_3(f) < 0.
\end{aligned}$$

Proof $\Longrightarrow$: If $f(x) \geq 0$ holds for all $x \in \mathbb{R}$, then the number of distinct real roots of $f(x)$ is less than 2. If it equals 2, then the roots of $f(x)$ are all real. If it equals 0, then $f(x)$ has no real root.

If $D_4(f) < 0$ and $D_2(f) > 0$, then the number of non-vanishing members of *revised sign list*, l, equals 4. Since $D_4(f)D_2(f) < 0$, then the number of the sign changes of *revised sign list*, v, equals 1, thus $l - 2v = 2$. By Lemma 1, the number of distinct real roots of $f(x)$ equals two and the number of the pairs of distinct conjugate imaginary root of $f(x)$, $v = 1$, which is impossible. Using function `RealTriangularize`, we can prove that the semi-algebraic system $a_4 > 0, D_4(f) < 0, D_2(f) \leq 0$ has no real solution. Therefore, $D_4(f) \geq 0$. Since $D_1(f) \geq 0$, the number of the sign changes of *revised sign list* $v \leq 2$.

If $D_4(f) > 0$, thus $l = 4$. Notice that the number of real roots of $f(x)$, namely $l - 2v \leq 2$, so $v \geq 1$, from which, we get

$$D_2(f) \leq 0 \vee D_3(f) \leq 0.$$

Using function `RealTriangularize`, we can prove that both the semi-algebraic system $a_4 > 0, a_0 > 0, D_4(f) > 0, D_2(f) \geq 0, D_3(f) = 0, a_1 \neq 0$ and the semi-algebraic system $a_4 > 0, a_0 > 0, D_4(f) > 0, D_3(f) \geq 0, D_2(f) = 0, a_1 \neq 0$ have no real solution. Hence, if $D_4(f) > 0$ and $D_2(f) = 0$, then $D_3(f) < 0$; if $D_4(f) > 0$ and $D_3(f) = 0$, then $D_2(f) < 0$. Thus, when $D_4(f) > 0$, either $D_2(f) < 0$ or $D_3(f) < 0$ holds.

If $D_4(f) = 0$ and $D_3(f) > 0$, then $l = 3$. The number of sign changes of *revised sign list* v equals either 2 or 0. From $0 \le l - 2v \le 2$, we have $v = 1$, which leads to contradiction. That implies if $D_4(f) = 0$, then $D_3(f) \le 0$. Using function `RealTriangularize`, we can prove that the semi-algebraic system $a_4 > 0$, $a_0 > 0$, $D_4(f) = 0$, $D_3(f) = 0$, $a_1 \ne 0$ has no real solution. Hence, when $D_4(f) = 0$, we have $D_3(f) < 0$.

$\Longleftarrow$: If $D_4(f) > 0 \wedge (D_2(f) < 0 \vee D_3(f) < 0)$, then the number of sign changes of *revised sign list* $v = 2$, so the number of distinct real roots of $f(x)$, $l - 2v$, equals 0, which means for any $x \in \mathbb{R}$, $f(x) > 0$.

If $D_4(f) = 0$ and $D_3(f) < 0$, then $l = 3$, the number of the sign changes of *revised sign list* $v = 2$. Thus, the number of distinct real roots of $f(x)$, $l - 2v$, equals 1, and the number of the pairs of distinct conjugate imaginary root of $f(x)$, v, equals 1, so f has a real root with multiplicity two, which means for any $x \in \mathbb{R}$, $f(x) \ge 0$. □

Now, we can provide a quantifier-free formula of the positive semidefinite cyclic ternary quartic form.

Theorem 2 *Given a cyclic ternary quartic form of real coefficients*

$$F(x, y, z) = \sum_{\text{cyc}} x^4 + k \sum_{\text{cyc}} x^2y^2 + l \sum_{\text{cyc}} x^2yz + m \sum_{\text{cyc}} x^3y + n \sum_{\text{cyc}} xy^3,$$

then

$$(\forall x, y, z \in \mathbb{R}) \qquad [F(x, y, z) \ge 0]$$

is equivalent to

$$\begin{aligned}
&(g_4 = 0 \wedge f_2 = 0 \wedge ((g_1 = 0 \wedge m \ge 1 \wedge m \le 4) \vee (g_1 > 0 \wedge g_2 \ge 0) \vee (g_1 > 0 \wedge g_3 \ge 0 \wedge g_5 \ge 0)))\\
&\vee (g_4^2 + f_2^2 > 0 \wedge f_1 > 0 \wedge f_3 = 0 \wedge f_4 \ge 0)\\
&\vee (g_4^2 + f_2^2 > 0 \wedge f_1 > 0 \wedge f_3 > 0 \wedge ((f_5 > 0 \wedge (f_6 < 0 \vee f_7 < 0)) \vee (f_5 = 0 \wedge f_7 < 0)))
\end{aligned}$$

where

$$\begin{aligned}
f_1 :=& 2 + k - m - n,\ f_2 := 4k + m + n - 8 - 2l,\\
f_3 :=& 1 + k + m + n + l,\ f_4 := 3(1 + k) - m^2 - n^2 - mn,\\
f_5 :=& -4k^3m^2 - 4k^3n^2 - 4k^2lm^2 + 4k^2lmn - 4k^2ln^2\\
&- kl^2m^2 + 4kl^2mn - kl^2n^2 + 8klm^3 + 6klm^2n + 6klmn^2\\
&+ 8kln^3 - 2km^4 + 10km^3n - 3km^2n^2 + 10kmn^3 - 2kn^4\\
&+ l^3mn - 9l^2m^2n - 9l^2mn^2 + lm^4 + 13lm^3n - 3lm^2n^2\\
&+ 13lmn^3 + ln^4 - 7m^5 - 8m^4n - 16m^3n^2 - 16m^2n^3 - 8mn^4\\
&- 7n^5 + 16k^4 + 16k^3l - 32k^2lm - 32k^2ln + 12k^2m^2\\
&- 48k^2mn + 12k^2n^2 - 4kl^3 + 4kl^2m + 4kl^2n - 12klm^2
\end{aligned}$$

$$
\begin{aligned}
&-60klmn-12kln^2+40km^3+48km^2n+48kmn^2+40kn^3\\
&-l^4+10l^3m+10l^3n-21l^2m^2+12l^2mn-21l^2n^2\\
&+10lm^3+48lm^2n+48lmn^2+10ln^3-17m^4-14m^3n\\
&-21m^2n^2-14mn^3-17n^4-16k^3+32k^2l-48k^2m\\
&-48k^2n+80kl^2-48klm-48kln+96km^2+48kmn+96kn^2\\
&-24l^3-24l^2m-24l^2n+24lm^2-24lmn+24ln^2-16m^3\\
&-48m^2n-48mn^2-16n^3-96k^2-64kl+64km+64kn+96l^2\\
&-32lm-32ln-16m^2-32mn-16n^2+64k-128l+64m+64n+128,
\end{aligned}
$$

$$
\begin{aligned}
f_6 :=&\,4k^2+2kl-4km-4kn+l^2-7lm-7ln+13m^2-mn+13n^2\\
&-40k+20l+8m+8n-32,
\end{aligned}
$$

$$
\begin{aligned}
f_7 :=&-768+352k^2-332l^2+180n^2+180m^2+56k^3-8k^4\\
&+14l^3+132n^3+132m^3+42n^4+42m^4-480k-60lmn-192n\\
&+32klmn-192m+912l+l^4-354kmn+158kln+158klm+26k^2mn\\
&-11kln^2+22k^2lm+22k^2ln-45kmn^2-90lm^2n-45km^2n\\
&-11klm^2+23l^2mn-90lmn^2+kl^2m+kl^2n+36mn-480km+592kl\\
&-480kn-60lm-60ln+8k^3m+8k^3n-20k^2l+32k^2n+32k^2m\\
&-12k^3l+234mn^2+234m^2n-192ln^2-258kn^2-192lm^2-258km^2\\
&+116l^2m+116l^2n+87m^3n+87mn^3-15kn^3+90m^2n^2-30ln^3\\
&-15km^3-30lm^3+25l^2m^2+25l^2n^2-14k^2m^2-14k^2n^2\\
&-146kl^2-10l^3m-10l^3n-2k^2l^2+3kl^3,
\end{aligned}
$$

$$
g_1 := k-2m+2,\ g_2 := 4k-m^2-8,\ g_3 := 8+m-2k,\ g_4 = m-n,\ g_5 = k+m-1.
$$

Proof By Theorem 1, it suffices to find a quantifier-free formula of

$$
\begin{aligned}
(\forall t\in\mathbb{R})[g(t) :=&\,3(2+k-m-n)t^4+3(4+m+n-l)t^2+k+1+m+n+l\\
&-\sqrt{27(m-n)^2+(4k+m+n-8-2l)^2}\,t^3\ge 0].
\end{aligned}
$$

Case 1 $\sqrt{27(m-n)^2+(4k+m+n-8-2l)^2}=0$, that is $m=n$ and $4k+m+n-8-2l=0$. Hence

$$
\begin{aligned}
g(t) =&\,3(2+k-2m)t^4+3(4+2m-l)t^2+k+1+2m+l\\
=&\,3(2+k-2m)t^4+3(8+m-2k)t^2+3(k+m-1).
\end{aligned}
$$

If $2+k-2m=0$, then

$$
\forall t\in\mathbb{R}\quad g(t)\ge 0 \Longleftrightarrow 1\le m\le 4.
$$

If $2+k-2m>0$, then

$$\forall t \in \mathbb{R} \quad g(t) \geq 0 \Longleftrightarrow (g_1 > 0 \wedge g_2 \geq 0) \vee (g_1 > 0 \wedge g_3 \geq 0 \wedge g_5 \geq 0).$$

Case 2 $\sqrt{27(m-n)^2+(4k+m+n-8-2l)^2} \neq 0$ and $1+k+m+n+l=0$. In this case, it is easy to show that $2+k-m-n>0$. Thus,

$$\begin{aligned}\forall t \in \mathbb{R}, \quad g(t) \geq 0 \Longleftrightarrow & \forall t \in \mathbb{R}, \quad 3(2+k-m-n)t^2+3(4+m+n-l) \\ & - \sqrt{27(m-n)^2+(4k+m+n-8-2l)^2}t \geq 0 \\ \Longleftrightarrow & 27(m-n)^2+(4k+m+n-8-2l)^2 \\ & \leq 36(2+k-m-n)(4+m+n-l) \\ \Longleftrightarrow & 3(1+k) \geq m^2+n^2+mn.\end{aligned}$$

Case 3 $\sqrt{27(m-n)^2+(4k+m+n-8-2l)^2} \neq 0$ and $1+k+m+n+l \neq 0$. In this case, by Lemma 4, we know that for all $x \in \mathbb{R}$, $g \geq 0$ holds if and only if

$$f_1 > 0 \wedge f_3 > 0 \wedge ((f_5 > 0 \wedge (f_6 \leq 0 \vee f_7 \leq 0)) \vee (f_5 = 0 \wedge f_7 < 0)).$$

To summarize, the theorem is proved. □

Acknowledgments This research was partly supported by President's Fund for Undergraduate Students of Peking University, NSFC-11271034 and the project SYSKF1207 from ISCAS. The author would like to thank the anonymous referees for their valuable comments on a previous version of this paper.

References

1. Arnon, D.S., Collins, G.E., McCallum, S.: Cylindrical algebraic decomposition II: An adjacency algorithm for the plane. SIAM J. Comput. **13**, 878–889 (1984)
2. Arnon, D.S., Collins, G.E., McCallum, S.: Cylindrical algebraic decomposition III. An adjacency algorithm for three-dimensional space. J. Symb. Comput. **5**, 163–187 (1988)
3. Arnon, D.S., Mignotte, M.: On mechanical quantifier elimination for elementary algebra and geometry. J. Symb. Comput. **5**, 237–260 (1988)
4. Brown, C.W.: Simple CAD construction and its applications. J. Symb. Comput. **31**, 521–547 (2001)
5. Brown, C.W.: Improved projection for cylindrical algebraic decomposition. J. Symb. Comput. **32**, 447–465 (2001)
6. Brown, C.W.: Fast simplifications for Tarski formulas based on monomial inequalities. J. Symb. Comput. **7**, 859–882 (2012)
7. Brown, C.W., McCallum, S.: On using bi-equational constraints in CAD construction. In: Kauers, M. (ed.) Proceedings of ISSAC2005, pp. 76–83. ACM Press, New York (2005)
8. Chen, C., Davenport, J.H., May, J.P., Moreno Maza, M., Xia, B., Xiao, R.: Triangular decomposition of semi-algebraic systems. In: Proceedings of ISSAC 2010, ACM Press (2010)
9. Caviness, B.F., Johnson, J.R. (eds.): Quantifier Elimination and Cylindrical Algebraic Decomposition. Springer, Berlin (1998)

10. Collins, G.E., Hong, H.: Partial cylindrical algebraic decomposition for quantifier elimination. J. Symb. Comput. **12**, 299–328 (1991)
11. Choi, M.D., Lam, T.Y., Reznick, B.: Even symmetric sextics. Math. Zeitschr. **195**, 559–580 (1987)
12. Collins, G.E.: Quantifier elimination for real closed fields by cylindrical algebraic decomposition. In: Lecture Notes in Computer Science 33, pp. 134–165. Springer, Berlin (1975)
13. Collins, G.E.: Quantifier elimination by cylindrical algebraic decomposition—20 years of progress. In: Caviness, B., Johnson, J. (eds.) Quantifier Elimination and Cylindrical Algebraic Decomposition, pp. 8–23. Springer, New York (1998)
14. González-Vega, L., Lombardi, H., Recio, T., Roy, M.F.: Sturm-habicht sequence. In: Proceedings of ISSAC89, pp. 136–146. ACM Press (1989)
15. González-Vega, L.: A combinatorial algorithm solving some quantifier elimination problems. In: Caviness, B., Johnson, J. (eds.) Quantifier Elimination and Cylindrical Algebraic Decomposition, pp. 365–375. Springer, New York (1998)
16. Han, J.J.: An introduction to the proving of elementary inequalities. Harbin: Harbin Institute of Technology Press, pp. 234–266 (2011) (in Chinese)
17. Harris, W.R.: Real even symmetric ternary forms. J. Algebra **222**, 204–245 (1999)
18. Hilbert, D.: Über die Darstellung definiter Formen als Summe von Formenquadraten. Math. Ann. **32**, 342–350 (1888)
19. Hong, H.: An improvement of the projection operator in cylindrical algebraic decomposition. In: Watanabe, S., Nagata, M. (eds.) Proceedings of ISSAC 90, pp. 261–264. ACM Press, New York (1990)
20. Hong, H.: Simple solution formula construction in cylindrical algebraic decomposition based quantifier elimination. In: Wang, P.S., (ed.) Proceedings of ISSAC '92, pp. 177–188. ACM Press, New York (1992)
21. Lazard, D.: Quantifier elimination: optimal solution for two classical examples. J. Symb. Comput. **5**, 261–266 (1988)
22. McCallum, S.: An improved projection operation for cylindrical algebraic decomposition of three-dimensional space. J. Symb. Comput. **5**, 141–161 (1988)
23. McCallum, S.: An improved projection operator for cylindrical algebraic decomposition. In: Caviness, B., Johnson, J. (eds.) Quantifier Elimination and Cylindrical Algebraic Decomposition, pp. 242–268. Springer, New York (1998)
24. McCallum, S., Brown, C.W.: On delineability of varieties in CAD-based quantifier elimination with two equational constraints. In: Proceedings of ISSAC' 2009: pp. 71–78
25. Tarski, A.: A decision method for elementary algebra and geometry. The RAND Corporation, SantaMonica (1948)
26. Timofte, V.: On the positivity of symmetric polynomial functions, Part I: General results. J. Math. Anal. Appl. **284**, 174–190 (2003)
27. Timofte, V.: On the positivity of symmetric polynomial functions, Part 2: Lattice general results and positivity criteria for degree 4 and 5. J. Math. Anal. Appl. **304**, 652–667 (2005)
28. Weispfenning, V.: Quantifier elimination for real algebra—the cubic case. In: Proceedings of ISSAC 94, pp. 258–263. ACM Press, Oxford (1994)
29. Weispfenning, V.: A new approach to quantifier elimination for real algebra. In: Caviness, B., Johnson, J. (eds.) Quantifier Elimination and Cylindrical Algebraic Decomposition, pp. 376–392. Springer, New York (1998)
30. Yan, A., Xia, S., Huang, L.: Single point creterion method for positive semidefinite bivariate polynomials over field of rationals. Comput. Appl. **21**, 183–184 (2001). (in Chinese)
31. Yang, L.: Recent advances on determining the number of real roots of parametric polynomials. J. Symbolic Comput. **28**, 225–242 (1999)
32. Yang, L., Hou, X., Zeng, Z.: A complete discrimination system for polynomials. Science in China (Ser. E) **39**, 628–646 (1996)
33. Yao, Y., Feng, Y.: Automated decision of positive of symmetric quintic forms. J. Sys. Scince Math. Sciences **28**(3), 313–324 (2008). (in Chinese)

The Vanishing Ideal of a Finite Set of Points with Multiplicity Structures

Na Lei, Xiaopeng Zheng and Yuxue Ren

Abstract Given a finite set of arbitrarily distributed points in affine space with multiplicity structures, we present an algorithm to compute the reduced Gröbner basis of the vanishing ideal under the lexicographic order. We split the problem into several smaller ones which can be solved by induction over variables and then use our new algorithm for intersection of ideals to compute the result of the original problem. The new algorithm for intersection of ideals is mainly based on the Extended Euclidean Algorithm. Our method discloses the essential geometric connection between the relative position of the points with multiplicity structures and the leading monomials of the reduced Gröbner basis of the vanishing ideal.

Keywords Vanishing ideal · Points with multiplicity structures · Reduced Gröbner basis · Intersection of ideals

1 Introduction

To describe the problem, first we give the definitions below.

Definition 1 $D \subseteq \mathbb{N}_0^n$ is called a lower set in n dimensional affine space as long as $\forall d \in D$ if $d_i \neq 0$, $d - e_i$ lies in D where $e_i = (0, \ldots, 0, 1, 0, \ldots, 0)$ with the 1 situated at the ith position $(1 \leq i \leq n)$. For a lower set D, we define its limiting set $E(D)$ to be the set of all $\beta \in \mathbb{N}_0^n - D$ such that whenever $\beta_i \neq 0$, then $\beta - e_i \in D$.

This work was supported by National Natural Science Foundation of China under Grant No. 11271156.

N. Lei (✉) · X. Zheng · Y. Ren
School of Mathematics, Jilin University, Changchun 130012, China
e-mail: lein@jlu.edu.cn

X. Zheng
e-mail: 490290756@qq.com

Y. Ren
e-mail: 1051221994@qq.com

© Springer-Verlag Berlin Heidelberg 2014

R. Feng et al. (eds.), *Computer Mathematics*, DOI 10.1007/978-3-662-43799-5_21

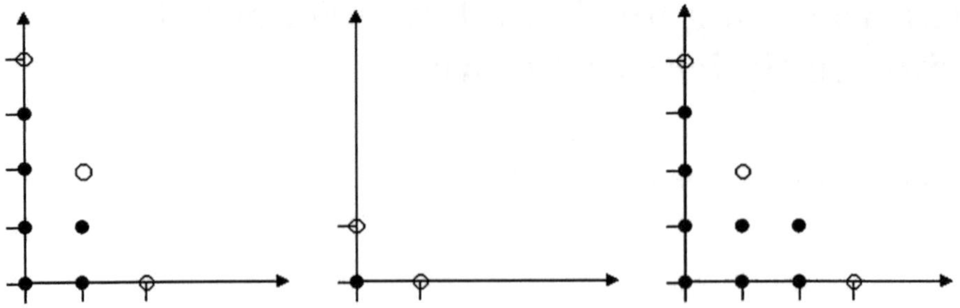

Fig. 1 Illustration of three lower sets and their limiting sets

As showed in Fig. 1, there are three lower sets and their limiting sets. The elements of the lower sets are marked by solid circles and the elements of the limiting sets are marked by blank circles.

Let k be a field and p be a point in the affine space k^n, i.e. $p = (p_1, \ldots, p_n) \in k^n$. Let $k[X]$ be the polynomial ring over k, where we write $X = (X_1, X_2, \ldots, X_n)$ for brevity's sake.

Definition 2 $\langle p, D\rangle$ represents a point p with multiplicity structure D, where p is a point in affine space k^n and $D \subseteq \mathbb{N}_0^n$ is a lower set. $\sharp D$ is called the multiplicity of point p (here we use the definition in [1]). For each $d = (d_1, \ldots, d_n) \in D$, we define a corresponding functional

$$L(f) = \frac{\partial^{d_1+\cdots+d_n}}{\partial x_1^{d_1}, \ldots, \partial x_n^{d_n}} f(p).$$

Hence for any given finite set of points with multiplicity structures $H = \{\langle p_1, D_1\rangle, \ldots, \langle p_t, D_t\rangle\}$, m functionals $\{L_i; i = 1, \ldots, m\}$ can be defined where $m \triangleq \sharp D_1 + \cdots + \sharp D_t$. We call

$$I(H) = \{f \in k[X]; L_i(f) = 0, \ i = 1, \ldots, m\}$$

the vanishing ideal of the set of the points H. The vanishing ideal problem we are focusing on is to compute the reduced Gröbner basis of the vanishing ideal for any given finite set of points H, which arises in several applications, for example, see [2] for statistics, [3] for biology, and [4–6] for coding theory.

A polynomial time algorithm for this problem was first given by Buchberger and Möller [7], then significantly improved by Marinari et al. [8], and Abbott et al. [9]. These algorithms perform Gauss elimination on a generalized Vandermonde matrix and have a polynomial time complexity in the number of points and in the number of variables. Jeffrey and Gao [10] presented a new algorithm that is essentially a generalization of Newton interpolation for univariate polynomial and has a good computational performance when the number of variables is small relative to the number of points.

In this paper the problem we consider is under the Lexicographical order with $X_1 \succ X_2 \succ \cdots X_n$ and a more transparent algorithm will be given. The ideas are summed-up as follows:

- Construct the reduced Gröbner basis of $I(H)$ and get the quotient basis by induction over variables (define $\{M;\, M$ is a monomial and it is not divisible by the leading monomial for any polynomial in $I(H)\}$ as the **quotient basis** for the vanishing ideal $I(H)$).
- Get the quotient basis of the vanishing ideal purely according to the geometric distribution of the points with multiplicity structures.
- Split the original n-variable problem into smaller ones which can be solved by converting them into $(n-1)$-variable problems.
- Compute the intersection of the ideals of the smaller problems by using Extended Euclidean Algorithm.

Our algorithm can get a lower set by induction over variables for any given set of points with multiplicity structures, and by constructing the reduced Gröbner basis at the same time we can prove that the lower set is the quotient basis. There are several publications which have a strong connection to the our work although they are all only focusing on the quotient basis, ignoring the reduced Gröbner basis of the vanishing ideal. Paper [11] gives a computationally efficient algorithm to get the quotient basis of the vanishing ideal over a set of points with no multiplicity structures and the authors introduce the interesting lex game to describe the problem and the algorithm. Paper [12] offers a purely combinatorial algorithm to obtain the quotient basis and the algorithm can handle the set of points with multiplicity structures as well.

The advantage of our method is insight rather than efficient computation. The computation cost depends closely on the structure of the given set of points and a full complexity analysis would be infeasible. Our method may not be particularly efficient, but is geometrically intuitive and appealing. The clear geometric meaning of our method reveals the essential connection between the relative position of the points with multiplicity structures and the quotient basis of the vanishing ideal, providing us a new perspective of view to look into the vanishing ideal problem and helping study the structure of the reduced Gröbner basis of zero dimensional ideal under lexicographic order. What's more, our method leads to the discovery of a new algorithm to compute the intersection of two zero dimensional ideals.

Since one important feature of our method is the clear geometric meaning, to demonstrate it we present an example in Sect. 2 together with some auxiliary pictures which can make the algorithms and conclusions in this paper easier to understand. In Sects. 3 and 4 some definitions and notions are given. Sections 5 and 6 are devoted to our main algorithms of computing the reduced Gröbner basis and the quotient basis together with the proofs. In Sect. 7 we demonstrate the algorithm to compute the intersection of two ideals and some applications.

2 Example

We will use two different forms to represent the set of points with multiplicity structures H in this paper.

For easier description, we introduce the matrix form which consists of two matrices $\langle \mathscr{P} = (p_{i,j})_{m\times n}, \mathscr{D} = (d_{i,j})_{m\times n}\rangle$ with $\mathscr{P}_i, \mathscr{D}_i$ denoting the ith row vectors of $\mathscr{P}$ and $\mathscr{D}$ respectively. Each pair $\{\mathscr{P}_i, \mathscr{D}_i\}$ $(1 \le i \le m)$ defines a functional in the following way.

$$L_i(f) = \frac{\partial^{d_{i,1}+\cdots+d_{i,n}}}{\partial x_1^{d_{i,1}} \ldots \partial x_n^{d_{i,n}}} f|_{x_1=p_{i,1},\ldots,x_n=p_{i,n}}.$$

And the functional set defined here is the same with that defined by the way in Sect. 1 with respect to H.

For example, given a set of three points with their multiplicity structures $\{\langle p_1, D_1\rangle, \langle p_2, D_2\rangle, \langle p_3, D_3\rangle\}$, where $p_1 = (1,1), p_2 = (2,1), p_3 = (0,2), D_1 = \{(0,0),(0,1),(1,0)\}, D_2 = \{(0,0),(0,1),(1,0),(1,1)\}, D_3 = \{(0,0),(1,0)\}$, the matrix form is like the follows.

$$\mathscr{P} = \begin{pmatrix} 1 & 1 \\ 1 & 1 \\ 1 & 1 \\ 2 & 1 \\ 2 & 1 \\ 2 & 1 \\ 2 & 1 \\ 0 & 2 \\ 0 & 2 \end{pmatrix}, \mathscr{D} = \begin{pmatrix} 0 & 0 \\ 1 & 0 \\ 0 & 1 \\ 0 & 0 \\ 1 & 0 \\ 0 & 1 \\ 1 & 1 \\ 0 & 0 \\ 1 & 0 \end{pmatrix}.$$

For intuition's sake, we also represent the points with multiplicity structures in a more intuitive way as showed in the left picture of Fig. 2 where each lower set that represents the multiplicity structure of the corresponding point p is also put in

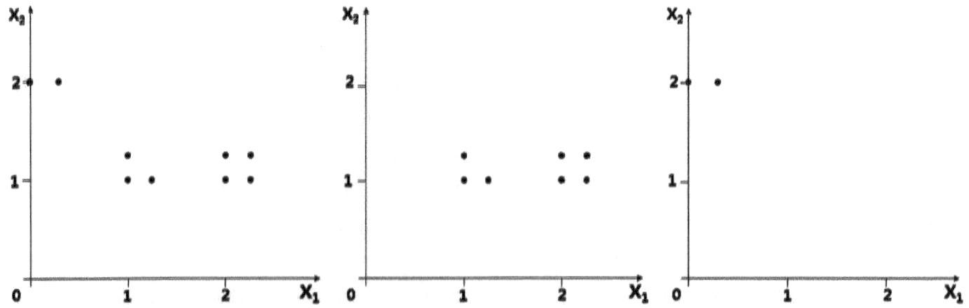

Fig. 2 The left picture represents H, the middle one is for H_1 and the right one for H_2

the affine space with the zero element (0,0) situated at p. This intuitive representing form is the basis of the geometric interpretation of our algorithm.

We take the example above to show how our method works and what the geometric interpretation of our algorithm is like:

Step 1: Define mapping $\pi : H \mapsto k$ such that $\langle p = (p_1, \ldots, p_n), D\rangle \in H$ is mapped to $p_n \in k$. So $H = \{\langle p_1, D_1\rangle, \langle p_2, D_2\rangle, \langle p_3, D_3\rangle\}$ consists of two π-fibres: $H_1 = \{\langle p_1, D_1\rangle, \langle p_2, D_2\rangle\}$ and $H_2 = \{\langle p_3, D_3\rangle\}$ as showed in the middle and the right pictures in Fig. 2. Each fibre defines a new problem, so we split the original problem defined by H into two small ones defined by H_1 and H_2 respectively.

Step 2: Solve the small problems. Take the problem defined by H_1 for example. First, it's easy to write down one element of $I(H_1)$:

$$f_1 = (X_2 - 1)(X_2 - 1) = (X_2 - 1)^2 \in I(H_1).$$

The geometry interpretation is: we draw two lines sharing the same equation of $X_2 - 1 = 0$ to cover all the points as illustrated in the left picture in Fig. 3 and the corresponding polynomial is f_1.

According to the middle and the right pictures in Fig. 3, we can write down another two polynomials in $I(H_1)$:

$$f_2 = (X_2 - 1)(X_1 - 1)(X_1 - 2)^2 \text{ and } f_3 = (X_1 - 1)^2(X_1 - 2)^2.$$

It can be checked that $G_1 = \{f_1, f_2, f_3\}$ is the reduced Gröbner basis of $I(H_1)$, and the quotient basis is $\{1, X_1, X_2, X_1X_2, X_1^2, X_2X_1^2, X_1^3\}$. In the following, we don't distinguish explicitly an n-variable monomial $X_1^{d_1}X_2^{d_2}\ldots X_n^{d_n}$ with the element $(d_1, d_2, \ldots, d_n)$ in $\mathbb{N}_0^n$. Hence this quotient basis can be written as a subset of $\mathbb{N}_0^n$: $\{(0,0), (1,0), (0,1), (1,1), (2,0), (2,1), (3,0)\}$, i.e. a lower set, denoted by D'.

In fact we can get the lower set in a more direct way by pushing the points with multiplicity structures leftward which is illustrated in the picture below (lower set D' is positioned in the right part of the picture with the (0,0) element situated at point (0,1)). The elements of the lower set D' in the right picture in Fig. 4 are marked by solid circles. The blank circles constitute the limiting set $E(D')$ and they are the leading terms of the reduced Gröbner basis $\{f_1, f_2, f_3\}$.

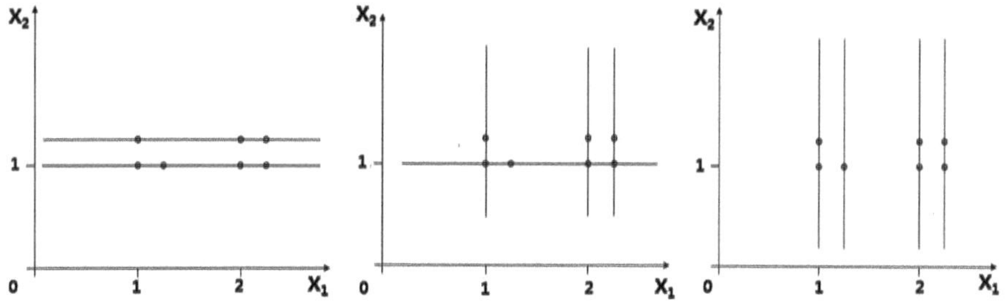

Fig. 3 Three ways to draw lines to cover the points

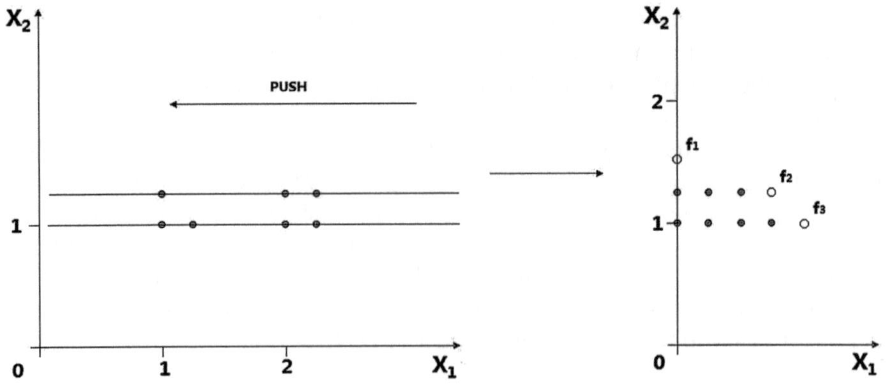

Fig. 4 Push the points leftward to get a lower set

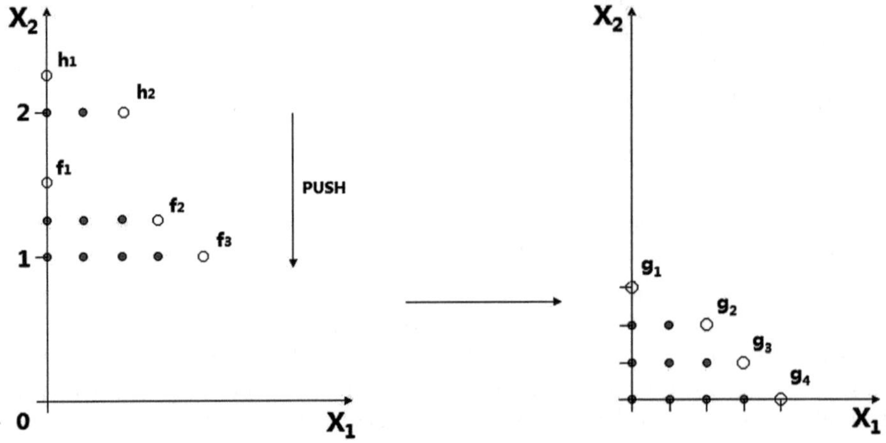

Fig. 5 Get the lower set D based on D' and D''

In the same way, we can get the Gröbner basis $G_2 = \{h_1, h_2\}$ and a lower set D'' for the problem defined by H_2, where $h_1 = (X_2 - 2)$, $h_2 = X_1^2$, $D'' = \{(0, 0), (1, 0)\}$.

Step 3: Compute the intersection of the ideals $I(H_1)$ and $I(H_2)$ to get the result for the problem defined by H.

First, we construct a new lower set D based on D', D'' in an intuitive way: let the solid circles fall down and the elements of D'' rest on the elements of D' to form a new lower set D which is showed in the right part of Fig. 5 and the blank circles represent the elements of the limiting set $E(D)$.

Then we need to find $\sharp E(D)$ polynomials vanishing on H with leading terms being the elements of $E(D)$. Take $X_1^3 X_2 \in E(D)$ for example to show the general way we do it.

We need two polynomials which vanish on H_1 and H_2 respectively, and their leading terms both have the **same degrees** of X_1 with that of the desired monomial $X_1^3X_2$ and both have the **minimal degrees** of X_2. Notice that f_2 and $X_1 \cdot h_2$ satisfy the requirement. And then we multiply f_2 and $X_1 \cdot h_2$ with h_1, f_1 respectively which are all univariate polynomials in X_2 to get two polynomials q_1, q_2 such that q_1 and q_2 both vanish on H. Obviously q_1 and q_2 still have the **same degrees** of X_1 with that of the desired monomial $X_1^3X_2$.

$$q_1 = f_2 \cdot h_1 = (X_2 - 1)(X_1 - 1)(X_1 - 2)^2(X_2 - 2),$$

$$q_2 = X_1 \cdot h_2 \cdot f_1 = X_1^3(X_2 - 1)^2.$$

Next try to find two univariate polynomials in X_2: r_1, r_2 such that $q_1 \cdot r_1 + q_2 \cdot r_2$ vanishes on H (which is obviously true already) and has the desired leading term $X_1^3X_2$.

$$\begin{aligned} q_1 &= (X_2 - 2)(X_2 - 1)X_1^3 - (5X_2^2 - 15X_2 + 10)X_1^2 \\ &\quad + (8X_2^2 - 24X_2 + 16)X_1 - 4X_2^2 + 12X_2 - 8, \\ q_2 &= (X_2 - 1)^2X_1^3. \end{aligned}$$

To settle the leading term issue, write q_1, q_2 as univariate polynomials in X_1 as above. Because $X_2 \prec X_1$ and the highest degrees of X_1 of the leading terms of q_1, q_2 are both 3, we know that as long as the leading term of $(X_2 - 2)(X_2 - 1)X_1^3 \cdot r_1 + (X_2 - 1)^2X_1^3 \cdot r_2$ is $X_1^3X_2$, the leading term of $q_1 \cdot r_1 + q_2 \cdot r_2$ is also $X_1^3X_2$.

$$\begin{aligned} &(X_2 - 2)(X_2 - 1)X_1^3 \cdot r_1 + (X_2 - 1)^2X_1^3 \cdot r_2 \\ &\quad = X_1^3(X_2 - 1)\left((X_2 - 2) \cdot r_1 + (X_2 - 1) \cdot r_2\right) \end{aligned}$$

Obviously if and only if $(X_2 - 2) \cdot r_1 + (X_2 - 1) \cdot r_2 = 1$ we can keep the leading term of $q_1 \cdot r_1 + q_2 \cdot r_2$ to be $X_1^3X_2$. In this case $r_1 = -1$ and $r_2 = 1$ will be just perfect. In our algorithm we use Extended Euclidean Algorithm to compute r_1, r_2.

Finally, we obtain

$$g_3 = q_1 \cdot r_1 + q_2 \cdot r_2$$

$$= (X_2 - 1)X_1^3 + (5X_2^2 - 15X_2 + 10)X_1^2 - (8X_2^2 - 24X_2 + 16)X_1 + 4X_2^2 - 12X_2 + 8$$

which vanishes on H and has $X_1^3X_2$ as its leading term.

In the same way, we can get $g_1 = (X_2 - 1)^2(X_2 - 2)$ for X_2^3, $g_2 = (X_2 - 1)^2X_1^2$ for $X_1^2X_2^2$ and $g_4 = X_1^4 + 6(X_2^2 - 2X_2)X_1^3 - 13(X_2^2 - 2X_2)X_1^2 + 12(X_2^2 - 2X_2)X_1 - 4(X_2^2 - 2X_2)$ for X_1^4. In fact we need to compute g_1, g_2, g_3 and g_4 in turn according to the lexicographic order because we need reduce g_2 by g_1, reduce g_3 by g_2 and g_1, and reduce g_4 by g_1, g_2 and g_3.

The reduced polynomial set can be proved in Sect. 6 to be the reduced Gröbner basis of the intersection of two ideals which is exactly the vanishing ideal over H, and D is the quotient basis.

This example shows what the geometric interpretation of our method is like: for any given point with multiplicity structure $\langle p_i, D_i \rangle$, we put the lower set D_i into the affine space with the (0,0) element situated at p_i to intuitively represent it, and it can be imagined as $\sharp D_i$ small balls in the affine space; for bivariate problem, we first push the balls along the X_1-axis, then along the X_2-axis to get a lower set as we did in the example above; the lower set is exactly the quotient basis and the limiting set of the lower set is the set of the leading monomials of the reduced Göbner basis. This intuitive understanding can be applied to the n-variable problem and can help us understand the algorithm better in the following.

3 Notions

First, we define the following mappings:

$$\begin{aligned}
\text{proj} : \mathbb{N}_0^n &\longrightarrow \mathbb{N}_0 \\
(d_1, \ldots, d_n) &\longrightarrow d_n. \\
\widehat{\text{proj}} : \mathbb{N}_0^n &\longrightarrow \mathbb{N}_0^{n-1} \\
(d_1, \ldots, d_n) &\longrightarrow (d_1, \ldots, d_{n-1}). \\
\text{embed}_c : \mathbb{N}_0^{n-1} &\longrightarrow \mathbb{N}_0^n \\
(d_1, \ldots, d_{n-1}) &\longrightarrow (d_1, \ldots, d_{n-1}, c).
\end{aligned}$$

Let $D \subset \mathbb{N}_0^n$, and naturally we define $\widehat{\text{proj}}(D) = \{\widehat{\text{proj}}(d) | d \in D\}$, and $\text{embed}_c(D') = \{\text{embed}_c(d) | d \in D'\}$ where $D' \subset \mathbb{N}_0^{n-1}$. In fact we can apply these mappings to any set $O \subset k^n$ or any matrix of n columns, because there is no danger of confusion. For example, let M be a matrix of n columns, and $\widehat{\text{proj}}(M)$ is a matrix of $n-1$ columns with the first $n-1$ columns of M reserved and the last one eliminated.

The embed_c mapping embeds a lower set of the $n-1$ dimensional space into the n dimensional space. When the embed_c operation parameter c is zero, we can get a lower set of $\mathbb{N}_0^n$ by mapping each element $d = (d_1, \ldots, d_{n-1})$ to $d = (d_1, \ldots, d_{n-1}, 0)$ as showed below.

Blank circles represent the elements of the limiting sets. Note that after the embed_c mapping, there is one more blank circle. In this case, the limiting set is always increased by one element $(0, \ldots, 0, 1)$.

In the case the embed_c operation parameter c is not zero, it is obvious that what we got is not a lower set any more. But there is another intuitive fact we should realize (Fig. 6).

Theorem 1 *Assume $D_0, D_1, \ldots, D_\ell$ are lower sets in $n-1$ dimensional space, and $D_0 \supseteq D_1 \supseteq \cdots \supseteq D_\ell$. Let $\hat{D}_i = embed_i(D_i), i = 0, \ldots, \ell$. Then $D = \bigcup_{i=0}^{\ell} \hat{D}_i$ is a lower set in n dimensional space, and $E(D) \subseteq C$ where $C = \bigcup_{i=0}^{\ell} embed_i(E(D_i)) \bigcup \{(0, \ldots, 0, \ell+1)\}$.*

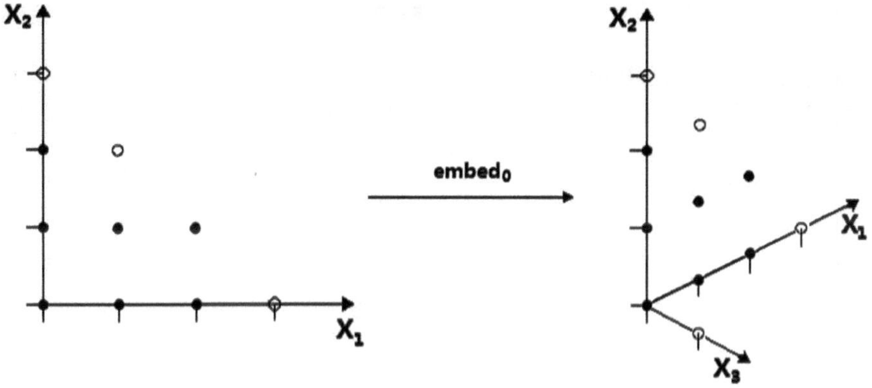

Fig. 6 Embed the lower set in 2-D space into 3-D space with parameter $c = 0$

Proof First to prove D is a lower set. $\forall d \in D$, let $i = \text{proj}(d)$, then $d \in \hat{D}_i$ i.e. $\widehat{\text{proj}}(d) \in \widehat{\text{proj}}(\hat{D}_i) = D_i$. Because D_i is a lower set, hence for $j = 1, \ldots, n-1$, if $d_j \neq 0$, then $\widehat{\text{proj}}(d) - \widehat{\text{proj}}(e_j) \in D_i$ where $e_j = (0, \ldots, 0, 1, 0, \ldots, 0)$ with the 1 situated at the jth position. So $d - e_j \in \hat{D}_i \subseteq D$. For $j = n$, if $i = 0$, we are finished. If $i \neq 0$, there must be $d - e_n \in \hat{D}_{i-1} \subseteq D$. Because if $d - e_n \notin \hat{D}_{i-1}$, we have $\widehat{\text{proj}}(d) \notin D_{i-1}$. Since we already have $\widehat{\text{proj}}(d) \in D_i$, this is contradictory to $D_i \subseteq D_{i-1}$.

Second, assume $\forall d \in E(D)$, $\widehat{\text{proj}}(d) \notin D_i, i = 0, \ldots, \ell$. If $\widehat{\text{proj}}(d)$ is a zero tuple, then d_n must be $\ell + 1$, that is $d \in C$. If $\widehat{\text{proj}}(d)$ is not a zero tuple, then we know $d_n < \ell + 1$. If $d_j \neq 0,\ j = 1, \ldots, n-1$, then $d - e_j \in \text{embed}_{d_n}(D_{d_n})$. Then $\widehat{\text{proj}}(d) - \widehat{\text{proj}}(e_j) \in D_{d_n}$, that is $\widehat{\text{proj}}(d) \in E(D_{d_n})$. Finally with the embed_{d_n} operation we have $d \in \text{embed}_{d_n}(E(D_{d_n}))$ where $d_n < \ell + 1$. So $d \in C$. □

4 Addition of Lower Sets

In this section, we define the addition of lower sets which is the same with that in [13], the following paragraph and Fig. 7 are basically excerpted from that paper with a little modification of expression.

To get a visual impression of what the addition of lower sets are, look at the example in Fig. 7. What is depicted there can be generalized to arbitrary lower sets D_1, D_2 and arbitrary dimension n. The process can be described as follows. Draw a coordinate system of $\mathbb{N}_0^n$ and insert D_1. Place a translate of D_2 somewhere on the X_2-axis. The translate has to be sufficiently far out, so that D_1 and the translate D_2 do not intersect. Then push the elements of the translate of D_2 down along the X_2-axis until on room remains between them and the elements of D_1. The resulting lower set is denoted by $D_1 + D_2$.

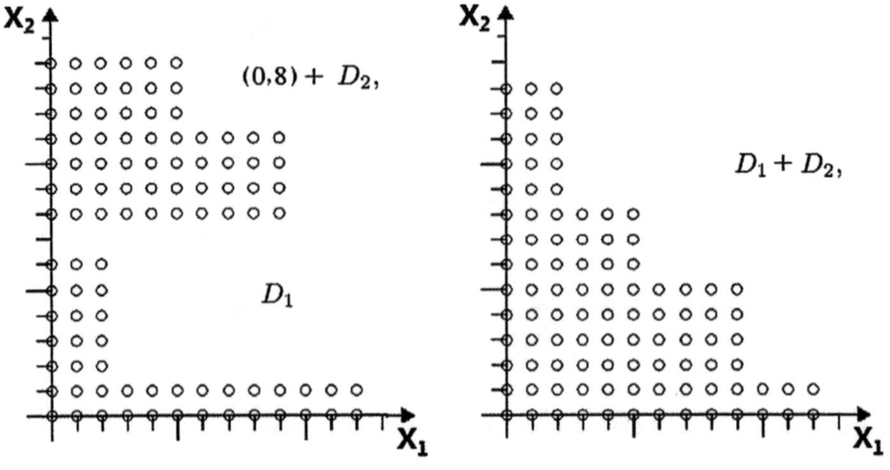

Fig. 7 Addition of D_1 and D_2

Intuitively, we define algorithm **ALS** (short for *Addition of Lower Sets*) to realize the addition of lower sets.

Algorithm **ALS:** Given two lower sets in n dimensional space D_1, D_2, determine another lower set as the addition of D_1, D_2, denoted by $D := D_1 + D_2$.

Step 1 $D := D_1$;
Step 2 If $\sharp D_2 = 0$ return D. Else pick $a \in D_2$, $D_2 := D_2 \setminus \{a\}$.

Step 2.1 If $a \in D$, add the last coordinate of a with 1. Go to Step 2.1.
Else
$D := D \bigcup \{a\}$, go to Step 2.

Given three lower sets D_1, D_2, D_3, the addition we defined satisfies:

1. $D_1 + D_2 = D_2 + D_1$,
2. $(D_1 + D_2) + D_3 = D_1 + (D_2 + D_3)$,
3. $D_1 + D_2$ is a lower set,
4. $\sharp(D_1 + D_2) = \sharp D_1 + \sharp D_2$.

These are all the same with that in [13]. And the proof can be referred to it.

As implied in the example of Sect. 2, when we want to get a polynomial with leading term d_3 showed in the right part of Fig. 8, we need two polynomials with the leading terms d_1, d_2 which are not the elements of the lower sets and have the same degrees of X_1 as d_3 and the minimal degrees of X_2 as showed in the left part of Fig. 8. In other words, $d_1 \notin D_1$, $d_2 \notin D_2$, $\widehat{\text{proj}}(d_1) = \widehat{\text{proj}}(d_2) = \widehat{\text{proj}}(d_3)$, $\text{proj}(d_1) + \text{proj}(d_2) = \text{proj}(d_3)$. It's easy to understand that these equations hold for the addition of three or even more lower sets.

We use algorithm **CLT** (short of *Computing the Leading Term*) to get the leading terms d_1 and d_2 from d_3 respectively.

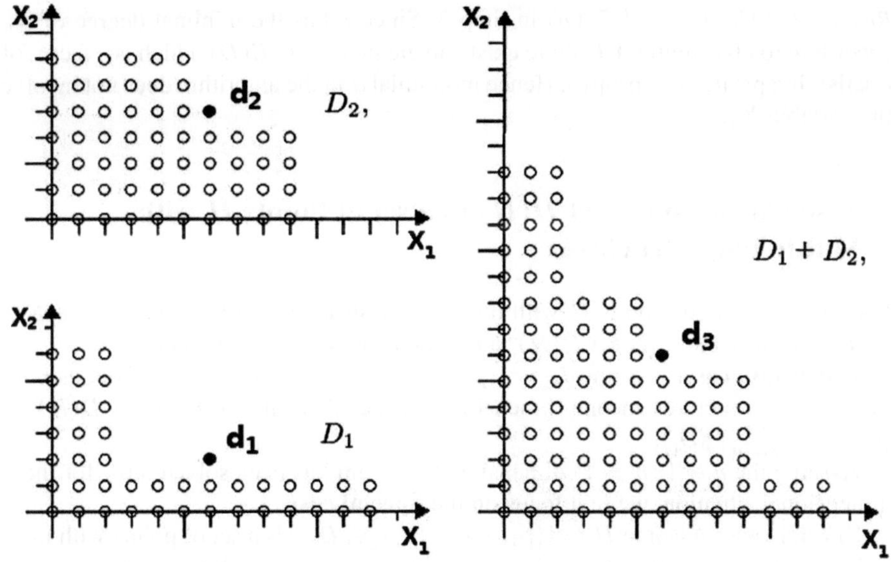

Fig. 8 $\widehat{\text{proj}}(d_1) = \widehat{\text{proj}}(d_2) = \widehat{\text{proj}}(d_3)$, $\text{proj}(d_1) + \text{proj}(d_2) = \text{proj}(d_3)$

Algorithm **CLT:** Given $a \in \mathbb{N}_0^n$, and a lower set D in n dimensional space satisfying $a \notin D$. Determine another $r = (r_1, \ldots, r_n) \in \mathbb{N}_0^n$ which satisfies that $r \notin D$, $\widehat{\text{proj}}(r) = \widehat{\text{proj}}(a)$ and $(r_1, \ldots, r_{n-1}, r_n - 1) \in D$, denoted by $r := \text{CLT}(a, D)$.

Step 1 Initialize r such as $\widehat{\text{proj}}(r) = \widehat{\text{proj}}(a)$ and $\text{proj}(r) = 0$.
Step 2 if $r \notin D$, return r, else $r_n := r_n + 1$, go to Step 2.

Then $d_1 = \text{CLT}(d_3, D_1)$, $d_2 = \text{CLT}(d_3, D_2)$.

Definition 3 For any $f \in k[X]$, view it as an element in $k(X_n)[X_1, \ldots, X_{n-1}]$ and define $\text{LC}_n(f)$ to be the leading coefficient of f which is a univariate polynomial in X_n.

Here is the algorithm **CP** (short for *Computing the Polynomial*) which can compute the polynomial with the leading term returned by algorithm **CLT**.

Algorithm **CP:** D is a lower set in n dimensional space, $a \in \mathbb{N}_0^n$ and $a \notin D$, $G := \{f_{ed} \in k[X];$ the leading term of f_{ed} is ed, $ed \in E(D)\}$, algorithm **CP** returns a polynomial p whose leading term is $\text{CLT}(a, D)$. Denoted by $p := \text{CP}(a, D, G)$.

Step 1 $c := \text{CLT}(a, D)$.
Step 2 Select $c' \in E(D)$, $s.t.$ c' is a factor of c. $d := \frac{c}{c'}$.
(d is well defined because $c \notin D$).
Step 3 $p := f_{c'} \cdot d$ where $f_{c'}$ is an element of G whose leading term is $c' \in E(D)$.
(p is well defined because d is well defined).

Remark 1 $\mathrm{LC}_n(f_{c'}) = \mathrm{LC}_n(p)$ in Step 3. Since c has the minimal degree of X_n according to algorithm **CLT**, there exists no element $c'' \in E(D)$ which is a factor of c satisfying $\mathrm{proj}(c'') < \mathrm{proj}(c)$. Hence monomial d in the algorithm does not involve the variable X_n.

5 Associate a Lower set $D(H)$ to a set of Points H with Multiplicity Structures

For any given set of points H with multiplicity structures in n dimensional space, we can construct a lower set $D(H)$ in n dimensional space by induction.

Univariate case: Assume $H = \{\langle p_1, D_1\rangle, \ldots, \langle p_t, D_t\rangle\}$ is a set of points with multiplicity structures in one dimensional space, then the lower set is $D(H) = \{0, 1, \ldots, \sum_{i=1}^{t} \sharp D_i\}$.

Assume the $n-1$ $(n > 1)$ dimensional problem has been solved, now for the n dimensional situation, we first focus on the **Special case**.

Special case: Assume $H = \{\langle p_1, D_1\rangle, \ldots, \langle p_t, D_t\rangle\}$ is a set of points with multiplicity structures in the n $(n > 1)$ dimensional space where all the points share the same X_n coordinate. Write H in matrix form as $\langle \mathscr{P}, \mathscr{D}\rangle$ and all the entries in the last column of matrix $\mathscr{P}$ have the same value. Classify the row vectors of $\langle \mathscr{P}, \mathscr{D}\rangle$ to get $\{\langle \mathscr{P}_0, \mathscr{D}_0\rangle, \ldots, \langle \mathscr{P}_w, \mathscr{D}_w\rangle\}$ according to the values of the entries in the last column of matrix $\mathscr{D}$ and we guarantee the corresponding relationship between the row vectors of matrix $\mathscr{P}$ and matrix $\mathscr{D}$ holds in $\langle \mathscr{P}_i, \mathscr{D}_i\rangle$ $(0 \le i \le w)$. All the entries in the last column of $\mathscr{D}_i$ are the same i and the entries of the last column of $\mathscr{P}_i$ stay the same too. Then eliminate the last columns of $\mathscr{P}_i$ and $\mathscr{D}_i$ to get $\langle \widehat{\mathrm{proj}}(\mathscr{P}_i), \widehat{\mathrm{proj}}(\mathscr{D}_i)\rangle$ which represents a set of points with multiplicity structures in $n-1$ dimensional space, by induction we get a lower set $\hat{D}_i$ in $n-1$ dimensional space. Then we set

$$D(H) = \bigcup_{i=0}^{w} \mathrm{embed}_i(\hat{D}_i).$$

Next we deal with the **General case**.

General case: Assume $H = \{\langle p_1, D_1\rangle, \ldots, \langle p_t, D_t\rangle\}$ is a set of points with multiplicity structures in the n $(n > 1)$ dimensional space. Split the set of points: $H = H_1 \bigcup H_2 \bigcup \cdots \bigcup H_s$ such that the points of each H_i are in the same π-fibre, i.e. they have the same X_n coordinate $c_i, i = 1, \ldots, s$,and $c_i \neq c_j, \forall i, j = 1, \ldots, s, i \neq j$. According to the **Special case**, for each $i = 1, \ldots, s$, we can get a lower set $D(H_i)$, then we set

$$D(H) = \sum_{i=1}^{s} D(H_i).$$

We now prove $D(H)$ is a lower set although it is easy to understand as long as the geometric interpretation involves. Since it is obviously true for **Univariate case**, induction over dimension would be helpful for the proof.

Proof Assume $D(H)$ is a lower set for the $n-1$ dimensional situation and now we prove the conclusion for n dimensional situation ($n > 1$).

First to prove $D(H)$ of the **Special case** is a lower set.

We claim that $\langle \widehat{\text{proj}(\mathscr{P}_i)}, \widehat{\text{proj}(\mathscr{D}_i)} \rangle$ represents a set of points with multiplicity structures in $n-1$ dimensional space ($i = 0, \ldots, w$). For any $D \subset \mathbb{N}_0^n$, define $F_a(D) = \{d \in D | \text{ proj}(d) = a\}$. Let $U = \{u | u \in \{1, \ldots, t\}, F_i(D_u) \neq \varnothing\}$. So $\langle \widehat{\text{proj}(\mathscr{P}_i)}, \widehat{\text{proj}(\mathscr{D}_i)} \rangle$ can be written in the form of $\{\langle \widehat{\text{proj}(p_u)}, \widehat{\text{proj}(F_i(D_u))} \rangle | u \in U\}$. It is apparent that $\widehat{\text{proj}(F_i(D_u))}$ is a lower set in $n-1$ dimensional space and can be viewed as the multiplicity structure of the point $\widehat{\text{proj}(p_u)}$. Hence $\langle \widehat{\text{proj}(\mathscr{P}_i)}, \widehat{\text{proj}(\mathscr{D}_i)} \rangle$ is a set of points with multiplicity structures in $n-1$ dimensional space.

What's more, we assert $\widehat{\text{proj}(\mathscr{P}_j)}$ is a sub-matrix of $\widehat{\text{proj}(\mathscr{P}_i)}$, and $\widehat{\text{proj}(\mathscr{D}_j)}$ is a sub-matrix of $\widehat{\text{proj}(\mathscr{D}_i)}$, $0 \leq i < j \leq w$. Because of the corresponding relationship between the row vectors in $\mathscr{P}$ and $\mathscr{D}$, we need only to prove $\widehat{\text{proj}(\mathscr{D}_j)}$ is a sub-matrix of $\widehat{\text{proj}(\mathscr{D}_i)}$. If it is not true, there exists a row vector g of $\widehat{\text{proj}(\mathscr{D}_j)}$ which is not a row vector of $\widehat{\text{proj}(\mathscr{D}_i)}$. That is, there exists b ($1 \leq b \leq t$) such that $\text{embed}_j(g)$ is an element of the lower set D_b, and $\text{embed}_i(g)$ is not included in any lower set D_a ($1 \leq a \leq t$). However since $i < j$ and $\text{embed}_j(g) \in D_b$, $\text{embed}_i(g)$ must be included in D_b. Hence our assertion is true.

Since $\widehat{\text{proj}(\mathscr{P}_j)}$ is a sub-matrix of $\widehat{\text{proj}(\mathscr{P}_i)}$, and $\widehat{\text{proj}(\mathscr{D}_j)}$ is a sub-matrix of $\widehat{\text{proj}(\mathscr{D}_i)}$, $0 \leq i < j \leq w$. According to the assumption of induction and the way we construct $D(H)$, we have $\hat{D}_i \supseteq \hat{D}_j, 0 \leq i < j \leq w$, where $\hat{D}_i, \hat{D}_j$ are both lower sets. Based on the Theorem 1 in Sect. 3, $D(H) = \bigcup_{i=0}^{w} \text{embed}_i(\hat{D}_i)$ is a lower set, and $E(D(H)) \subseteq \bigcup_{i=0}^{w} \text{embed}_i(E(\hat{D}_i)) \bigcup \{(0, \ldots, 0, w+1)\}$.

Then to prove $D(H)$ of **General case** is a lower set. Since $D(H_i), i = 1, \ldots, s$ are lower sets, and the addition of lower sets is also a lower set according to Sect. 4, $D(H)$ is obviously a lower set. □

6 Associate a set of Polynomials poly(H) to $D(H)$

For every lower set constructed during the induction procedure showed in the last section, we associate a set of polynomials to it.

We begin with the univariate problem as we did in the last section.

6.1 Univariate Problem

P-Univariate case:

Assume $H = \{\langle p_1, D_1 \rangle, \ldots, \langle p_t, D_t \rangle\}$ is a set of points with multiplicity structures in one dimensional space, and $D(H) = \{0, 1, \ldots, \sum_{i=1}^{t} \sharp D_i\}$. Then the set of univariate polynomials associated to $D(H)$ is $poly(H) = \{\prod_{i=1}^{t} (X_1 - p_i)^{\sharp D_i}\}$.

Obviously $poly(H)$ of **P-Univariate case** satisfies the following **Assumption**.

Assumption For any given set of points with multiplicity structures H in the $n-1$ $(n>1)$ dimensional space, there are the following properties. For any $\lambda \in E(D(H))$, there exists a polynomial $f_\lambda \in k[X]$ where $X=(X_1,\ldots,X_{n-1})$ such that

- The leading term of f_λ under lexicographic order is X^λ.
- The exponents of all lower terms of f_λ lies in $D(H)$.
- f_λ vanishes on H.
- $\text{poly}(H)=\{f_\lambda|\lambda \in E(D(H))\}$.

Now assume the $(n-1)$-variable $(n>1)$ problem has been solved i.e. for any given set of points with multiplicity structures H in $n-1$ dimensional space, we can construct a set of polynomial $\text{poly}(H)$ which satisfies the **Assumption**. And then to tackle the n-variable problem, we still begin with the special case.

6.2 Special Case of the n-variable (n > 1) Problem

P-Special case:
Given a set of points with multiplicity structures in n $(n>1)$ dimensional space $H=\{\langle p_1,D_1\rangle,\ldots,\langle p_t,D_t\rangle\}$ or in matrix form $\langle \mathscr{P}=(p_{ij})_{m\times n}, \mathscr{D}=(d_{ij})_{m\times n}\rangle$. All the given points have the same X_n coordinate, i.e. the entries in the last column of $\mathscr{P}$ are the same. We compute $\text{poly}(H)$ with the following steps.

Step 1 $c:=p_{1n}$; $w=\max\{d_{in}; i=1,\ldots,m\}$.

Step 2 $\forall i=0,\ldots,w$, define $\mathscr{S}\mathscr{D}_i$ as a sub-matrix of $\mathscr{D}$ containing all the row vectors whose last coordinates equal i. Extract the corresponding row vectors of $\mathscr{P}$ to form matrix $\mathscr{S}\mathscr{P}_i$, and the corresponding relationship between the row vectors in $\mathscr{P}$ and $\mathscr{D}$ holds for $\mathscr{S}\mathscr{P}_i$ and $\mathscr{S}\mathscr{D}_i$.

Step 3 $\forall i=0,\ldots,w$, eliminate the last columns of $\mathscr{S}\mathscr{P}_i$ and $\mathscr{S}\mathscr{D}_i$ to get $\langle \tilde{\mathscr{S}\mathscr{P}}_i, \tilde{\mathscr{S}\mathscr{D}}_i\rangle$ which represents a set of points in $n-1$ dimensional space with multiplicity structures.According to the induction assumption, we have the polynomial set $\tilde{G}_i=\text{poly}(\langle \tilde{\mathscr{S}\mathscr{P}}_i,\tilde{\mathscr{S}\mathscr{D}}_i\rangle)$ associated to the lower set $\tilde{D}_i=D(\langle \tilde{\mathscr{S}\mathscr{P}}_i,\tilde{\mathscr{S}\mathscr{D}}_i\rangle)$.

Step 4 $D:=\bigcup_{i=0}^{w}\text{embed}_i(\tilde{D}_i)$. Multiply every element of $\tilde{G}_i$ with $(X_n-c)^i$ to get G_i. $\tilde{G}:=\bigcup_{i=0}^{w}G_i\bigcup\{(X_n-c)^{w+1}\}$.

Step 5 Eliminate the polynomials in $\tilde{G}$ whose leading term is not included in $E(D)$ to get $\text{poly}(H)$.

Theorem 2 *The* $\text{poly}(H)$ *obtained in* ***P-Special case*** *satisfies the* ***Assumption***.

Proof According to the Sect. 5, $\langle \tilde{\mathscr{S}\mathscr{P}}_i,\tilde{\mathscr{S}\mathscr{D}}_i\rangle$ represents a set of points with multiplicity structures in $n-1$ dimensional space for $i=0,\ldots,w$. And $\tilde{D}_j \supseteq \tilde{D}_i$, $0\le j\le i\le w$. D is a lower set and $E(D)\subseteq \bigcup_{i=0}^{w}\text{embed}_i(E(\tilde{D}_i))\bigcup \{(0,\ldots,0,w+1)\}$.

For $\lambda = (0, \ldots, 0, w+1) \in E(D)$, we have $f_\lambda = (X_n - c)^{w+1}$. It is easy to check that it satisfies the first three terms of the **Assumption**.

For any other element ed of $E(D)$, $\exists \ell\ s.t.\ ed \in \text{embed}_\ell E(\tilde{D}_\ell)$. So let $\tilde{ed}$ be the element in $E(\tilde{D}_\ell)$ such that $ed = \text{embed}_\ell(\tilde{ed})$. We have $f_{\tilde{ed}}$ vanishes on $\langle \mathscr{S}\tilde{\mathscr{P}}_\ell, \mathscr{S}\tilde{\mathscr{D}}_\ell\rangle$ whose leading term is $\tilde{ed} \in E(\tilde{D}_\ell)$ and the lower terms belong to $\tilde{D}_\ell$. According to the algorithm $f_{ed} = (X_n - c)^\ell \cdot f_{\tilde{ed}} \in \text{poly}(H)$.

First it is easy to check that the leading term of f_{ed} is ed since $ed = \text{embed}_\ell(\tilde{ed})$.

Second, the lower terms of f_{ed} are all in the set $S = \bigcup_{j=0}^{\ell} \text{embed}_j(\tilde{D}_\ell)$ because all the lower terms of $f_{\tilde{ed}}$ are in the set $\tilde{D}_\ell$. $\tilde{D}_0 \supseteq \tilde{D}_1 \supseteq \ldots \tilde{D}_\ell$, so $\text{embed}_j(\tilde{D}_\ell) \subset \text{embed}_j(\tilde{D}_j)$ $(0 \le j \le \ell)$, hence $S \subseteq D = \bigcup_{j=0}^{w} \text{embed}_j(\tilde{D}_j)$ and the second term of the **Assumption** is satisfied.

Third, we are going to prove that f_{ed} vanishes on all the functionals defined by $\langle \mathscr{P}, \mathscr{D}\rangle$, i.e. all the functionals defined by $\langle \mathscr{S}\mathscr{P}_i, \mathscr{S}\mathscr{D}_i\rangle$ $(i = 0, \ldots, w)$. Write all the functionals defined by $\langle \mathscr{S}\mathscr{P}_i, \mathscr{S}\mathscr{D}_i\rangle$ in this form: $L' \cdot \frac{\partial^i}{\partial X_n^i}|_{X_n=c}$ where L' is an $n-1$ variable functional. Substitute the zeroes and use the fact that $f_{\tilde{ed}}$ vanishes on $\langle \mathscr{S}\tilde{\mathscr{P}}_\ell, \mathscr{S}\tilde{\mathscr{D}}_\ell\rangle$, it's apparent that $f_{ed} = (X_n - c)^\ell \cdot f_{\tilde{ed}}$ vanishes on these functionals.

So f_{ed} vanishes on H, and satisfies the first three terms of the **Assumption**.

In summary $\text{poly}(H)$ satisfies the **Assumption**. □

Remark 2 For $f_\lambda \in \text{poly}(H)$, $\lambda \in E(D)$ where $\text{poly}(H)$ is the result gotten in the algorithm above, we have the conclusion that $\text{LC}_n(f_\lambda) = (X_n - c)^{\text{proj}(\lambda)}$.

6.3 General Case of the n-variable ($n > 1$) Problem

P-General case:
Given a set of points with multiplicity structures in n $(n > 1)$ dimensional space $H = \{\langle p_1, D_1\rangle, \ldots, \langle p_t, D_t\rangle\}$ or in matrix form $\langle \mathscr{P} = (p_{ij})_{m\times n}, \mathscr{D} = (d_{ij})_{m\times n}\rangle$, we are going to get $\text{poly}(H)$.

Step 1 Write H as $H = H_1 \bigcup H_2 \bigcup \cdots \bigcup H_s$ where H_i $(1 \le i \le s)$ is a π-fibre ($\pi : H \mapsto k$ such that $\langle p = (p_1, \ldots, p_n), D\rangle \in H$ is mapped to $p_n \in k$) i.e. the points of H_i have the same X_n coordinate c_i, $i = 1, \ldots, s$, and $c_i \ne c_j, \forall i, j = 1, \ldots, s, i \ne j$.

Step 2 According to the **P-Special case**, we have $D_i' = D(H_i)$, $G_i = \text{poly}(H_i)$. Write H_i as $\langle \mathscr{P}_i, \mathscr{D}_i\rangle$, and define w_i as the maximum value of the elements in the last column of $\mathscr{D}_i$.

Step 3 $D := D_1'$, $G := G_1$, $i := 2$.

Step 4 If $i > s$, go to Step 5. Else

Step 4.1 $D := D + D_i'$; $\hat{G} := \varnothing$. View $E(D)$ as a monomial set $MS := E(D)$.

Step 4.2 If $\sharp MS = 0$, go to Step 4.7, else select the minimal element of MS under lexicographic order, denoted by LT. $MS := MS \setminus \{LT\}$.

Step 4.3

$$f_1 := \mathrm{CP}(LT, D, G),\ f_2 := \mathrm{CP}(LT, D_i', G_i).$$

$$v_\wp := \mathrm{proj}(g_\wp),\ \text{where}\ g_\wp := \mathrm{CLT}(LT, D_\wp'),\ \wp = 1, \ldots, i.$$

Step 4.4

$$q_1 := f_1 \cdot (X_n - c_i)^{w_i+1};\quad q_2 := f_2 \cdot \prod_{\wp=1}^{i-1}(X_n - c_\wp)^{w_\wp+1}.$$

$$pp1 := (X_n - c_i)^{w_i+1-v_i};\quad pp2 := \prod_{\wp=1}^{i-1}(X_n - c_\wp)^{w_\wp+1-v_\wp}.$$

Step 4.5 Use Extended Euclidean Algorithm to compute r_1 and r_2 s.t. $r_1 \cdot pp_1 + r_2 \cdot pp_2 = 1$.

Step 4.6 $f := r_1 \cdot q_1 + r_2 \cdot q_2$. Reduce f with the elements in $\hat{G}$ to get f'; $\hat{G} := \hat{G} \bigcup \{f'\}$. Go to Step 4.2.

Step 4.7 $G := \hat{G}$. $i := i + 1$. Go to Step 4.

Step 5 $\mathrm{poly}(H) := G$.

Theorem 3 *The* $poly(H)$ *obtained in* ***P-General case*** *satisfies the* ***Assumption***.

Proof It is easy to know $v_\wp \leq w_\wp + 1$ according to their definitions, so the polynomials $pp1$ and $pp2$ in Step 4.4 do make sense. And to prove Theorem 3, we need only to prove the situation that $s \geq 2$ in Step 1.

For $i = 2, D = D_1' + D_2'$, $\forall ed \in E(D), v := \mathrm{proj}(ed)$ and $X_0 := \frac{X^{ed}}{X_n^v}$. According to Sect. 4, we have $v = v_1 + v_2$. Based on the Remarks 1 and 2, f_1 and f_2 can be written as polynomials of $k(X_n)[X_1, \ldots, X_{n-1}] : f_1 = X_0 \cdot (X_n - c_1)^{v_1} + \textit{the rest}$ and $f_2 = X_0 \cdot (X_n - c_2)^{v_2} + \textit{the rest}$ and none of the monomials in *the rest* is greater than or equal to X_0. Because f_1 and $(X_n - c_1)^{w_1+1}$ vanish on H_1, f_2 and $(X_n - c_2)^{w_2+1}$ vanish on H_2, we know that $q_1 = f_1 \cdot (X_n - c_2)^{w_2+1}$ and $q_2 = f_2 \cdot (X_n - c_1)^{w_1+1}$ both vanish on $H_1 \bigcup H_2$. Then f vanishes on $H_1 \bigcup H_2$ where

$$\begin{aligned}
f &= r_1 \cdot q_1 + r_2 \cdot q_2 \\
&= X_0 \cdot (X_n - c_1)^{v_1} \cdot (X_n - c_2)^{v_2}(r_1 \cdot (X_n - c_2)^{w_2+1-v_2} + r_2 \cdot (X_n - c_1)^{w_1+1-v_1}) \\
&\quad + \textit{the rest} \\
&= X_0 \cdot (X_n - c_1)^{v_1} \cdot (X_n - c_2)^{v_2}(r_1 \cdot pp1 + r_2 \cdot pp2) + \textit{the rest} \\
&= X_0 \cdot (X_n - c_1)^{v_1} \cdot (X_n - c_2)^{v_2} + \textit{the rest}.
\end{aligned}$$

None monomial in *the rest* is greater than or equal to X_0 , so the leading term of f is obviously $X_0 \cdot X_n^v$ which is equal to ed. Naturally $\mathrm{LC}_n(f) = \prod_{j=1}^{i}(X_n - c_j)^{v_j}$ for $i = 2$.

We **assert** that for any i, the polynomial f in [step 4.6] satisfies that $\mathrm{LC}_n(f) = \prod_{j=1}^{i}(X_n - c_j)^{v_j}$.

When $i > 2$, assume the **assertion** above holds for $i-1$. $\forall\, ed \in E(D)$, $v :=$ proj(ed) and $X_0 := \frac{X^{ed}}{X_n^v}$. According to Sect. 4, we have $v = v_1 + \cdots + v_i$. Based on the **assertion** for $i = 1$, Remarks 1 and 2, f_1 and f_2 can be written as polynomials of $k(X_n)[X_1, \ldots, X_{n-1}]$:

$$f_1 = X_0 \cdot \prod_{j=1}^{i-1}(X_n - c_j)^{v_j} + \textit{the rest}$$

$$f_2 = X_0 \cdot (X_n - c_i)^{v_i} + \textit{the rest}$$

and none of the monomials in *the rest* is greater than or equal to X_0. Because f_1 and $\prod_{j=1}^{i-1}(X_n - c_j)^{w_j+1}$ vanish on $\bigcup_{j=1}^{i-1} H_j$, f_2 and $(X_n - c_i)^{w_i+1}$ vanish on H_i, we know that $q_1 = f_1 \cdot (X_n - c_i)^{w_i+1}$ and $q_2 = f_2 \cdot \prod_{j=1}^{i-1}(X_n - c_j)^{w_j+1}$ both vanish on $\bigcup_{j=1}^{i} H_j$. Then f vanishes on $\bigcup_{j=1}^{i} H_j$ where

$$\begin{aligned} f &= r_1 \cdot q_1 + r_2 \cdot q_2 \\ &= X_0 \cdot \prod_{j=1}^{i}(X_n - c_j)^{v_j}\left(r_1 \cdot (X_n - c_i)^{w_i+1-v_i} + r_2 \cdot \prod_{j=1}^{i-1}(X_n - c_j)^{w_j+1-v_j}\right) \\ &\quad + \textit{the rest} \\ &= X_0 \cdot \prod_{j=1}^{i}(X_n - c_j)^{v_j}(r_1 \cdot pp1 + r_2 \cdot pp2) + \textit{the rest} \\ &= X_0 \cdot \prod_{j=1}^{i}(X_n - c_j)^{v_j} + \textit{the rest.} \end{aligned}$$

None monomial in *the rest* is greater than or equal to X_0 and the leading term of f is obviously $X_0 \cdot X_n^v$ which is equal to ed. Hence the **assertion** holds for arbitrary i.

Therefore we have proved that for any element $ed \in E(D)$, $f_{ed} := f$ vanishes on H and the leading term is ed. In the algorithm, we compute f_{ed} in turn according to the lexicographic order of the elements of $E(D)$. Once we get a polynomial, we use the polynomials obtained previously to reduce it (refer to Step 4.6). Now to prove the lower terms of the polynomial are all in D after such a reduction operation.

Let D be a lower set, a be a monomial, define $L(a, D) = \{b \in \mathbb{N}_0^n; b \prec a, b \in D\}$. Given any $d \notin D$, there exist only two situations: $d \in E(D)$ or $d \notin E(D)$ but $\exists d' \in E(D)$, *s.t.* d' is a factor of d. Of course $d' \prec d$.

Consider the sequence $\mathfrak{R} = \{T_1, T_2, T_3, \ldots\}$ of all the monomials with the elements of D discarded and all the elements are arranged according to the lexicographic order, use induction on it to prove that for every element T_t $(t > 0)$ we can construct a vanishing polynomial with the leading term T_t and all the lower terms are in D, i.e. T_t can be represented as the linear combination of the elements of $L(T_t, D)$.

The very first vanishing polynomial we got in the algorithm is a univariate polynomial in X_n whose leading term is exactly T_1. It is obvious that the lower terms are in D, i.e. T_1 can be represented as the combination of the elements of $L(T_1, D)$.

Assume that T_{m-1} $(m \geq 2)$ can be written as the combination of the elements of $L(T_{m-1}, D)$, now to prove it is true for T_m.

If $T_m \in E(D)$, the algorithm provides us a vanishing polynomial whose leading term is T_m, i.e. T_m can be represented as the combination of the terms which are all smaller than T_m. According to the induction assumption, for any lower term $T \notin D$ of the polynomial, T can be represented as the linear combination of the elements of $L(T, D)$, then T_m can be represented as the linear combination of the elements of $L(T_m, D)$.

If $T_m \notin E(D)$, there exists $d' \in E(D)$ s.t. $T_m = T'_m \cdot d'$. Since $d' \prec T_m$, according to the assumption, we can substitute d' with the linear combination of the elements of $L(d', D)$. Since all the elements in $L(d', D)$ are smaller than d', then T_m can be represented as the combination of elements which are all smaller than T_m. Then for the same reason described in the last paragraph, T_m can be represented as the linear combination of the elements of $L(T_m, D)$.

Therefore for every element T_t $(t > 0)$ we can construct a vanishing polynomial with the leading term T_t and all the lower terms are in D. Particularly for any $ed \in E(D)$, all the lower terms of the polynomial f_{ed} we got in the algorithm after the reduction operation are in D. □

Remark 3 According to the proof of Theorem 3, f and f' in Step 4.6 for arbitrary i satisfy that $\mathrm{LC}_n(f) = \mathrm{LC}_n(f') = \prod_{j=1}^{i}(X_n - c_j)^{\nu_j}$.

Theorem 4 *Given a set of points H with multiplicity structures, $poly(H)$ is the reduced Gröbner basis of the vanishing ideal $I(H)$ and $D(H)$ is the quotient basis under lexicographic order.*

Proof Let m be the number of functionals defined by H and then $m = \dim(k[X]/I(H))$. Denote by J the ideal generated by $\mathrm{poly}(H)$. According to the **Assumption**, $\mathrm{poly}(H) \subseteq I(H)$. So $\dim(k[X]/I(H)) \leq \dim(k[X]/J)$. Let C be the set of the leading terms of polynomials in J under lexicographic order, then $C \supseteq \bigcup_{\beta \in E(D(H))}(\beta + \mathbb{N}_0^n)$ where the latter union is equal to $\mathbb{N}_0^n \backslash D(H)$. Then we can get $C' = \mathbb{N}_0^n \backslash C \subseteq D(H)$. Because $k[X]/J$ is isomorphic as a k-vector space to the k-span of C', here C' is viewed as a monomial set. So $\dim(k[X]/J) \leq \sharp D(H) = m$. Hence we have

$$m = \dim(k[X]/I(H)) \leq \dim(k[X]/J) \leq m.$$

Therefore $J = I(H)$, where $J = \langle \mathrm{poly}(H) \rangle$. Hence it is easy to know that $\mathrm{poly}(H)$ is exactly the reduced Gröbner basis of the vanishing ideal under lexicographic order, and $D(H)$ is the quotient basis. □

Based on Remark 3 and Theorem 4, we can naturally get the following lemma.

Lemma 1 *Assume G is the reduced Gröbner basis of some zero dimensional n-variable polynomial ideal under lexicographic order with $X_1 \succ X_2 \succ \cdots \succ X_n$. Define $p_0(G)$ as the univariate polynomial in X_n of G. View $g \in G$ as polynomial of $k(X_n)[X_1, \ldots, X_{n-1}]$ and define $LC_n(g)$ to be the leading coefficient of g which is a univariate polynomial in X_n and we have the conclusion that $LC_n(g)$ is always a factor of $p_0(G)$.*

Proof In fact for any given zero dimensional n-variable polynomial ideal, its reduced Gröbner basis G can be constructed from its zeros in the way our algorithm provides. Because the reduced Gröbner basis under lexicographic order is unique, Remark 3 holds for all the elements of G i.e. $\forall g \in G, \mathrm{LC}_n(g) = \prod_{j=1}^{s}(X_n - c_j)^{v_j}$ and particularly $p_0(G) = \prod_{j=1}^{s}(X_n - c_j)^{w_j+1}$ (refer the algorithm of **P-General case** for the symbols c_j, v_j, w_j). Because $v_j \leq w_j + 1$, $\mathrm{LC}_n(g)$ is a factor of $p_0(G)$. □

7 Intersection of Ideals

Based on Lemma 1 and the algorithm of **P-General case** in Sect. 6, we present a new algorithm named **Intersection** to compute the intersection of two ideals I_1 and I_2 satisfying that the greatest common divisor of $p_0(G_1)$ and $p_0(G_2)$ equals 1 where G_1 and G_2 are respectively the reduced Gröbner bases of I_1 and I_2 under the lexicographic order i.e. satisfying that the zeros of I_1 and that of I_2 does not share even one same X_n coordinate.

Denote by $Q(G)$ the quotient basis where G is the reduced Gröbner basis. The following algorithm **CPI** (short for *Computing the Polynomial for Intersection*) is a sub-algorithm called in algorithm **Intersection**.

Algorithm **CPI:** G is a reduced Gröbner basis, for any given monomial LT which is not in $Q(G)$, we get a polynomial p in $\langle G \rangle$ whose leading term is a factor of LT: the $X_1, \ldots, X_{n-1}$ components of the leading term are the same with that of LT and the X_n component has the lowest degree. Denoted by $p := \mathrm{CPI}(LT, G)$.

Step 1 $G' := \{g \in G |$ the leading monomial of g is a factor of LT $\}$.
Step 2 $G'' := \{g \in G' |$ the leading monomial of g has the smallest degree of X_n for that of all the elements in $G'\}$.
Step 3 Select one element of G'' and multiply it by a monomial of $X_1, \ldots, X_{n-1}$ to get p whose leading monomial is LT.

Algorithm **Intersection:** G_1 and G_2 are the reduced Gröbner bases of two different ideals satisfying that $\mathrm{GCD}(p_0(G_1), p_0(G_2)) = 1$. Return the reduced Gröbner basis of the intersection of these two ideals, denoted by $G := \mathrm{Intersection}(G_1, G_2)$.

Step 1 $D := Q(G_1) + Q(G_2)$. View $E(D)$ as a monomial set. $G := \varnothing$.

Step 2 If $E(D) = \varnothing$, the algorithm is done. Else select the minimal element of $E(D)$, denoted by T. $E(D) := E(D) \backslash \{T\}$.

Step 3

$$f_1 := \mathrm{CPI}(T, G_1),\ \ f_2 := \mathrm{CPI}(T, G_2).$$

$$q_1 := f_1 \cdot p_2,\ q_2 := f_2 \cdot p_1.$$

Step 4

$$t_1 := \frac{p_0(G_2)}{\mathrm{LC}_n(f_2)},\ t_2 := \frac{p_0(G_1)}{\mathrm{LC}_n(f_1)}.$$

Step 5 Use Extended Euclidean Algorithm to find r_1, r_2 s.t.

$$r_1 \cdot t_1 + r_2 \cdot t_2 = 1.$$

Step 6 $f := q_1 \cdot r_1 + q_2 \cdot r_2$. Reduce f with G to get f', and $G := G \bigcup \{f'\}$. Go to Step 2.

Remark 4 This algorithm is essentially the same with Step 4.1–Step 4.7 of **P-General case** in Sect. 6, so it is obvious that r_1, r_2 in Step 5 do exist and the polynomials in the algorithm are all well defined. Besides, D in Step 1 is not empty, so it is easy to know the result of this algorithm can never be empty.

Because this algorithm is essentially the same with Step 4.1–Step 4.7 of **P-General case** in Sect. 6, here we omit the proof. And in return, the algorithm of **P-General case** in Sect. 6 can be simplified according to this **Intersection** algorithm: we can delete the last sentence in Step 2 and replace Step 4.3 and Step 4.4 respectively by:

Step 4.3′

$$f_1 := \mathrm{CP}(LT, D, G),\ f_2 := \mathrm{CP}(LT, D_i', G_i).$$

Step 4.4′

$$q_1 := f_1 \cdot p_0(G_i);\ \ q_2 := f_2 \cdot p_0(G).$$

$$pp1 := \frac{p_0(G_i)}{\mathrm{LC}_n(f_2)};\ \ pp2 := \frac{p_0(G)}{\mathrm{LC}_n(f_1)}.$$

8 Conclusion

During the induction of the algorithm in Sect. 6, we can record the leading coefficients for later use to save the computation cost and the computation cost is mainly on the Extended Euclidean Algorithm. But it's hard to compute how many times we need to use the Extended Euclidean Algorithm for a given problem, and the computation cost depends closely on the structures of the given set of points.

The benefit is the explicit geometric interpretation. For any given point with multiplicity structure $\langle p, D\rangle$, we put the lower set D into the affine space with the $(0, 0)$ element situated at p to intuitively represent it, and it can be imagined as $\sharp D_i$ small balls in the affine space. Given a set of points with multiplicity structures in n dimensional space, in the way showed in Sect. 2, we push the balls first along X_1-axis, then along X_2-axis, and so on, at last along X_n-axis to finally get a lower set which turns out to be exactly the quotient basis under the lexicographic order with $X_1 \succ X_2 \succ \cdots \succ X_n$. In the future, we will try to apply this geometric interpretation to our previous work on Birkhoff problem [14].

The geometric interpretation in this paper reveals the essential connection between the relative position of the points with multiplicity structures and the quotient basis of the vanishing ideal. It provides us a new perspective of view to look into the vanishing ideal problem and helps study the structure of the reduced Gröbner basis of zero dimensional ideal under lexicographic order. The new algorithm **Intersection** which computes the intersection of two ideals and Lemma 1 are the direct byproducts of our algorithm. Lemma 1 reveals important property of the reduced Gröbner basis under lexicographic order, which is necessary for a set of polynomials to be a reduced Gröbner basis. Lemma 1 can also help us to solve the polynomial system. It is well-known that the Gröbner basis of an ideal under lexicographic order holds good algebraic structures and hence is convenient to use for polynomial system solving [15]. Once we get the reduced Gröbner basis G of a zero dimensional ideal, to solve the polynomial system, we need first compute the roots of $p_0(G)$. Since $\mathrm{LC}_n(g)$ $(g \neq p_0(G),\ g \in G)$ is a factor of $p_0(G)$, computing the roots of $\mathrm{LC}_n(g)$ which has a smaller degree would be helpful for saving the computation cost.

Lederer [13] presented an algorithm to compute the reduced Gröbner basis of the vanishing ideal over a set of points with no multiplicity structures. The author splits the problem into several small ones and combines the results of the small problems by using Lagrange interpolation method to get the final result and the idea really inspired us a lot. Because the problem considered here concerning the points with multiplicity structures, we have to consider **P-Special case** and **P-General case**, and the Lagrange interpolation method is not available any more, we use the Extended Euclidean Algorithm instead.

References

1. Stetter, H.J.: Numerical Polynomial Algebra. SIAM, Philadelphia (2004)
2. Pistone, G., Riccomagno, E., Wynn, H.P.: Algebraic Statistics: Computational Commutative Algebra in Statistics. Monographs on statistics & applied probability. Chapman & Hall/CRC, Boca Raton (2001)
3. Laubenbacher, R., Stigler, B.: A computational algebra approach to the reverse engineering of gene regulatory networks. J. Theor. Biol. **229**, 523–537 (2004)
4. Guruswami, V., Sudan, M.: Improved decoding of Reed-Solomon and algebraic geometric codes. IEEE Trans. Inf. Theory **46**, 1757–1767 (1999)
5. Koetter, R., Vardy, A.: Algebraic soft-decision decoding of Reed-Solomon codes. IEEE Trans. Inf. Theory **49**, 2809–2825 (2003)

6. Sudan, M.: Decoding of Reed-Solomon codes beyond the error-correction bound. J. Complex. **13**, 180–193 (1997)
7. Buchberger, B., Möller, H.M.: The construction of multivariate polynomials with preassigned zeros. In: Computer Algebra, EUROCAM'82. Lecture Notes in Computer Science, vol. 144, Springer, Berlin, pp. 24–31 (1982)
8. Marinari, M.G., Möller, H.M., Mora, T.: Gröbner basis of ideals defined by functionals with an application to ideals of projective points. Appl. Algebra Eng. Commun. Comput. **4**, 103–145 (1993)
9. Abbott, J., Bigatti, A., Kreuzer, M., Robbiano, L.: Computing ideals of points. J. Symb. Comput. **30**, 341–356 (2000)
10. Farr, J., Gao, S.: Computing Gröbner bases for vanishing ideals of finite sets of points. In: Proceedings of the Conference, ACA (2003)
11. Felszeghy, B., Ráth, B., Rónyai, L.: The lex game and some applications. J. Symb. Comput. **41**, 663–681 (2006)
12. Cerlienco, L., Mureddu, M.: From algebraic sets to monomial linear bases by means of combinatorial algorithms. Discret. Math. **139**, 73–87 (1995)
13. Lederer, M.: The vanishing ideal of a finite set of closed points in affine space. J. Pure Appl. Algebra **212**, 1116–1133 (2008)
14. Lei, N., Chai, J., Xia, P., Li, Y.: A fast algorithm for the multivariate Birkhoff interpolation problem. J. Comp. Appl. Math. **236**, 1656–1666 (2011)
15. Cox, D., Little, J., OShea, D.: Ideals, Varieties, and Algorithms. Springer, New York (1997)

Signature-Based Method of Deciding Program Termination

Yaohui Li, Yuqing Song and Zhifeng Wu

Abstract We present a method based on the discriminant sequence and Gröbner bases to verify the termination of a class of linear program. This method relates the program termination to the existence of real zeros of polynomial system with constraint conditions. To avoid the wrong determination due to approximate computation, we use Gröbner bases and revised sign list to count sign changes of characteristic polynomial of a trace matrix. This method need not solve the equations of polynomial system but count the number of real zeros which satisfy the constraint condition by using symbolic computation. The number of real zeros of polynomial in the linear program can always be computed correctly. Therefore, the termination of the program can be decided accurately.

Keywords Linear loop program · Termination · Program verification · Gröbner basis · Sturm sequence

1 Introduction

Analysis of program termination plays a very important role in program verification. However, the total program verification problem is undecidable, it is doomed to be impossible to find an universal approach to mechanically verify the correctness of program without any simplification or restriction [1]. Considering the linear program of the following form

$$P : while(\boldsymbol{c}^T \boldsymbol{x} > 0)\{\boldsymbol{x} := A\boldsymbol{x}\} \tag{1}$$

This paper is supported partially by No. 20071311 of Education Committee of Tianjin Municipality. The research is also supported by Talent Foundation in TUTE through grant KYQD06005 and Natural Science Foundation of TUTE under grant No. KJY12-09 and KJ20080039.

Y. Li (✉) · Y. Song · Z. Wu
Department of Computer Science, Tianjin University of Technology and Education, Tianjin 300222, People's Republic of China
e-mail: philliplee@163.com

© Springer-Verlag Berlin Heidelberg 2014

R. Feng et al. (eds.), *Computer Mathematics*, DOI 10.1007/978-3-662-43799-5_22

where $x \in R^n$ and A is a $M \times N$ matrix. Tiwari proves that the termination of a class of single-path loop with linear guards is decidable and provides a decision procedure via constructive proofs [2]. Yang et al. convert the program verification to a semi-algebraic system, then use DISCOVERER to solve it [1, 3] so as to decide the termination of the program. If the system has nontrivial zero points, then it is nonterminating. Li et al. in their paper discuss the termination of a linear program with nonlinear loop condition [4]. In Yang and Li's method, they have to solve algebraic equations so as to decide the termination of linear program. In order to guarantee the accuracy of solution, they isolate the real root in the tiny interval rather than the approximate value.

In this paper, we present new results on using signature of characteristic polynomial of a matrix to decide the termination of a class of linear programs. When checking the existence of real zeros of polynomial system, we do not solve the semi-algebraic system but using improved Sturm sequence and signature theory to determine whether the matrix have positive eigenvalue and eigenvector such that $c^T v > 0$. It can be seen that there is no approximate computation in our method. Hence, all procedures of computation are accurate and the result of decision is trustworthy.

2 Determination of Positive Eigenvalues

2.1 Linear Loop Program

As to the simple loop program P, Tiwari analyzes the decidability of its termination. His decision procedure is based on the observation that only the eigenvectors (and the generalized eigenspace) corresponding to positive real eigenvalues of the assignment matrix are relevant for program termination.

Theorem 1 *If the linear loop program P, defined by an $(N \times N)$ matrix A and a nonzero $N \times 1$-vector $\mathbf{c}$, is nonterminating then there exists a real eigenvector $\mathbf{v}$ of A, corresponding to positive eigenvalue, such that $\mathbf{c}^T \mathbf{v} \geq 0$.*

For the proof of Theorem 1 refer to [2] for further detail. Usually, in order to decide the termination of program many papers compute the eigenvalues of A directly after converting this problem to algebraic expression. However, this may lead to the wrong result because of the error in approximate computation. To overcome this, we count the number of positive eigenvalues in the characteristics polynomial of A by using Sturm sequence and improved Descart's rule of sign.

According to the definition of linear algebra, suppose λ is the eigenvalue of matrix A, then $(A - \lambda E)x = 0$ and λ is the root of $|A - \lambda E| = 0$. Assume that $f_{cp}(\lambda) = a_0\lambda^m + a_1\lambda^{m-1} + \cdots + a_{m-1}\lambda + a_m$, where m is the order of A, is the characteristic polynomial. To decide the termination of the program accurately, we do not solve this univariate equation for checking the positivity of eigenvalue. In our method, the idea is counting the number N of positive eigenvalues over R by using sturm sequence or revised sturm sequence in real algebra. If number N is 0, this means that program P is terminating undoubtedly. Our idea is to check whether λ has positive root directly by using Sturm-Tarski theorem and number of sign changes in real algebra.

2.2 Real Roots Discrimination System of Polynomials

In this section, we review some definitions and theorems which mainly come from [3, 7]. Yang in [7] presented a method to determine the number of real roots of univariate polynomial. Let $f(x) \in R[x]$ with $f(x) = a_0x^m + a_1x^{m-1} + \cdots + a_m$ and $g(x)$ be another polynomial. Then, consider $r(x)$ as

$$r(x) = \text{rem}(f'g, f) = b_1x^{m-1} + \cdots + b_m,$$

where f' is the derivation of f.

Definition 1 The $2n \times 2n$ matrix $M =$

$$\begin{pmatrix} a_0 & a_1 & a_2 & \cdots & a_m & & & \\ 0 & b_1 & b_2 & \cdots & b_m & & & \\ & a_0 & a_1 & a_2 & \cdots & a_m & & \\ & 0 & b_1 & b_2 & \cdots & b_m & & \\ & & \vdots & \vdots & \vdots & & & \\ & & & a_0 & a_1 & a_2 & \cdots & a_m \\ & & & 0 & b_1 & b_2 & \cdots & b_m \end{pmatrix}$$

is called as discriminant matrix of f and g, and denoted it by Discr(f,g). If $g = 1$, the matrix is called as discriminant matrix of f.

Definition 2 For $1 \leqslant k \leqslant 2n$, let M_k be the kth principal minor of M, let $D_0 = 1$, $D_k(f, g) = M_{2k}$. The n-tuple $D = [D_0, D_1(f, g), D_2(f, g), \ldots, , D_n(f, g)]$ is called the discriminant sequence of f with respect to g, denoted by $GDL(f, g)$

In practice, we usually get the discriminant sequence from the Bezout matrix of $f(x)$ and $g(x)$, which is the same as that getting from the Sylvester matrix, read [3, 7] for details.

Definition 3 If sgn(x) is the signum function, sgn(0) = 0, then the list $[s_0, s_1, s_2, \ldots, s_n]$ = [sgn(D_0), sgn(D_1), sgn(D_2),..., sgn(D_n)] is called the sign list of the discriminant sequence D.

Definition 4 Given a sign list $[s_1, s_2, \ldots, s_n]$, we construct a new list $[\varepsilon_1, \varepsilon_2, \ldots, \varepsilon_n]$ as follows:

If $[s_i, s_{i+1}, \ldots, s_{i+j}]$ is a section of the given list, where $s_i \neq 0, s_{i+1} = s_{i+2} = \cdots = s_{i+j-1} = 0, s_{i+j} \neq 0$, then, we replace the subsection $[s_{i+1}, s_{i+2}, \ldots, s_{i+j-1}]$

by $[-s_i, -s_i, s_i, s_i, -s_i, -s_i, si, si, -s_i, \ldots]$, i.e. $\varepsilon_{i+r} = (-1)^{\lfloor (r+1)/2 \rfloor} \cdot s_i$ for $r = 1, 2, \ldots, j-1$. Otherwise, $\varepsilon_k = s_k$. We call this new list as revised sign list(*RSL*).

Based on the symbol and definition above, there is the following theorem:

Theorem 2 *Given real coefficient polynomials $f = f(x)$ and $g = g(x)$, if the sign changes number of $RSL(f, g)$ is n and $D_\eta \neq 0$ but $D_t = 0(t > \eta)$, then*

$$\eta - 2n = \# f_{g_+} - \# f_{g_-},$$

where

$$\# f_{g_+} = \text{card}(\{x \in R | f(x) = 0, g(x) > 0\}),$$
$$\# f_{g_-} = \text{card}(\{x \in R | f(x) = 0, g(x) < 0\}).$$

If there is only one variate in $\boldsymbol{x}$ of P, regard the constraint condition as $g(x)$ and then count the number of real zeros of characteristic polynomial such that $g(x) > 0$ according to Theorem 2. If there is no positive eigenvalue, termination of the program can be decided easily according to Theorem 1. While the positive eigenvalue exists, we have to consider whether there is eigenvector v corresponding to it such that $\boldsymbol{c}^T \boldsymbol{v} \geq 0$.

2.3 Existence of Positive Eigenvalues of Matrix in Linear Program

By t we represent the variable in character polynomial of the matrice in the paper. Let $f(t) = f_{cp}(t)$ and the number of positive eigenvalue be N. According to Theorem 1, if N is greater than 0 and there exists at least an eigenvector v, which a positive eigenvalue corresponds to, such that $c^T v > 0$ then program p is not termination. So, the termination of linear program can be converted to the problem whether matrix A has positive eigenvalues at first. We determine the existence of positive real eigenvalues by using the result in the above subsection.

To determine the number of positive eigenvalues, we choose a suitable $g(t)$ and let $g(t) = t$. We construct discriminant matrix by using Bezout method but not Sylvester method because the size of Bezout matrix is small. Assume that the revised sign list RSL of $(f_{cp}(t), t)$ has $m+1$ nonzero elements in total, we count the sign changes number of RSL and let the number be n. As $g(t) = t$, f_{g_+} represents the number of eigenvalues such that $t > 0$, i.e., positive eigenvalues; similarly, f_{g_-} represents the number of negative eigenvalues. Therefore, $\eta - 2n$ is the difference between them. It is meaningful in counting real zeros even if we do not know the numbers of positive eigenvalues and negative ones at present.

The next step is to count the total number of nontrivial eigenvalues. In this step, let $g(t) = t^2$, then we construct the revised sign list similarly to the above and count the number of sign changes. According to Theorem 2, we still count the number of sign changes, and suppose it is n'. In fact, n' is the number of real roots of $f(t)$ in the first step such that $t^2 > 0$ no matter whatever $t < 0$ or $t > 0$. Thus, we count the total number of nonzero real eigenvalues according to Theorem 2. On the other hand, the number of elements in revised sign list is decided by the size of Bezout matrix. The order of Bezout matrix is $\max(\deg(f), \deg(r))$. Although $f'g(t) = t^2 f'$, $\deg(f'g/f) \le m - 1$. From this, $\max(\deg(f), \deg(r)) = m$. Therefore, the length of sign list is still m and $m - 2n'$ is the total number of real eigenvalues which is not equal to 0.

$$\begin{pmatrix} 1 & -1 \\ 1 & 1 \end{pmatrix} \begin{pmatrix} \# f_{t_>} \\ \# f_{t_<} \end{pmatrix} = \begin{pmatrix} m - 2n \\ m - 2n' \end{pmatrix}$$

$\# f_{t_{(\cdot)}}$ represents card($\{t \in R | f(t) = 0, g(t) \cdot 0\}$). Finally, the number of positive eigenvalues is $a = \# f_{t_>} = (m - 2n + m - 2n')/2 = m - (n + n')$.

Theorem 3 *Let the characteristic polynomial of A in program P be $f_{cp}(t)$, its discriminant sequence, and the discriminant sequences with respect to $g = 1$, $g = t$, $g = t^2$ be D , D',D'', respectively. The numbers of sign changes in RSLs they corresponding to are n, n' and n''. If $m = 2n''$ or $m = (n' + n'')$, then program p is terminating.*

Proof According to the definition, the sequences of D is constructed when $g = 1$. No matter what real zero of $f_{cp}(t)$ is, g is always greater than 0. Therefore, $m - 2n$ is the number of all real zeros, where n is the sign changes in RSL formed by the discriminant matrix Discr(f).

$$\begin{pmatrix} 1 & 1 & 1 \\ 0 & 1 & -1 \\ 0 & 1 & 1 \end{pmatrix} \begin{pmatrix} \# f_{t_=} \\ \# f_{t_>} \\ \# f_{t_<} \end{pmatrix} = \begin{pmatrix} m - 2n \\ m - 2n' \\ m - 2n'' \end{pmatrix},$$

By this, there is

$$\begin{pmatrix} \# f_{t_-} \\ \# f_{t_>} \\ \# f_{t_<} \end{pmatrix} = \begin{pmatrix} 1 & 0 & -1 \\ 0 & \frac{1}{2} & \frac{1}{2} \\ 0 & -\frac{1}{2} & \frac{1}{2} \end{pmatrix} \begin{pmatrix} m - 2n \\ m - 2n' \\ m - 2n'' \end{pmatrix}.$$

Characteristic polynomial f_{cp} has trivial eigenvalues only if $m = 2n''$. On the other hand, $\# f_{t_+} = m - (n' + n'') = 0$ means there is no positive eigenvalues either. Therefore, the theorem holds.

In paper [2, 4], they discussed several kinds of linear program with different loop conditions—linear or nonlinear conditions. Our method can be used always because all of them are determined also by using positive eigenvalue and eigenvector.

2.4 Signature Method of Determining Positive Eigenvalues

When using the method in Sect. 2.3 to determine program termination, we have to compute the sequence of principal minors. However, it is expensive to compute the determinant of different size for sign list. To overcome this, we propose a signature method to determine if there exists positive eigenvalues in the matrix of linear loop program.

Same with last subsection, we construct discriminant matrix by using Bezout method. It is lucky that discriminant matrix D of A is a real symmetry one. Hence, all eigenvalues of D are real according to the theory in linear algebra. The number of real eigenvalues of D is related to A in linear loop program. Suppose D is the discriminant matrix of $f_{cp}(t)$ with respect to polynomial $g(t)$, there exists the following Theorem for the number of eigenvalues of $f_{cp}(t)$.

Theorem 4 *The symbols are defined as above. Let D be discriminant matrix of $f(x)$ with respect to $g(x)$, its characteristic polynomial $f_{cp}(t)$. The difference σ between the number of sign changes of $f_{cp}(t)$ and $f_{cp}(-t)$ satisfies*

$$\sigma(f_{cp}) = \# f_{g_+} - \# f_{g_-}.$$

Proof According to the definition of Bezoutian, D is as follows:

$$\frac{f(x)r(y) - f(y)r(x)}{x - y} = \sum_{i,j=1}^{m} b_{ij} x^{i-1} y^{j-1} = \left(1\ x\ \cdots\ x^{m-1}\right)(b_{ij}) \begin{pmatrix} 1 \\ y \\ \vdots \\ y^{m-1} \end{pmatrix},$$

where (b_{ij}) is $M \times M$ matrix produced by coefficients b_{ij}, it is called as Bezout matrix. On the other hand, there exists [6]

$$\frac{f(x)r(y) - f(y)r(x)}{x - y} = \begin{cases} \frac{f(x)r(y) - f(y)r(x)}{x - y}, x \neq y, \\ r(x)f'(x) - r'(x)f(x), x = y. \end{cases}$$

Let $\alpha_1, \alpha_2, \ldots, \alpha_m$ be the eigenvalues of D, construct Vandermonde matrix

$$\Gamma(\alpha) = \begin{pmatrix} 1 & 1 & \cdots & 1 \\ \alpha_1 & \alpha_2 & \cdots & \alpha_m \\ \vdots & \vdots & \ddots & \vdots \\ \alpha_1^{m-1} & \alpha_2^{m-1} & \cdots & \alpha_m^{m-1} \end{pmatrix}$$

Consider a new matrix $S = \Gamma^T(b_{ij})\Gamma(\Gamma^T$ is the transpose of $\Gamma)$, the elements in matrix satisfy

$$\left(1\ \alpha_i\ \cdots\ \alpha_i^{m-1}\right)(b_{ij})\begin{pmatrix}1\\ \alpha_j\\ \vdots\\ \alpha_j^{m-1}\end{pmatrix} = \begin{cases}\dfrac{f(\alpha_i)r(\alpha_j)-f(\alpha_j)r(\alpha_i)}{x-y}=0, \alpha_i\neq\alpha_j,\\ r(\alpha_i)f'(\alpha_j)-r'(\alpha_i)f(\alpha_i)=r(\alpha_i)f'(\alpha_j),\\ \qquad\qquad\qquad\qquad \alpha_i=\alpha_j.\end{cases}$$

Therefore,

$$S=\begin{pmatrix} r(\alpha_1)f'(\alpha_1) & & & \\ & r(\alpha_2)f'(\alpha_2) & & \\ & & \ddots & \\ & & & r(\alpha_m)f'(\alpha_m)\end{pmatrix}$$

Because $f'g(\alpha)=r(\alpha)$, hence

$$S=\begin{pmatrix} g(\alpha_1)f'(\alpha_1)^2 & & & \\ & g(\alpha_2)f'(\alpha_2)^2 & & \\ & & \ddots & \\ & & & g(\alpha_m)f'(\alpha_m)^2\end{pmatrix}$$

The quadratic form $Q(g)$ is defined by S as follows

$$Q(g)=\sum_{i=1}^{m} g(\alpha_i)f'(\alpha_i)^2x_m^2=\sum_{\alpha\in R} g(\alpha_i)f'(\alpha_i)^2x_m^2+\sum_{\alpha\notin R} g(\alpha_i)f'(\alpha_i)^2x_m^2$$

The signature of

$$\sum_{\alpha\notin R} g(\alpha_i)f'(\alpha_i)^2x_m^2$$

is 0, since for α and $\bar{\alpha}$ complex eigenvalues of cp,

$$g(\alpha)f'(\alpha)^2x_m^2+g(\bar{\alpha})f'(\bar{\alpha})^2x_m^2$$

is a difference of two real squares. So the signature of $cp(x)$ agrees with the ones of $\sum_{\alpha\in R} g(\alpha_i)f'(\alpha_i)^2x_m^2$. Clearly, the signature of $\sum_{\alpha\in R} g(\alpha)f'(\alpha)^2x_m^2$ is equal to that of $g(\alpha)x_m^2$.

Furthermore, $S \equiv \Gamma^T(b_{ij})\Gamma$ indicates Bezout matrix is congruent to diagonal matrix S. Therefore, both S and (b_{ij}) have same signature. Meanwhile, the signature of S is equal to the difference of the number of positive eigenvalues and negative ones. Hence, the theorem holds.

According to this theorem, we construct discriminant matrix D of $f(t)$ with respect to $g(t)=t$. Then, derive its characteristic polynomial $f_{cp}(t)$ and count the number σ_+ and σ_- of sign changes of coefficients when $t = t$ and $t = -t$. From this, it is easy to compute the signature of $\sigma(t) = \sigma_+ - \sigma_-$, that is, the difference of the number between positive eigenvalues and negative ones. After this, Let $g(t) = t^2$ and count the signature $\sigma(t^2)$, i.e., the total number of eigenvalues except zeros. Last, we compute the number of positive eigenvalues and the result is $(\sigma(t) + \sigma(t^2))/2$.

Remark 1 Usually, the coefficients in $cp(t)$ are very large. We concern with the positivity or negativity of these numbers but not their value when counting number of sign changes. Therefore, we replace those coefficients that are greater than 0 by 1, that are less than 0 by -1 and ignore zero coefficients. The characteristic polynomial whose coefficients replaced by 1 or -1 is called as *sign polynomial.*

By using Theorem 4, unlike the method in above subsection, we can determine the existence of eigenvalues of A in program P in one step once having constructed discriminant matrix D. It is more efficient because we need not compute the principal minor sequence.

3 Positivity of Eigenvectors

The method in Sect. 2 is only suitable to the circumstance that there is no positive eigenvalue in $\boldsymbol{R}^n$. But if positive eigenvalue in A exists, the problem of checking program termination is complicated. Furthermore, we should also check whether its corresponding eigenvector $\boldsymbol{v}$ such that $\boldsymbol{c}^T\boldsymbol{v} > 0$ exists. Only such a condition satisfies, nontermination of the program P can be decided. In the following, we determine the existence of such a eigenvector but not solving equations directly.

Definition 5 In matrix A of program P, if there exists eigenvector $\boldsymbol{v}$ such that $\boldsymbol{c}^T\boldsymbol{v} > 0$, then we call it as positive eigenvector.

Regarding λ as variable, we still use $x_1, x_2, \ldots, x_m$ to represent all x_i and λ. Suppose the polynomial system that the equations corresponds to is $ps \in R(x_1, x_2, \ldots, x_m)$ and its ideal is I. Construct finite-dimensional algebra $A = R(x_1, x_2, \ldots, x_m)/I$. Clearly, all polynomials in A are linear combinations of the monomials $x^\alpha \notin \langle I \rangle$. Since this set of monomials is linearly independent in A, it can be regarded as a basis of A and is called as S-basis [5]. In order to obtain S-basis of polynomials, compute Gröbner basis G for total degree order of ideal I. Let $G = \{g_1, g_2, \ldots, g_n\}$, the leading items of $g_1, g_2, \ldots, g_n$ be $x^{\alpha_{g_1}}, x^{\alpha_{g_2}}, \ldots, x^{\alpha_{g_n}}$ respectively, where $x^{\alpha_{g_i}} = x_1^{\alpha_{g_{i1}}} x_2^{\alpha_{g_{i2}}} \ldots x_m^{\alpha_{g_{im}}}$ and $g_{i1} + g_{i2} + \cdots + g_{im} = g_i$. All of these terms are the generators of $\langle LT(I) \rangle$. Those monomials not lying in the ideal $\langle LT(I) \rangle$ are supposed to be $\{m_1, m_2, \ldots, m_k\}$ and $m_i \notin LT(I)(1 \le i \le k)$. All m_i form the S-basis of A. Then, we use multiplication to define a linear map m_f from $A = C[x_1, \ldots, x_n]/I$ to itself, where f is any polynomial. Using f multiply with m_i, the result is mapping to A and let it be $fm_i = a_{1i}m_1 + a_{2i}m_2 + \cdots + a_{ki}m_k$. Therefore, for polynomial f there is a matrix

$$m = \begin{bmatrix} a_{11} & a_{12} & \cdots & a_{1k} \\ a_{21} & a_{22} & \cdots & a_{2k} \\ \vdots & \vdots & \ddots & \vdots \\ a_{k1} & a_{k2} & \cdots & a_{kk} \end{bmatrix}$$

and $Tr(m) = \sum_{i=1}^{m} a_{ii}$. According to the theory in Ref. [5], any eigenvalue of m is a value of the function f on $V(I)$. Same with polynomial f, there is a matrix corresponding to each monomial $m_{ij} = m_i \cdot m_j \in A$ respectively, we represent it as m_{ij} still. The traces of matrix m_{ij} constructs an another matrix

$$M = \begin{bmatrix} \mathrm{Tr}(m_{11}) & \mathrm{Tr}(m_{12}) & \cdots & \mathrm{Tr}(m_{1m}) \\ \mathrm{Tr}(m_{21}) & \mathrm{Tr}(m_{22}) & \cdots & \mathrm{Tr}(m_{2m}) \\ \vdots & \vdots & \ddots & \vdots \\ \mathrm{Tr}(m_{m1}) & \mathrm{Tr}(m_{m2}) & \cdots & \mathrm{Tr}(m_{mm}) \end{bmatrix}$$

Clearly, M is a symmetry matrix. Hence, all of its eigenvalues are real. Moreover, $Tr(m_{ij})$ equals to the sum of m_{ij}'s eigenvalues and it is $\sum_{p \in V(I)} \mu(p) p^{\alpha(i)} p^{\alpha}(j)$, where $p^{\alpha(i)}$ denotes the value of the monomial $x^{\alpha(i)}$ at the point $p \in V(I)$ and $\mu(p)$ is the multiplicity of p. Similarly, given a polynomial $h \in k[x_1, x_2, \ldots, x_m]$ and $hm_{ij} = hm_i m_j$, there exists a matrix M_h whose entry is the form of $\sum_{p \in V(I)} h(p)\mu(p) p^{\alpha(i)} p^{\alpha(j)}$. From this, we obtain a matrix factorization $M_h = W \Delta h W^t$, where W is a matrix whose entry is $m_i(p_j)$ and Δh is the diagonal matrix with entries $h(p_1), h(p_2) \ldots h(p_k)$. According to linear algebra, there are two fundamental invariants under such changes of basis—the signature $\sigma(hm_{ij})$ and the rank $\rho(hm_{ij})$. There is the following theorem.

Theorem 5 *[5] Let I be a zero-dimensional ideal generated by polynomials in $k[x_1, \ldots, x_n](k \subset R)$, so that $V(I) \subset C^n$ is finite. Then, for $h \in k[x_1, \ldots, x_n]$, the signature and rank of the bilinear form $S_h(f, g) = Tr(m_{hfg})$ satisfy: $\sigma(S_h) = \#\{a \in V(I) \cap R^n : h(a) > 0\} - \#\{a \in V(I) \cap R^n : h(a) < 0\}$ $\rho(S_h) = \#\{a \in V(I) : h(a) \neq 0\}$.*

If $h = h_1, h_2, \ldots, h_s$ are the constraint conditions in the linear program P, we use Theorem 5 to determine whether there exists positive v such that $h(v) > 0$. The detail is as follows: Let λ be the eigenvalue of A and $h = C^T x$, now the problem converts to proving the existence of $\lambda > 0$ and $h > 0$ if the program P is nonterminating. Therefore, once computing the number of real zeros satisfies the above constraints, we determine the termination of P immediately. We use the signature of M_h to count the number of positive vectors.

In order to compute $m_1, \ldots, m_k$, we compute the Gröbner basis of ideal that polynomial system $X - AX$ generates. As the system is linear, it is easy to compute Gröbner basis $(g_1, \ldots, g_m)$ of the ideal generated by $p_1, \ldots, p_k$.

Definition 6 The remainder on division of polynomial f by a Gröbner basis G for I is called as Normal form of f, denoted by $\overline{f}^G$.

Let the normal forms of $p_1, \ldots, p_k$ w.r.t. G be $\overline{p_1}^G, \ldots, \overline{p_k}^G$, the leading terms of these normal forms are $(x^{\alpha_i} | x^{\alpha_i} \notin \langle LT(I) \rangle, 1 \leq i \leq s)$, where $x^{\alpha_i} = x_1^{\alpha_{i1}} x_2^{\alpha_{i12}} \ldots x_s^{\alpha_{is}}$ and $\alpha_{i1} + \cdots + \alpha_{is} = \alpha_i$. Because G is Gröbner basis of I in some order, therefore x_i^{α} is not always 1 and let them be $m_1, m_2, \ldots, m_m$. Then, we use any constraint condition, say h_k, to multiply $m_i m_j$ and construct a matrix $m_{hm_{ij}}$. Its trace is the i, j entry in M_{h_k}. After getting M_{h_k}, we compute its characteristic polynomial $cp(t)$. The number of positive eigenvalues of M_{h_k} is equal to the number of sign changes in the sequence of coefficients of $cp(t)$. Let $t = -t$, we count the number of sign changes and it is the negative eigenvalues. Assume they are n_1 and n_2 respectively, then we calculate the signature $\sigma(M_h) = n_1 - n_2$ and it is the signature of Δh_k too. According to Theorem 5, $n_1 - n_2$ is the difference of number of real zeros such that $h_k > 0$ and number of those such that $h_k < 0$. Similarly, let $h' = h_k^2$, we obtain the signature of $\sigma(M_h')$ and let it be $n_1' - n_2'$. It is the total number of real zeros such that $h_k \neq 0$. On the base of this, it is easy to compute the number of real zeros such that $h_k > 0$.

Usually, h is a set of n constraint conditions and those conditions form a "box". We should determine whether there exists positive eigenvector in this box. We count the number of sign changes of $cp(t)$ for any $h_k > 0$. $H||m$ represents the multiplication of any m constraint conditions, for example $H||2 = h_i h_j (i = 1..k, j = j + 1..k)$. Then, we compute the number of sign changes of each $H||m_k (k = 1..C_n^m)$. Finally, according to Theorem 2 we obtain the number of real zeros such that $h > 0$ and it is

$$N_R = \frac{1}{2^n} \sum_{i=1}^{m} \sum_{j=1}^{C_n^i} S(H||i_j), m = 1..n. \tag{2}$$

4 Experimental Result

In this section, we use three examples to show the use of signature-based method discussed above. It can be seen that the termination of the program is determined by using RSL or number of sign changes in characteristic polynomial of discriminant matrix directly.

Example 1 [2] The effect of two sequential assignments $x := x - y; y := x + 2y$ is captured by the simultaneous assignment

$$\begin{pmatrix} x \\ y \end{pmatrix} = \begin{pmatrix} 1 & -1 \\ 1 & 2 \end{pmatrix} \begin{pmatrix} x \\ y \end{pmatrix}$$

In the example,

$$A = \begin{pmatrix} 1 & -1 \\ 1 & 2 \end{pmatrix}$$

Its characteristic polynomial is $f_{cp} = t^2 - 3t + 3$. Let $g(t) = 1$, we compute the revised sign list of discriminant sequence and it turns out to be $[1, 1, -1]$. Clearly, the number of sign change is 1, which means the number of real eigenvalues is $2 - 2 \times 1 = 0$. Therefore, no matter what the loop condition is, this program is terminating always.

Example 2 While $(7x + 9y - 10z > 0)\{X := AX\}$ [8], where $X = [x \ \ y \ \ z]^T$ and

$$A = \begin{pmatrix} 7 & -10 & 17 \\ 3 & 9 & 2 \\ -4 & 5 & 13 \end{pmatrix}$$

At first, we construct the equations $(A - \lambda E)X = 0$ in eigenvalues λ. As a result, the system becomes three equations with four variables. However, as the system is homogenous, the solution of eigenvector is sure on the unit ball, that is, $x^2 + y^2 + z^2 - 1 = 0$. In the following, we compute the Gröbner basis G in total degree ordering of ideal that these four polynomial generates and $G = [\lambda^3 - 29\lambda^2 + 359\lambda - 2086, 106020x^2 + 761\lambda^2 - 11100\lambda - 29096, 106020yx - 469\lambda^2 + 11775\lambda - 108266, 106020y^2 - 382\lambda^2 + 5955\lambda - 43838, 106020zx + 56\lambda^2 - 6285\lambda + 41674, 106020zy - 139\lambda^2 + 2190\lambda - 38756, 106020z^2 - 33086 + 5145\lambda - 379\lambda^2, -7x + x\lambda + 10y - 17z, -3x - 9y + y\lambda - 2z, 4x - 5y - 13z + z\lambda]$. According to G, those monomials $m_m \notin \langle I \rangle$ are $A = [1, x, y, z, \lambda, \lambda^2]$. Let $h = 1$, then compute the entries of matrix of $H = hm_i m_j (i = 1..k, j = 1..k)$. when $i = j = 1$, $hm_i m_j = 1$ and $\overline{Hm_1}^G = 1 = 1 + 0x + 0y + 0z + 0a + 0a^2$, so the first column in matrix $M_{b_{11}}$ is $(1\ 0\ 0\ 0\ 0\ 0)^T$; Hm_2 is x, its coefficients in A are $(0\ 1\ 0\ 0\ 0\ 0)^T$..., so $M_{b_{11}}$ is identity matrix and $tr(M_{b_{11}}) = 6$. This is the $1, 1$ entry in M_h. In the second loop, when $i = 1, j = 2, H = x, \overline{Hm_1}^G = x$ and the coefficients in A are $(0\ 1\ 0\ 0\ 0\ 0)^T$; As $Hm_2 = 1 \times x \times x = x^2, \overline{Hm_2}^G = -\frac{761}{106020}\lambda^2 + \frac{185}{1767}\lambda + \frac{7274}{26505}$. The coefficients of $\overline{Hm_2}^G$ w.r.t. A is $\frac{7274}{26505}, 0, 0, 0, \frac{185}{1767}, -\frac{761}{106020}$. Same with this, $Hm_3 = xy$, the coefficients of $\overline{Hm_3}^G$ w.r.t. A is $54133/53010, 0, 0, 0, -785/7068, 469/106020$... From this, we obtain $M_{b_{11}}$ and it is

$$M_{b_{12}} = \begin{bmatrix} 0 & \frac{7274}{26505} & \frac{54133}{53010} & -\frac{20837}{53010} & 0 & 0 \\ 1 & 0 & 0 & 0 & 7 & -49 \\ 0 & 0 & 0 & 0 & -10 & -75 \\ 0 & 0 & 0 & 0 & 17 & 320 \\ 0 & \frac{185}{1767} & -\frac{785}{7068} & \frac{419}{7068} & 0 & 0 \\ 0 & -\frac{761}{106020} & \frac{469}{106020} & -\frac{14}{26505} & 0 & 0 \end{bmatrix}$$

Clearly, the 1, 2 entry of M_h is $tr(M_{b_{12}}) = 0$. Finally, all traces of $M_{b_{ij}}$ form M_h and is

$$M_{h=1}\begin{bmatrix} 6 & 0 & 0 & 0 & 58 & 246 \\ 0 & \frac{7013}{1178} & \frac{1367}{1767} & \frac{1119}{1178} & 0 & 0 \\ 0 & \frac{1367}{1767} & \frac{129}{1178} & \frac{4657}{3534} & 0 & 0 \\ 0 & \frac{1119}{1178} & \frac{4657}{3534} & -\frac{37}{589} & 0 & 0 \\ 58 & 0 & 0 & 0 & 246 & -1172 \\ 246 & 0 & 0 & 0 & -1172 & -1314 \end{bmatrix}$$

The characteristic polynomial is $f_{cp}(t) = t^6 + 1056t^5 - \frac{37605102449}{21204}t^4 + \frac{38103207691}{589}t^3 - \frac{1692609345794}{5301}t^2 - \frac{103356487836}{589}t + \frac{270918780300}{589}$. For our purpose, the exact values of the coefficients are much less important than their signs. The sign characteristic polynomial of $M_{h=1}$ is $f_{scp} = t^6+t^5-t^4+t^3-t^2-t+1$. Its number of sign changes is $s_1 = 2$. Then, let $h = 7x+9y-10z$, we get the number of sign changes is $s_h = 0$. At last, $N_R = \frac{1}{2}(2+0) = 1$, it means that there exists a positive eigenvalue and its eigenvector such that $\boldsymbol{c}^T\boldsymbol{v} > 0$. Therefore, this linear loop program is nonterminating.

Example 3 While $(3x_1^5 - 5x_2x_3^4 + 4x_4^5 - x_3^2x_4^3 > 0 \wedge 5x_3^3x_2^2 + 45x_1x_2^4 - 6x_2^3x_1^2 + x_4^5 > 0)\{X := AX\}$ [4], where

$$A = \begin{pmatrix} -55 & -94 & 87 & -56 \\ 0 & -62 & 97 & -73 \\ -4 & -83 & -10 & 62 \\ -82 & 80 & -44 & 71 \end{pmatrix}$$

Its characteristic polynomial is $f_{cp}(t) = t^4+56t^3+7938t^2+708976t-42287841$. At first, we construct discriminant matrix of fc with respect to $g(t) = t$, it is

$$D = \begin{bmatrix} -56 & -15876 & -2126928 & 169151364 \\ -15876 & -2571456 & 89746052 & 7104357288 \\ -2126928 & 89746052 & 1476505800 & 671361763716 \\ 169151364 & 7104357288 & 671361763716 & 29981064360816 \end{bmatrix}$$

Then, compute principal minor sequence ps from D and it is $ps = [-56, -108045840, 17985254002673987840, 2757698689534963996815770632971264]$. Further, the sign list is $sl = [1, -1, -1, 1, 1]$. As there is no zero in the list, so we need not revise it. From sign list sl, we obtain the number of sign changes is $sc_t = 2$.

Besides this, we have to know the total number of positive eigenvalues and negative ones. Let $g(t) = t^2$. Repeat the method above and get sign list $sl = [1, -1, -1, -1, -1]$ of f_{cp} with respect to t^2. The number of sign changes is $sc_t^2 = 1$. According to Theorem 3, the number of positive eigenvalues is $a = \deg(f_{cp}) - (sc_t + sc_t^2) = 4 - (2+1) = 1$. It means that matrix A has one positive eigenvalue.

Next, we discuss how to use signature method by using this example. On the basis of discriminant matrix of fc with respect to t, compute its characteristic polynomial and it is $f_{cp} = t^4 - a_1 t^3 - a_2 t^2 - a_3 t + a_4$, where $a_1 = 29982538295104$, $a_2 = 40658701138810637296 4416$, $a_3 = 15846586055162164781998 65724928$, $a_4 = 2757698689534963996815770632971264$. In fact, the coefficients in f_{cp} is not important because we only concern with the number of sign changes. Simplify f_{cp} as $f_{cp} = t^4 - t^3 - t^2 - t + 1$. The number of sign changes sc_+ is 2. Then, let $t = -t$. Count sign changes sc_- and its number is 2. So, when $g(t) = t, \sigma = 0$. It means that the number of the positive eigenvalues and that of negative ones is equal. Similar to the RSL method, then let $g(t) = t^2$. Sign polynomial is $f_{cp} = t^4 - t^3 + t^2 - t - 1$. The signature of $f_{cp}(t)$ and $f_{cp}(-t)$ is 2. It is easy to compute the number of positive eigenvalues of A and the result is 1.

Because this program has only one positive eigenvalue accurately, therefore we have to check whether there still exists positive eigenvector. The semi-algebraic system that the linear program corresponds to is as follows:

$$\begin{cases} f_1 = -(\lambda + 55)x_1 - 94x_2 + 87x_3 - 56x_4 = 0, \\ f_2 = -(\lambda + 62)x_2 + 97x_3 - 73x_4 = 0, \\ f_3 = -4x_1 - 83x_2 - (\lambda + 10)x_3 + 62x_4 = 0, \\ f_4 = -82x_1 + 80x_2 - 44x_3 + (71 - \lambda)x_4 = 0, \\ f_5 = x_1^2 + x_2^2 + x_3^2 + x_4^2 - 1 = 0, \\ \lambda > 0, \\ g_1 = 3x_1^5 - 5x_2x_3^4 + 4x_4^5 - x_3^2x_4^3 > 0, \\ g_2 = 5x_3^3x_2^2 + 45x_1x_2^4 - 6x_2^3x_1^2 + x_4^5 > 0 \end{cases}$$

We use software $maple$ to compute the gröbner basis G of ideal that f_1, f_2, f_3, f_4, f_5 generates in total degree order. In G, there are 15 polynomials and they are very long, so we do not write it out here. According to G, we obtain S-basis of A is $m = \{1, x_1, x_2, x_3, x_4, x_4^2, \lambda, \lambda^2\}$. Then, let $h = 1$ and compute M_h according to Sect. 4 for further. As M_h is a numerical matrix. We compute its characteristic polynomial very quickly. When $h = 1$, the sign style of characteristic polynomial is $f_{cp}(t) = t^8 - t^7 - t^6 + t^5 + t^4 - t^3 + t^2 - t + 1$. The number of sign changes in the sequence of coefficients of $f_{cp}(t)$ is 6. Let $t = -t$, the number of sign changes is 2. So, the signature is $\sigma(M_1) = 4$. Same with this, we compute the signature of M_h when $h = \lambda, g_1, g_2, \lambda g_1, \lambda g_2, g_1 g_2, \lambda g_1 g_2$. They are 0,2,0,2,0,2,−2, respectively. Finally, we obtain $N_R = 1$ according to formula (2). It means that there exists a positive eigenvalue and its eigenvector satisfies $\boldsymbol{c}^T \boldsymbol{v} > 0$. From this, it can be seen that this program is nonterminating.

5 Conclusion

In this paper, we present a symbolic method to decide the termination of a class of linear program. When the program is univariate, we use Sylvester resultant of discriminant matrix of the polynomial in the program to compute the number of positive eigenvalues. If there is no positive eigenvalue, the linear program is terminating. When positive eigenvalue exists or the program is multivariate, it is complicated to determine whether the program is terminating. In this case, we convert the terminating problem to the number of real zeros of polynomial system by using number of sign changes of characteristic polynomial. This method counts the number of real zeros by using the number of sign changes of characteristic polynomial but not computing the value of real zeros. Therefore, it avoids the error of approximate computation and can always determine the termination of the program trustworthily.

References

1. Yang, L., Zhan, N., Xia, B., et al. Program verification by Using DISCOVERER. Proc.VSTTE'05, Zurich. (2005)
2. Tiwari, A.: Termination of Linear Programs, pp. 70–82. CAV'04, USA (2004). LNCS
3. Yang, L., Xia, B.C.: Automated Proving and Discovery of Inequalities. Science Press, China (2008)
4. Li, J., Li, Y., Feng, Y.: Termination of a class of nonlinear loop programs. J. Sichuan Univ. (Eng. Sci. Edition) **41**(1), 129–133 (2009)
5. Cox, D.A., Little, J.B., O'Shea, Donald, B.: Using Algebric Geometry. Springer, New York (1998)
6. Mirshra, B.: Algorithmic Algebra. Springer, New York (1993)
7. Yang, L., Zhang, J.Z., Hou, X.R.: Nonlinear Algebraic Equation System and Automated Theorem Proving. ShangHai Scientific and Technological Education Publishing House, Shanghai (1996)
8. Zhou, C.Z.: Formal Techniques for software Embeded System. East-China Normal University, Shanghai (2008)

High-Precision Eigenvalue Bound for the Laplacian with Singularities

Xuefeng Liu, Tomoaki Okayama and Shin'ichi Oishi

Abstract For the purpose of bounding eigenvalues of the Laplacian over a bounded polygonal domain, we propose an algorithm to give high-precision bound even in the case that the eigenfunction has singularities around reentrant corners. The algorithm is a combination of the finite element method and the Lehmann–Goerisch theorem. The interval arithmetic is adopted in floating point number computation. Since all the error in the computation, e.g., the function approximation error, the floating point number rounding error, are exactly estimated, the result can be mathematically correct. In the end of the chapter, there are computational examples over an L-shaped domain and a square-minus-square domain that demonstrate the efficiency of our proposed algorithm.

Keywords Eigenvalue problem · Elliptic operator · Finite element method · Verified computation · Lehmann-Goerisch's theorem

1 Introduction

This chapter is a continued work of [10] on evaluating the eigenvalues of the Laplacian over a bounded polygonal domain Ω,

X. Liu (✉)
Research Institute for Science and Engineering, Waseda University, 3-4-1 Okubo, Shinjuku-ku,Tokyo 169-8555, Japan
e-mail: xfliu.math@gmail.com

T. Okayama
Graduate School of Economics, Hitotsubashi University, 2-1 Naka, Kunitachi, Tokyo 186-8601, Japan

S. Oishi
Faculty of Science and Engineering, Waseda University, 3-4-1 Okubo, Shinjuku-ku, Tokyo 169-8555, Japan

S. Oishi
CREST/JST, Saitama, Japan

© Springer-Verlag Berlin Heidelberg 2014
R. Feng et al. (eds.), *Computer Mathematics*, DOI 10.1007/978-3-662-43799-5_23

$$-\Delta u = \lambda u \text{ in } \Omega, \; u = 0 \text{ on } \partial\Omega.$$

The above eigenvalue problem has a spectrum of infinitely many eigenvalues (see, e.g., [5]), $0 < \lambda_1 \le \lambda_2 \le \lambda_3 \cdots$. To give explicit bounds for the eigenvalues is not an easy work, especially when the eigenfunction associated to the eigenvalue has a singularity around the reentrant corner.

In [10], based on the finite element method (FEM), X. Liu et al. proposed a robust method to bound the eigenvalues of the Laplacian, where the difficulty caused by the singularity of the eigenvalue problem is solved by introducing a hyper-circle equation originating from Prager–Synge's theorem. For detailed literature on approximate and rigorous eigenvalue estimation, refer to [10] and the references therein. A drawback of the method proposed in [10] is that it can only provide relatively rough bounds for the eigenvalues; a high-precision bound requires mesh refinement and the resulting huge matrices are very difficult to process.

In this chapter, we aim to develop a method that provides rigorous high-precision bounds for the leading eigenvalues of the Laplacian while only light computation is needed. The Lehmann–Goerisch theorem and Kato's bound [2, 7, 8] are well known for being able to provide precise bounds for up to the nth eigenvalue if the lower bound ν for the $(n+1)$th eigenvalue is available. As an example, the efficiency of such bound has been demonstrated by solving the eigenvalue problems over an L-shaped domain [18], where a pseudo lower bound ν is used. However, finding an exact *a priori* estimation ν is not an easy task. The homotopy method developed by Goerisch [6] and Plum [13] independently is a good choice for this purpose, which sets up a connection between a base problem with an explicit spectrum and the objective eigenvalue problem. When the domain in consideration is not so regular, for example, a dumbbell-shaped domain, the base problem in need is not so easy to construct and a slightly complex domain deformation homotopy becomes necessary; see e.g., [15].

Since the proposed method of [10] can provide the a priori bound needed by the Lehmann–Goerisch theorem, it is natural to combine the two methods to obtain high-precision bounds. The Lehmann–Goerisch theorem is an extension of Lehmann's theorem [8] by weakening the strict preconditions while the precision is a little dropped. In this chapter, we only apply Lehmann's theorem to bound the eigenvalues of the Laplacian.

The remainder of this chapter is organized as follows: In Sect. 2, we provide some preliminary background and introduce the result of [10]. In Sect. 3, Lehmann's theorem and the way to construct the approximate function needed by this theorem are introduced. In Sect. 4, we explain how to calculate the integration of singular base functions. In Sect. 5, two computation examples are presented.

2 Eigenvalue Bound Based on FEM

First, let us introduce the function spaces to be used. The space $L_2(\Omega)$ contains all the real-square integrable functions over Ω with standard norm $\|\cdot\|_{L_2}$, and $H^k(\Omega)$ $(k = 1, 2, \ldots,)$ are the kth order Sobolev function spaces that the function has up to the kth derivative to be in $L_2(\Omega)$. Let $|\cdot|_{H^k}$ and $\|\cdot\|_{H^k}$ be the semi-norm and norm of $H^k(\Omega)$, respectively. Denote by $(\cdot,\cdot)_\Omega$ the inner product in $L_2(\Omega)$ or $(L_2(\Omega))^2$. The inner product is abbreviated to be $(\cdot,\cdot)$ if the domain is obvious. Define $V := H_0^1(\Omega)$ by $H_0^1(\Omega) = \{v \in H^1(\Omega) \mid v = 0 \text{ on } \partial\Omega\}$ associated with the norm $\|\cdot\|_V := |\cdot|_{H^1}$.

The following Poisson's equation plays an important role in considering the eigenvalue problem of the Laplacian: For $f \in L_2(\Omega)$, find u in V such that

$$(\nabla u, \nabla v) = (f, v), \quad \forall v \in V. \tag{1}$$

In the case that the domain has a reentrant corner, the solution of problem (1) may have a singularity around the reentrant corner. Suppose that the angle of the inner corner is π/ω , and set up the polar coordinate (r,θ) with the origin point at the reentrant corner. Then, we have (see [11]),

$$u(r,\theta) = c_0 r^\omega \sin(\omega\theta) + O(r^{2\omega}) + O(r^{\omega+2}) \quad (1/2 < w < 1). \tag{2}$$

Due to the leading term r^ω, such a solution does not belong to $H^2(\Omega)$ if $c_0 \neq 0$.

Let $\mathcal{T}_h$ be a proper triangulation of the domain Ω and h the mesh size of the triangulation. Denote by $V^h (\subset V)$ the conforming FEM space over $\mathcal{T}_h$, in which we seek an approximation solution of (1). The FEM solution $u_h \in V^h$ is given by solving the above weak formulation (1) in V^h:

$$(\nabla u_h, \nabla v_h) = (f, v_h) \quad \forall v_h \in V^h. \tag{3}$$

By proper selection of the finite element space, the following a priori error estimate is expected,

$$\|u - u_h\|_V \leq M_h \|f\|_{L_2},$$

where M_h is a quantity with an explicit value and $M_h \to 0$ as $h \to 0$.

Generally, due to the singularity of solution u, i.e., $u \notin H^2(\Omega)$, the a priori error estimation with explicit M_h is not an easy task. In [10], X. Liu et al. successfully constructed an explicit a priori bound for the eigenvalues by adopting several FEM spaces (see, e.g., [3], for detailed description of the spaces):

- V^h: the conforming piecewise linear FEM space;
- X^h: the piecewise constant function space;
- W^h: the lowest order Raviart-Thomas mixed FEM space;
- $W^h_{f_h} (\subset W^h)$: each $w_h \in W^h_{f_h}$ satisfies $\operatorname{div} w_h + f_h = 0$ for $f_h \in X^h$.

In the rest of this section, we review the result of [10] in a brief way. Let π_0^h be the L_2 orthogonal projection that maps $L_2(\Omega)$ to X^h and

$$\|f - \pi_0^h f\|_{L_2} \leq C_0^h |f|_{H^1} \text{ if } f \in H^1(\Omega),$$

where C_0^h is a constant independent on f. There are several results on providing an explicit upper bound for C_0^h; see the references in [10]. Here, we follow the way of [9] to define $C_0(K)$ on each element K of the triangulation,

$$C_0(K) := \frac{L}{\pi}\sqrt{1 + |\cos\theta|},$$

where θ is the maximal inner angle of K and L is the second longest edge length of K. Thus C_0^h can be taken as

$$C_0^h := \max_{K \in \mathcal{T}_h} C_0(K).$$

To give an explicit value of M_h, a quantity κ_h is introduced in [10],

$$\kappa_h := \max_{f_h \in X^h \setminus \{0\}} \min_{v_h \in V^h} \min_{\mathbf{p_h} \in W_{f_h}^h} \|\mathbf{p_h} - \nabla v_h\|_{L_2} / \|f_h\|_{L_2}. \quad (4)$$

The quantities C_0^h and κ_h can help to construct an a priori error estimate for FEM solution u_h.

Theorem 1 *(A priori* error estimation [10]) *For any $f \in L_2(\Omega)$, let $u \in V$ and $u_h \in V^h$ be the solutions of variational problems* (1) *and* (3), *respectively. Let $M_h := \sqrt{{C_0^h}^2 + {\kappa_h}^2}$. Then, we have*

$$|u - u_h|_{H^1} \le M_h \|f\|_{L_2}, \quad \|u - u_h\|_{L_2} \le M_h |u - u_h|_{H^1} \le M_h^2 \|f\|_{L_2}. \quad (5)$$

By using the quantity M_h, we can obtain lower bounds for eigenvalues of the Laplacian.

Theorem 2 (Explicit eigenvalue bounds [10]) *Consider the variational form of the eigenvalue problem in V_h: find $\lambda^h \in R$ and $u_h \in V^h$ such that*

$$(\nabla u_h, \nabla v_h) = \lambda^h (u_h, v_h) \quad \forall v_h \in V^h. \quad (6)$$

Denote by $\{\lambda_k^h\}_{k=1}^n (n = Dim(V^h))$ the eigenvalues of (6) in an increasing order. Suppose that the kth eigenvalue λ_k of the Laplacian satisfies $\lambda_k M_h^2 < 1$. Then a bound for λ_k is given as,

$$\lambda_k^h / (1 + M_h^2 \lambda_k^h) \le \lambda_k \le \lambda_k^h. \quad (7)$$

Remark 1 In practical computation, we consider a loosed condition for $\lambda_k M_h^2 < 1$, that is, $\lambda_k^h M_h^2 < 1$, since $\lambda_k^h (\ge \lambda_k)$ can be easily computed using FEMs.

Remark 2 The computation of κ_h and λ_k^h is in fact to solve generalized eigenvalue problems for matrices. In particular, to obtain high-precision verified evaluation of

κ_h, a relatively large-scale of full matrix must be processed. For detailed discussion the computation of κ_h, refer to [10].

3 Sharpen the Eigenvalue Bounds by Applying Lehmann's Theorem

The smaller value of M_h, the higher precision of eigenvalue bound is possible. To refine the value of M_h, we need a dense mesh and have to deal with large-scale matrices. It can be even worse if the domain has a reentrant corner, which is due to the missing of H^2-regularity of eigenfunction around the corner. Another try is to utilize higher degree FEM and even the singular function characterizing the singularity of the eigenfunction. However, for such kinds of trial spaces, it is hard to obtain explicit error estimates.

To give a high-precision bound with light computation, we introduce Lehmann's theorem [8], which can greatly sharpen the eigenvalue bounds by using a rough *a priori* bound. We display Lehmann's theorem in a compact form; see [14] for a concise proof.

Theorem 3 (Section 5 of [14]) *Let H be a pre-Hilbert space with inner product $\langle\cdot,\cdot\rangle$, $D(L) \subset H$ a dense subspace, and $L : D(L) \to H$ a self-adjoint operator. Suppose that the real eigenvalues of $Lu = \lambda u$ to be $\lambda_1 \le \lambda_2 \le \cdots$ and H possesses an orthonormal basis $\{\phi_k\}_{k\in\mathbb{N}}$, where $\phi_k \in D(L)$ is the eigenfunction of L corresponding to eigenvalue λ_k. Take m linearly independent function $v_1, \dots, v_m$ from $D(L)$. Define $m \times m$ matrices A_1, A_2, A_3, the components of which are*

$$A_1(i,j) := \langle Lv_i, v_j\rangle,\ A_2(i,j) := \langle v_i, v_j\rangle,\ A_3(i,j) := \langle Lv_i, Lv_j\rangle \quad (i,j = 1,\dots,m).$$

Denoted by Λ_m the upper bound of λ_m obtained by Rayleigh–Ritz's method, that is, the maximum eigenvalue of $A_1 x = \lambda A_2 x$. Suppose ν is a quantity satisfying

$$\Lambda_m < \nu \le \lambda_{m+1}. \tag{8}$$

Define B_1, B_2 by

$$B_1 := A_1 - \nu A_2,\ B_2 := A_3 - 2\nu A_1 + \nu^2 A_2.$$

Thus, $(-B_1)$ and B_2 must be positive definite matrices. Let the eigenvalues of eigenproblem $B_1 x = \mu B_2 x$ be $\mu_1 \le \mu_2 \le \cdots \le \mu_m < 0$. Then lower bound for eigenvalue problem $Lu = \lambda u$ is given by

$$\lambda_{m+1-k} \ge \nu + \frac{1}{\mu_k} \quad (k = 1,\dots,m). \tag{9}$$

Remark 3 The test function $\{v_i\}$ should be well designed to approximate the exact eigenfunction accurately. Once $\{v_i\}$ is fixed, the lower bound is monotonically increasing on variable ν. However, the selection of ν is not so sensible; even rough lower bound ν of λ_{m+1} can produce precise lower bounds through (9).

A proper selection of ν satisfying (8) can be obtained in three steps. First, index m should be selected that λ_m and λ_{m+1} are not close to each other, which can be done through approximate computation based on FEMs. Second, by applying the conforming piecewise linear FEM with proper mesh size, a lower bound ν of λ_{m+1} can be obtained through Theorem 2,

$$\nu := \lambda^h_{m+1}/(1 + M_h^2 \lambda^h_{m+1}) \leq \lambda_{m+1}.$$

Third, with well-designed trial functions space, for example, the FEM space of higher order, we can take the approximate eigenfunction as the v_i needed in Theorem 3. If the quantity ν satisfies $\Lambda_m < \nu$, then we can apply Lehmann's theorem to evaluate the eigenvalues with high precision. Otherwise, we may need to select new index m, refine the mesh to get improved lower bound of λ_{m+1} and choose better v_i's to make Λ_m smaller.

Let us take uniform rectangular mesh $\mathcal{T}_h$ of L-shape domain as an example. The origin point of the coordinate is located at the reentrant corner of the domain, c.f., Fig. 1. Denote by Ω_s the sub-domain that contains the elements neighbor to the reentrant corner. Let h be the size of the element. The trial function $v \in C^1(\Omega)$ is defined as follows:

$$v = \sum_{K \in \mathcal{T}_h} \sum_{i,j=0}^{N} a_{K,ij} B_i^N(x) B_j^N(y) + b_1\eta_1 + b_2\eta_2. \tag{10}$$

Here $B_i^N(t) = \binom{N}{i} t^i (1-t)^{(N-i)}$ is Bernstein polynomial with degree N; η_k's $(k = 1, 2)$ are special functions presented in polar coordinate (r, θ) with the support being Ω_s,

$$\eta_k := r^{2k/3} \sin \frac{2k}{3}(\theta + \pi/2)\, \Psi \quad (k = 1, 2),$$

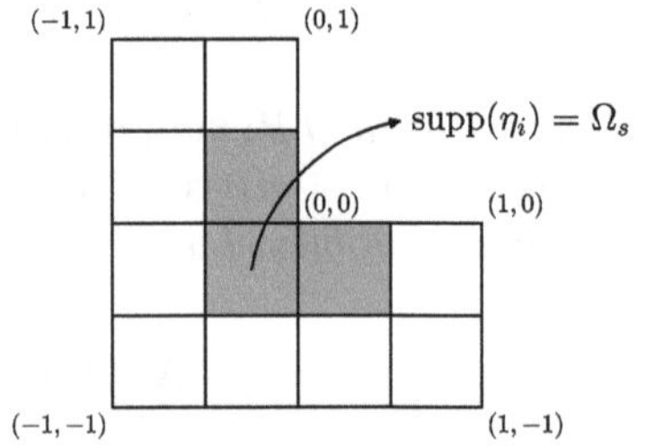

Fig. 1 Uniform rectangle mesh of L-shape domain and support of η_i $(h = 1/2)$

where Ψ is a C^1 continuous cut-off function that makes η_k vanish outside Ω_s:

$$\Psi := (1 - x^2/h^2)^2(1 - y^2/h^2)^2 \quad \text{for } |x|, |y| \leq h; \;\; \Psi := 0, \text{ otherwise.}$$

The C^1 continuous condition of v and boundary condition will add constraint conditions to the coefficients $a_{K,ij}$. We point out that $\eta_1 \notin H^2(\Omega)$ but $\Delta\eta_1 \in L_2(\Omega)$.

Denote by S^h the finite dimensional space constructed by the C^1 functions in the form (10). Thus, the m test function $\{v_i\}_{i=1,\dots,m}$ in Theorem 3 can be taken as the eigenfunctions associated to the leading mth eigenvalues of the following variational eigenvalue problem over S^h: find $\lambda \in R$ and $u_h \in S^h$ such that,

$$(\nabla u_h, \nabla v_h) = \lambda(u_h, v_h) \quad \forall v_h \in S^h.$$

The above way to construct trial functions has the flexibility to deal with domain of general shape. For a domain of general shape, Bernstein polynomials over elements of a triangulation of given domain, which is also called Bézier patch, can be used to construct C^1 base functions.

4 Integration of Singular Base Function

For the purpose of bounding the eigenvalues correctly, the matrices A_i $(i = 1, 2, 3)$ in Lehmann's theorem should be constructed exactly or enclosed by interval matrices, where the integration of the singular base function η_k is involved. In this section, we show the techniques used in calculating these integrations.

Let us consider a sample rectangle element which is divided into two right triangles, K and K'; see Fig. 2. Suppose that the singular point is located at $(0, 0)$. To illustrate the main technique used in the integration evaluation, we focus on the integration over the triangle K. Let $\psi_{i,j} := B_i^N(x)B_j^N(y)$. The inner products between $\{\psi_{i,j}\}$'s can be easily done. We need pay efforts to evaluate the terms involving η_k,

$$\begin{aligned}&(\psi_{i,j}, \eta_k)_K,\; (\nabla\psi_{i,j}, \nabla\eta_k)_K,\; (\Delta\psi_{i,j}, \Delta\eta_k)_K,\; (\eta_k, \eta_{k'})_K\\ &(\nabla\eta_k, \nabla\eta_{k'})_K,\; (\Delta\eta_k, \Delta\eta_{k'})_K \quad (i, j = 1, \dots, N;\; k, k' = 1, 2).\end{aligned} \tag{11}$$

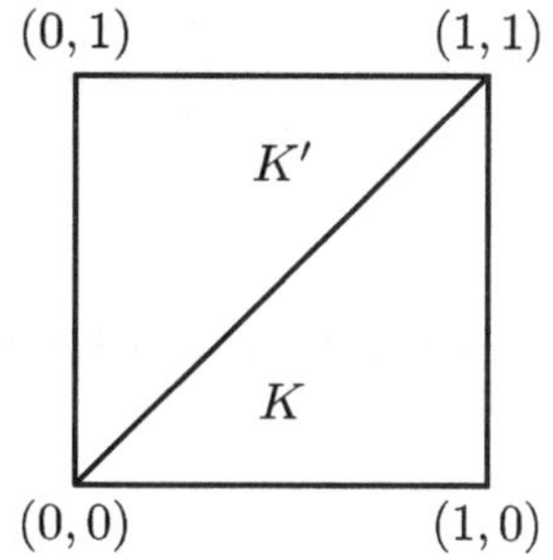

Fig. 2 A sample rectangle element

As $\psi_{i,j}$, η_k and their derivatives are productions of polynomials of x and y, power of r, and trigonometric functions of θ, each term in (11) can be presented by linear combinations of terms like $\int_K r^\beta g(\theta) \mathrm{d}\, r \mathrm{d}\, \theta$, where β is a certain positive fraction number and $g(\theta)$ is a function depending only on θ. Substituting $r = x/\cos\theta$, we have

$$\int_{\theta=0}^{\pi/4} \int_{r=0}^{1/\cos\theta} r^\beta g(\theta) \mathrm{d}\, r \mathrm{d}\, \theta = \int_{x=0}^{1} x^\beta dx \cdot \int_{\theta=0}^{\pi/4} (1 + \tan^2\theta)^{\beta/2} \frac{g(\theta)}{\cos\theta} d\theta. \qquad (12)$$

Thus the 2-dimensional integrations are reduced to be the 1-dimensional ones. Thanks to the powerful symbolic computation ability of Mathematica [12], all of the integrations can be given with explicit form. Particularly, the integration on θ is a little complicated, which may contain the hypergeometric function ${}_2F_1$. As an example, let us consider the integration of η_k defined over above element,

$$\eta_k = r^{2k/3} \sin\frac{2k}{3}(\theta)\,(1 - x^2)^2 (1 - y^2)^2 \quad (k = 1, 2).$$

The integration values $I_{k,k'} := (\Delta\eta_k, \Delta\eta_{k'})_K$ $(k, k' = 1, 2)$ are given as follows: where the underlined numbers denote the same digits shared by the left side and the right side of an interval that contains the integration value.

$$\begin{aligned}
I_{11} &= -\frac{6 \cdot 2^{2/3} \left(124570561120\, {}_2F_1\left(\frac{1}{6}, 1; \frac{3}{2}; -1\right) + 40223569113\sqrt{3} - 192961512624\right)}{201713136625} \\
&\in (\underline{0.40063755034929}09, \underline{0.40063755034929}10) \\
I_{12} = I_{21} &= -\frac{2\left(-369380860536\sqrt{3} + 1063726624119 + 706947619840\sqrt{3}\tanh^{-1}\left(3 - 2\sqrt{3}\right)\right)}{580054565475} \\
&\in (\underline{0.659895609882747}4, \underline{0.659895609882747}5) \\
I_{22} &= \frac{446104569792\, {}_2F_1\left(\frac{1}{2}, \frac{2}{3}; \frac{3}{2}; -1\right) + 36\sqrt[3]{2}\left(6306697864 - 3823310403\sqrt{3}\right)}{296527606375} \\
&\in (\underline{1.2262039192334}79, \underline{1.2262039192334}80).
\end{aligned}$$

When the domain Ω is of general shape, we need to adopt triangular meshes. In this case, the integrations on each element K can still be separated into two 1-dimensional integrations as in (12), while the one of θ may not have an explicit value. Thus, verified numerical quadrature algorithms, e.g., [4], will be applied to give verified enclosure of the integrations of θ.

5 Computation Examples

In this section, we show two computation examples: one is the eigenvalue problem over an L-shaped domain; the other over a square domain containing a square hole.

In the practical computation, the inner products involving singular functions is computed by using Mathematica [12]. The numerical computation of floating point number is executed with interval arithmetic; the INTLAB toolbox for MATLAB (developed by Rump [16]) are used for interval computation. To give sharp bound for the matrices eigenvalues, we adopt the method of Behnke [1]. As the rounding error in floating point number computation is estimated, the desired results can be mathematically accurate if there is no programming bug in the computation codes.

5.1 L-Shaped Domain

Let us consider the eigenvalue problem over an L-shaped domain: $(0,2) \times (0,2) \setminus [1,2] \times [1,2]$. A uniform triangulation of the domain is performed, as is displayed in Fig. 3.

The rough bounds for the leading 6 eigenvalues are obtained by applying the result of Theorem 2 along with FEM spaces V^h, X^h and W^h; the result is displayed in Table 1. The several important quantities in the computation are,

$$\#\text{Elements} = 6144,\ C_0^h = 0.010,\ \kappa_h = 0.033,\ M_h = 0.035.$$

The relative error for the lower bound λ_{low} and the upper bound λ_{upper} is estimated by ReErr:= $2(\lambda_{\text{upper}} - \lambda_{\text{low}})/(\lambda_{\text{upper}} + \lambda_{\text{low}})$. To apply Lehmann's theorem, we take $\lambda_5 < \nu := 39 < \lambda_6$. Thus we obtain more precise bound for $\lambda_1, \ldots, \lambda_5$, where

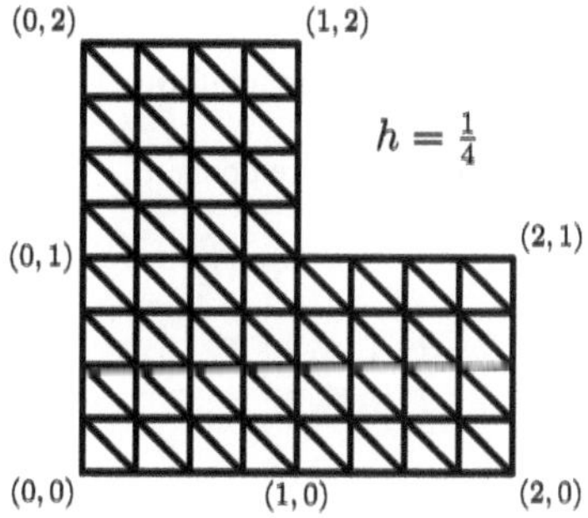

Fig. 3 Uniform mesh for L-shaped domain

Table 1 Eigenvalue evaluation for L-shaped domain (uniform mesh with $h = 1/32$) Note the third eigenpair is known:$\{\lambda_3 = 2\pi^2, u_3 = \sin \pi x \sin \pi y\}$

λ_i	Lower bound	Upper bound	ReErr
1	9.5585	9.6699	0.012
2	14.950	15.225	0.018
3	19.326	19.787	0.024
4	28.605	29.626	0.035
5	30.866	32.058	0.038
6	39.687	41.680	0.049

Table 2 High-precision bound for eigenvalues over L-shaped domain

λ_i	Yuan-He [18]	Liu-Oishi ($N = 6$)	Liu-Oishi ($N = 10$)
1	9.63972384^{44}_{04}	9.639^{73}_{55}	9.6397^{24}_{17}
2	15.19725192^{66}_{59}	15.1972^{53}_{30}	15.197251^{93}_{75}
3	$19.739208802178^{95}_{31}$	19.7329208^{81}_{65}	19.7392088021^{80}_{25}
4	29.52148111^{42}_{38}	29.521^{49}_{34}	29.5214^{82}_{77}
5	31.9126359^{59}_{37}	31.912^{64}_{18}	31.9126^{36}_{21}

Bernstein function up to degree 10 and singular basis introduced in §3 is used. The high-precision eigenvalue bounds are displayed in Table 2, together with the ones from Q. Yuan and Z. He [18].

The result of [18] is also based on Lehmann's theorem, where the trial functions $\{v_i\}$ are taken as Bessel functions and quite accurate eigenvalue bounds become possible. However, there have been no efforts on verified computations of the integrations of these trial functions. Also, the quantity ν in [18] is upon the result of [17], which is not a guaranteed estimation and only available for special domains.

5.2 Square-Minus Domain

Let us consider the eigenvalue problem over a square-minus-square domain: $(0, 8) \times (0, 8) \setminus [1, 7] \times [1, 7]$. To apply the FEM to obtain rough bounds for the eigenvalues, we triangulate the domain by a nonuniform mesh; see Fig. 4. The quantities in the

Fig. 4 A sample nonuniform mesh for square-minus-square domain

computation are

$$\#\text{Elements} = 48920,\ C_0^h = 0.033,\ \kappa_h = 0.027,\ M_h = 0.043.$$

The rough bounds for the leading several eigenvalues are displayed in Table 3.

From Theorem 2, we obtain an a priori estimate as: $\lambda_{17} < \nu := 13.71 \le \lambda_{18}$. Then, we turn to Lehmann's theorem to sharpen the eigenvalue bounds. To set polynomial functions space, we subdivide the domain into 16 small rectangles. As the domain has four reentrant corners, around each of the corners, we introduce singular base functions, the support of which is plotted in gray color; see the graph around Table 4. In the computation, the degree of the polynomial base function is taken to be 10. The high-precision eigenvalue bounds are displayed in Table 4. The eigenfunctions associated to the leading 4 eigenvalues are plotted in Fig. 5.

Table 3 Eigenvalue evaluation for square-minus-square domain (uniform mesh with $h = 1/32$)

λ_i	Lower bound	Upper bound	ReErr
1	9.029	9.182	0.017
2	9.039	9.192	0.017
3	9.039	9.193	0.017
4	9.050	9.203	0.017
5	9.920	10.104	0.018
17	13.080	13.403	0.024
18	13.710	14.064	0.026

Table 4 Bounds for the eigenvalues over square-minus-square domain

λ_i	Eigenvalue bound	ReErr	
1	9.16021^{64}_{37}	2.8E-7	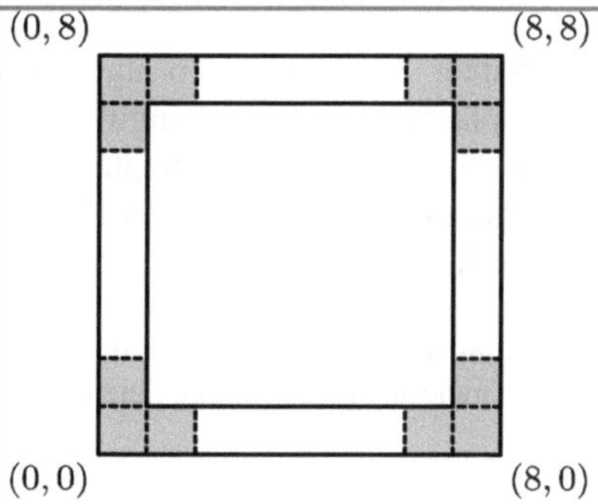
2	9.17008^{89}_{61}	2.9E-7	
3	9.17008^{89}_{61}	2.9E-7	
4	9.18056^{80}_{52}	3.0E-7	
5	10.089843^{37}_{14}	2.2E-8	
17	13.3751^{79}_{00}	5.8E-6	

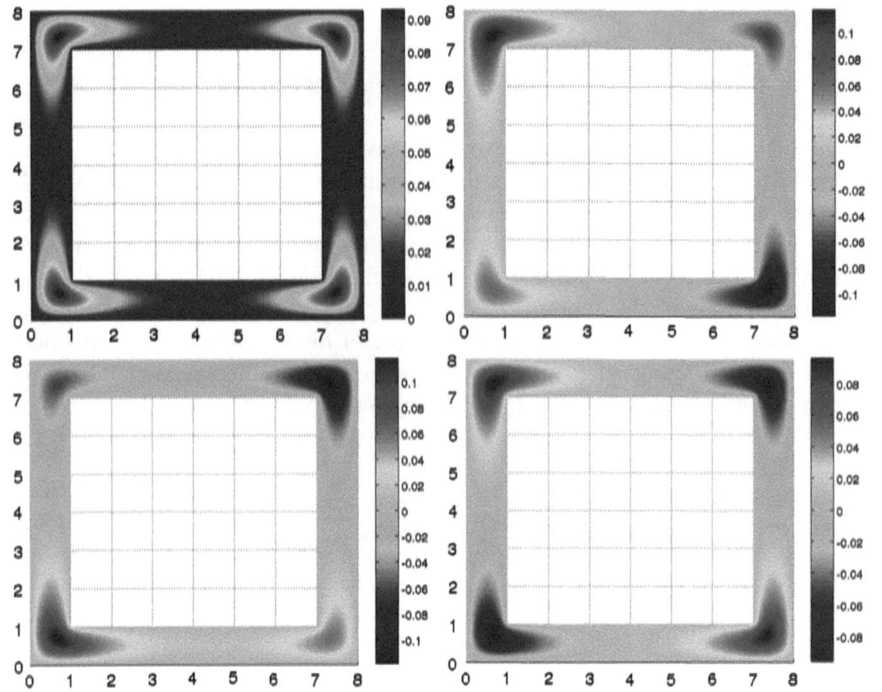

Fig. 5 Eigenfunctions corresponding to the leading 4 eigenvalues

6 Conclusion

In this chapter, we propose an algorithm to give high-precision eigenvalue bound for the Laplacian even with singularity. In the future, we would like to apply the technique developed in this chapter to verify the solution existence of semilinear elliptic differential equation, where the investigation of the spectrum of certain elliptic operator is the essential task.

Acknowledgments The first author of this chapter is supported by Japan Society for the Promotion of Science, Grand-in-Aid for Young Scientist (B) 23740092.

References

1. Behnke, H.: The calculation of guaranteed bounds for eigenvalues using complementary variational principles. Computing **47**, 11–27 (1991)
2. Behnke, H., Goerisch, F.: Inclusions for eigenvalues of selfadjoint problems. Topics in Validated Computations, pp. 277–322. Elsevier, Amsterdam (1994)
3. Brezzi, F., Fortin, M.: Mixed and Hybrid Finite Element Methods. Springer, Heidelberg (1991)

4. Chen, C.-Y.: On the properties of sard kernels and multiple error estimates for bounded linear functionals of bivariate functions with application to non-product cubature. Numerische Mathematik **122**, 603–643 (2012)
5. Dautray, R., Lions, J., Artola, M., Cessenat, M.: Mathematical Analysis and Numerical Methods for Science and Technology: Spectral Theory and Applications. Springer, Heidelberg (2000)
6. Goerisch, F.: Ein stufenverfahren zur berechnung von eigenwertschranken, in Numerical Treatment of Eigenvalue Problems, Vol. 4. In: Albrecht, J., Collatz, L., Velte, W., Wunderlich, W. (eds.) International Series of Numerical Mathematics, vol. 83, pp. 104–114. Birkhäuser, Basel (1987)
7. Goerisch, F., He, Z.: The determination of guaranteed bounds to eigenvalues with the use of variational methods I. Computer Arithmetic and Self-validating Numerical Methods, pp. 137–153. Academic Press Professional Inc, San Diego (1990)
8. Lehmann, N.: Optimale Eigenwerteinschließungen. Numer. Math. **5**, 246–272 (1963)
9. Liu, X., Kikuchi, F.: Analysis and estimation of error constants for interpolations over triangular finite elements. J. Math. Sci. Univ. Tokyo **17**, 27–78 (2010)
10. Liu, X., Oishi, S.: Verified eigenvalue evaluation for the Laplacian over polygonal domains of arbitrary shape. SIAM J. NUMER. ANAL. **51**, 1634–1654 (2013)
11. Mason, J.: Chebyshev polynomial approximations for the l-membrane eigenvalue problem. SIAM J. Appl. Math. **15**, 172–186 (1967)
12. Mathematica, Ver. 9.0, Wolfram Research Inc, Champaign, IL (2013)
13. Plum, M.: Eigenvalue inclusions for second-order ordinary differential operators by a numerical homotopy method. Zeitschrift für Angewandte Mathematik und Physik (ZAMP) **41**, 205–226 (1990)
14. Plum, M.: Guaranteed numerical bounds for eigenvalues. In: Hinton, D., Schaefer, P. (eds.) Spectral Theory and Computational Methods of Sturm, pp. 313–332. Marcel Dekker, New York (1997)
15. Plum, M., Wieners, C.: New solutions of the Gelfand problem. J. Math. Anal. Appl. **269**, 588–606 (2002)
16. Rump, S.: INTLAB - INTerval LABoratory. In: Csendes, T. (ed.) Developments in Reliable Computing, pp. 77–104. Kluwer Academic Publishers, Dordrecht (1999). http://www.ti3.tu-harburg.de/rump/
17. Still, G.: Approximation theory methods for solving elliptic eigenvalue problems. ZAMM Z. Angew. Math. Mech. **83**, 468–478 (2003)
18. Yuan, Q., He, Z.: Bounds to eigenvalues of the Laplacian on L-shaped domain by variational methods. J. Comput. Appl. Math. **233**, 1083–1090 (2009)

POLY: A New Polynomial Data Structure for Maple 17

Michael Monagan and Roman Pearce

Abstract We demonstrate how a new data structure for sparse distributed polynomials in the Maple 17 kernel significantly accelerates several key Maple library routines. The POLY data structure and its associated kernel operations are programmed for compactness, scalability, and low overhead. This allows polynomials to have tens of millions of terms, increases parallel speedup, and improves the performance of Maple library routines.

1 Introduction

Figure 1 shows the default polynomial data structure in Maple 16 and previous versions of Maple for the polynomial $f = 9xy^3z - 4y^3z^2 - 6xy^2z - 8x^3 - 5$. It is a "sum-of-products" where each monomial is stored a separate Maple object, a `PROD`. To compute the degree of f, a coefficient in x, test for a subexpression, or do almost anything else, the Maple kernel must recursively descend through multiple levels of dags. This involves extensive branching and random memory access, which prevents Maple from achieving high-performance on modern computer processors.

For example, to compute the degree of f in x, when Maple sees x in the monomial [PROD 7 | x | 1 | y | 3 | z | 1], because Maple does not have a dedicated polynomial data structure, Maple does not know that the next factor does not have x in it.

This work was supported by Maplesoft and the National Science and Engineering Research Council (NSERC) of Canada.

M. Monagan (✉) · R. Pearce
Department of Mathematics, Simon Fraser University, Burnaby, BC V5A 1S6, Canada
e-mail: mmonagan@cecm.sfu.ca

R. Pearce
e-mail: rpearcea@cecm.sfu.ca

© Springer-Verlag Berlin Heidelberg 2014

R. Feng et al. (eds.), *Computer Mathematics*, DOI 10.1007/978-3-662-43799-5_24

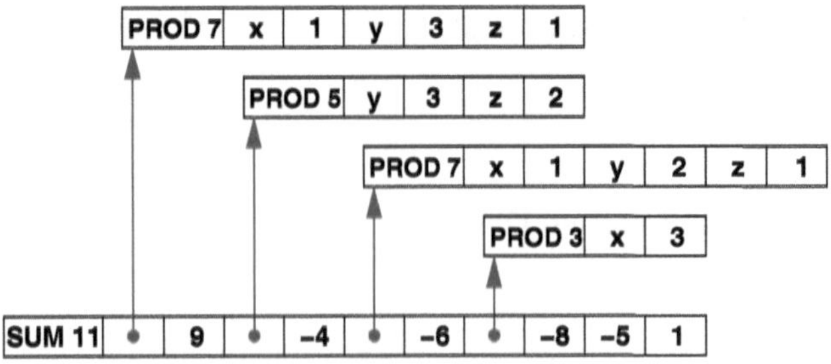

Fig. 1 Maple's sum-of-products representation encodes each object in an array of words where the first word encodes the type and length (in words) of the object

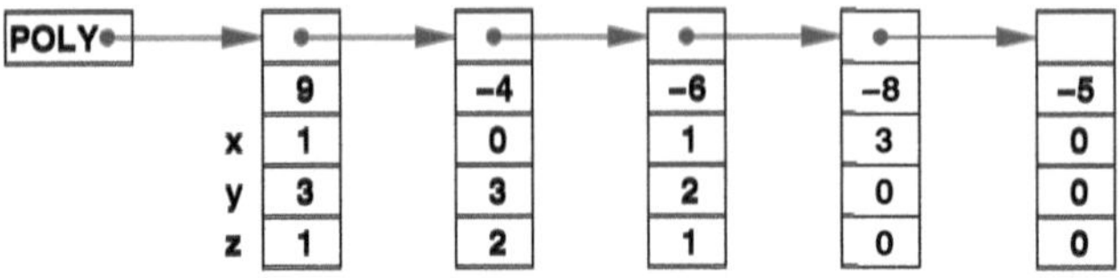

Fig. 2 Singular's polynomial representation for $f = 9xy^3z - 4y^3z^2 - 6xy^2z - 8x^3 - 5$

The monomial could be $x(x+y)^3z$ or $x\sin(x)^3 z$. Thus, computing the degree of a polynomial with t terms in n variables is $O(nt)$ in Maple.

Monomial operations are particularly slow. To multiply xy^3z by xz Maple forms the product [PROD 11 | x | 1 | y | 3 | z | 1 | x | 1 | z | 1] . It then "simplifies" this product to obtain [PROD 7 | x | 2 | y | 3 | z | 2]. Since the variables in a PROD are not sorted, Maple uses an $O(n^2)$ double loop to add the exponents of like variables. Finally, because Maple stores unique copies of objects, the resulting PROD object is hashed and inserted in an internal table if it not already present. In all, there are many function calls and many loops. We estimate that Maple takes more than 200 clock cycles to multiply monomials in 3 variables.

For comparison, Fig. 2 shows Singular's data structure for the same polynomial. Singular uses a linked list of terms and dense exponent vectors to represent monomials.

Singular is representative of several computer algebra systems which have a dedicated distributed polynomial representation with obvious advantages. Monomial multiplication consists of a memory allocation and a loop of fixed length to add exponents. This is the main reason Singular is faster than Maple at polynomial multiplication.

The relatively slow performance of Maple for polynomials lead us to develop a high performance C library for sparse polynomial arithmetic which we integrated into Maple in [1]. To multiply two polynomials, we first convert the Maple inputs to a special polynomial representation, then we call our library to compute the product using our parallel algorithm from [2], then we convert the result to Maple's sum-of-products representation. This substantially improved Maple's performance for large polynomials. However, for very sparse polynomials, and small polynomials, the overhead of constructing the sum-of-products representation often negated the parallel speedup achieved by our library. Furthermore, algorithms in the Maple library continued to use the sum-of-products structure, so the cost of ancillary operations like computing the degree or coefficients in a variable could easily overshadow the cost of arithmetic. These problems are addressed in this paper by introducing a new polynomial data structure in the Maple kernel. The new data structure, called POLY, was integrated into Maple 17 which was released in April 2013.

Our paper is organized as follows. In Sect. 2 we describe POLY, list some of its obvious advantages, and detail how it is integrated into Maple. In Sect. 3 we describe how we implemented various Maple kernel operations and give benchmarks demonstrating the improvement in their performance. In Sect. 4 we consider the impact of the POLY data structure and improved kernel routines on Maple library codes and measure improved parallel speedup in polynomial arithmetic. We end with a conclusion.

2 The POLY Data Structure

Figure 3 shows our new data structure for sparse distributed polynomials. The first word is a header word, which encodes the length and type of the object. The second word points to the variables, which are sorted in Maple's canonical ordering for sets. This is followed by the monomials and coefficients, where the monomials encode the exponents and the total degree in a single machine word. For example, for xy^2z^3 we encode the values $(6, 1, 2, 3)$ as the integer $6 \cdot 2^{48} + 2^{32} + 2 \cdot 2^{16} + 3$, using 16 bits each on a 64-bit machine. Terms are sorted into graded lexicographical order by comparing the monomials as unsigned integers. This gives a canonical representation for the polynomial. Small integer coefficients $-2^{62} < x < 2^{62}$ are encoded as $2x+1$, so that the rightmost bit is 1. Larger integers are a pointer (with rightmost bit 0) to a GMP multiprecision integer [3]. The current implementation requires all coefficients to be integers.

Five advantages of the new representation are readily apparent.

1. It is much more compact. Polynomials use two words per term instead of $2n + 3$ words, where n is the number of variables. For polynomials in three variables this saves over a factor of four.

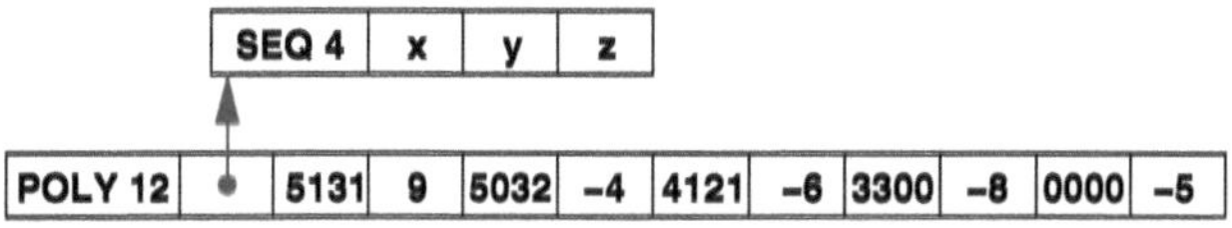

Fig. 3 The new packed representation for $f = 9xy^3z - 4y^3z^2 - 6xy^2z - 8x^3 - 5$

2. Monomial comparisons become machine word comparisons and monomial multiplication becomes machine word addition (provided there is no overflow), and monomial division becomes subtraction with a bitwise test for failure. This dramatically speeds up polynomial arithmetic.
3. Explicitly storing variables and sorting the terms lets us perform many common Maple idioms without looking at all of the terms, e.g. `degree`(f) (total degree), `indets`(f) (extract the set of variables), `has`(f, x), and `type`$(f, \mathit{polynom})$.
4. Other operations such as `degree`(f, x), `diff`(f, x), or `coeff`(f, x, i) (extract the coefficient of x^i) access memory sequentially to make good use of cache. We can isolate groups of exponents using masks. This eliminates branching and loops at the level of the exponents.
5. For large polynomials, we avoid creating many small Maple objects (the PRODs) that must be simplified by Maple's internal simplifier and stored in Maple's `simpl` table, an internal hash table of all Maple objects. They fill the `simpl` table and slow down Maple's garbage collector.

The idea of packing monomials in one or more computer words is not new; the ALTRAN computer algebra system [4] allowed the user to pack monomials in lexicographical order to conserve memory. Bruno Buchberger [5] experimented with packed monomials using the gradex lexicogrpahical ordering when he was computing Gröbner bases in the early 1960s because of the scarcity of main memory. A number of special purpose computer algebra systems have supported monomial packing. For example, the McCaulay2 computer algebra system [6] for computing Gröbner bases supports packing for the lexicographical and graded reverse lexicographical monomial orderings and the Magma computer algebra system [7] dynamically packs monomials using bytes. In [8], Bachmann and Schönemann compared the graded packing with packings for other monomial orderings for Gröbner basis computation.

We have integrated the POLY data structure into the Maple kernel and it is used by default as of Maple 17. When Maple parses input from the user or a file it creates the sum of products structure which is then simplified. When the simplifier detects an expanded polynomial with integer coefficients and at least two terms it creates a POLY structure when: (i) the number variables $n < \beta/2$ for β-bit machines, (ii) the total degree d satisfies $1 < d < 2^b$ where $b = \lfloor \beta/(n+1) \rfloor$ for $n > 1$ and $b = \beta - 2$ for $n = 1$, (iii) the variables are Maple *names* with regular evaluation rules, e.g. $\{x, y_1, \pi\}$ but not *infinity* or *undefined*. Otherwise the sum-of-products format is retained. On output, and for other cases where it was convenient, Maple recreates a sum-of-products structure for a POLY and caches it. Thus Maple 17 uses two representations for polynomials, automatically selecting between the two. All conversions between representations are automatic and invisible to the Maple user.

Note, the monomial encoding is determined solely by the number of variables in the polynomial. We use b bits for the exponent in each variable and for the total degree. It follows that arithmetic operations between polynomials in the same variables do not need to repack monomials.

Note, for polynomials with total degree $d = 1$, we do not store them in as a POLY dag because Maple's sum-of-products representation is better in this case.

For example $f = 2x + 3y + 4z + 5$ is represented as | SUM 9 | x | 2 | y | 3 | z | 4 | 1 | 5 |. This is compact as monomials are not explicitly represented. Maple's sparse linear solver uses this polynomial representation.

We chose the graded lex ordering rather than pure lexicographical ordering for several reasons. First, the graded ordering appears more natural for output. Second, when multiplying $a \times b$, if the total degree $d = \deg a + \deg b$ does not overflow, that is, $d < 2^b$, then the entire product can be computed without overflow and with no overflow detection. This allows us to look at only the leading terms of polynomials and predict overflow in $O(1)$ time. Third, in the division algorithm, if one uses pure lexicographical order, degrees in the remainder can increase and overflow the exponents. For example, consider the following division

$$x^2y^5 + y^3 \div x^2y + xy^5$$

in lexicographical order with $x > y$. The quotient is y^4 and the remainder is $-xy^9 + y^3$. If we had 3 bits per variable, the y^9 would overflow. In contrast, when a graded ordering is used the total degree of the monomials in the division algorithm always decreases. In our example the leading term of the divisor would be xy^5 and the division

$$x^2y^5 + y^3 \div xy^5 + x^2y$$

would result in the quotient x and remainder $-x^3y + y^3$.

We encode the monomial $x^iy^jz^k$ as | $d\ i\ j\ k$ | where $d = i + j + k$. Notice, however, that we could store | $d\ i\ j$ | and recover $k = d - i - j$ as needed to allow for more bits. We rejected this idea because it adds too much code complexity, both in the kernel routines and in external libraries that must support POLY. It also makes the cost of some monomial operations $O(n)$ instead of $O(1)$.

We expect to pack many practical problems into 64-bit words, which is the norm for today's computers. For example, if a polynomial has 8 variables then we store 9 integers for each monomial using $\lfloor 64/9 \rfloor = 7$ bits each. So the POLY dag can accommodate polynomials in 8 variables of total degree up to 127. Table 1 below shows the number of bits per exponent for a polynomial in n variables assuming a 64-bit word. Column lex shows how many bits would be available if lexicographical order were used and one bit (at the top) were left unset to test for overflow. To check for overflow in $a + b$ we would compute $(a + b) \oplus a \oplus b$. This sets the bits that are carried into (see [9]) after which we can apply a mask.

Column grlex is our packing. It shows how many bits are available for each variable and for the total degree. For univariate polynomials we do not store the degree because it simply duplicates the exponent. Instead we restrict the degree to the range of Maple immediate integers (62 bits signed x is stored as $2x + 1$) to avoid handling multiprecision exponents in conversions. The table also shows the number of unused bits at the top of each word.

Table 1 The number of bits per exponent for n variables in the two orderings

n	lex	grlex	Unused	n	lex	grlex	Unused	n	lex	grlex	Unused	n	lex	grlex	Unused
1	63	62	2	7	9	8	0	13	4	4	8	19	3	3	4
2	31	21	1	8	7	7	1	14	4	4	4	20	3	3	1
3	21	16	0	9	7	6	4	15	4	4	0	21	3	2	20
4	15	12	4	10	6	5	9	16	3	3	13	22	2	2	18
5	12	10	4	11	5	5	4	17	3	3	10	...	...	...	...
6	10	9	1	12	5	4	12	18	3	3	7	31	2	2	0

When this work was presented at the Asian Symposium on Computer Mathematics meeting in Beijing in October 2012, Joris van der Hoven asked "Why don't you use the extra bits for increased total degree?". For example, in Table 1 we see that for $n = 10$ variables, we allocate $b = 5$ bits per exponent and 5 bits for the total degree leaving 9 unused bits. The answer we gave was that doing so would increase the complexity of the code significantly. We also just did not know how much difference those extra bits would make. Over the following year, when computing determinants of various matrices of polynomials, we encountered three determinants which do not fit in POLY. In all three cases, however, they would fit if we used the unused bits for total degree. Consider the $n \times n$ Vandermonde matrix V_n where the (i, j)th entry of V_n is x_i^{j-1}. For example

$$V_4 = \begin{bmatrix} 1 & x_1 & x_1^2 & x_1^3 \\ 1 & x_2 & x_2^2 & x_2^3 \\ 1 & x_3 & x_3^2 & x_3^3 \\ 1 & x_4 & x_4^2 & x_4^3 \end{bmatrix}$$

The determinant D_n of V_n is a polynomial in n variables of total degree $0 + 1 + 2 + 3 + \cdots + n - 1 = n(n-1)/2$. Looking at Table 1, we can see that D_{10} which has total degree 45 does not fit in POLY because there are only 5 bits for the total degree. However, it would easily fit if the extra 9 bits were used for the total degree. In fact, D_n would fit for $n = 11, 12, 13$ and 14 as well. This example, and the significant number of unused bits in the range $9 \leq n \leq 14$ has convinced us that it will be worthwhile using them for the total degree. We are presently implementing this.

3 Operations in the Maple Kernel

The new representation allowed us to design new high performance algorithms for the Maple kernel. In the old data structure most operations are $O(nt)$ where n is the number of variables and t is the number of terms. Maple must examine the entire sum of products structure because its contents are unknown.

With the new data structure we can often avoid doing expensive operations on all of the terms, or we can do them much more efficiently. Our first example is the command diff(f, x). To compute the derivative with respect to x in the sum of products representation, Maple searches for x in each PROD. If x is found it copies the PROD, decreases the exponent of x, and scales the coefficient. Our algorithm for diff(f, x) first locates x in the set of variables. If x is not there it returns zero. Otherwise let x be variable k of n, let $s = \lfloor 64/(n+1) \rfloor$ be the width of the exponents, $sk = s(n-k)$ is the shift to find the exponent of x, and $b = 2^s - 1$ is a mask of s bits. The core of diff(f, x) is a loop that updates the exponent of x and the total degree (if present) using one subtraction.

```
/* subtracted from monomial */
d = 1 << sk;
if (n > 1) d += 1 << (s*n);
for (i=j=2; i < LENGTH(f); i+=2) {
        m = f[i];                /* monomial */
        e = (m >> sk) & b;       /* exponent */
        if (!e) continue;        /* skip the constant */
        r[j+0] = m - d;          /* new monomial */
        r[j+1] = mulint(f[i+1], IMMEDIATE(e), NULL);
        j += 2; }
```

The new monomials in the derivative remain sorted in the graded ordering. For example, consider $f = 9xy^3z - 4y^3z^2 - 6xy^2z - 8x^3 - 5$. If we differentiate with respect to x we obtain $f' = 9y^3z + 0 - 6y^2z - 24x^2 + 0$ and the nonzero terms remain sorted in the graded monomial ordering. Thus we can differentiate in $O(n+t)$ instead of $O(nt)$.

As a second example, consider degree(f, $\{x, z\}$); for $f \in \mathbb{Z}[x, y, z]$. We exploit a trick from Hacker's Delight [9]. If a and b are packed monomials $(a - b) \oplus a \oplus b$ sets the bits that are borrowed to subtract $a - b$. The absence of underflow in the exponent fields indicates that two monomials divide and $a - b$ is their quotient. To test for division in $\{x, z\}$ only, we construct a mask m with 1 bits above the x and z fields and compute `((a-b) ^ a ^ b) & m`. If this is zero we know b divides a in $\{x, z\}$, so degree(b, $\{x, z\}$) < degree(a, $\{x, z\}$) and we can skip over b without computing its degree. This optimization is also used for lcoeff which computes the leading terms of polynomials. In the case of degree we can stop when the total degree of the monomials is less than or equal to the largest degree already found.

Another optimization that we use is binary search. Consider has(f, x^5) and coeff($f, x, 5$). In both cases it is pointless to examine terms of total degree less than five, but rather than test each monomial with an additional branch in the main loop, we use binary search to adjust the endpoint for linear search.

Our results are summarized in Table 2 which reports the speedup for kernel operations on a polynomial with 9,90,000 terms in 3 variables. Note that none of these operations are parallelized, and the cost for evaluation is added to the other commands if you are using Maple interactively.

Table 2 Improvements for Maple kernel operations

Command	Description	Maple 16	Maple 17	Speedup
`f;`	top level evaluation evaluate the variables, interactive statements incur this cost	0.162 s	0.000 s	$\to O(n)$
`coeff`$(f, x, 20)$	coefficient of x^{20} linear search to degree 19, binary search for univariate f	2.140 s	0.005 s	420x
`coeffs`(f, x)	extract all coefficients in x reorder exponents $[d, x, y, z] \to [x, d, y, z]$, sort to collect in x	0.979 s	0.119 s	8x
`frontend`$(g, [f])$	subs functions for variables there's nothing to do since POLY has no functions	3.730 s	0.000 s	$\to O(n)$
`degree`(f)	total degree in all variables	0.175 s	0.000 s	$\to O(1)$
`degree`(f, x)	degree in x	0.073 s	0.003 s	24x
`diff`(f, x)	differentiate wrt x	0.956 s	0.031 s	30x
`eval`$(f, x = 6)$	compute $f(6, y, z)$ reorder exponents $[d, x, y, z] \to [d, y, z, x]$, sort, Horner form	3.760 s	0.175 s	21x
`expand`$(2xf)$	multiply by a term	1.190 s	0.066 s	18x
`has`(f, x^{101})	search for subexpression linear search to degree 100, $O(n)$ for names, $O(\log t)$ for terms	0.040 s	0.002 s	20x
`indets`(f)	set of indeterminates	0.060 s	0.000 s	$\to O(1)$
`lcoeff`(f, x)	leading coefficient in x monomial division avoids comparisons, stop early using degree	0.058 s	0.005 s	11x

(continued)

Table 2 (continued)

Command	Description	Maple 16	Maple 17	Speedup
`op`(f)	extract terms of f create old structure, cached until the next garbage collection	0.634 s	2.420 s	0.26x
`subs`($x = y$, f)	replace variable add exponents of like variables, sort and combine like terms	1.160 s	0.076 s	15x
`taylor`($f, x, 50$)	Taylor series to $O(x^{50})$ reorder exponents $[d, x, y, z] \rightarrow [x, d, y, z]$, sort to collect in x	0.668 s	0.055 s	12x
`type`(f, *polynom*)	type check type check the variables, other types are optimized as well	0.029 s	0.000 s	$\rightarrow O(n)$

The polynomial f is constructed as follows f := expand(mul(randpoly(i,degree=100,dense), i=[x,y,z]))

Table 3 Profiles for executing `collect(f,x)` in Maple 16 and Maple 17

Function	Calls	Maple 16		Maple 17	
		Time	Time (%)	Time	Time (%)
frontend	1	3.672	59.26	0.004	3.60
indets	2	1.617	26.10	0.000	0.00
series	1	0.747	12.06	0.094	84.68
collect	1	0.160	2.58	0.013	11.71
Total:	5	6.196	100.00	0.111	100.00

For some algorithms we need to sort polynomials after modifying the monomial data. We sort by treating the monomials as unsigned 64-bit integers and using an inplace MSD American flag radix sort [10]. For example, the `coeffs(f,x)`, `eval(f,x=6)` and `taylor(f,x,n)` commands all need the coefficients of f in x. Suppose f is a polynomial in $\{w, x, y\}$. For each monomial $w^i x^j y^k$ in f, encoded as `dijk` where $d = i + j + k$ is its degree, with a constant number of masks and bit operations (seven suffices) we move j, the degree of x to the front to obtain the `jdik`. We sort this modified data to group terms by power of x with ties broken by the monomial ordering on $\{w, y\}$. From that we extract the coefficients in x, already sorted, in linear time.

The biggest improvement we have seen for a Maple library command is the `collect` command, which is used to write a polynomial in recursive form. For example, if $f = xy^3 + x^2y - x^2z + xyz - 2$ then `collect(f,x)` will rewrite f as $(y - z)x^2 + (y^3 + yz)x - 2$. The Maple code for the `collect` command uses the `series(f,x,3)` command to implement this. Since the `series` command is 8x faster, we did not expect collect to be 6.196/0.111 = 56 times faster. In Table 3 is a profile showing that most of the time in Maple 16 was not in `series` at all, but rather in the `frontend` and `indets` commands.

In Maple 17, the cost of `frontend` and `indets` are now negligible since they no longer need to descend into the sum-of-products dag. In Maple 16, `frontend` recursively descends the sum-of-products dag looking for objects like $x^{1/2}$, $\sin(x)$, 2^n, etc. But our polynomial has none; it only has variables x, y and z in it.

3.1 Unpacking

The one case where we lose is when we must unpack the POLY dag and convert to the old data structure. The Maple command `op(f)` constructs a sequence of all the terms of f. Other Maple commands which effectively do the same thing include the common Maple programming idioms:

Table 4 Calls to `indets`(f, t) for f a POLY in the Maple library test suite

Type t	Number	Type t	Number
name	11937973	{rtable, table}	1509366
nonreal	7081486	specfunc(anything, RootOf)	1429737
float	6930777	radical	1101539
function	6678146	indexed	1089504
Or(RootOf, radical)	1863699	'^'	1047368
{name, function}	1861368	Or('+', '*', '^')	828257

`map(g, f)`	apply the function g to each term of f
`for t in f do`	iterate over the terms of f
`indets(f, t)`	extract all subexpressions in f of type t

Each term, e.g. $8xy^2$, is stored as [SUM | $\uparrow P$ | 8] where P is the monomial stored as [PROD | x | 1 | y | 2]. Thus Maple 17 must build a SUM and a PROD for each term in f whereas Maple 16 only builds a SUM and the PROD already exists. Theoretically, Maple 17 is $O(nt)$ compared with $O(t)$ for Maple 16. We have tried to improve the speed of unpacking by creating the PROD objects in simplified form, but the slowdown for `op(f)` remains a factor of 4.

However, we observed a tendency of Maple library code to either frequently unpack POLY or rarely unpack, often in the course of checking high level types. Alongside other internal caches in the Maple kernel (e.g. for `subs` and `indets`) we added a cache for unpacked POLY dags. This cache is cleared out on every garbage collection, so its practical effect is restricted to small polynomials that would be repeatedly unpacked.

For `indets(f,t)` we can avoid unpacking in many cases by detecting types that do not appear in POLY or appear only in the variables. Table 4 shows the most common types in `indets(f,t)` when f is a POLY when the entire Maple library test suite is run. For products or powers we must unpack the terms of f to create the result, but in the top ten cases (and many others) that is avoided.

3.2 Repacking

A number of the Maple kernel operations require us to repack monomials. For example, when adding $x^2+y^2+z^2$ and $x^2+y^2-z^2$, the result $2x^2+2y^2$ does not have the variable z. When Maple simplifies the result it must detect that z is missing and repack the polynomial into $\mathbb{Z}[x,y]$. Repacking also occurs in the `coeff`, `coeffs`, `eval`, `lcoeff`, and `taylor` commands that remove one or more variables, or `expand` and `divide`, which convert polynomials to a common ring. Critically, these operations do not permute the variables; they insert or remove exponent fields and change the sizes of the remaining fields.

In Maple 17 this is coded as a straightforward $O(n)$ loop, but we digress on the topic because new Intel microprocessors (codenamed Haswell, see [11]) have added two new instructions with exciting possibilities. The PEXT instruction is short for parallel extract, and it flushes masked bits to the bottom of the word.

monomial $x^3y^5z^7$	00001111 00000011 00000101 00000111
mask for $\{x, z\}$	00000000 11111111 00000000 11111111
result from PEXT	00000000 00000000 00000011 00000111

Its inverse is PDEP, short for parallel deposit, which distributes bits to various locations in a word, starting from the lowest bit. Both operations are $O(1)$.

input data	00000000 00000000 00000011 00000111
mask of locations	00000000 11111111 00000000 11111111
result from PDEP	00000000 00000011 00000000 00000111

The parallel design for these operations was originally proposed by Hilewitz and Lee in [12]. With these instructions, we can repack monomials without any branches or loops, e.g. to convert from $\mathbb{Z}[x, z]$ to $\mathbb{Z}[x, y, z]$ we would do:

3×21 bits	extract	4×16 bits	deposit	4×16 bits
d | i | k	$\longrightarrow$	0 | d | i | k	$\longrightarrow$	d | i | 0 | k

This would be needed, for example, to multiply $f(x, z)$ and $g(x, y)$ in $\mathbb{Z}[x, y, z]$. In fact Maple 17 does use the PEXT operation to reorder multiple variables for `coeffs`, but we coded this in C using the algorithm in Hacker's Delight [9]. After precomputation it uses 24 bitwise instructions per word, and it is called twice per term by `coeffs` to reorder the exponents. We attempted to use a C routine for PDEP to repack monomials but it was hard to get a gain. Nevertheless this should be our approach as soon as there is widespread hardware support.

3.3 Hashing and Simplification

When an algebraic expression is created in Maple, it is simplified recursively by the kernel. For example, consider the polynomial $f = 9xy^3z - 4y^3z^2 - 6xy^2z - 8x^3 - 5$ from Fig. 1. Each object in f that is not a small integer is simplified and hashed to see if it already exists in memory. In Maple 16, this is first done for the variables x, y, z then for the monomials (PRODs in Fig. 1) xy^3z, y^3z^2, xy^2z and x^3, and finally for the whole expression f. These objects are hashed and stored in the internal `simpl` table, which maps each object to a unique copy in memory. This feature of Maple allows it to identify equal objects by address.

Table 5 Real time in seconds for computing and simplifying a large power

	Expand power (s)	Allocate dag (s)	Simplify dag (s)	Total time (s)
Maple 16	0.133	0.080	1.180	1.420
Maple 17	0.128	0.000	0.010	0.139

In Maple 17 (see Fig. 3), because monomials are encoded immediately as machine integers, they are not stored in the `simpl` table. Rather, only the `POLY` object, the `SEQ` object (the sequence of variables), and multiprecision coefficients are stored in the `simpl` table.

What is gained from not having to create, simplify, and hash each monomial as a `PROD` object? The following benchmark gives us a clue. Consider

```
> f := expand((1+s+t+u+v+w+x+y+z)^16):
```

which creates a polynomial f in 8 variables with 7,35,471 terms. By first issuing the command `sdmp:-info(1):` we can obtain profiling information from the C library that computes this result. Table 5 shows that most of the time is spent simplifying the result in Maple 16, whereas in Maple 17 this time is fairly small. The C routine itself is identical with only a tiny difference due to the compiler.

The C library uses the same monomial representation as Maple 17, so it can copy the term data to a new POLY object. Maple 16 has to allocate the sum of products structure, which is almost as expensive as the computation. Maple 17 simplifies the POLY by checking that its terms and variables are sorted and all variables have a nonzero exponent. This takes 0.01 s or 7.2 % of the time. Maple 16 must do considerably more work to simplify the sum of products dag. For each `PROD`, it checks that the variables are distinct (they are) using an $O(n^2)$ loop, then it hashes the `PROD` and inserts it into the `simpl` table. Then it has to sort the `SUM` to check that all the `PROD`s are distinct (they are), because hashing has destroyed any previous order. Finally, it hashes the `SUM`.

A hidden cost is that the code to simplify `SUM`s and `PROD`s is quite expensive. It handles special objects like *infinity* and *undefined*, complex and floating point arithmetic, operator overloading, binary relations like $<$ or $=$ because equations can be added or scaled, matrix arithmetic, etc. These routines implement much of the expressive power of the Maple language, and this is not free. To simplify `POLY`, we have the luxury of analyzing the variables and calling algorithms that work in restricted domains.

4 Benchmarks

What is the impact of the POLY data structure on Maple's overall performance? This was difficult to predict in advance. One goal was to reduce sequential overhead in polynomial algorithms so that parallel speedup in multiplication and division (see [2, 13]) would speed up the Maple library. Was that achieved? To this end we developed two benchmarks; expanding determinants of polynomial matrices and factoring multivariate polynomials. Both are higher level algorithms.

4.1 Determinant Benchmark

Our first benchmark computes the determinant of the $n \times n$ symmetric Toeplitz matrix A for $6 \leq n \leq 11$. This is a matrix with n variables $\{x_1, \ldots, x_n\}$ with x_i along the ith diagonal and ith subdiagonal. To compute $\det(A)$ we use our own implementations of the Bareiss algorithm (see [14]), which we provide in Appendix A. At the kth elimination step, ignoring pivoting, the Bareiss algorithm computes

$$A_{i,j} := \frac{A_{k,k}A_{i,j} - A_{i,k}A_{k,j}}{A_{k-1,k-1}} \text{ for } i = k+1, \ldots, n \text{ and } j = k+1, \ldots, n \quad (1)$$

where the division is exact. At the end of the algorithm $A_{n,n} = \pm\det(A)$. Thus the Bariess algorithm does a sequence of $O(n^3)$ polynomial multiplications and divisions that grow in size, with the largest one occurring in the last step when $k = n - 1$.

In Table 6 below, #det is the number of terms in the determinant which has degree n, and #num is the number of terms in $A_{n-1,n-1}A_{n,n} - A_{n,n-1}A_{n-1,n}$ which has degree $2n - 2$ and is much larger than $\det(A)$.

We used a quad core Intel Core i7 920 2.66 GHz CPU running 64-bit Linux. Timings are real times in seconds. With four cores we achieve a factor of 3 to 4 speedup over Maple 16, which is large. That gain is entirely from reducing the overhead of Maple data structures; Maple 16 and Maple 17 use the same C library routines for polynomial arithmetic. The reduction of overhead increases parallel speedup from 1.6x to 2.59x over Maple 16. For comparison we include times for Maple 13 (which does not use our C library) and Magma 2.17.

Table 6 Real time in seconds to compute $\det(A)$ using the Bareiss algorithm

n	#det	#num	Maple 13	Maple 16		Maple 17		Magma 2.17
			1 core	1 core	4 cores	1 core	4 cores	1 core
6	120	575	0.015	0.008	0.009	0.002	0.002	0.000
7	427	3277	0.105	0.030	0.030	0.010	0.010	0.020
8	1628	21016	1.123	0.181	0.169	0.047	0.037	0.200
9	6090	128530	19.176	1.450	1.290	0.482	0.294	2.870
10	23797	813638	445.611	14.830	12.240	5.680	2.940	77.020
11	90296	5060172	–	151.200	94.340	78.500	26.890	2098.790

By default, Maple and Magma do not use the Bareiss algorithm to compute these determinants. Instead, they use the method of minor expansion of Gentleman and Johnson [15]. Recall that given an $n \times n$ matrix A

$$\det(A) = \sum_{i=1}^{n} (-1)^{n+1} A_{i,1} \det(M(1,i)) \quad (2)$$

where $M(1,i)$ is the $n = 1$ by $n - 1$ matrix obtained from A by deleting column 1 and row i. Applied naively, this formula recomputes the determinants of sub-matrices many times. Gentleman and Johnson avoided that by computing from the bottom up; they compute all 2×2 determinants then all 3×3 determinants and so on. This is still exponential in n. It computes $\binom{n}{k}$ determinants of $k \times k$ sub-matrices for a total of $\sum_{k=1}^{n} \binom{n}{k} = 2^n - 1$ determinants.

For our Toeplitz matrices, the multiplications in (2) are of the form *variable times polynomial*. For example, to multiply $f = 9xy^3z - 4y^3z^2 - 6xy^2z - 8x^3 - 5$ by y, we add the monomial representation for y to each monomial in f, namely y = `1010` $= 2^{48} + 2^{16}$ to the encodings of `5131`, `5032`, `4121`, `3300`, `0000`. Notice how the result remains sorted in the monomial ordering. The additions in (2) are n-ary, for which we copy terms to a new POLY and apply radix sort. The improvement shown in Table 7 was huge. It surprised us.

4.2 Factorization Benchmark

Our second benchmark is multivariate factorization. For perspective we include timings for Magma [7], Mathematica, Maxima [16], Sage [17], Singular [18], and Trip [19] which is a computer algebra system for celestial mechanics.

Table 8 reports the real times for multiplication, division, and factorization on a hyperthreaded quad core Intel Core i7 920 2.66 GHz running 64-bit Linux. For each timing we report the median of three test runs. Maple 16 and 17 start up to four threads depending on the size of each multiplication or division. The factorization

Table 7 Real times in seconds for determinants using minor expansion

n	#det	Maple 16	Maple 17	Magma 2.17
6	120	0.002	0.002	0.001
7	427	0.010	0.004	0.003
8	1628	0.049	0.013	0.019
9	6090	0.305	0.047	0.116
10	23797	1.991	0.252	0.770
11	90296	19.370	1.322	6.210
12	350726	274.990	6.737	44.500
13	1338076	2024.370	37.570	337.770

Table 8 Real times in seconds for polynomial multiplication, division and factorization

	Maple 13	Maple 16		Maple 17		Magma	Sage	Singular		Maxima	Mathematica		Trip 1.2.26	
		1 core	4 cores	1 core	4 cores	2.19-1	5.8	3-1-4	3-1-6	5.25.0	8.0	9.0	(RS)	(RD)
multiply														
$p_1 := f_1 \cdot (f_1 + 1)$	1.561	0.063	0.030	0.041	0.012	0.33	1.09	0.57	0.56	0.31	5.92	0.120	0.011	0.008
$p_2 := f_2 \cdot (f_2 + 1)$	1.544	0.063	0.032	0.041	0.012	0.33	1.08	0.58	0.56	0.31	5.86	0.115	0.031	0.015
$p_3 := f_3 \cdot (f_3 + 1)$	26.501	0.521	0.171	0.403	0.102	3.99	10.86	6.77	6.60	3.95	59.32	0.855	0.094	0.080
$p_4 := f_4 \cdot (f_4 + 1)$	98.351	2.180	0.649	1.814	0.416	13.70	61.77	30.99	30.77	13.85	306.10	5.732	0.488	0.388
$p_5 := f_5 \cdot g_5$	13.666	1.588	0.384	0.153	0.154	13.24	32.73	18.22	17.41	10.00	196.28	1.526	0.177	0.145
$p_6 := f_6 \cdot g_6$	11.486	0.772	0.628	0.204	0.082	0.89	3.04	2.75	1.75	1.44	22.45	1.338	0.064	0.047
divide														
$q_1 := p_1/f_1$	1.451	0.065	0.033	0.042	0.015	0.36	1.30	0.40	0.18	0.31	5.00	0.197	0.238	0.191
$q_2 := p_2/f_2$	1.435	0.065	0.033	0.042	0.015	0.36	1.30	0.39	0.18	0.31	5.04	0.194	0.225	0.208
$q_3 := p_3/f_3$	25.054	0.524	0.184	0.411	0.117	4.14	17.81	3.64	4.65	4.07	47.01	1.510	2.340	1.776
$q_4 := p_4/f_4$	92.867	2.253	0.736	1.842	0.483	18.54	80.39	14.96	11.21	14.29	228.83	5.662	9.540	7.374
$q_5 := p_5/f_5$	5.570	1.636	0.417	1.445	0.333	12.48	50.16	10.80	10.27	10.58	35.94	2.789	8.827	5.670
$q_6 := p_6/f_6$	10.421	0.769	0.627	0.215	0.095	7.90	4.87	1.84	1.45	1.56	30.28	6.800	1.181	1.016

(continued)

Table 8 (continued)

	Maple 13	Maple 16		Maple 17		Magma	Sage	Singular		Maxima	Mathematica		Trip 1.2.26	
		1 core	4 cores	1 core	4 cores	2.19-1	5.8	3-1-4	3-1-6	5.25.0	8.0	9.0	(RS)	(RD)
factor														
p_1 : 12341 terms	31.330	2.792	2.658	0.790	0.650	6.51	1.51	2.01	0.84	3.18	11.70	18.478	NA	NA
p_2 : 12341 terms	275.508	3.240	3.071	0.991	0.850	7.09	1.58	2.10	0.91	3.52	67.83	112.243	NA	NA
p_3 : 38711 terms	360.862	16.714	14.110	6.927	4.399	119.32	18.14	12.48	9.81	34.02	166.10	276.161	NA	NA
p_4 : 135751 terms	2856.388	59.009	46.151	24.345	12.733	320.04	68.32	61.85	38.70	110.97	604.74	951.725	NA	NA
p_5 : 12552 terms	302.453	26.435	16.152	12.131	6.800	105.55	14.63	13.83	9.45	792.60	563.54	935.149	NA	NA
p_6 : 417311 terms	1359.473	51.702	48.808	8.295	6.330	369.12	37.56	42.08	20.28	95.70	290.07	423.032	NA	NA

$f_1 = (1 + x + y + z)^{20} + 1$
1771 terms
$f_2 = (1 + x^2 + y^2 + z^2)^{20} + 1$
1771 terms
$f_3 = (1 + x + y + z)^{30} + 1$
5456 terms
$f_4 = (1 + x + y + z + t)^{20} + 1$
10626 terms
$f_5 = (1 + x)^{20}(1 + y)^{20}(1 + z)^{20} + 1$
$g_5 = (1 - x)^{20}(1 - y)^{20}(1 - z)^{20} + 1$
9261 terms
$f_6 = (1 + u^2 + v + w^2 + x - y)^{10} + 1$
$g_6 = (1 + u + v^2 + w + x^2 - y)^{10} + 1$
3003 terms

routine is sequential Maple code, which gains parallelism from our multiplication and division routines.

For Mathematica 9 we timed the internal functions `Algebra'IPExpand` and `Algebra'IPExactQuotient` for multiplication and division. The additional time for the top level `Expand` was small, and we found no suitable command for exact division at the top level. The `Factor` command did not make use of parallelism.

We report two timings for Trip: the (RS) time is for the optimized recursive sparse polynomial data structure POLYV, while the (RD) time is the optimized recursive dense data structure POLPV. Both use Trip's parallel routines (see [20]) with 8 threads and rational arithmetic, including a fast representation for small machine integers similar to Maple's.

There are some anomalies in Table 8. Maple's times for division are close to those for multiplication, except on Problem 5 where Maple 17 uses a sequential dense method to multiply. Singular's timings for division are often twice as fast as multiplication. This is because Singular multiplies in the distributed representation and divides in a recursive representation. Trip's times for division are much slower than those for multiplication. This is partly because division in Trip is not parallelized.

In comparing the timings for factoring p_1 and p_2 we see that factoring p_2 is much slower in Maple 13 and Mathematica, but not in Maple 16 or 17, Magma, or Singular. The fast systems apply the substitution $p_2(x^2 = u, y^2 = v, z^2 = w)$ to reduce the degree of the input polynomial before factoring it. This halves the number of Hensel lifting steps in each variable.

We note that Singular's timings for factorization have improved enormously since version 3-1-0. Times for version 3-1-0 on the first four factorizations were 12.28, 23.67, 97.10, 404.86 s. The factorization code in 3-1-4 was changed by Michael Lee to use a recursive representation for polynomials instead of the default distributed representation shown in Fig. 2.

Our first improvement from Maple 13 to Maple 16 was due to our C library for polynomial multiplication and division described in [1, 2, 13, 21]. The speedup in multiplication and division produces a speedup in multivariate factorization, because most of the time is spent in "Hensel lifting" which consists of many multiplications and some exact divisions.

Our second improvement was to parallelize the algorithms for multiplication and division. In many cases we obtain superlinear speedup in our C library but the top level `expand` and `divide` have lower speedup, because of the extra time to import and export Maple data structures. For higher level algorithms such as `factor`, parallel speedup is further reduced by the need to perform many small operations in sequence. The cost of `degree`, `indets`, and `type(f,polynom)` also reduce parallelism in higher level code.

With the introduction of the POLY dag in Maple 17, we have substantially reduced data structure overhead and the cost of almost all supporting routines. Table 9 shows the improvements to parallel speedup which come on top of the large gains achieved for sequential time. The speedup for `expand` and `divide` is now much closer to our C library, and the speedup for `factor`, while modest, is respectable for a sequential algorithm.

Table 9 Parallel speedup (1 core)/(4 cores) in Maple 17 versus Maple 16

	Maple 17							Maple 16				
expand	3.41	3.41	3.95	4.36	1.00	2.49	2.10	1.97	3.04	3.36	4.13	1.23
divide	2.80	2.80	3.51	3.81	4.34	2.26	1.97	1.97	2.85	3.06	3.92	1.23
factor	1.21	1.16	1.57	1.91	1.78	1.31	1.05	1.05	1.18	1.28	1.63	1.06

Table 10 Observed speedup (Maple 17)/(Maple 16) on 4 cores versus 1 core

	4 cores							1 core				
expand	2.50	2.66	1.67	1.56	2.49	7.65	1.53	1.53	1.29	1.20	10.38	3.78
divide	2.20	2.20	1.57	1.52	1.25	6.60	1.54	1.54	1.27	1.22	1.13	3.57
factor	4.09	3.61	3.20	3.62	2.37	7.71	3.53	3.27	2.41	2.42	2.18	6.32

Table 10 shows the speedup of Maple 17 over Maple 16, on 4 cores versus 1. Notice how the gain is larger in parallel. This is just a pleasant consequence of Amdahl's Law when you reduce sequential overhead in parallel algorithms. The sole exception (Problem 5 expand) uses a sequential dense method in Maple 17.

With the POLY dag in Maple 17, the time for factorization on one core has been reduced by more than 50 %, but the parallel speedups are even greater:

The savings are entirely sequential time, as can be seen by subtracting the parallel times from the sequential times. The reduction in overhead improves parallel speedup significantly, from 1.28x to 1.91x in the case of factoring p_4.

To see where the improvements in the factorization have come from we have profiled the main parts of the factorization code. The profile (see Table 11) shows the %age of the time in the main parts of the factorization algorithm for Maple 16 and Maple 17. The data under *improved coeftayl* includes a further algorithmic improvement. The data shows we have eliminated $0.599 - 0.377 = 0.222$ s of overhead from the polynomial multiplications (see row `expand`) or 37 %. The biggest speedup is division (see row `divide`). This is because the divisions are mostly trial divisions which fail quickly. In such cases almost all the time is in conversion which is wasted.

The biggest absolute gain is for the routine `coeftayl(f,x-a,k)` which computes the coefficient of f in $(x - a)^k$. This computation is not done by expanding f as a Taylor series about $x = a$, but instead, by using the formula $g(x = a)/k!$ where $g = \frac{\mathrm{d}f}{\mathrm{d}^k x}$, the kth derivative of f. Referring back to Table 2, we can see that the speedup is due to the improvement of differentiation and polynomial evaluation. We also tried the following formula to compute the coefficient: $\sum_{i=k}^{\deg_x f} \mathrm{coeff}(f, x^i) a^i \binom{i}{k}$. We can see that this is 3x faster again (see improved coeftayl). The total real time is reduced from 2.59 s to 1.07 s to 0.790 s.

Table 11 Profile for `factor(p1);` (1 core)

function	Maple 16		Maple 17		Improved coeftayl	
	Time (s)	Time (%)	Time (s)	Time (%)	Time (s)	Time (%)
coeftayl	1.086	41.06	0.310	28.21	0.095	12.03
expand	0.506	19.13	0.263	23.93	0.255	32.28
diophant	0.424	16.03	0.403	34.94	0.299	37.85
divide	0.256	9.68	0.034	3.09	0.035	4.43
factor	0.201	7.60	0.011	1.00	0.010	1.27
factor/hensel	0.127	4.80	0.064	5.82	0.063	7.97
factor/unifactor	0.045	1.70	0.033	3.00	0.033	4.18
Total:	2.645	100.00	1.099	100.00	0.790	100.00

5 Conclusion

Maple, Mathematica, Magma and Singular all use a distributed representation for multivariate polynomials. Maple's sum-of-products data structure and Singular's linked list data structure are illustrated in Figs. 1 and 2 in the introduction. We ask the the reader take another good look at them. Mathematica's data structure is similar to Maple's and Magma's data structure is similar to Singular's. These data structures, which were designed in the 1980s when memory access was constant time, will not yield high-performance on todays computers because memory access is not sequential.

One way to speed up polynomial multiplication, division, or factorization would be to convert the input to a more suitable data structure, compute the result, then convert back. This is what we did in [1] for Maple 14 for polynomial multiplication and division. Singular 3-1-4 does this for polynomial division and factorization. It switches to using a recursive representation for division and factorization. However, conversion overhead will limit parallel speedup. Amdahl's law states that if the sequential proportion of a task is S then parallel speedup on N cores is limited to

$$\text{speedup} \leq \frac{1}{S + (1 - S)/N}.$$

When S is large (50 % or more say), then we cannot get good parallel speedup.

What we have done in this work for Maple is to make our POLY data structure the default data structure in Maple. The POLY data structure is used when all monomials in a polynomial can be packed into a single word. This enabled us to eliminate conversion overhead in multiplication and division. The data in Table 6 shows improved parallel speedup for polynomial multiplication and division. We also implemented highly efficient algorithms for many Maple kernel operations for POLY. The data in Table 7 shows a speedup of a factor of 50 over Maple 16 for a routine polynomial determinant computation. The data in Table 8 shows speedups of factors of between 2 and 3 for large multivariate polynomial factorizations which is a huge gain. Although not reported here, we also find speedups of a factor of 2 for large multivariate polynomial gcd computations.

The cost incurred is mainly in code complexity. We must manage two data structures for polynomials, one where the coefficients are integers and the monomials can be packed into a single machine word, and one, Maple's sum-of-products data structure, which does not have these restrictions. A substantial programming effort was required to support the new data structure in the Maple kernel. The gains suggest this is worthwhile.

In closing, the reader may have wondered why we only use one word of memory to encode monomials, and not two, or more? For if we use two words, we could encode polynomials in twice as many variables or of much higher degree. With 128 bits, one will cover almost all applications. We would like to see how far 64-bits takes us before considering such an extension. For supporting two word exponents potentially doubles the amount of code. Another desirable extension is to allow the coefficients in the POLY dag to be fractions or floating point numbers as well as integers.

Appendix A

Maple code (no pivoting) for the Bareiss algorithm.

```
ffge := proc(A,n) local d,i,j,k,t;
   d := 1;
   for k to n-1 do
       for i from k+1 to n do
           for j from k+1 to n do
               t := expand(A[k,k]*A[i,j]-A[i,k]*A[k,j]);
               divide(t, d, evaln(A[i,j]));
           od;
           A[i,k] := 0;
       od;
       d := A[k,k];
   od:
   A[n,n];
end;
n := 8;
T := linalg[toeplitz]([seq(x[i],i=1..n)]);
A := array(1..n,1..n):
for i to n do for j to n do A[i,j] := T[i,j] od od:
det := CodeTools[Usage]( ffge(A,n) ):
```

Magma code for the Bareiss algorithm.

```
Z := IntegerRing();
P<x,y,z,u,v,w,p,q,r,s,t,a> := PolynomialRing(Z,12);
X := [x,y,z,u,v,w,p,q,r,s,t,a];
n := 8;
A := Matrix(P,n,n,[0 : i in [1..n^2]]);
for i in [1..n] do
    for j in [1..n] do
        A[i,j] := X[AbsoluteValue(j-i)+1];
    end for;
end for;
d := 1;
time for k in [1..n-1] do
         for i in [k+1..n] do
            for j in [k+1..n] do
                t := A[k,k]*A[i,j]-A[i,k]*A[k,j];
                A[i,j] := ExactQuotient(t,d);
            end for;
         end for;
         d := A[k,k];
     end for;
det := A[n,n];
```

Appendix B

Maple code for timing benchmarks.

```
f := expand( (1+x+y+z)^20 )+1:
p := CodeTools[Usage]( expand( f*(f+1) ) ):
CodeTools[Usage]( divide(p,f,'q') );
CodeTools[Usage]( factor(p) ):
```

Magma code for timing benchmarks.

```
Z := IntegerRing();
P<x,y,z> := PolynomialRing(Z,3);
f := (1+x+y+z)^20+1;
g := f+1;
time h := f*g;
time q := ExactQuotient(h,f);
time ff := Factorization(h);
```

Mathematica code for timing benchmarks. Note, for the fifth benchmark, we used Cancel[p/f] which was much faster than PolynomialQuotient.

```
f = Expand[(1+x+y+z)^20]+1;
AbsoluteTiming[p = Expand[f*(f+1)];]
AbsoluteTiming[q = PolynomialQuotient[p,f,x];]
AbsoluteTiming[h = Factor[p];]
```

Maxima code for timing benchmarks.

```
showtime : true;
f : rat( (1+x+y+z)^20 ) +1 $
h : f*(f+1)$
qr : divide( h, f )$
f : factor( h )$
```

Sage code for timing benchmarks

```
Q = RationalField()
P.<x,y,z> = PolynomialRing(Q,3,order='deglex')
f = (1+x+y+z)^20+1
%time p = f*(f+1)
%time q,r = p.quo_rem(f)
%time h = factor(p)
```

Singular code for timing benchmarks.

```
ring R=0,(x,y,z),lp;
poly f = (1+x+y+z)^20+1;
poly g = f+1;
int TIMER;
TIMER = timer; poly p = f*g; timer-TIMER;
TIMER = timer; poly q = p/f; timer-TIMER;
TIMER = timer; list L = factorize(p); timer-TIMER;
```

Trip code for timing benchmarks. POLYV means recursive sparse, POLPV means recursive dense.

```
reset; _cpu=8$ _mode=POLYV$ _modenum=NUMDBL$
f=(1+x+y+z)^20+1$ g=f+1$ p = 0$
time_s; p = f*g$ time_t(usertime, realtime); realtime;
time_s$ div(p,f,q,r)$ time_t(ctime,rtime)$ rtime;
```

References

1. Monagan, M., Pearce, R.: Sparse polynomial multiplication and division in Maple 14. ACM Commun. Comput. Algebra **44**(4), 205–209 (2010)
2. Monagan, M., Pearce, R.: Parallel sparse polynomial multiplication using heaps. In: Proceedings of ISSAC 2009, pp. 295–315. ACM Press (2009).
3. Granlund, T.: The GNU Multiple Precision Arithmetic Library, version 4.2.2. http://www.gmplib.org/ (2008)
4. Hall Jr, A.D.: The ALTRAN system for rational function manipulation - a survey. Commun. ACM 14, 517–521 (1971). (ACM Press).
5. Buchberger, B.: Private communication, May 2013.
6. Grayson, D.R., Stillman, M.E.: Macaulay2, a software system for research in algebraic geometry. http://www.math.uiuc.edu/Macaulay2/
7. Bosma, W., Cannon, J., Playoust, C.: The magma algebra system I: the user language. J. Symbol Comput. 24(3–4), 235–265. http://magma.maths.usyd.edu.au/magma
8. Bachmann, O., Schönemann, H.: Monomial representations for Grobner bases computations. In: Proceedings of ISSAC '98, pp. 309–316 (1998).
9. Warren, H.S.: Hacker's Delight. Addison-Wesley, Boston (2003)
10. McIlroy, P.M., Bostic, K., McIlroy, M.D.: Engineering radix sort. Comput. Syst. **6**(1), 5–27 (1993)
11. Intel corporation. Advanced vector extensions programming reference. http://software.intel.com/sites/default/files/m/8/a/1/8/4/36945-319433-011.pdf (2011)
12. Hilewitz, Y., Lee, R.B.: Fast bit compression and expansion with parallel extract and parallel deposit instructions. In: Proceedings of ASAP '06, IEEE, pp. 65–72, (2006).
13. Monagan, M., Pearce, R.: Parallel sparse polynomial division using heaps. In: Proceedings of PASCO 2010, pp. 105–111. ACM Press (2010).
14. Bariess, E.: Sylvester's identity and multistep integer-preserving Gaussian elimination. Math. Comput. **22**(102), 565–578 (1968)
15. Gentleman, W.M., Johnson, S.C.: Analysis of algorithms, a case study: determinants of matrices with polynomial entries. ACM Trans. Math. Softw. **2**(3), 232–241 (1976)
16. The maxima computer algebra system. http://maxima.sourceforge.net/
17. Stein W.A., et al.: Sage Mathematics Software (Version 5.0), The sage development team. http://www.sagemath.org (2012)
18. Greuel, G.-M., Pfister, G., Schönemann, H.: Singular 3.0: A computer algebra system for polynomial computations. Centre for Computer Algebra, University of Kaiserslautern. http://www.singular.uni-kl.de (2005)
19. Gastineau, M., Laskar, J.: Development of TRIP: fast sparse multivariate polynomial multiplication using burst tries. In: Proceedings of ICCS 2006, pp. 446–453. Springer LNCS 3992 (2006).
20. Gastineau, M.: Parallel operations of sparse polynomials on multicores - I. Multiplication and Poisson bracket. In: Proceedings of PASCO '2010, pp. 44–52. ACM Press (2010).
21. Monagan, M., Pearce, R.: Sparse polynomial division using heaps. J. Symbol Comput. **46**(7), 807–822 (2011)

Degree and Dimension Estimates for Invariant Ideals of P-Solvable Recurrences

Marc Moreno Maza and Rong Xiao

Abstract Motivated by the generation of polynomial loop invariants of computer programs, we study P-solvable recurrences. While these recurrences may contain non-linear terms, we show that the solutions of any such relation can be obtained by solving a system of linear recurrences. We also study invariant ideals of P-solvable recurrences (or equivalently of while loops with no branches). We establish sharp degree and dimension estimates of those invariant ideals.

1 Introduction

In many applications, such as program verification, non-linear recurrence relations, like the following one, may arise:

$$\begin{cases} x(n+1) = x(n) + 1 \\ y(n+1) = y(n) + x(n)^2 + 1 \end{cases}, \text{ with } \begin{array}{l} x(0) = 1 \\ y(0) = 1. \end{array}$$

In these recurrences, some variables may appear non-linearly, but not in a completely arbitrary way. A fundamental case is that of the so called P-solvable recurrences. This paper focuses on P-solvable recurrences with rational coefficients, that we define formally below.

Definition 1 (*P-solvable recurrence*) Let $n_1, \ldots, n_k$ be positive integers and define $s := n_1 + \cdots + n_k$. Let M be a square matrix over $\mathbb{Q}$ and with order s. We assume that M is block-diagonal with the following shape:

M. Moreno Maza (✉) · R. Xiao (✉)
University of Western Ontario, London, ON, Canada
e-mail: moreno@csd.uwo.ca

R. Xiao
e-mail: akelux@gmail.com

© Springer-Verlag Berlin Heidelberg 2014

R. Feng et al. (eds.), *Computer Mathematics*, DOI 10.1007/978-3-662-43799-5_25

$$M := \begin{pmatrix} \mathbf{M}_{n_1 \times n_1} & & & \\ & \mathbf{M}_{n_2 \times n_2} & & \\ & & \ddots & \\ & & & \mathbf{M}_{n_k \times n_k} \end{pmatrix}.$$

Consider an s-variable recurrence relation R in the variables $x_1, x_2, \ldots, x_s$ and with the following form:

$$\begin{pmatrix} x_1(n+1) \\ x_2(n+1) \\ \vdots \\ x_s(n+1) \end{pmatrix} = M \times \begin{pmatrix} x_1(n) \\ x_2(n) \\ \vdots \\ x_s(n) \end{pmatrix} + \begin{pmatrix} \mathbf{f}_1 \\ \mathbf{f}_2 \\ \vdots \\ \mathbf{f}_k \end{pmatrix},$$

where $\mathbf{f}_1$ is a vector of length n_1 with coordinates in $\mathbb{Q}$ and where $\mathbf{f}_i$ is a tuple of length n_i with coordinates in the polynomial ring $\mathbb{Q}[x_1, \ldots, x_{n_1+\cdots+n_{i-1}}]$, for $i = 2, \ldots, k$. Then, the recurrence relation R is called *P-solvable* over $\mathbb{Q}$ and the matrix M is called the *coefficient matrix* of R.

Example 6 in Sect. 4 illustrates the above definition with a 2-block matrix and non-linear terms, while Example 1 below is a simpler case with a 1-block matrix. Our study of P-solvable recurrences originates in a previous work [7] dedicated to the computation of loop invariants of while-loops of the following shape:

```
while cond do
    X := A(X);
end do
```

where the recurrence $X(n+1) = A(X(n))$ induced by the assignments in the loop body is a P-solvable recurrence. We call *P-solvable* such while-loops.

Example 1 Consider the following code segment:

```
a, b := 0, 1 ;
while true do
    a, b := b, a + b;
end do
```

At each iteration, the variables a, b hold two consecutive elements in the Fibonacci sequence. To be more precise, let us associate a counter variable n to the variables a, b. Let us initialize n to 0 before entering the loop and let us increase n by 1 after each loop iteration. Then we have

$$\begin{cases} a(n+1) := b(n) \\ b(n+1) := a(n) + b(n) \end{cases}, \text{ with } \begin{array}{l} a(0) := 0 \\ b(0) := 1 \end{array}.$$

which is clearly a P-solvable recurrence.

There are several variants of the notion of a P-solvable recurrence, see the concept of *solvable mapping* in [12] or that of *solvable loop* in [6].

For a recurrence $X(n+1) = A(X(n))$, or equivalently for a P-solvable while loop, an *invariant* is a condition on the recurrence variables which holds for all values of n. In this paper, we are mainly interested by invariants which are polynomial equations, as defined formally below.

Definition 2 Given an s-variable P-solvable recurrence R with recurrence variables $x_1, x_2, \ldots, x_s$, a polynomial p in $\mathbb{Q}[x_1, x_2, \ldots, x_s]$ is called a *polynomial invariant* of R if for all n, we have $p(x_1(n), x_2(n), \ldots, x_s(n)) = 0$. All polynomial invariants of R form an ideal of the polynomial ring $\mathbb{Q}[x_1, x_2, \ldots, x_s]$; this ideal is called the (polynomial) *invariant ideal* of R.

It is known that, for instance in [12], that P-solvable recurrences have poly-geometrical expressions (which are defined formally in Definition 3) as closed form solutions. However, solving P-solvable recurrences, even linear ones, is a computationally hard problem, since many algebraic numbers could be involved.

Returning to the question of computing polynomial invariants of P-solvable recurrences, there are approaches based on solving those recurrences explicitly. See the work of Kauers and Zimmermann in [5] and that of Kovács in [6]. In contrast, our goal in [7] as that of Kapur and Rodriguez–Carbonell in [4], is to compute polynomial invariants of P-solvable recurrences without explicitly solving those recurrences. In [7], we proposed a method, based on interpolating polynomials at finitely many points on the trajectory (i.e. point sequence) of the recurrence under study. This interpolation process yields "candidate invariants" which are then checked by a criterion performing a polynomial ideal membership test.

The objective of the present paper is to provide degree and dimension estimates for invariant ideals. These results are clearly needed by the interpolation method of [7] and can benefit any methods for computing polynomial invariants that require a degree bound as input, such as the method by Kapur and Rodriguez–Carbonell [11].

Our paper proposes the following original results. We show that P-solvable recurrences and linear recurrences are equivalent in the sense that every P-solvable recurrence can be obtained by solving a system of linear recurrences (Theorem 1). We also supply a sharp degree bound (Theorems 2 and 3) for *invariant ideals* as well as a dimension analysis (Theorem 4) of those ideals. In addition, Corollary 2 states a sufficient condition for a given invariant ideal to be trivial.

The paper is organized as follows. In Sect. 2, we review some results on symbolic summation; those results are related to the properties of closed form solutions of P-solvable recurrences. We also include a brief review on the notion of a degree of a polynomial ideal. In Sect. 3, we show how solving P-solvable recurrence reduces to solving linear recurrences; thus we refer to that process as "linearizing" a P-solvable recurrence. Finally, in Sect. 4, we exhibit degree and dimension estimates for invariant ideals of P-solvable recurrences. We conclude this introduction with an example illustrating the notion of an invariant ideal together with our results on dimension and degree of invariant ideals.

Example 2 Consider the following P-solvable recurrence relation with x, y as recurrence variables:

$$x(n+1) = y(n), \; y(n+1) = x(n) + y(n), \; \text{ with } x(0) = 0, \; y(0) = 1.$$

Closed form formulas for $x(n)$ and $y(n)$ are easily obtained:

$$\begin{aligned} x(n) &= \frac{(\frac{\sqrt{5}+1}{2})^n}{\sqrt{5}} - \frac{(\frac{-\sqrt{5}+1}{2})^n}{\sqrt{5}}, \\ y(n) &= \frac{\sqrt{5}+1}{2} \frac{(\frac{\sqrt{5}+1}{2})^n}{\sqrt{5}} - \frac{-\sqrt{5}+1}{2} \frac{(\frac{-\sqrt{5}+1}{2})^n}{\sqrt{5}}. \end{aligned}$$

Let a, u, v be 3 variables. Replace $(\frac{\sqrt{5}+1}{2})^n$ (resp. $(\frac{-\sqrt{5}+1}{2})^n$) by u (resp. by v) and replace $\sqrt{5}$ by a. Taking into account the dependencies $u^2\, v^2 = 1, a^2 = 5$, one can check that the invariant ideal is given by:

$$\langle x - \frac{au}{5} + \frac{av}{5}, y - a\frac{a+1}{2}\frac{u}{5} + a\frac{-a+1}{2}\frac{v}{5}, a^2 - 5, u^2v^2 - 1\rangle \cap \mathbb{Q}[x, y],$$

which turns out to be $\langle 1 - y^4 + 2xy^3 + x^2y^2 - 2x^3y - x^4\rangle$. Observe that this ideal has dimension 1 and degree 4.

Now, we use Theorem 3 to estimate the degree of this invariant ideal. Denote by $A := \frac{-\sqrt{5}+1}{2}, \frac{\sqrt{5}+1}{2}$, the eigenvalues of the coefficient matrix of the input P-solvable recurrence. One can easily check that the set A is weakly multiplicatively independent, see Definition 6 for this notion. Note that the multiplicative relation ideal, See Definition 5, of A associated with the variables u, v is generated by $u^2v^2 - 1$ and thus has degree 4 and dimension 1 in $\mathbb{Q}[u, v]$. Therefore, by Theorem 3, the degree of the invariant ideal is bounded by 4. This implies that the degree bound given by Theorem 3 is sharp. Meanwhile, Theorem 4 estimates the dimension as 1, which is also sharp.

2 Preliminaries

Let $\overline{\mathbb{Q}}$ be the algebraic closure of $\mathbb{Q}$. Let $\mathbb{Q}^*$ (resp. $\overline{\mathbb{Q}}^*$) denote the non zero elements in $\mathbb{Q}$ (resp. $\overline{\mathbb{Q}}$) .

2.1 Poly-Geometric Summation

In this subsection, we recall several well-known notions together with related results around the topic of P-solvable recurrences. Those notions and results are adapted to our needs and could be stated in a more general context. For instance, the notion of multiplicative relation can be defined among elements of an arbitrary Abelian group, whereas we define it for a multiplicative group of algebraic numbers.

Definition 3 Let $\alpha_1, \ldots, \alpha_k$ be k pairwise distinct elements of $\overline{\mathbb{Q}}^* \setminus \{1\}$. Let n be a variable taking non-negative integer values. We regard $n, \alpha_1^n, \ldots, \alpha_k^n$ as independent variables and we call $\alpha_1^n, \ldots, \alpha_k^n$ n-exponential variables. Any polynomial of $\overline{\mathbb{Q}}[n, \alpha_1^n, \ldots, \alpha_k^n]$ is called a *poly-geometrical expression in n over $\overline{\mathbb{Q}}$* w.r.t. $\alpha_1, \ldots, \alpha_k$.

Let f, g be two poly-geometrical expressions n over $\overline{\mathbb{Q}}$ w.r.t. $\alpha_1, \ldots, \alpha_k$. Given a non-negative integer number i, we denote by $f|_{n=i}$ the *evaluation* of f at i, which is obtained by substituting all occurrences of n by i in f. We say that f and g are *equal* whenever $f|_{n=i} = g|_{n=i}$ holds for all non-negative integer i.

We say that $f(n)$ is in *canonical form* if there exist

(i) finitely many numbers $c_1, \ldots, c_m \in \overline{\mathbb{Q}}^*$, and
(ii) finitely many pairwise different couples $(\beta_1, e_1), \ldots, (\beta_m, e_m)$ all in $(\overline{\mathbb{Q}}^* \setminus \{1\}) \times \mathbb{Z}_{\geq 0}$, and
(iii) a polynomial $c_0(n) \in \mathbb{Q}[n]$,

such that each $\beta_1, \ldots, \beta_m$ is a product of some of the $\alpha_1, \ldots, \alpha_k$ and such that the poly-geometrical expressions $f(n)$ and $\sum_{i=1}^{m} c_i\, \beta_i^n\, n^{e_i} + c_0(n)$ are equal. When this holds, the polynomial $c_0(n)$ is called the *exponential-free part* of $f(n)$.

Remark 1 Note that sometime when referring to poly-geometrical expressions, for simplicity, we allow n-exponential terms with base 0 or 1, that is, terms with 0^n or 1^n as factors. Such terms will always be evaluated to 0 or 1 respectively.

Proving the following result is routine.

Lemma 1 *With the notations of Definition 3, let f a poly-geometrical expression in n over $\overline{\mathbb{Q}}$ w.r.t. $\alpha_1, \ldots, \alpha_k$. There exists a unique poly-geometrical expression c in n over $\overline{\mathbb{Q}}$ w.r.t. $\alpha_1, \ldots, \alpha_k$ such that c is in canonical form and such that f and c are equal. We call c the* canonical form *of f.*

Example 3 The closed form $f := \frac{(n+1)^2\, n^2}{4}$ of $\sum_{i=0}^{n} i^3$ is a poly-geometrical expression in n over $\overline{\mathbb{Q}}$ without n-exponential variables. The expression $g := n^2\, 2^{(n+1)} - n\, 2^n\, 3^{\frac{n}{2}}$ is a poly-geometrical in n over $\overline{\mathbb{Q}}$ w.r.t. 2, 3. Some evaluations are: $f|_{(n=0)} = 0$, $f|_{n=1} = 1$, $g|_{n=0} = 0$, $g|_{n=2} = 8$.

Notation 1 Let x be an arithmetic expression and let $k \in \mathbb{N}$. Following [13], we call *k-th falling factorial of x* and denote by $x^{\underline{k}}$ the product

$$x\,(x-1)\,\cdots\,(x-k+1).$$

For $i = 1, \ldots, k$, we denote by $\left\{ {k \atop i} \right\}$ the number of ways to partition k into i non-zero summands, that is, the *Stirling number of the second kind* also denoted by $S(n, k)$. We define $\left\{ {k \atop 0} \right\} := 0$.

Example 4 Consider a fixed non-negative integer k. The sum $\sum_{i=1}^{n-1} i^k$ has $n-1$ terms while its closed form [13] below

$$\sum_{i=1}^{k} \left\{ {k \atop i} \right\} \frac{n^{\underline{i+1}}}{i+1}$$

has a fixed number of terms and thus is poly-geometrical in n over $\overline{\mathbb{Q}}$.

The following result is well-known and one can find a proof in [13].

Lemma 2 *Let x be an arithmetic expression and let $k \in \mathbb{N}$. Then we have*

$$x^k = \sum_{i=1}^{k} \left\{ {k \atop i} \right\} x^{\underline{i}}.$$

Notation 2 Let $r \in \overline{\mathbb{Q}}$ and let $k \in \mathbb{N}$. We denote by $H(r, k, n)$ the following symbolic summation

$$H(r, k, n) := \sum_{i=0}^{n-1} r^i\, i^{\underline{k}}.$$

Let's denote by $H(r, 0, n)$ the symbolic summation $\sum_{i=0}^{n-1} r^i$. One can easily check that $H(r, 0, n) = \frac{r^n - 1}{r-1}$ holds for $r \neq 1$. Moreover, we have the following result.

Lemma 3 *Assume $r \neq 0$. Then, we have*

$$(r-1)\, H(r, k, n) = (n-1)^{\underline{k}}\, r^n - r\, k\; H(r, k-1, n-1). \qquad (1)$$

In addition, we have

(i) if $r = 1$, then $H(r, k, n)$ equals to $\frac{n^{\underline{k+1}}}{k+1}$, which is a polynomial in n over $\overline{\mathbb{Q}}$ of degree $k+1$.
(ii) if $r \neq 1$, then $H(r, k, n)$ has a closed form like $r^n\, f(n) + c$, where $f(n)$ is a polynomial in n over $\overline{\mathbb{Q}}$ of degree k and c is a constant in $\overline{\mathbb{Q}}$.

Proof We can verify Relation (1) by expanding $H(r, k, n)$ and $H(r, k-1, n-1)$. Now let us show the rest of the conclusion. First, assume $r = 1$. With Relation (1), we have

$$k\,H(r, k-1, n-1) = (n-1)^{\underline{k}}.$$

Therefore, we deduce

$$H(r, k, n) = \frac{n^{\underline{k+1}}}{k+1}.$$

One can easily check that $\frac{n^{\underline{k+1}}}{k+1}$ is a polynomial in n over $\overline{\mathbb{Q}}$ and the degree of n is $k+1$.

From now on assume $r \neq 1$. We proceed by induction on k. When $k = 0$, we have $H(r, 0, n) = \frac{r^n - 1}{r-1}$. We rewrite $\frac{r^n-1}{r-1}$ as

$$r^n \frac{1}{r-1} - \frac{1}{r-1},$$

which is such a closed form. Assume there exists a closed form $r^{n-1} f_{k-1}(n-1) + c_{k-1}$ for $H(r, k-1, n-1)$, where $f_{k-1}(n-1)$ is a polynomial in $n-1$ over $\overline{\mathbb{Q}}$ of degree $k-1$. Substitute $H(r, k-1, n-1)$ by $r^{n-1} f_{k-1}(n-1) + c_{k-1}$ in Relation (1) and solve $H(r, k, n)$, we have

$$H(r, k, n) = \frac{(n-1)^{\underline{k}}\, r^n - r\,k\,(r^{n-1} f_{k-1}(n-1) + c_{k-1})}{r-1}.$$

We rewrite the right hand side of the above equation as

$$r^n \frac{(n-1)^{\underline{k}} - k\, f_{k-1}(n-1)}{r-1} - \frac{r\,k\,c_{k-1}}{r-1},$$

from which one can easily check it satisfies the requirements of (ii) in the conclusion. This completes the proof. □

Lemma 4 *Let $k \in \mathbb{N}$ and let λ be a non zero algebraic number over $\mathbb{Q}$. Consider the symbolic summation*

$$S := \sum_{i=1}^{n} i^k \lambda^i.$$

1. *if $\lambda = 1$, then there exists a closed form $s(n)$ for S, where s is a polynomial in n over $\overline{\mathbb{Q}}$ of degree $k+1$.*
2. *if $\lambda \neq 1$, then there exists a closed form $\lambda^n\, s(n) + c$ for S, where s is a polynomial in n over $\overline{\mathbb{Q}}$ of degree k and $c \in \overline{\mathbb{Q}}$ is a constant.*

Proof By Lemma 2, we deduce

$$\begin{aligned}\sum_{i=1}^{n} i^k \lambda^i &= \sum_{i=1}^{n} \left(\sum_{j=1}^{k} \left\{ {k \atop j} \right\} i^{\underline{j}} \right) \lambda^i \\ &= \sum_{j=1}^{k} \left(\left\{ {k \atop j} \right\} \sum_{i=1}^{n} i^{\underline{j}} \lambda^i \right) \\ &= \sum_{j=1}^{k} \left(\left\{ {k \atop j} \right\} H(\lambda, j, n) \right)\end{aligned}$$

Then, the conclusions on each case follow from the corresponding results in Lemma 3.
□

The following definition is a specialization of the general definition of multiplicative relation to the case of non-zero algebraic numbers.

Definition 4 (*Multiplicative relation*) Let k be a positive integer. Let $A := (\alpha_1, \ldots, \alpha_k)$ be a sequence of k non-zero algebraic numbers over $\mathbb{Q}$ and $\mathbf{e} := (e_1, \ldots, e_k)$ be a sequence of k integers. We say that $\mathbf{e}$ is a *multiplicative relation on* A if $\prod_{i=1}^{k} \alpha_i^{e_i} = 1$ holds. Such a multiplicative relation is said *non-trivial* if there exists $i \in \{1, \ldots, n\}$ such that $e_i \neq 0$ holds. If there exists a non-trivial multiplicative relation on A, then we say that A is *multiplicatively dependent*; otherwise, we say that A is *multiplicatively independent*.

All multiplicative relations of A form a lattice, called the *multiplicative relation lattice* on A, which can effectively be computed, for instance with the algorithm proposed by Ge in his Ph.D. thesis [2].

For simplicity, we need the following generalized notion of multiplicative relation ideal, which is defined for a sequence of algebraic numbers that may contain 0 and repeated elements.

Definition 5 Let $A := (\alpha_1, \ldots, \alpha_k)$ be a sequence of k algebraic numbers over $\mathbb{Q}$. Assume w.l.o.g. that there exists an index ℓ, with $1 \leq \ell \leq k$, such that $\alpha_1, \ldots, \alpha_\ell$ are non-zero and $\alpha_{\ell+1}, \ldots, \alpha_k$ are all zero. We associate each α_i with a variable y_i, where $y_1, \ldots, y_k$ are pairwise distinct. We call the *multiplicative relation ideal of A associated with variables* $y_1, \ldots, y_k$, the binomial ideal of $\mathbb{Q}[y_1, y_2, \ldots, y_k]$ generated by

$$\left\{ \prod_{j \in \{1,\ldots,\ell\},\, v_j > 0} y_j^{v_j} - \prod_{i \in \{1,\ldots,\ell\},\, v_i < 0} y_i^{-v_i} \mid (v_1, \ldots, v_\ell) \in Z \right\}$$

and $\{y_{\ell+1}, \ldots, y_k\}$, denoted by $\mathrm{MRI}(A; y_1, \ldots, y_k)$, where Z is the multiplicative relation lattice on $(\alpha_1, \ldots, \alpha_\ell)$. When no confusion is possible, we shall not specify the associated variables $y_1, \ldots, y_k$.

Lemma 5 *Let* $\alpha_1, \ldots, \alpha_k$ *be k multiplicatively independent elements of* $\overline{\mathbb{Q}}$ *and let n be a non-negative integer variable. Let* $f(n)$ *be a poly-geometrical expression in*

n w.r.t. $\alpha_1, \ldots, \alpha_k$. Assume that $f|_{(n=i)} = 0$ holds for all $i \in \mathbb{N}$. Then, f is the zero polynomial of $\overline{\mathbb{Q}}[n, \alpha_1^n, \ldots, \alpha_k^n]$.

The following definition will be convenient in later statements.

Definition 6 (*Weak multiplicative independence*) Let $A := (\alpha_1, \ldots, \alpha_k)$ be a sequence of k non-zero algebraic numbers over $\mathbb{Q}$ and let $\beta \in \overline{\mathbb{Q}}$. We say that β is *weakly multiplicatively independent* w.r.t. A, if there exist no non-negative integers $e_1, e_2, \ldots, e_k$ such that $\beta = \prod_{i=1}^{k} \alpha_1^{e_i}$ holds. Furthermore, we say that A is *weakly multiplicatively independent* if

(i) $\alpha_1 \neq 1$ holds, and
(ii) α_i is weakly multiplicatively independent w.r.t. $\{\alpha_1, \ldots, \alpha_{i-1}, 1\}$, for all $i = 2, \ldots, s$.

Lemma 7 is a structural result for the closed form solutions of single-variable linear recurrences involving poly-geometrical expressions. For the proof, we need Lemma 6, which is easy to check, see for instance [10].

Lemma 6 *Let n a variable holding non-negative integer values. Let a and b be two sequences in $\mathbb{Q}$ indexed by n. Consider the following recurrence equation of variable x:*

$$x(n) = a(n-1)\, x(n-1) + b(n-1).$$

Then we have

$$x(n) = \prod_{i=0}^{n-1} a(i) \left(x(0) + \sum_{j=0}^{n-1} \frac{b(j)}{\prod_{s=0}^{j} a(s)} \right).$$

Lemma 7 *Let $\alpha_1, \ldots, \alpha_k$ be k elements in $\overline{\mathbb{Q}}^* \setminus \{1\}$. Let $\lambda \in \overline{\mathbb{Q}}^*$. Let $h(n)$ be a poly-geometrical expression in n over $\overline{\mathbb{Q}}$ w.r.t. $\alpha_1, \ldots, \alpha_k$. Consider the following single-variable recurrence relation R:*

$$x(n+1) = \lambda x(n) + h(n).$$

Then, there exists a poly-geometrical expression $s(n)$ in n over $\overline{\mathbb{Q}}$ w.r.t. $\alpha_1, \ldots, \alpha_k$ such that we have

$$\deg(s(n), \alpha_i^n) \leq \deg(h(n), \alpha_i^n) \ \text{ and } \ \deg(s(n), n) \leq \deg(h(n), n) + 1,$$

and such that

- *if $\lambda = 1$ holds, then $s(n)$ solves R,*
- *if $\lambda \neq 1$ holds, then there exists a constant c depending on $x(0)$ (that is, the initial value of x) such that $c\, \lambda^n + s(n)$ solves R.*

Moreover, in both cases, if the exponential-free part of the canonical form of $(\frac{1}{\lambda})^n\, h(n)$ is 0, then we can further require that $\deg(s(n), n) \leq \deg(h(n), n)$ holds.

Proof By Lemma 6, we have

$$x(n) = \lambda^n \left(x(0) + \sum_{j=0}^{n-1} \frac{h(j)}{\lambda^{j+1}} \right). \tag{2}$$

Denote by $\texttt{terms}(h)$ all the terms of the canonical form of $h(n)$. Assume each $t \in \texttt{terms}(h)$ is of form

$$c_t\, n^{q_t}\, \beta_t^n,$$

where c_t is a constant in $\overline{\mathbb{Q}}$, q_t is a non-negative integer and β_t is a product of finitely many elements (with possible repetitions) from $\{\alpha_1, \ldots, \alpha_k\}$. Define $g(n) := \frac{h(n)}{\lambda^{n+1}}$. Then $g(n)$ is a poly-geometrical expression in n w.r.t. $\{\beta_t\}_{t\in\texttt{terms}(h)}, \frac{1}{\lambda}$. Clearly we have

$$g(n) = \sum_{t\in\texttt{terms}(h(n))} \frac{c_t}{\lambda}\, n^{q_t} \left(\frac{\beta_t}{\lambda}\right)^n.$$

Therefore, we have

$$\sum_{j=0}^{n-1} \frac{h(j)}{\lambda^{j+1}} = \sum_{t\in\texttt{terms}(h)} \sum_{j=0}^{n-1} \frac{c_t}{\lambda}\, j^{q_t} \left(\frac{\beta_t}{\lambda}\right)^j. \tag{3}$$

According to Lemma 4, for each $t \in \texttt{terms}(h)$, we can find a poly-geometrical expression

$$s_t := \left(\frac{\beta_t}{\lambda}\right)^n f_t(n) + a_t$$

in n over $\overline{\mathbb{Q}}$ w.r.t. $\frac{\beta_t}{\lambda}$ satisfying

1. $s_t = \sum_{j=0}^{n-1} \frac{c_t}{\lambda}\, j^{q_t}\, (\frac{\beta_t}{\lambda})^j$;
2. f_t is a polynomial in n over $\overline{\mathbb{Q}}$ of degree q_t (if $\beta_t \neq \lambda$) or $q_t + 1$ (if $\beta_t = \lambda$), and a_t is a constant in $\overline{\mathbb{Q}}$; note in the later case, $c_t\, n^{q_t}\, (\frac{\beta_t}{\lambda})^n$ is a summand of the constant term of the canonical form of $(\frac{1}{\lambda})^n\, h(n)$ is 0 when viewed as a polynomial of the n-exponential variables.

Therefore, using s_t, for $t \in \texttt{terms}(h)$, we can simplify the right hand side of Eq. (2) to

$$\left(x(0) + \sum_{t\in\texttt{terms}(h)} a_t \right) \lambda^n + \sum_{t\in\texttt{terms}(h)} f_t(n)\, \beta_t^n. \tag{4}$$

Assume that, for each $t \in \texttt{terms}(h)$, we have $\beta_t = \alpha_1^{e_{t,1}}\, \alpha_1^{e_{t,2}} \cdots \alpha_1^{e_{t,k}}$. Define

$$\beta_t(n) := (\alpha_1^n)^{e_{t,1}}\, (\alpha_1^n)^{e_{t,2}} \cdots (\alpha_1^n)^{e_{t,k}},$$

$$c := x(0) + \sum_{t \in \mathtt{terms}(h)} a_t \text{ and } s(n) := \sum_{t \in \mathtt{terms}(h)} f_t(n)\,\beta_t(n).$$

We easily deduce $\deg(s(n), \alpha_i^n) = \max_{t\in\mathtt{terms}(h)}(\deg(\beta_t(n), \alpha_i^n) \leq \deg(h(n), \alpha_i^n)$. Finally, one can easily verify that c and $s(n)$ satisfy the requirements of the conclusion. □

Remark 2 In Lemma 7, if λ is weakly multiplicatively independent w.r.t. $\alpha_1, \ldots, \alpha_k$, then we know that the exponential-free part of the canonical form of $(\frac{1}{\lambda})^n\, h(n)$ is 0, without computing the canonical form explicitly.

2.2 Degree of a Polynomial Ideal

In this subsection, we review some notions and results on the degree of a polynomial ideal. Up to our knowledge, Proposition 1 is a new result which provides a degree estimate for an ideal of a special shape and which can be applied to estimate the degree of invariant ideals of P-solvable recurrences. Throughout this section, $\mathbb{K}$ is an algebraically closed field. Let F be a set of polynomials of $\mathbb{K}[x_1, x_2, \ldots, x_s]$. We denote by $V_{\mathbb{K}^s}(F)$ (or simply by $V(F)$ when no confusion is possible) the zero set of the ideal generated by $F \subset \mathbb{K}[x_1, x_2, \ldots, x_s]$ in $\mathbb{K}^s$.

Definition 7 Let $V \subset \mathbb{K}^s$ be an r-dimensional equidimensional algebraic variety. The number of points of intersection of V with an $(s-r)$-dimensional generic linear subspace $L \subset \mathbb{K}^s$ is called the *degree* of V [1], denoted by $\deg(V)$. The degree of a non-equidimensional variety is defined to be the sum of the degrees of its equidimensional components. The degree of an ideal $I \subseteq \mathbb{K}[x_1, x_2, \ldots, x_s]$ is defined to be the degree of the variety of I in $\mathbb{K}^s$.

We first recall a few well-known results. Note that, for a zero-dimensional algebraic variety, the degree is just the number of points in that variety.

Lemma 8 *Let $V \subset \mathbb{K}^s$ be an r-dimensional equidimensional algebraic variety of degree δ. Let L be an $(s-r)$-dimensional linear subspace. Then, the intersection of L and V is either of positive dimension or consists of no more than δ points.*

Lemma 9 *Let $V \subset \mathbb{K}^s$ be an algebraic variety. Let L be a linear map from $\mathbb{K}^s$ to $\mathbb{K}^k$, for some integer $k > 0$. Then, we have $\deg(L(V)) \leq \deg(V)$.*

Lemma 10 ([3]) *Let $I \subset \mathbb{Q}[x_1, x_2, \ldots, x_s]$ be a radical ideal of degree δ. Then there exist finitely many polynomials in $\mathbb{Q}[x_1, x_2, \ldots, x_s]$ generating I and such that each of these polynomials has total degree less than or equal to δ.*

The following Lemma is a generalized form of Bézout's Theorem.

Lemma 11 *Let $V, W, V_1, V_2, \ldots, V_e$ be algebraic varieties in $\mathbb{K}^s$ such that we have $V = W \cap \cap_{i=1}^{e} V_i$ Define $r := \dim(W)$. Then, we have*

$$\deg(V) \leq \deg(W) \max(\{\deg(V_i) \mid i = 1 \cdots e\})^r.$$

Proposition 1 *Let $X = x_1, x_2, \ldots, x_s$ and $Y = y_1, y_2, \ldots, y_t$ be pairwise different $s + t$ variables. Let M be an ideal in $\mathbb{Q}[Y]$ of degree d_M and dimension r. Let $f_1, f_2, \ldots, f_s$ be s polynomials in $\mathbb{Q}[Y]$, each with maximum total degree d_f. Denote by I the ideal $\langle x_1 - f_1, x_2 - f_2, \ldots, x_s - f_s\rangle$. Then the ideal $J := I + M$ has degree upper bounded by $d_M\, {d_f}^r$.*

Proof We assume first that M is equidimensional. Let $L := l_1, l_2, \ldots, l_r$ be r linear forms in X, Y such that the intersection of the corresponding r hyperplanes and $V(J)$ consists of finitely many points, i.e. $H_L := J + \langle L\rangle$ is zero-dimensional. By virtue of Lemma 8, the degree of J equals the maximum degree of H_L among all possible choices of linear forms $l_1, l_2, \ldots, l_r$ satisfying the above conditions.

Let $L^* := l_1^*, l_2^*, \ldots, l_r^*$, where each l_j^* $(j = 1 \cdots r)$ is the polynomial obtained by substituting x_i with f_i, for $i = 1 \cdots s$, in the polynomial l_j. Consider the ideal $L^* + M$ in $\mathbb{Q}[Y]$. It is easy to show that the canonical projection map Π_Y onto the space of Y coordinates is a one-one-map between $V_{\mathbb{C}^t}(M + L^*)$ and $\Pi_Y(V_{\mathbb{C}^{t+s}}(H_L))$. Therefore, $V_{\mathbb{C}^t}(M + L^*)$ is zero-dimensional and $\deg(M + L^*) = \deg(H_L)$. Hence, viewing $V_{\mathbb{C}^t}(M + L^*)$ as

$$V_{\mathbb{C}^t}(M) \bigcap_{j=1}^{r} V_{\mathbb{C}^t}(l_j^*)$$

and thanks to Lemma 11, we have $\deg(V_{\mathbb{C}^t}(M + L^*)) \leq d_M\, d_f^r$. Therefore, we deduce that $\deg(J) = \max_L \deg(M + L^*) \leq d_M\, d_f^r$ holds, by Lemma 8.

Assume now that $V_{\mathbb{C}^t}(M)$ is not necessarily equidimensional. Let $V_1, V_2, \ldots, V_k$ be an irredundant equidimensional decomposition of $V_{\mathbb{C}^t}(M)$, with corresponding radical ideals $P_1, P_2, \ldots, P_k$. Then, applying the result proved in the first part of the proof to each $I + P_i$ $(i = 1, \ldots, k)$, we deduce

$$\begin{aligned}
\deg(J) &= \sum_{i=1}^{k} \deg(I + P_i) \\
&\leq \sum_{i=1}^{k} \deg(P_i)\, d_f^{r_i} \\
&\leq \sum_{i=1}^{k} \deg(P_i)\, d_f^{r} \\
&= d_M\, d_f^r,
\end{aligned}$$

where r_i is the dimension of P_i in $\mathbb{Q}[Y]$. This completes the proof. □

Remark 3 For J in Proposition 1, a less tight degree bound, namely

$$d_M\, d_f^{r+s},$$

can easily be deduced from a generalized form of Bezout's bound, since $V_{\mathbb{C}^{t+s}}(M)$ has degree d_M and is of dimension $r + s$ in $\mathbb{C}^{t+s}$.

Example 5 Consider $M := \langle n^2 - m^3 \rangle$, $g_1 := x - n^2 - n - m$, $g_2 := y - n^3 - 3n + 1$, and the ideal $J := M + \langle g_1, g_2 \rangle$. The ideal M has degree 3, and is of dimension 1 in $\mathbb{Q}[n, m]$. The degree of J is 9, which can be obtained by computing the dimension of

$$\mathbb{Q}(a, b, c, d, e)[x, y, m, n]/(J + \langle a\,x + b\,y + c\,n + d\,m + e \rangle),$$

where a, b, c, d, e are indeterminates. The degree bound estimated by Proposition 1 is 3×3, which agrees with the actual degree.

3 Linearization of P-Solvable Recurrences

In this section, we show that every P-solvable recurrence can be "linearized", that is, given an s-variable P-solvable recurrence R, there exists an affine recurrence L, such that the first s components of the solution to L solves R. In other words, although non-linear terms are allowed in P-solvable recurrences, these recurrences are essentially linear ones.

We will first show that, every poly-geometrical expression is a component of the solution of some affine recurrence.

Lemma 12 *Given a positive integer k, there exists a k-variable affine recurrence A with rational coefficients such that $(n, n^2, \ldots, n^{k-1}, n^k)$ is the solution to A.*

Proof We proceed by induction on k. The case $k = 1$ is easy: n solves the recurrence $x(n) = x(n-1) + 1$. Now assume that there exists a $(k-1)$-variable affine recurrence B of variables $x_1, x_2, \ldots, x_{k-1}$ with rational coefficients, whose solution is $(n, n^2, \ldots, n^{k-1})$. Let $x_k(n) = n^k$ and consider $x_k(n) - x_k(n-1)$, which is a polynomial in n of degree $k-1$. Therefore, $x_k(n) - x_k(n-1)$ can be written as a linear form with basis $1, n-1, (n-1)^2, \ldots, (n-1)^{k-1}$ (say by Taylor expansion) with coefficients $c_0, c_1, \ldots, c_{k-1}$. We deduce that:

$$x_k(n) = c_0 + x_k(n-1) + \sum_{i=1}^{k-1} c_i\, x_i(n-1) \tag{5}$$

Let A be the affine recurrence of recurrence variables $x_1, x_2, \ldots, x_k$ defined by the recurrence equations from B and Eq. (5). Clearly $(n, n^2, \ldots, n^k)$ is the solution to the k-variable affine recurrence A, which coefficients are all rational. □

Similarly, for hyper-geometrical terms in n, we have the following result.

Lemma 13 *Given a non-negative integer k and an algebraic number λ ($\lambda \neq 1$), there exists an $(k+1)$-variable affine recurrence A such that $(\lambda^n, n\,\lambda^n, \ldots, n^k\,\lambda^n)$ is the solution of A.*

Proof We proceed by induction on k as well. The case of $k = 0$ is trivial: the recurrence $x(n) = \lambda x(n-1)$ has the properties specified in the conclusion. Now assume there exists a k-variable affine recurrence B with recurrence variables $x_0, x_1, \ldots, x_{k-1}$, whose solution is $(\lambda^n, n\,\lambda^n, n^2, \ldots, n^{k-1}\,\lambda^n)$. Consider $x_k(n) = n^k \lambda^n$, which can be rewritten as $(n^k \lambda)\,\lambda^{n-1}$. Now consider $n^k\,\lambda$, which can be rewritten as a linear combination $c_0 + c_1\,(n-1) + \cdots + c_k(n-1)^k$ where $c_0, c_1, \ldots, c_k$ are constants. Therefore, we have:

$$x_k(n) = \sum_{i=0}^{k} c_i\, x_i(n-1) \tag{6}$$

Let A be the affine recurrence with recurrence variables $x_0, x_1, \ldots, x_k$ defined by the recurrence equations from B and Eq. (6). Clearly $(\lambda^n, n\,\lambda^n, \ldots, n^k\,\lambda^n)$ is the solution to the $(k+1)$-variable affine recurrence A. □

Next, as a consequence, we shall show that every poly-geometrical expression is a component of the solution of some affine recurrence.

Proposition 2 *Given a poly-geometrical expression h in n, there exists an affine recurrence A such that h equals the first component of the solution to A.*

Proof Assume w.l.o.g. that h is in canonical form and has m terms, say $h = \sum_{i=1}^{m} c_i\, t_i(n)$. We know that each term $t_i(n)$ of h is either of the form n^k or the form $n^k\,\lambda^n$. According to Lemmas 12 and 13, we can substitute each term $t_i(n)$ by a recurrence variable x_i from some affine recurrence, say A_i. We can assume w.l.o.g the variables in those recurrences are all pairwise different.

Then, we can form a new affine recurrence A by putting together $x_0(n) = \sum_{i=0}^{m} c_i\, x_i(n)$ and the equations in A_j, for $j = 1, \ldots, m$, yielding a system where h will be the first component of the solution to A. □

Since the solutions to P-solvable recurrences consist of poly-geometrical expressions, Proposition 2 implies that there exists a "linearization" procedure for P-solvable recurrences. Next, we show that we can "linearize" P-solvable recurrences without knowing their solutions. More precisely, Theorem 1 states that if one has a P-solvable recurrence R with rational coefficients, we can always find an affine recurrence A with rational coefficients such that each component of a solution of R is a component of a solution of A. One of the key point is the construction made in Lemma 14.

Lemma 14 *Given any two recurrence variables x_1 and x_2 from an affine recurrence A with rational coefficients, there exists an affine recurrence A^* with rational coefficients such that $x_1(n)\, x_2(n)$ is a component of the solution to A^*.*

Proof Let $x_1, x_2, \ldots, x_s$ be all the recurrence variables in A. For each pair (i, j), with $1 \le i \le j \le s$, we define a new recurrence variable $y_{i,j} = x_i x_j$. It is easy to check that $y_{i,j}(n)$ can be represented as a linear combination of $y_{k,\ell}(n-1)$ with rational coefficients, for $1 \leq k, \ell \leq s$. Indeed, each $x_i(n)$, $i = 1, \ldots, s$, is a linear combination of $x_j(n-1)$, for $j = 1, \ldots, s$. □

Theorem 1 *Given a P-solvable recurrence R with rational coefficients, we can find an affine recurrence A^* with rational coefficients, without solving the recurrence R, such that each component of the solution of R is a component of the solution of A^*.*

Proof Assume R has k blocks. If $k = 1$, then R is an affine recurrence and nothing needs to be done for this case.

From now on, we assume $k > 1$. As we shall see, however, treating the case of two blocks is sufficient to raise the key argument in the construction. Thus, for clarity, we assume that the coefficient matrix M of R is 2-block diagonal.

Let $x_{i_1}(n-1)x_{i_2}(n-1), \ldots, x_{i_j}(n-1)$ be a non-linear term occuring in the second block. Note that $x_{i_1}, x_{i_2}, \ldots, x_{i_j}$ are actually variables of the affine recurrence induced by the first block. According in Lemma 14, there exists an affine recurrence A with rational coefficients, such that $x_{i_1} x_{i_2}$ is solution to some variable y of A. Substitute in R each occurrence of $x_{i_1}(n-1)x_{i_2}(n-1)$ by $y(n-1)$ we obtain recurrence equations A_y. Let A^1 be the recurrence defined by the equations in A_y and R. Note that

(1) A^1 is a P-solvable recurrence with rational coefficients and allows a 2 block coefficient matrix;
(2) each component of the solution of R is a component of the solution of A^1.

If A^1 still has non-linear terms, we apply again the above trick to A^1, yielding a recurrence A^2. It is easy to check, that this "linearization" process will be completed in a finite number of steps. Finally, we obtain an affine recurrence A^* with rational coefficients, such that each component of the solution of R is a component of the solution of A^*. □

Actually, the proof of Theorem 1 implies an algorithm for "linearizing" any P-solvable recurrence. However, the resulting affine recurrence by this an algorithm will have exponentially many (roughly $\binom{n+d+1}{d}$) variables, which is hardly of practical use). An interesting problem would be to find an "optimal linearization", with a minimum number of recurrence variables.

4 Invariant Ideal of P-Solvable Recurrences

We will first formalize the notion of a P-solvable recurrence. Then in the rest of this section, we will investigate the shape of the closed form solutions of a P-solvable recurrence equation, for studying the degree and the dimension of invariant ideal. We will provide degree estimates for the invariant ideal, which is useful for all invariant generation methods which need a degree bound, like the proposed polynomial interpolation based method and those in [4, 8, 9]. Last but not least, we will investigate the dimension of the invariant ideal. So that we can get a sufficient for non-trivial polynomial invariants of a given P-solvable recurrence to exist. Note that in our invariant generation method, we do not need (thus never compute) the closed form solutions explicitly.

It is known that the solutions to P-solvable recurrences are poly-geometrical expressions in n w.r.t. the eigenvalues of the matrix M, see for example [12]. However, we need to estimate the "shape", e.g. the degree of those poly-geometrical expression solutions, with the final goal of estimating the "shape" (e.g. degree, height, dimension) of the invariant ideal. In this paper, we focus on degree and dimension estimates.

We first generalize the result of Lemma 7 to the multi-variable case.

Proposition 3 *Let $\alpha_1, \ldots, \alpha_m \in \overline{\mathbb{Q}}^* \setminus \{1\}$. Let $\lambda \in \overline{\mathbb{Q}}$ and $M \in \overline{\mathbb{Q}}^{s\times s}$ be a matrix in the following Jordan form*

$$\begin{pmatrix} \lambda & 0 & 0 & \cdots & 0 & 0 \\ 1 & \lambda & 0 & \cdots & 0 & 0 \\ 0 & 1 & \lambda & 0 & 0 & 0 \\ 0 & \ddots & \ddots & \ddots & \ddots & 0 \\ 0 & 0 & 0 & \cdots & \lambda & 0 \\ 0 & 0 & 0 & \cdots & 1 & \lambda \end{pmatrix}.$$

Consider an s-variable recurrence R defined as follows:

$$X(n+1)_{s\times 1} = M_{s\times s}\, X(n)_{s\times 1} + F(n)_{s\times 1}, \text{ where}$$

(a) $X := x_1, x_2, \ldots, x_s$ are the recurrence variables;
(b) $F := (f_1, f_2, \ldots, f_s)$ is a list of poly-geometrical expression in n w.r.t. $\alpha_1, \ldots, \alpha_m$, with maximal total degree d.

Then we have:

1. *if $\lambda = 0$, then $(f_1, f_1 + f_2, \ldots, f_1 + f_2 + \cdots + f_s)$ solves R.*
2. *if $\lambda = 1$, then there exist s poly-geometric expressions $(g_1, g_2, \ldots, g_s)$ in $\alpha_1, \ldots, \alpha_m$ such that for each $i \in 1, \ldots, s$, g_i is a poly-geometrical expression in n w.r.t. $\alpha_1, \ldots, \alpha_m$ with total degree less or equal than $d + i$.*
3. *if $\lambda \notin \{0, 1\}$, then there exists a solution of R, say $(y_1, y_2, \ldots, y_s)$, such that for each $i = 1, \ldots, s$ we have*

$$y_i := c_i \lambda_i^n + g_i, \text{ where} \tag{7}$$

for each $i \in 1, \ldots, s$: (a) c_i is a constant depending only on the initial value of the recurrence; and (b) g_i is like in the case of $\lambda = 1$. Moreover, assume furthermore that the following conditions hold:

(i) λ is weakly multiplicatively independent w.r.t. $\alpha_1, \ldots, \alpha_m$;
(ii) $\deg(f_j, n) = 0$ holds for all $j \in \{1, 2, \ldots, s\}$.

Then, for all $i = 1, \ldots, s$, *we can further choose* g_i *such that* $\deg(g_i, n) = 0$ *holds and the total degree of* g_i *is less or equal than* $\max(d, 1)$.

Proof We observe that the recurrence variables of R can be solved one after the other, from x_1 to x_s. When $\lambda = 0$, the conclusion is easy to verify. The case $\lambda \neq 0$ is easy to prove by induction on s with Lemma 7. □

Proposition 4 *Let* $\lambda_1, \ldots, \lambda_s, \alpha_1, \ldots, \alpha_m \in \overline{\mathbb{Q}}^* \setminus \{1\}$. *Let* $M \in \overline{\mathbb{Q}}^{s \times s}$ *be a matrix in the following Jordan form*

$$\begin{pmatrix} \lambda_1 & 0 & 0 & \cdots & 0 & 0 \\ \epsilon_{2,1} & \lambda_2 & 0 & \cdots & 0 & 0 \\ 0 & \epsilon_{3,2} & \lambda_3 & 0 & 0 & 0 \\ 0 & \ddots & \ddots & \ddots & \ddots & 0 \\ 0 & 0 & 0 & \cdots & \lambda_{s-1} & 0 \\ 0 & 0 & 0 & \cdots & \epsilon_{s,s-1} & \lambda_s \end{pmatrix},$$

where for $i = 2, \ldots, s$, $\epsilon_{i,i-1}$ *is either* 0 *or* 1. *Consider an* s-*variable recurrence* R *defined as follows:*

$$X(n+1)_{s\times 1} = M_{s\times s}\, X(n)_{s\times 1} + F(n)_{s\times 1},$$

where

1. $X := x_1, x_2, \ldots, x_s$ *are the recurrence variables;*
2. $F := (f_1, f_2, \ldots, f_s)$ *is a list of poly-geometrical expression in* n *w.r.t.* $\alpha_1, \ldots, \alpha_m$, *with maximal total degree* d.

Then there exists a solution of R, *say* $(y_1, y_2, \ldots, y_s)$, *such that for each* $i = 1, \ldots, s$ *we have*

$$y_i := c_i \lambda_i^n + g_i, \tag{8}$$

where

(a) c_i *is a constant depending only on the initial value of the recurrence and*
(b) g_i *is a poly-geometrical expression in* n *w.r.t.* $\lambda_1, \ldots, \lambda_{i-1}, \alpha_1, \ldots, \alpha_m$, *with total degree less or equal than* $d + i$.

Assume furthermore that the following conditions hold:

(i) the sequence consisting of $\lambda_1, \lambda_2, \ldots, \lambda_s$ *is weakly multiplicatively independent;*
(ii) $\deg(f_j, n) = 0$ *holds for all* $j \in \{1, 2, \ldots, s\}$.

Then, for all $i = 1, \ldots, s$, *we can further choose* y_i *such that* $\deg(g_i, n) = 0$ *holds and the total degree of* g_i *is less or equal than* $\max(d, 1)$.

Proof We observe that the recurrence variables of R can be solved one after the other, from x_1 to x_s. We proceed by induction on s. The case $s = 1$ follows directly from

Lemma 7. Assume from now on that $s > 1$ holds and that we have found solutions $(y_1, y_2, \ldots, y_{s-1})$ for the first $s - 1$ variables satisfying the requirements, that is, Relation (8) with (a) and (b). We define

$$\tilde{f}(n) = f_s(n) - \epsilon_{s,s-1}\, y_{s-1}(n+1). \tag{9}$$

Note that $\tilde{f}(n)$ is a poly-geometrical expression in n w.r.t. $\lambda_1, \ldots, \lambda_{s-1}, \alpha_1, \ldots, \alpha_m$ with total degree less than or equal to $d + s - 1$. Moreover, for $v \in \{n, \lambda_1^n, \ldots, \lambda_{s-1}^n, \alpha_1^n, \ldots, \alpha_m^n\}$ we have

$$\deg(\tilde{f}(n), v) \leq \max\left(\deg(f_s(n), v), \deg(y_{s-1}(n), v)\right). \tag{10}$$

It remains to solve x_s from

$$x_s(n+1) = \lambda_s\, x_s(n) + \tilde{f}(n) \tag{11}$$

in order to solve all the variables $x_1, \ldots, x_s$. Again, by Lemma 7, there exists a poly-geometrical expression

$$y_s := c_s\, \lambda_s^n + g_s(n),$$

where $g_s(n)$ is poly-geometrical expression in n w.r.t. $\lambda_1, \ldots, \lambda_{s-1}, \alpha_1, \ldots, \alpha_m$, of total degree upper bounded by $d + s$. This completes the proof of the properties (a) and (b) for y_s.

Now we assume that (i), (ii) hold and we prove the second half of the conclusion. Observe that we have $\deg(g_s(n), n) = \deg(\tilde{f}(n), n)$, which is 0, according to Relation (10) and the fact that we can choose y_{s-1} such that $\deg(y_{s-1}(n), n) = 0$ holds. Next, we observe that for each

$$v \in \{n, \lambda_1^n, \ldots, \lambda_{s-1}^n, \alpha_1^n, \ldots, \alpha_m^n\},$$

we have $\deg(g_s(n), v) = \deg(\tilde{f}(n), v)$, which is less or equal to $\deg(y_{s-1}(n), v)$ by Relation (10). Therefore, the total degree of g_s is less or equal than the total degree of y_{s-1}, which is less or equal than $\max(d, 1)$ by our induction hypothesis. This completes the proof. □

Theorem 2 *Let R be a P-solvable recurrence relation. Using the same notations $M, k, s, F, n_1, n_2, \ldots, n_k$ as in Definition 1. Assume M is in a Jordan form. Assume the eigenvalues $\lambda_1, \ldots, \lambda_s$ of M (counted with multiplicities) are different from* 0, 1, *with λ_i being the i-th diagonal element of M. Assume for each block j the total degree of any polynomial in $\mathbf{f}_j$ (for $i = 2, \ldots, k$) is upper bounded by d_j. For each i, we denote by $b(i)$ the block number of the index i, that is,*

$$\sum_{j=1}^{b(i)-1} n_j < i \leq \sum_{j=1}^{b(i)} n_j. \tag{12}$$

Let $D_1 := n_1$ and for all $j \in \{2, \ldots, k\}$ let $D_j := d_j\, D_{j-1} + n_j$. Then, there exists a solution $(y_1, y_2, \ldots, y_s)$ for R of the following form:

$$y_i := c_i \lambda_i^n + g_i, \tag{13}$$

for all $i \in 1, \ldots, s$, where

(a) c_i is a constant depending only on the initial value of the recurrence;
(b) g_i is a poly-geometrical expression in n w.r.t. $\lambda_1, \ldots, \lambda_{i-1}$, and with total degree less or equal than $D_{b(i)}$.

Moreover, if the sequence consisting $\{\lambda_1, \ldots, \lambda_s\}$ is weakly multiplicatively independent, then, for all $i = 1, \ldots, k$, we can further choose y_i such that $\deg(g_i, n) = 0$ holds and the total degree of g_i is less or equal than $\prod_{2 \leq t \leq b(i)} \max(d_t, 1)$.

Proof We proceed by induction on the number of blocks, that is, k. The case $k = 1$ follows immediately from Proposition 4. Assume from now on that the conclusion holds for a value $k = \ell$, with $\ell \geq 1$ and let us prove that it also holds for $k = \ell+1$. We apply the induction hypothesis to solve the first ℓ blocks of variables, and suppose that $\mathbf{y}_\ell$ is a solution satisfying the properties in the conclusion. For solving the variables in the $(\ell + 1)$th block, we substitute $\mathbf{y}_\ell$ to $f_{\ell+1}$ and obtain a tuple of poly-geometrical expressions in n w.r.t the eigenvalues of the first ℓ blocks and with total degree bounded by $d_\ell\, D_\ell$. Therefore, applying again Proposition 4, we can find solutions for the variables in the $(\ell + 1)$th block satisfying the properties required in the conclusion. This completes the proof. □

Note that the degree estimate in Theorem 2 depends on how the block structure of the recurrence is exploited, for example, a 2×2 diagonal matrix can be viewed as a matrix with a single block or a matrix with two 1×1 diagonal blocks.

In practice, one might want to decouple the recurrence first, and then study the recurrence variable one by one (after a linear coordinate change) to get better degree estimates for the poly-geometrical expression solutions, regarded as polynomials of n-exponential terms as the eigenvalues of the coefficient matrix. We will just use a simple example to illustrate this idea.

Example 6 Consider the recurrence:

$$\begin{pmatrix} x(n+1) \\ y(n+1) \\ z(n+1) \end{pmatrix} := \begin{pmatrix} 2\,0\,0 \\ 0\,3\,0 \\ 0\,0\,3 \end{pmatrix} \times \begin{pmatrix} x(n) \\ y(n) \\ z(n) \end{pmatrix} + \begin{pmatrix} 0 \\ x(n)^2 \\ x(n)^3 \end{pmatrix}$$

Viewing the recurrence as two blocks corresponding to variables (x) and (y, z) respectively, the degree estimate according to Theorem 2 would be bounded by $5 = 3 \times 1 + 2$.

If we decouple the (y, z) block to the following two recurrences

$$y(n+1) = 3\,y(n) + x(n)^2 \text{ and } z(n+1) = 3\,z(n) + x(n)^3,$$

then we can easily deduce that the degree of the poly-geometrical expression for y and z are upper bounded by 2 and 3 respectively, again according to Theorem 2.

It is easy to generalize the previous results to the case of a matrix M which is not in Jordan form. Let Q be a non-singular matrix such that $J := Q\,M\,Q^{-1}$ is a Jordan form of M. Let the original recurrence R be

$$X(n+1) = M\,X(n) + F.$$

Consider the following recurrence R_Q

$$Y(n+1) = J\,Y(n) + QF.$$

It is easy to check that if

$$(y_1(n), y_2(n), \ldots, y_s(n))$$

solves R_Q, then

$$Q^{-1}\;(y_1(n), y_2(n), \ldots, y_s(n))$$

solves R. Note that an invertible matrix over $\overline{\mathbb{Q}}$ maps a tuple of poly-geometrical expressions to another tuple of poly-geometrical expressions; moreover it preserves the highest degree among the expressions in the tuple.

We turn now our attention to the question of estimating the degree of the invariant ideal of a P-solvable recurrence relation.

Proposition 5 *Let R be an s-variable P-solvable recurrence relation, with recurrence variables $(x_1, x_2, \ldots, x_s)$. Let $\mathcal{I} \subset \mathbb{Q}[x_1, x_2, \ldots, x_s]$ be the invariant ideal of R. Denote by $\mathcal{I}^e$ the extension of $\mathcal{I}$ in $\overline{\mathbb{Q}}[x_1, x_2, \ldots, x_s]$. Let $A = \alpha_1, \alpha_2, \ldots, \alpha_s$ be the eigenvalues (counted with multiplicities) of the coefficient matrix of R. Let $\mathcal{M}$ be the multiplicative relation ideal of A associated with variables $y_1, \ldots, y_s$. Then, there exists a sequence of s poly-geometrical expressions in n w.r.t. $\alpha_1, \alpha_2, \ldots, \alpha_s$, say*

$$f_1(n, \alpha_1^n, \ldots, \alpha_k^n), \ldots, f_s(n, \alpha_1^n, \ldots, \alpha_k^n),$$

which solves R. Moreover, we have

$$\mathcal{I}^e \;=\; (\mathcal{S} + \mathcal{M}) \;\cap\; \overline{\mathbb{Q}}[x_1, x_2, \ldots, x_s],$$

where $\mathcal{S}$ is the ideal generated by $\langle x_1 - f_1(n, y_1, \ldots, y_s), \ldots, x_s - f_s(n, y_1, \ldots, y_s)\rangle$ in $\overline{\mathbb{Q}}[x_1, x_2, \ldots, x_s, n, y_1, \ldots, y_s]$.

Proof The existence of $f_1, f_2, \ldots, f_s$ follows by Theorem 2 and the fact that linear combination of poly-geometrical expressions w.r.t. n are still poly-geometrical expressions. The conclusion follows from Lemma 5. □

The following lemma is not hard to prove and one can find a proof in [5].

Lemma 15 *Let R be a P-solvable recurrence relation defining s sequences in $\mathbb{Q}^s$, with recurrence variables $(x_1, x_2, \ldots, x_s)$. Let $\mathcal{I}$ be the invariant ideal of R in $\mathbb{Q}[x_1, x_2, \ldots, x_s]$; let $\overline{\mathcal{I}}$ be the invariant ideal of R in $\overline{\mathbb{Q}}[x_1, x_2, \ldots, x_s]$. Then $\overline{\mathcal{I}}$ equals to $\mathcal{I}^e$, the extension of $\mathcal{I}$ in $\overline{\mathbb{Q}}[x_1, x_2, \ldots, x_s]$.*

With Propositions 5 and 1, we are able to estimate the degree of polynomials in a generating system of the invariant ideals. Now we are able to estimate the total degree of closed form solutions of a P-solvable recurrence without solving the recurrence explicitly.

Theorem 3 *Let R be a P-solvable recurrence relation defining s sequences in $\mathbb{Q}^s$, with recurrence variables $(x_1, x_2, \ldots, x_s)$. Let $\mathcal{I} \subset \mathbb{Q}[x_1, x_2, \ldots, x_s]$ be the invariant ideal of R. Let $A = \alpha_1, \alpha_2, \ldots, \alpha_s$ be the eigenvalues (counted with multiplicities) of the coefficient matrix of R. Let $\mathcal{M}$ be the multiplicative relation ideal of A associated with variables $y_1, \ldots, y_k$. Let r be the dimension of $\mathcal{M}$. Let $f_1(n, \alpha_1^n, \ldots, \alpha_k^n), \ldots, f_s(n, \alpha_1^n, \ldots, \alpha_k^n)$ be a sequence of s poly-geometrical expressions in n w.r.t. $\alpha_1, \alpha_2, \ldots, \alpha_s$ that solves R. Suppose R has a k block configuration as $(n_1, 1), (n_2, d_2), \ldots, (n_k, d_k)$. Let $D_1 := n_1$; and for all $j \in \{2, \ldots, k\}$, let $D_j := d_j\, D_{j-1} + n_j$. Then we have*

$$\deg(\mathcal{I}) \leq \deg(\mathcal{M})\, D_k^{r+1}.$$

Moreover, if the degrees of n in f_i $(i = 1 \cdots s)$ are 0, then we have

$$\deg(\mathcal{I}) \leq \deg(\mathcal{M})\, D_k^{r}.$$

Proof Denoting by Π the standard projection from $\overline{\mathbb{Q}}^{s+1+s}$ to $\overline{\mathbb{Q}}^{s}$:

$$(x_1, x_2, \ldots, x_s, n, y_1, \ldots, y_s) \mapsto (x_1, x_2, \ldots, x_s),$$

we deduce by Proposition 5 that

$$V(\mathcal{I}) = \overline{\Pi(V(\mathcal{S} + \mathcal{M}))}, \tag{14}$$

where $\mathcal{S}$ is the ideal generated by $\langle x_1 - f_1(n, y_1, \ldots, y_s), \ldots, x_s - f_s(n, y_1, \ldots, y_s)\rangle$ in $\overline{\mathbb{Q}}[x_1, x_2, \ldots, x_s, n, y_1, \ldots, y_s]$. Thus, by Lemma 9, we have

$$\deg(\mathcal{I}) \leq \deg(\mathcal{S} + \mathcal{M}).$$

It follows from Proposition 1 that

$$\deg(\mathcal{S} + \mathcal{M}) \leq \deg(\mathcal{M})\, D_k^{r+1},$$

since the total degree of f_i of R is bounded by D_k according to Theorem 2 and the dimension of $\mathcal{M}$ is $r + 1$ is in $\mathbb{Q}[n, y_1, \ldots, y_s]$.

With similar arguments, the second part of the conclusion follows from the fact that $S + M$ can be viewed as an ideal in in $\overline{\mathbb{Q}}[x_1, x_2, \ldots, x_s, n, y_1, \ldots, y_s]$, where M has dimension r. □

Indeed, the degree bound in Theorem 3 is "sharp" in the sense that it is reached by many of the examples (Example 2) we have considered.

In the rest of this section, we are going to investigate the dimension of the invariant ideal of a P-solvable recurrence. This can help checking whether or not the invariant ideal of a P-solvable recurrence over $\mathbb{Q}$ is the trivial ideal of $\mathbb{Q}[x_1, \ldots, x_s]$. Note that it is obvious that the invariant ideal is not the whole polynomial ring.

Theorem 4 *Using the same notations as in Definition 1. Let $\lambda_1, \lambda_2, \ldots, \lambda_s$ be the eigenvalues of M counted with multiplicities. Let $\mathcal{M}$ be the multiplicative relation ideal of $\lambda_1, \lambda_2, \ldots, \lambda_s$. Let r be the dimension of $\mathcal{M}$. Let $\mathcal{I}$ be the invariant ideal of R. Then $\mathcal{I}$ is of dimension at most $r + 1$. Moreover, for generic initial values,*

1. *the dimension of $\mathcal{I}$ is at least r;*
2. *if 0 is not an eigenvalue of M and the sequence consisting of $\lambda_1, \lambda_2, \ldots, \lambda_s$ is weakly multiplicatively independent, then $\mathcal{I}$ has dimension r.*

Proof Assume without loss of genericity that M is in Jordan form. By Theorem 2, we deduce that R has a solution $(f_1, f_2, \ldots, f_s)$ as follows

$$\left(c_1\, \lambda_1^n + h_1(n), c_2\, \lambda_2^n + h_2(n), \ldots, c_s\, \lambda_s^n + h_s(n)\right),$$

where for each $i \in 1, \ldots, s$, c_i is a constant in $\overline{\mathbb{Q}}$ depending only on the initial value of R, and h_i is a poly-geometrical expression in n w.r.t. $\lambda_1, \ldots, \lambda_{i-1}$. Moreover, we have

1. for generic initial values, none of $c_1, c_2, \ldots, c_s$ is 0;
2. if the eigenvalues of M can be ordered in $\lambda_1, \lambda_2, \ldots, \lambda_s$ s.t. $\lambda_1 \neq 1$ and for each $i \in 2, \ldots, s$, λ_i is weakly multiplicatively independent w.r.t. $\lambda_1, \lambda_2, \ldots, \lambda_{i-1}$, then we can require that, for all $i \in 1, \ldots, s$, we have $\deg(f_i, n) = 0$.

Viewing n, λ_i^n (for $i = 1, \ldots, s$) as indeterminates, let us associate coordinate variable u_0 to n, u_i to λ_i^n (for $i = 1, \ldots, s$). Denote by V the variety of $\mathcal{I}$ in $\overline{\mathbb{Q}}^s$ (with coordinates $x_1, x_2, \ldots, x_s$). Note that we have

$$\dim(V) = \dim(\mathcal{I}).$$

Denote by W_1, W_2 respectively the variety of $\mathcal{M}$ in $\overline{\mathbb{Q}}^s$ (with coordinates $u_1, u_2, \ldots, u_s$) and in $\overline{\mathbb{Q}}^{s+1}$ (with coordinates $u_0, u_1, u_2, \ldots, u_s$). Note that we have

$$\dim(W_1) = r \text{ and } \dim(W_2) = r + 1.$$

Consider first the map F_0 defined below:

$$F_0 : \overline{\mathbb{Q}}^{s+1} \mapsto \overline{\mathbb{Q}}^{s+1}$$
$$(u_0, u_1, \ldots, u_s) \to (c_1 u_1 + f_1, \ldots, c_s u_s + f_s).$$

By Theorem 3, we have $V = \overline{F_0(W_2)}$. Therefore, we have we have $\dim(\mathcal{I}) = \dim(V) \leq \dim(W_2) = r + 1$.

Now assume the initial value of R is generic, thus we have $c_i \neq 0$, for all $i \in 1, \ldots, s$. Let us consider the map F_1 defined below:

$$F_1 : \overline{\mathbb{Q}}^{s+1} \mapsto \overline{\mathbb{Q}}^{s+1}$$
$$(u_0, u_1, \ldots, u_s) \to (u_0, c_1 u_1 + f_1, \ldots, c_s u_s + f_s).$$

Let us denote by V_2 the variety $\overline{F_1(W_2)}$. By virtue of Theorem 3, we have $\dim(V_2) = \dim(W_2) = r + 1$. Denote by Π the standard projection map that forgets the first coordinate, that is, u_0. We observe that $V = \overline{\Pi(V_2)}$. Therefore, we have $\dim(V) \geq \dim(\overline{\Pi(V_2)}) - 1 = r$.

Now we further assume $\lambda_1 \neq 1$ and for each $i \in 2, \ldots, s$, λ_i is weakly multiplicatively independent w.r.t. $\lambda_1, \lambda_2, \ldots, \lambda_{i-1}$ the invariant ideal of R. In this case, we have that for all $i \in 1, \ldots, s$, $\deg(f_i, n) = 0$. Let us consider the map F_2 defined below:

$$F_2 : \overline{\mathbb{Q}}^{s} \mapsto \overline{\mathbb{Q}}^{s}$$
$$(u_1, \ldots, u_s) \to (c_1 u_1 + f_1, c_2 u_2 + f_2, \ldots, c_s u_s + f_s).$$

By Theorem 3, we have $V = \overline{F_2(W_1)}$. Therefore, we have $\dim(\mathcal{I}) = \dim(V) = \dim(W_1) = r$. This completes the proof. □

The following result, which is a direct consequence of Theorem 4, can serve as a sufficient condition for the invariant ideal to be non-trivial. This condition is often satisfied when there are eigenvalues with multiplicities or when 0 and 1 are among the eigenvalues.

Corollary 1 *Using the same notations as in Theorem 4. If $r + 1 < s$ holds, then $\mathcal{I}$ is not the zero ideal in $\mathbb{Q}[x_1, x_2, \ldots, x_s]$.*

The following corollary indicates that, the fact that the invariant ideal of a given P-solvable recurrence is trivial could be determined by just investigating the multiplicative relation among the eigenvalues of the underlying recurrence.

Corollary 2 *Using the same notations as in Theorem 4, consider an s variable P-solvable recurrence R with initial value $x_1(0) := a_1, \ldots, x_s(0) := a_s$, where $a_1, \ldots, a_s$ are indeterminates. If the eigenvalues of R are multiplicatively independent, then the invariant ideal of R is $\langle 0 \rangle$ in $\mathbb{Q}(a_1, \ldots, a_s)[x_1, x_2, \ldots, x_s]$.*

Proof The assumption implies that the multiplicative relation ideal of the eigenvalues is of dimension s. By Theorem 4, the dimension of the invariant ideal of R must be at least s, thus the invariant ideal of R must be zero ideal in $\mathbb{Q}(a_1, \ldots, a_s)[x_1, \ldots, x_s]$. □

Example 7 Consider the recurrence:

$$(x(n+1), y(n+1)) := (3\,x(n) + y(n), 2\,y(n)) \text{ with } x(0) = a,\, y(0) = b.$$

The two eigenvalues f the coefficient matrix are 2 and 3 which are multiplicatively independent. Therefore, by Corollary 2, the invariant ideal of the recurrence is trivial.

Note in Theorem 4, if we drop the "generic" assumption on the initial values, then the conclusion might not hold. The following example illustrate this for the case when all the eigenvalues are different and multiplicatively independent, but the invariant ideal is not trivial.

Example 8 Consider the linear recurrence $x(n+1) = 3\,x(n) - y(n), y(n+1) = 2\,y(n)$ with $(x(0), y(0)) = (a, b)$. The eigenvalues of the coefficient matrix are $2, 3$, which are multiplicatively independent. One can check that, when $a = b$, the invariant ideal is generated by $x - y$. However, generically, that is when $a \neq b$ holds, the invariant ideal is the zero ideal.

5 Concluding Remarks

In this article, we study the equivalence between P-solvable recurrences and linear recurrences, and supply sharp estimate on the degree and dimension of invariant ideals of P-solvable recurrences. As future work, we would be interested on finding simple linearizations of P-solvable recurrences, which could help obtaining more precise estimates on the degree of the invariant ideal.

References

1. Cox, D., Little, J., O'Shea, D.: Using Algebraic Geometry. Graduate Text in Mathematics. Springer, New York (1998)
2. Ge, G.: Algorithms related to multiplicative representations of algebraic numbers. Ph.D. thesis, U.C. Berkeley (1993).
3. Heintz, Joos: Definability and fast quantifier elimination in algebraically closed fields. Theor. Comput. Sci. **24**(3), 239–277 (1983)
4. Kapur, D., Rodriguez-Carbonell, E.: Automatic generation of polynomial invariants of bounded degree using abstract interpretation. Sci. Comput. Program. **64**(1), 54–75 (2007)
5. Kauers, M., Zimmermann, B.: Computing the algebraic relations of c-finite sequences and multisequences. J. Symb. Comput. **43**, 787–803 (2008)

6. Kovács, L.: Invariant generation for p-solvable loops with assignments. In: Proceedings of the 3rd international conference on computer science: theory and applications. CSR'08, pp. 349–359. Springer, Berlin (2008)
7. Moreno Maza, M., Xiao, R.: Generating program invariants via interpolation. CoRR, abs/1201.5086, 2012.
8. Müller-Olm, M., Seidl, H.: A Note on Karr's Algorithm. In: Díaz, J., Karhumäki, J., Lepistö, A., Sannella, D. (eds.) Automata, Languages and Programming, vol. 3142 of Lecture Notes in Computer Science, pp. 1016–1028. Springer, Turku (2004)
9. Müller-Olm, M., Seidl, H.: Computing polynomial program invariants. Inf. Process. Lett. **91**(5), 233–244 (2004)
10. Osborne, M.J.: Math tutorial: first-order difference equations (2000).
11. Rodríguez-Carbonell, E., Kapur, D.: An Abstract Interpretation Approach for Automatic Generation of Polynomial Invariants. In International Symposium on Static Analysis (SAS 2004), vol. 3148 of Lecture Notes in Computer Science, pp. 280–295. Springer, Heidelberg (2004).
12. Rodríguez-Carbonell, E., Kapur, D.: Automatic generation of polynomial loop invariants: algebraic foundations. ISSAC '04, pp. 266–273. ACM (2004).
13. von zur Gathen, J., Gerhard, J.: Modern Computer Algebra. Cambridge University Press, Cambridge (1999).

Real Root Isolation of Polynomial Equations Based on Hybrid Computation

Fei Shen, Wenyuan Wu and Bican Xia

Abstract A new algorithm for real root isolation of zero-dimensional nonsingular square polynomial systems based on hybrid computation is presented in this paper. First, approximate the (complex) roots of the given polynomial equations via homotopy continuation method. Then, for each approximate root, an initial box relying on the Kantorovich theorem is constructed, which contains the corresponding accurate root. Finally, the Krawczyk interval iteration with interval arithmetic is applied to the initial boxes so as to check whether or not the corresponding approximate roots are real and to obtain the real root isolation boxes. Moreover, an empirical construction of initial box is provided for speeding-up the computation in practice. Our experiments on many benchmarks show that the new hybrid method is very efficient. The method can find all real roots of any given zero-dimensional nonsingular square polynomial systems provided that the homotopy continuation method can find all complex roots of the equations.

1 Introduction

The *Real Root Isolation* of polynomial equations is a procedure that produces disjoint regions to isolate all the distinct real roots of polynomial equations, with only one root in each region. Formally speaking, let $\mathbf{F} = (f_1, f_2, \ldots, f_n)^{\mathrm{T}}$ be polynomial equations defined on $\mathbf{R}^n$, i.e. $f_i \in \mathbf{R}[x_1, x_2, \ldots, x_n]$. Suppose $\mathbf{F}(\mathbf{x}) = 0$ has only finite many real roots, say $\xi^{(1)}, \xi^{(2)}, \ldots, \xi^{(m)}$. The target of real root isolation is

F. Shen (✉) · B. Xia (✉)
LMAM & School of Mathematical Sciences,
Peking University, Beijing, China
e-mail: shenfei@pku.edu.cn

B. Xia
e-mail: xbc@math.pku.edu.cn

W. Wu (✉)
Chongqing Institute of Green and Intelligent Technology,
Chinese Academy of Sciences, Chongqing, China
e-mail: wuwenyuan@cigit.ac.cn

© Springer-Verlag Berlin Heidelberg 2014
R. Feng et al. (eds.), *Computer Mathematics*, DOI 10.1007/978-3-662-43799-5_26

to compute a family of regions $S_1, S_2, \ldots, S_m,\ S_j \subset \mathbf{R}^n (1 \le j \le m)$, such that $\xi^{(j)} \in S_j$ and $S_i \cap S_j = \emptyset\ (1 \le i, j \le m)$. Usually, we use rectangular boxes to denote the regions above. So we often call these isolated boxes *intervals* in this paper. Theoretically, the width of intervals for some special problems can be very small. Hence, we assume that the accuracy of numerical computation in this paper can be arbitrarily high. However, it is also important to point out that such case rarely happens in nearly 50 examples we computed and double-precision is usually enough to obtain very small intervals.

Real root isolation is an important problem in symbolic computation. It is an algorithm for solving equations accurately since no root formula is available in general situation. It is also a critical part of some other important algorithms, such as CAD and real root classification for semi-algebraic systems, etc. Improvement on real root isolation will benefit all of these algorithms.

We impose some hypothesis on the problem discussed here. The first is that the system is square, i.e., the number of equations is the same as that of variables. Then we only handle the systems with finitely many complex roots. Positive dimensional complex solution is beyond the scope of this paper. Moreover, we suppose that the Jacobian matrix of $\mathbf{F}$ is nonsingular at each root of $\mathbf{F}(\mathbf{x}) = 0$. So we only deal with the simple root cases. For the singular situation, the deflation method [12, 16, 17, 32] can be applied, which is one of our ongoing work.

Most of the previous real root isolation algorithms are based on symbolic computations. For instance, the Collins–Akritas algorithm [10, 11] based on Descartes' rule of signs is for polynomials in one variable. In multi-variable scenario, there are many work using different theories and techniques with focuses on complexity and/or solving benchmarks, see for example [6–9, 13, 15, 21, 24, 29, 34, 35].

An advantage of those symbolic methods is that exact results can be obtained since they use symbolic computation and some of them can be extended to semi-algebraic systems. However, there are also some disadvantages. Some of these methods could only handle the isolation of complex roots. And some of them need to triangularize the system first, which is very time-consuming in computation when the scale (such as the number of variables or the degrees of polynomials) of the system is big. While some methods that do not use triangularization have to give a huge initial interval to include all the real roots [37, 38], which is extremely inefficient.

In order to avoid these problems and design a new algorithm that could efficiently solve more complicated systems and provide accurate interval results, we employ hybrid computation to take the advantages of both symbolic and numerical methods.

The basic idea of this paper is to use a numerical method to obtain all the approximate roots of polynomial systems, including possible nonreal ones. With these approximations, small initial intervals which contains the corresponding real roots are constructed. Then, we apply a symbolic method to these initial intervals to verify whether there is a real root in it or not. The main method we use during numerical computation is homotopy continuation, and for symbolic process we use the Krawczyk iteration.

Most of the work in this paper comes from [27]. In Sect. 2, we will introduce some preliminaries, including homotopy continuation and interval arithmetic. A new real root isolation algorithm is discussed in Sect. 3. To test our new method, our experimental results on benchmarks together with comparison and analysis will be presented in Sect. 4. Finally, there is a summary in Sect. 5 and some future work will also be discussed.

2 Preliminary

We introduce in this section the basic theory and tools that would be used in our algorithm.

2.1 Homotopy Continuation Method

Homotopy continuation method is an important numerical computation method, which is used in various fields. We only treat it as an "algorithm black box" here, where the input is a polynomial system, and the output is its approximate roots. Please find the details about the theory in [18, 28].

Our method to find all real roots of any given zero-dimensional nonsingular square polynomial equations relies on the homotopy continuation methods which theoretically obtain all complex roots of the equations. For the real homotopies, we refer to the work by Li and Wang [20]. For the certified numerical homotopy tracking and its complexity, we first refer to the work by M. Shub and S. Smale on complexity of Bezout's theorem [5] and the readers can also find the recent results on this topic by Beltran and Leykin in [1–3].

For our purpose, it is convenient to utilize some existing software, such as Hom4ps-2.0 [19], PHCpack [30] and HomLab [31].

In our implementation, we use Hom4ps-2.0, which could return all the approximate complex roots of a given polynomial system efficiently, along with residues and condition numbers.

2.2 Interval Arithmetic

Interval arithmetic plays an important role in real root isolation algorithms [34, 37, 38]. The two main differences between our new algorithm and the traditional ones in [37, 38] are: (1) Verification only carry out on the localized "small" intervals; (2) symbolic computation is replaced with floating point numerical computation.

Most of the interval operations in this paper's algorithms are based on Rump's floating point verification work [26] and accomplished by using the Matlab package Intlab [25], including interval arithmetic operations and Jacobian matrix, Hessian matrix calculations.[1]

[1] See reference [26], Sect. 11, *Automatic differentiation.*

2.2.1 Basic Concepts

We introduce some basic interval arithmetic theory in this section. See reference [22] for more details.

For given numbers $\underline{x}, \overline{x} \in \mathbf{R}$, if $\underline{x} \leq \overline{x}$, we call

$$X = [\underline{x}, \overline{x}] = \{x \in \mathbf{R} | \underline{x} \leq x \leq \overline{x}\}$$

a *bounded closed interval*, or *interval* for short. Denote by $I(\mathbf{R})$ the set of all the bounded close intervals on $\mathbf{R}$, and $I(A) = \{X \in I(\mathbf{R}) | X \subseteq A\}$ all the intervals on $A \subseteq \mathbf{R}$. Especially, if $\underline{x} = \overline{x}$, we call X a *point interval.*

For intervals, there are some common quantities:

midpoint $\mathrm{mid}(X) = (\underline{x} + \overline{x})/2$
width $\mathrm{W}(X) = \overline{x} - \underline{x}$
radius $\mathrm{rad}(X) = \frac{1}{2}\mathrm{W}(X)$
low end point $\inf(X) = \underline{x}$
high end point $\sup(X) = \overline{x}$

Obviously we have $X = [\mathrm{mid}(X) - \mathrm{rad}(X), \mathrm{mid}(X) + \mathrm{rad}(X)]$. An interval is usually expressed by its midpoint and radius. For example, if $m = \mathrm{mid}(X)$, $r = \mathrm{rad}(X)$, then we can write the formula above as $X = \mathrm{midrad}(m, r)$.

We can also define the arithmetic operations over intervals. Let $X = [\underline{x}, \overline{x}]$, $Y = [\underline{y}, \overline{y}] \in I(\mathbf{R})$,

- $X + Y = [\underline{x} + \underline{y}, \overline{x} + \overline{y}]$
- $X - Y = [\underline{x} - \overline{y}, \overline{x} - \underline{y}]$
- $X \cdot Y = [\min(\underline{x}\underline{y}, \underline{x}\overline{y}, \overline{x}\underline{y}, \overline{x}\overline{y}), \max(\underline{x}\underline{y}, \underline{x}\overline{y}, \overline{x}\underline{y}, \overline{x}\overline{y})]$
- $X/Y = [\underline{x}, \overline{x}] \cdot [1/\overline{y}, 1/\underline{y}],\ 0 \notin Y$

A vector is called an *interval vector* if all its components are intervals. *Interval matrix* can be similarly defined. For interval vectors and interval matrices, the concepts such as midpoint, width, radius, etc., and the arithmetic operations are defined in components.

Let $f : \mathbf{R}^n \to \mathbf{R}$ be a function, if there exists an interval map

$$F : I(\mathbf{R}^n) \to I(\mathbf{R})$$

such that for all $x_i \in X_i (i = 1, 2, \ldots, n)$,

$$F([x_1, x_1], [x_2, x_2], \ldots, [x_n, x_n]) = f(x_1, x_2, \ldots, x_n)$$

holds, then we call F an *interval expansion* of f.

We call $F : I(\mathbf{R}^n) \to I(\mathbf{R})$ an interval map with *inclusive monotonicity* if $\mathbf{X} \subseteq \mathbf{Y}$ implies $F(\mathbf{X}) \subseteq F(\mathbf{Y})$ for any given intervals $\mathbf{X}$ and $\mathbf{Y}$. The definitions above can all be extended to the situations in $I(\mathbf{R}^n) \to I(\mathbf{R}^n)$. And it is easy to prove that all the polynomial operations satisfy the inclusive monotonicity.

2.2.2 Krawczyk Operator

The Krawczyk operator plays a key role in the real root verification of interval arithmetic. The main accomplishment comes from the work of Krawczyk and Moore. We only list some important results here. Complete proofs can be found in [22].

Suppose $\mathbf{f} : D \subseteq \mathbf{R}^n \rightarrow \mathbf{R}^n$ is continuous differentiable on D. Consider the equation

$$\mathbf{f}(\mathbf{x}) = 0. \tag{1}$$

Let $\mathbf{f}'$ be the Jacobi matrix of $\mathbf{f}$, $\mathbf{F}$ and $\mathbf{F}'$ be the interval expansion of $\mathbf{f}$ and $\mathbf{f}'$ with inclusive monotonicity, respectively. For $\mathbf{X} \in I(D)$ and any $\mathbf{y} \in \mathbf{X}$, define the *Krawczyk operator* as:

$$K(\mathbf{y}, \mathbf{X}) = \mathbf{y} - Y\mathbf{f}(\mathbf{y}) + (\mathbf{I} - Y\mathbf{F}'(\mathbf{X}))(\mathbf{X} - \mathbf{y}) \tag{2}$$

where $\mathbf{I}$ is the $n \times n$ unit matrix and Y is any $n \times n$ nonsingular matrix.

Especially, we assign $\mathbf{y} = \mathrm{mid}(\mathbf{X})$, so Formula (2) becomes

$$K(\mathbf{X}) = \mathrm{mid}(\mathbf{X}) - Y\mathbf{f}(\mathrm{mid}(\mathbf{X})) + (\mathbf{I} - Y\mathbf{F}'(\mathbf{X}))\mathrm{rad}(\mathbf{X})[-1, 1]. \tag{3}$$

Formula (3) is often used in practice.

The reason why the Krawczyk operator is so important is that it has some nice properties.

Proposition 1 *Suppose $K(\mathbf{y}, \mathbf{X})$ is defined as Formula* (2)*, then*

1. *If $\mathbf{x}^* \in \mathbf{X}$ is a root of Eq.* (1)*, then for any $\mathbf{y} \in \mathbf{X}$, we have $\mathbf{x}^* \in K(\mathbf{y}, \mathbf{X})$;*
2. *For any $\mathbf{y} \in \mathbf{X}$, if $\mathbf{X} \cap K(\mathbf{y}, \mathbf{X}) = \emptyset$ holds, then there is no roots in $\mathbf{X}$;*
3. *For any $\mathbf{y} \in \mathbf{X}$ and any nonsingular matrix Y, if $K(\mathbf{y}, \mathbf{X}) \subseteq \mathbf{X}$ holds, then Eq.* (1) *has a solution in $\mathbf{X}$;*
4. *Moreover, for any $\mathbf{y} \in \mathbf{X}$ and any nonsingular matrix Y, if $K(\mathbf{y}, \mathbf{X})$ is strict inclusive in $\mathbf{X}$, then Eq.* (1) *has only one root in $\mathbf{X}$.*

With the properties above, we can easily develop a real root verification method which will be explained later in this paper.

Meanwhile, with the hypothesis we set in introduction, all the systems considered here are nonsingular ones with only simple roots. So the Jacobian matrix at the roots are all invertible. Thus, we often set $Y = (\mathrm{mid}\,\mathbf{F}'(\mathbf{X}))^{-1}$ and the Krawczyk operator becomes

$$\begin{aligned} K(\mathbf{X}) = &\ \mathrm{mid}(\mathbf{X}) - (\mathrm{mid}\,\mathbf{F}'(\mathbf{X}))^{-1}\mathbf{f}(\mathrm{mid}(\mathbf{X})) \\ &+ (\mathbf{I} - (\mathrm{mid}\,\mathbf{F}'(\mathbf{X}))^{-1}\mathbf{F}'(\mathbf{X}))\mathrm{rad}(\mathbf{X})[-1, 1]. \end{aligned} \tag{4}$$

This is also called the Moore form of the Krawczyk operator.

3 Real Root Isolation Algorithm

In this section, we will present our new algorithm for real root isolation based on hybrid computation. As mentioned before, we first construct initial intervals for the approximate roots obtained by homotopy continuation, and then find out those intervals containing real roots via the Krawczyk interval iteration.

3.1 Construction of Initial Intervals

To apply the Krawczyk interval iteration, obviously the construction of initial intervals is a key procedure. We should guarantee both the *correctness* and *efficiency*, that is, make sure the initial box contains the corresponding accurate real root, and meanwhile keep the interval radius as small as possible so as to shorten the iteration time.

Thus a valid error estimate for the initial approximate roots should be established. And we discuss this issue in both theoretical and practical aspects here.

3.1.1 Error Estimate Theory

The core problem of the construction of initial box is the choice of interval radius, which is essentially an error estimate for the approximate root. There are many work on error analysis, from classical results to modern ones. For example, in [5], Smale et al. gave a systemic method which is now often called alpha theory.

Here we employ the Kantorovich Theorem to give our error estimate.

Theorem 2 (Kantorovich) *Let X and Y be Banach spaces and $F : D \subseteq X \to Y$ be an operator, which is Fréchet differentiable on an open convex set $D_0 \subseteq D$. For equation $F(x) = 0$, if the given approximate root $x_0 \in D_0$ meets the following three conditions:*

1. *$F'(x_0)^{-1}$ exists, and there are real numbers B and η such that*

$$\|F'(x_0)^{-1}\| \leq B, \quad \|F'(x_0)^{-1}F(x_0)\| \leq \eta,$$

2. *F' satisfies the Lipschitz condition on D_0:*

$$\|F'(x) - F'(y)\| \leq K\|x - y\|, \ \forall x, y \in D_0,$$

3. *$h = BK\eta \leq \frac{1}{2}$, $O(x_0, \frac{1-\sqrt{1-2h}}{h}\eta) \subset D_0$,*

then we claim that:

1. *$F(x) = 0$ has a root x^* in $\overline{O(x_0, \frac{1-\sqrt{1-2h}}{h}\eta)} \subset \overline{D_0}$, and the sequence $\{x_k : x_{k+1} = x_k - F'(x_k)^{-1}F(x_k)\}$ of Newton method converges to x^*;*
2. *For the convergence of x^*, we have:*

$$\|x^* - x_{k+1}\| \leq \frac{\theta^{2^{k+1}}(1-\theta^2)}{\theta(1-\theta^{2^{k+1}})}\eta \tag{5}$$

where $\theta = \frac{1-\sqrt{1-2h}}{1+\sqrt{1-2h}}$;

3. *The root* x^* *is unique in* $\overline{D_0} \cap \overline{O(x_0, \frac{1+\sqrt{1-2h}}{h}\eta)}$.

In the theorem, $O(x, r)$ denotes the ball neighborhood whose center is x and radius is r, and $\overline{O(x, r)}$ refers to the closure of the ball neighborhood. The proof can be found in [14].

Since the approximation x_0 is already the result of a homotopy process, what we care about is the initial interval w.r.t. x_0, i.e. the proper upper bound for $\|x^* - x_0\|$. So we have the following proposition, which is a direct corollary of the Kantorovich Theorem.

Proposition 3 *Let* $\mathbf{F} = (f_1, \ldots, f_n)^{\mathrm{T}}$ *be a polynomial system, where* $f_i \in \mathbf{R}[x_1, \ldots, x_n]$. *Denote by* $\mathbf{J}$ *the Jacobian matrix of* $\mathbf{F}$. *For an approximation* $\mathbf{x}_0 \in \mathbf{C}^n$, *if the following conditions hold:*

1. $\mathbf{J}^{-1}(\mathbf{x}_0)$ *exists, and there are real numbers* B *and* η *such that*

$$\|\mathbf{J}^{-1}(\mathbf{x}_0)\| \leq B, \quad \|\mathbf{J}^{-1}(\mathbf{x}_0)\mathbf{F}(\mathbf{x}_0)\| \leq \eta,$$

2. *There exists a ball neighborhood* $O(\mathbf{x}_0, \omega)$ *such that* $\mathbf{J}(\mathbf{x})$ *satisfies the Lipschitz condition on it:*

$$\|\mathbf{J}(\mathbf{x}) - \mathbf{J}(\mathbf{y})\| \leq K\|\mathbf{x} - \mathbf{y}\|, \quad \forall \mathbf{x}, \mathbf{y} \in O(\mathbf{x}_0, \omega)$$

3. *Let* $h = BK\eta$,

$$h \leq \frac{1}{2}, \quad \text{and} \quad \omega \geq \frac{1-\sqrt{1-2h}}{h}\eta,$$

then $\mathbf{F}(\mathbf{x}) = 0$ *has only one root* $\mathbf{x}^*$ *in* $\overline{O(\mathbf{x}_0, \omega)} \cap \overline{O(\mathbf{x}_0, \frac{1+\sqrt{1-2h}}{h}\eta)}$.

Proof We consider F as an operator on $\mathbf{C}^n \to \mathbf{C}^n$, obviously it is Fréchet differentiable, and from

$$\mathbf{F}(\mathbf{x} + \mathbf{h}) = \mathbf{F}(\mathbf{x}) + \mathbf{J}(\mathbf{x})\mathbf{h} + o(\mathbf{h})$$

we can get

$$\lim_{\mathbf{h} \to 0} \frac{\|\mathbf{F}(\mathbf{x} + \mathbf{h}) - \mathbf{F}(\mathbf{x}) - \mathbf{J}(\mathbf{x})\mathbf{h}\|}{\|\mathbf{h}\|} = 0.$$

Thus the first order Fréchet derivative of $\mathbf{F}$ is just the Jacobian matrix $\mathbf{J}$, i.e. $\mathbf{F}'(\mathbf{x}) = \mathbf{J}(\mathbf{x})$. So by Theorem 2, the proof is completed immediately after checking the situation of $\|\mathbf{x}^* - \mathbf{x_0}\|$.

It is easy to know that $\frac{1-\sqrt{1-2h}}{h} \leq 2$. So we can just assign $\omega = 2\eta$. Then we need to check whether $BK\eta \leq \frac{1}{2}$ in the neighborhood $O(\mathbf{x}_0, 2\eta)$. Even though the initial $\mathbf{x}_0$ does not satisfy the conditions, we can still find a proper $\mathbf{x}_k$ after several Newton iterations, since B and K are bounded and η will approach zero. And we only need to find an upper bound for the Lipschitz constant K.

3.1.2 Constructive Algorithm

Now we will give a constructive procedure for the Lipschitz constant K.

Let $J_{ij} = \partial f_i/\partial x_j$, apply mean value inequality [23] to each element of $\mathbf{J}$ on $O(\mathbf{x}_0, \omega)$ to get

$$\|J_{ij}(\mathbf{y}) - J_{ij}(\mathbf{x})\| \leq \sup_{\kappa_{ij} \in line(\mathbf{x},\mathbf{y})} \|\nabla J_{ij}(\kappa_{ij})\| \cdot \|\mathbf{y} - \mathbf{x}\|, \ \forall \mathbf{x}, \mathbf{y} \in O(\mathbf{x}_0, \omega) \quad (6)$$

where $\nabla = (\partial/\partial x_1, \partial/\partial x_2, \ldots, \partial/\partial x_n)$ is the gradient operator and $line(\mathbf{x}, \mathbf{y})$ refers to the line connecting $\mathbf{x}$ with $\mathbf{y}$. Since ∇J is continous, we can find a $\zeta_{ij} \in line(\mathbf{x}, \mathbf{y})$ such that $\|\nabla J_{ij}(\zeta_{ij})\| = \sup_{\kappa_{ij} \in line(\mathbf{x},\mathbf{y})} \|\nabla J_{ij}(\kappa_{ij})\|$. So we get

$$\|J_{ij}(\mathbf{y}) - J_{ij}(\mathbf{x})\| \leq \|\nabla J_{ij}(\zeta_{ij})\| \cdot \|\mathbf{y} - \mathbf{x}\|, \quad \forall \mathbf{x}, \mathbf{y} \in O(\mathbf{x}_0, \omega) \quad (7)$$

Setting $\left(\|\nabla J_{ij}(\zeta_{ij})\| \cdot \|\mathbf{y} - \mathbf{x}\|\right)_{n \times n} = \Delta\mathbf{J}$, then $\|J(y) - J(x)\| \leq \|\Delta\mathbf{J}\|$. And for $\Delta\mathbf{J}$ we have

$$\begin{aligned} \|\Delta\mathbf{J}\|_\infty &= \|\left(\|\nabla J_{ij}(\zeta_{ij})\|_\infty \|\mathbf{y} - \mathbf{x}\|_\infty\right)_{n \times n}\|_\infty \\ &\leq \|\left(\|\nabla J_{ij}(\zeta_{ij})\|_\infty\right)_{n \times n}\|_\infty \cdot \|\mathbf{y} - \mathbf{x}\|_\infty \\ &= \max_{1 \leq i \leq n} \sum_{j=1}^{n} \|\nabla J_{ij}(\zeta_{ij})\|_\infty \cdot \|\mathbf{y} - \mathbf{x}\|_\infty \end{aligned} \quad (8)$$

Note that $\nabla J_{ij}(\zeta_{ij})$ is a vector, so if we use $|\cdot|_{\max}$ to denote the maximum module component of a vector, then we have

$$\|\Delta\mathbf{J}\|_\infty \leq \max_{1 \leq i \leq n} \sum_{j=1}^{n} |\nabla J_{ij}(\zeta_{ij})|_{\max} \cdot \|\mathbf{y} - \mathbf{x}\|_\infty. \quad (9)$$

Let $H_i = (\frac{\partial^2 f_i}{\partial x_j \partial x_k})_{n \times n}$ be the Hessian matrix of f_i, and let $H_i = (h_1^{(i)}, \ldots, h_n^{(i)})$, where $h_j^{(i)}$ are the column vectors. Then we have

$$\|\Delta\mathbf{J}\|_\infty \leq \max_{1 \leq i \leq n} \sum_{j=1}^{n} |h_j^{(i)}(\zeta_{ij})|_{\max} \cdot \|\mathbf{y} - \mathbf{x}\|_\infty. \quad (10)$$

For convenience, we construct $\mathbf{X}_0 = \text{midrad}(\mathbf{x}_0, \omega)$ with $\mathbf{x}_0$ as centre and $\omega = 2\eta$ as radius.

Now we have $|h_j^{(i)}(\zeta_{ij})|_{\max} \le |h_j^{(i)}(\mathbf{X}_0)|_{\max}$. So

$$\|\Delta \mathbf{J}\|_\infty \le \max_{1\le i\le n} \sum_{j=1}^{n} |h_j^{(i)}(\mathbf{X}_0)|_{\max} \cdot \|\mathbf{y} - \mathbf{x}\|_\infty. \tag{11}$$

Therefore

$$K = \max_{1\le i\le n} \sum_{j=1}^{n} |h_j^{(i)}(\mathbf{X}_0)|_{\max} \tag{12}$$

is the Lipschitz constant w.r.t. $\mathbf{J}$.

Now we give an algorithm for computing initial intervals in Algorithm 1.

Algorithm 1 init_width

Input: Equation F; Approximation x_0; Number of variables n
Output: Initial interval's radius r

repeat
  $\mathbf{x}_0 = \mathbf{x}_0 - \mathbf{J}^{-1}(\mathbf{x}_0)\mathbf{F}(\mathbf{x}_0)$;
  $B = \|\mathbf{J}^{-1}(\mathbf{x}_0)\|_\infty$; $\eta = \|\mathbf{J}^{-1}(\mathbf{x}_0)\mathbf{F}(\mathbf{x}_0)\|_\infty$;
  $\omega = 2\eta$;
  $\mathbf{X}_0 = \text{midrad}(\mathbf{x}_0, \omega)$;
  $K = 0$;
  for $i = 1$ **to** n **do**
    Compute the Hessian matrix $H_i = (h_1^{(i)}, h_2^{(i)}, \ldots, h_n^{(i)})$ of $\mathbf{F}$ on $\mathbf{X}_0$;
    if $\sum_{j=1}^{n} |h_j^{(i)}(\mathbf{X}_0)|_{\max} > K$ **then**
      $K = \sum_{j=1}^{n} |h_j^{(i)}(\mathbf{X}_0)|_{\max}$;
    end if
  end for
  $h = BK\eta$;
until $h \le 1/2$
return $r = \frac{1-\sqrt{1-2h}}{h}\eta$

3.2 Empirical Estimate

As so far, we have established a rigorous method to construct initial intervals. This method takes a complex approximate root as input to obtain an initial box. But in practice we often find many approximations with "large" imaginary parts which strongly indicate that they are nonreal. A natural question is

Can we detect these nonreal roots without using interval arithmetic?

Let $\mathbf{z}$ be an approximation of the real root ξ. Because

$$\|\mathscr{R}e(\mathbf{z}) - \xi\| \le \|\mathbf{z} - \xi\|,$$

then we can see the real part $\mathscr{R}e(\mathbf{z})$ is also an approximation of this root and is even closer. So we can simply replace $\mathbf{x_0}$ by $\mathscr{R}e(\mathbf{x_0})$ in Algorithm 1 to construct the initial box.

The other consideration is the efficiency of numerical computation. When we use Proposition 3, lots of interval matrix operations would be executed, which cost much more time than the point operations. So if we can find an empirical estimate radius, which can be computed much faster, but is still valid for most of the equations, then that will be a good choice in practice.

We now give one such empirical estimate.

For $\mathbf{F} = 0$, let $\mathbf{x}^*$ be an accurate root and $\mathbf{x}_0$ be its approximation. Although the mean value theorem is not valid in complex space, the Taylor expansion is still valid. And the polynomial systems considered here are all continuous, so we suppose the equation satisfies the mean value theorem approximately:

$$0 = \mathbf{F}(\mathbf{x}^*) \approx \mathbf{F}(\mathbf{x}_0) + \mathbf{J}(\xi)(\mathbf{x}^* - \mathbf{x}_0) \tag{13}$$

where ξ is between $\mathbf{x}^*$ and $\mathbf{x}_0$. So we have

$$\mathbf{x}^* - \mathbf{x}_0 \approx -\mathbf{J}^{-1}(\xi)\mathbf{F}(\mathbf{x}_0).$$

Let $\mathbf{J}(\xi) = \mathbf{J}(\mathbf{x}_0) + \triangle\mathbf{J}$, then

$$\begin{aligned} \mathbf{J}(\xi) &= \mathbf{J}(\mathbf{x}_0)(\mathbf{I} + \mathbf{J}^{-1}(\mathbf{x}_0)\triangle\mathbf{J}), \\ \mathbf{J}^{-1}(\xi) &= (\mathbf{I} + \mathbf{J}^{-1}(\mathbf{x}_0)\triangle\mathbf{J})^{-1}\mathbf{J}^{-1}(\mathbf{x}_0). \end{aligned} \tag{14}$$

For $\triangle\mathbf{J}$, we can get an estimate similar to Formula (10):

$$\|\triangle\mathbf{J}\|_\infty \le \max_{1\le i\le n} \sum_{j=1}^{n} |h_j^{(i)}(\zeta_{ij})|_{\max} \cdot \|\mathbf{x}^* - \mathbf{x}_0\|_\infty. \tag{15}$$

From our hypothesis, $\mathbf{x}^*$ and $\mathbf{x}_0$ are very close, so are ζ_{ij} and $\mathbf{x}_0$. Thus, we approximate $\mathbf{x}_0$ with ζ_{ij}. Meanwhile, from $\mathbf{x}_0$, after a Newton iteration, we get $\mathbf{x}_1 = \mathbf{x}_0 - \mathbf{J}^{-1}(\mathbf{x}_0)\mathbf{F}(\mathbf{x}_0)$. Thus we may consider that the distance between $\mathbf{x}^*$ and $\mathbf{x}_0$ is more or less the same with that of $\mathbf{x}_0$ and $\mathbf{x}_1$, so we replace $\|\mathbf{x}^* - \mathbf{x}_0\|$ with $\|\mathbf{x}_1 - \mathbf{x}_0\| = \|\mathbf{J}^{-1}(\mathbf{x}_0)\mathbf{F}(\mathbf{x}_0)\|$ for approximation.

So we get

$$\|\triangle\mathbf{J}\|_\infty \le \max_{1\le i\le n} \sum_{j=1}^{n} |h_j^{(i)}(\mathbf{x}_0)|_{\max} \cdot \|\mathbf{J}^{-1}(\mathbf{x}_0)\mathbf{F}(\mathbf{x}_0)\|_\infty. \tag{16}$$

Let $\lambda = \max_{1\le i\le n} \sum_{j=1}^{n} |h_j^{(i)}(\mathbf{x}_0)|_{\max}$, then

$$\|\mathbf{J}^{-1}(\mathbf{x}_0)\triangle\mathbf{J}\|_\infty \leq \lambda\|\mathbf{J}^{-1}(\mathbf{x}_0)\|_\infty^2\|\mathbf{F}(\mathbf{x}_0)\|_\infty.$$

Because $\|\mathbf{F}(\mathbf{x}_0)\|_\infty \ll 1$, the last formula is also much less than 1. So substitute that into Formula (14) we can get

$$\|\mathbf{J}^{-1}(\xi)\|_\infty \leq \frac{\|\mathbf{J}^{-1}(\mathbf{x}_0)\|_\infty}{1-\lambda\|\mathbf{J}^{-1}(\mathbf{x}_0)\|_\infty^2\|\mathbf{F}(\mathbf{x}_0)\|_\infty}. \tag{17}$$

Finally we obtain the empirical estimate

$$\begin{aligned}\|\mathbf{x}^*-\mathbf{x}_0\|_\infty &\approx \|\mathbf{J}^{-1}(\xi)\mathbf{F}(\mathbf{x}_0)\|_\infty \\ &\leq \frac{\|\mathbf{J}^{-1}(\mathbf{x}_0)\|_\infty\|\mathbf{F}(\mathbf{x}_0)\|_\infty}{1-\lambda\|\mathbf{J}^{-1}(\mathbf{x}_0)\|_\infty^2\|\mathbf{F}(\mathbf{x}_0)\|_\infty}. \end{aligned}\tag{18}$$

The inequality (18) is only an empirical estimate. As stated at the beginning of this subsection, an empirical estimate is only a way expected to speed-up computation in some cases and may not be valid for all examples. We test this empirical estimate by many numerical experiments and report the results in Sect. 4, which show that this empirical estimate performs very well on all those examples. An algorithm based on inequality (18) is described as Algorithm 2.

Algorithm 2 IsComplex

Input: Equation F; Approximation $\mathbf{z}$;
Output: `true` or `false`
1: Compute Formula (18), denote the result by r';
2: **if** `any(` $\mathscr{I}(\mathbf{z}) > r'$ `)` **then**
3: **return** `true`; // not a real root
4: **else**
5: **return** `false`; // may be a real root
6: **end if**

In Algorithm 2, `any()` is a default function in Matlab, which returns true if there is nonzero component in a vector.

3.3 Krawczyk–Moore Interval Iteration

We now discuss about the real root verification with a given interval. In Sect. 2.2.2, we have introduced the Krawczyk operator. With the properties in Proposition 1, we can determine whether an interval contains a real root by the relationship of the original interval and the one after the Krawczyk iteration.

However, in practice, we cannot expect the intervals to be entire inclusion or disjoint after just one iteration. Partly intersection is the most common cases that we encounter. Since the real root is still in the interval after the Krawczyk iteration, a normal method is to let $\mathbf{X} \cap K(\mathbf{X})$ be the new iteration interval. So suppose $\mathbf{X}^{(0)}$ is the initial interval, the iteration rule is $\mathbf{X}^{(k+1)} = \mathbf{X}^{(k)} \cap K(\mathbf{X}^{(k)})$, where $K(\mathbf{X}^{(k)})$ is defined by Formula (4). This update rule can make sure that the size of $\mathbf{X}^{(k)}$ is nonincreasing. But a problem is once we encounter $K(\mathbf{X}^{(k)}) \cap \mathbf{X}^{(k)} == \mathbf{X}^{(k)}$, the iteration will be trapped into endless loop. So we have to divide $\mathbf{X}^{(k)}$ if this happened.

Thus, we introduce a bisection function `divide()`. To speedup the convergence of our algorithm, we divide the longest dimension of an interval vector. This strategy may not be the optimal choice when the system's dimension is high. Greedy method or optimization algorithm will be studied in the future.

We now give a formal description of `divide` function in Algorithm 3 and the Krawczyk–Moore iteration process in Algorithm 4.

Algorithm 3 divide

Input: Interval vector $\mathbf{X}$
Output: $\mathbf{X}^{(1)}$ and $\mathbf{X}^{(2)}$, a decomposition of $\mathbf{X}$

Let $\mathbf{X}_l$ be the coordinate with the largest width in $\mathbf{X}$
$\mathbf{X}^{(1)} = \mathbf{X};\ \mathbf{X}^{(2)} = \mathbf{X};$
$\mathbf{X}^{(1)}_l = [\inf(\mathbf{X}_l), \mathrm{mid}(\mathbf{X}_l)];$
$\mathbf{X}^{(2)}_l = [\mathrm{mid}(\mathbf{X}_l), \sup(\mathbf{X}_l)];$
return $\mathbf{X}^{(1)},\ \mathbf{X}^{(2)}$

Algorithm 4 Krawczyk

Input: F; initial box $\mathbf{X}$; isolation boxes $real\_roots$; number of real roots $nreal$
Output: symbol of whether there is a real root $flag$; $real\_roots$; $nreal$

$Y = \mathrm{mid}(\mathbf{F}'(\mathbf{X}))^{-1}$; $\mathbf{X}_t = K(\mathbf{X})$, where $K(\mathbf{X})$ is define by Formula (4);
if $\mathbf{X}_t \cap \mathbf{X} == \emptyset$ **then**
  return $flag = $ `false`;
end if
while not($\mathbf{X}_t \subseteq \mathbf{X}$) **do**
  if $\mathbf{X}_t \cap \mathbf{X} == \mathbf{X}$ **then**
    $[\mathbf{X}^{(1)}, \mathbf{X}^{(2)}] = $ `divide`$(\mathbf{X})$;
    $[f1, real_roots, nreal] = $ `Krawczyk`$(F, \mathbf{X}^{(1)}, real\_roots, nreal)$;
    if $f1 ==$ `false` **then**
      $[f2, real_roots, nreal] = $ `Krawczyk`$(F, \mathbf{X}^{(2)}, real\_roots, nreal)$;
    end if
    return $f1$ **or** $f2$;
  end if
  $\mathbf{X} = \mathbf{X}_t \cap \mathbf{X};$
  $Y = (\mathrm{mid}\mathbf{F}'(\mathbf{X}))^{-1}.$
  $\mathbf{X}_t = K(\mathbf{X});$
  if $\mathbf{X}_t \cap \mathbf{X} == \emptyset$ **then**
    return flag = `false`;
  end if
end while
$nreal = nreal + 1;$
$real\_roots[nreal] = \mathbf{X}_t$
return $flag = $ `true`, $real\_roots, nreal$;

3.4 *Verification and Refinement*

After the Krawczyk iteration, we already have all the real root isolation intervals, but these are not the final results. Since we require an isolation of disjoint intervals, we have to check for possible overlaps.

On the other hand, some intervals may not be as small as required by users, so we can narrow them via bisection method until they match the requirement.

We discuss these details in this subsection.

3.4.1 Remove the Overlaps

There is a basic hypothesis: for nonsingular systems, each root has an approximation, and from this approximation, the iteration will end up in its corresponding accurate root, not any other root. So we only have to remove the overlaps, and the number of real roots will not change.

However, we want to expand our algorithm into multi-roots cases. And in that situation, it is possible that two isolated intervals contain the same real root. So whether or not the overlap part contains a real root, our algorithm has its corresponding processes. See Algorithm 5 for details.

Algorithm 5 disjoint_process

Input: Isolated intervals *real_roots*; number of real roots *nreal*; *F*
Output: Checked isolated intervals *real_roots*; *nreal*

```
k = 0;
for i = 1 to nreal do
  X = real_roots[i] ; new_root = true;
  for j = 1 to k do
    Y = real_roots[j];
    Z = X ∩ Y;
    if Z == ∅ then
      continue;
    end if
    flag = Krawczyk(F,Z);
    if flag == true then
      new_root = false ; break;
    else
      X = X \ Z;
      real_roots[j] = real_roots[j] \ Z;
    end if
  end for
  if new_root == true then
    k = k + 1 ; real_roots[k] = X
  end if
end for
return real_roots ,nreal = k;
```

The function Krawczyk() in Algorithm 5 is a little bit different from that in the Krawczyk–Moore iteration. In the Krawczyk–Moore iteration, we have to store the information of isolated real root intervals, so the $real\_roots$ and $nreal$ are in the function arguments. However, we only need to know whether there is a real root here, so only the symbol variable $flag$ is returned. The situation is the same in Algorithm 6.

3.4.2 Narrow the Width of Intervals

The user may require that the width of isolation intervals be less than or equal to a prescribed number. Different from symbolic algorithms which can get any precision they want in theory, our floating point number calculation cannot beat the machine precision. In fact, in the Matlab environment that we implement our algorithm, the width cannot be smaller than the system zero threshold.[1]

Algorithm 6 narrowing

Input: Isolated intervals $real\_roots$; Number of real roots $nreal$; F; Threshold τ
Output: $real\_roots$ after bisection

for $i = 1$ **to** $nreal$ **do**
  $\mathbf{X} = real\_roots[i]$;
  while any(rad($\mathbf{X}$) $> \tau$) **do**
    $[\mathbf{Y}^{(1)}, \mathbf{Y}^{(2)}] =$ divide($\mathbf{X}$);
    $flag =$ Krawczyk($F, \mathbf{Y}^{(1)}$);
    if $flag ==$ true **then**
      $\mathbf{X} = \mathbf{Y}^{(1)}$;
    else
      $\mathbf{X} = \mathbf{Y}^{(2)}$
    end if
  end while
  $real\_roots[i] = \mathbf{X}$;
end for
return $real\_roots$

We also use bisection to do the narrowing job. Since there is only one root in the interval, we only have to continue dividing and checking the half that contains that root. Formal description of this procedure is in Algorithm 6.

3.5 Algorithm Description

Up to now, we have discussed all the parts of real root isolation algorithm in detail. We give the final main program in Algorithm 7.

[1] In Matlab2008b that we do the experiments, the zero threshold is 2.2204e−016.

Algorithm 7 real_root_isolate

Input: Equation $F(\mathbf{x})$; number of variables n; Threshold τ;
Output: Isolated intervals of $F(\mathbf{x}) = 0$ and number of real roots $nreal$

$[complex\_roots, ncomplex] =$ `hom4ps(F,n)`;
Initialize $real\_roots$ to be empty; $nreal = 0$;
for $i = 1$ **to** $ncomplex$ **do**
  $\mathbf{z} = complex\_roots[i]$;
  if `IsComplex`$(F, \mathbf{z})$ **then**
    continue;
  end if
  $r =$ `init_width`$[F, \mathbf{z}, n]$;
  $X_0 = \text{midrad}(\mathscr{R}(\mathbf{z}), r)$;
  $[flag, real\_roots, nreal] =$ `Krawczyk`$(F, X_0, real\_roots, nreal)$;
end for
$[real\_roots, nreal] =$ `disjoint_process`$(real\_roots, nreal, F)$;
$real\_roots =$ `narrowing`$(real\_roots, nreal, F, \tau)$;
return $real\_roots,\ nreal$;

4 Experiments

Now we apply our new method to some polynomial systems and do some comparison with some former algorithms.

All the experiments are undertaken in Matlab2008b, with Intlab [25] of Version 6. For arbitrarily high accuracy, we can call Matlab's vpa (variable precision arithmetic), but in fact all the real roots of the examples below are isolated by using Matlab's default double-precision floating point. We use Hom4ps-2.0 [19] as our homotopy continuation tool to obtain initial approximate roots.

4.1 Demo Example

We begin our illustration with a simple example.

Example 4 Consider the real root isolation of the system below.

$$\begin{cases} x^3y^2 + x + 3 = 0 \\ 4yz^5 + 8x^2y^4z^4 - 1 = 0 \\ x + y + z - 1 = 0 \end{cases}$$

The homotopy program tells us this system has 28 complex roots in total. And we get the following results after calling our `real_root_isolate` program.

```
intval =
[ - 0.94561016957416, - 0.94561016957415]
[ 1.55873837303161, 1.55873837303162]
[ 0.38687179654254, 0.38687179654255]
```

```
intval =
[ - 1.18134319868123, - 1.18134319868122]
[ - 1.05029487815439, - 1.05029487815438]
[ 3.23163807683560, 3.23163807683561]
intval =
[ - 2.99999838968782, - 2.99999838968781]
[ 0.00024421565895, 0.00024421565896]
[ 3.99975417402886, 3.99975417402887]
intval =
[ - 0.79151164911096, - 0.79151164911095]
[ 2.11038450699949, 2.11038450699950]
[ - 0.31887285788855, - 0.31887285788854]
The order of variables:
'x'
'y'
'z'
The number of real roots: 4
```

We verify the answers above with the DISCOVERER [33] package under Maple, which also return 4 isolated real roots. Here we show its output in floating point number format, i.e.

[[−2.999998391, −2.999998389], [0.0002442132, 0.0002442180], [3.999754090, 3.999754249]],
[[−1.181343199, −1.181343199], [−1.050294975, −1.050294818],[3.231637836, 3.231638372]],
[[−.9456101805, −.9456101656], [1.558738033, 1.558738728], [.3868716359, .3868719935]],
[[−.7915116549, −.7915116400], [2.110384024, 2.110385000], [−.3188729882, −.3188727498]].
And we can see the answers perfectly match the ones of our program.

We list some information during the calculation of our algorithm here for reference. Only the 4 real ones are given, and the other nonreal ones are all detected by our empirical estimate method. We mention that all the imaginary parts of complex roots are significant larger than the initial radius of our algorithm in order of magnitude in this example.

We give some remarks on Table 1. In the first row, *root1* to *root4* are refer to the 4 real roots mentioned above respectively. And B, K, η, h are exactly the same as they

Table 1 Key quantities comparison

	Root1	Root2	Root3	Root4
B	1.060227	1.192159	2.000864	0.874354
K	14.941946	7.198937e+003	4.095991e+003	16.988990
η	4.260422e−016	4.20807e−016	8.882333e−016	5.764449e−016
h	2.024791e−014	1.083446e−011	2.183861e−011	2.568823e−014
Estimate-rad	4.274976e−016	4.208067e−016	8.882344e−016	5.779921e−016
Empirical-rad	1.015249e−015	1.29164e−012	2.156138e−015	1.559270e−015

are defined in Algorithm 1. The *estimate-rad* are the radius obtained via Algorithm 1, while the *empirical-rad* are refer to the ones calculated by Formula (18).

We say a little more words about the *empirical-rad*. Firstly, although the empirical ones are basically larger than the rigorous error radius, they are still small enough, which hardly have any influence on the efficiency of interval iteration. We will see this in the comparison experiments later. But avoiding of interval matrix computation is very helpful to the algorithm. Secondly, the radius obtained from Algorithm 1 are so small that they are even comparable to the zero threshold of Matlab system.[2] And this could bring some uncertainty of floating point operation to our algorithm, such as misjudgement of interval inclusion in Intlab, etc. So we intend to use empirical estimate bound in next experiments.

For system *cyclic6*, the classic symbolic algorithm can do nothing due to the difficulty of triangularization. Meanwhile, we can easily get the 24 isolated real roots intervals with our `real_root_isolate` program.

4.2 Comparison Experiment

Many benchmarks have been checked with our `real_root_isolate` program. Since no results on the time complexity of our method have been obtained, we mainly focus on the isolation results and the program execution time.

We investigate over 130 benchmarks provided by Hom4ps [4], among which about 40 equations are nonsingular systems. Due to space limitation, we only list 11 of them in Tables 2, 4 and 5. Comparison to the symbolic tool DISCOVERER on the

Table 2 Comparison to symbolic method

Problem	Total roots	Real roots	DISCOVERER	Complex roots detected
barry	20	2	2	18
cyclic5	70	10	10	60
cyclic6	156	24	N/A	132
des18 3	46	6	N/A	40
eco7	32	8	8	24
eco8	64	8	N/A	56
geneig	10	10	N/A	0
kinema	40	8	N/A	32
reimer4	36	8	8	28
reimer5	144	24	N/A	120
virasoro	256	224	N/A	32

2 As mentioned before, the zero threshold in Matlab2008b is 2.2204e−016, which is almost the same order of magnitude of those radiuses.

Table 3 Comparison to a hybrid method

	Verifyrealroot0(M)			Verifyrealroot0(H)			Real_root_isolate		
Problem	Time	Sol	Width	Time	Sol	Width	Time	Sol	Width
comb3000	1.56	1	2.0e−20	1.38	4	2.7e−20	1.11	4	5.3e−10
d1	52.3	2	1.7e−14	6.24	16	1.8e−14	6.19	16	7.3e−13
boon	27.6	1	5.1e−15	1.98	8	2.9e−15	1.14	8	1.8e−14
des22_24	1.79	1	2.5e−14	1.73	10	1.2e−08	4.46	10	1.5e−01
geneig	6.53	2	6.7e−15	4.63	10	2.7e−13	4.21	10	3.6e−13
heart	24.9	2	5.3e−15	1.40	2	4.9e−15	1.54	2	8.5e−15
kin1	52.3	2	1.8e−14	5.91	16	1.8e−14	6.44	16	1.2e−12
ku10	37.8	1	4.7e−14	0.96	2	6.7e−14	0.41	2	6.8e−13
noon3	1.88	1	1.6e−16	11.7	8	1.6e−15	0.94	7	2.6e−15
noon4	9.70	1	3.6e−15	30.2	22	3.9e−15	2.96	15	1.0e−05
puma	5.85	2	2.9e−14	3.99	16	1.8e−13	2.27	16	1.8e−13
quadfor2	1.48	2	5.6e−16	0.71	2	2.2e−16	1.11	2	1.2e−15
rbpl	5.59	1	2.6e−15	23.2	4	8.4e−14	7.78	4	4.1e−12
redeco5	0.95	1	8.3e−17	1.07	4	1.3e−15	0.48	4	1.0e−05
reimer5	26.7	3	8.4e−14	5.83	24	3.2e−13	10.5	24	7.8e−13

11 systems in Table 2 just tries to verify the results and to show the advantage of our hybrid method. Comparison to another hybrid tool in [36] on 15 zero-dimensional nonsingular systems is listed in Table 3.

The 11 systems in Table 2 were computed on a computer with OS: Windows Vista, CPU: Inter®Core 2 Duo T6500 2.10 GHz, Memory: 2G. The column *real roots* in Table 2 tells the number of intervals that our program isolated. Compared with the results of DISCOVERER, the new algorithm indeed works out all equations that are beyond the capability of classic symbolic algorithm. Moreover, the last column show that our empirical estimate method detects all the nonreal roots successfully.

The 15 problems in Table 3 are all from Table 1 of [36]. There are 20 problems in Table 1 of [36], of which 5 are either positive dimensional (e.g., cohn2) or singular (e.g., katsura5). They are beyond the scope of this paper and thus have not been listed here. The data of Columns 2–7 in Table 3 is copied from [36], which was obtained with Matlab(2011R) on a computer with Intel(R) Core(TM) at 2.6 GHz under Windows. The data of `real_root_isolate` (Columns 8–10) was obtained with Matlab(2008b) on a computer with intel(R) core(TM)i3 at 2.27 GHz under Windows 7.

From Table 3, one could see that our program is faster on most of the examples. The widths of the output intervals of our program are usually much larger than those of verifyrealroot0(H) and verifyrealroot0(M). It should be pointed out that our results on systems "noon3" and "noon4" are different from that of verifyrealroot0(H). On the other hand, verifyrealroot0 is a tool applicable also to positive dimensional systems but our tool is only effective on zero-dimensional nonsingular systems.

Table 4 Execution time comparison, unit:s

Problem	Total time	Homotopy time	Interval time
barry	0.421203	0.093601	0.327602
cyclic5	2.948419	0.218401	2.652017
cyclic6	9.984064	0.639604	9.063658
des18_3	4.180827	0.702004	3.385222
eco7	2.371215	0.265202	2.012413
eco8	3.946825	0.499203	3.354022
geneig	4.243227	0.249602	3.868825
kinema	3.946825	1.014006	2.808018
reimer4	2.480416	0.374402	2.059213
reimer5	12.963683	3.073220	9.578461
virasoro	137.124879	4.570829	109.996305

Table 4 shows that interval iterations consume more time than homotopy continuation. The reason is complicated and we enumerate some here:

1. The homotopy continuation focuses only on floating-point number, while the Krawczyk iteration cares about intervals;
2. Hom4ps-2.0 is a software complied from language C, which is much more efficient than the tool that we use to implement our algorithm, say Matlab.
3. The interval iteration time increases as roots number grows since we examine the approximate roots one by one. So the parallel computation of homotopy is much faster.

We believe that with efficient language such as C/C++, and parallel computation, the implementation of our algorithm will be much faster.

In order to verify our idea and see whether parallelization could help, we go into every approximate root's iteration process. Some critical data are recorded in Table 5. The *avg. rad. of ans* is the average radius of the final isolated intervals, while the *avg. rad. of init.* indicates the average radius of the initial intervals. The average time of each root's interval iteration is shown in column *avg. time of iteration* along with the max interval iteration time in *max time of iter*. We think the consumption for each root's process is acceptable.

From Table 5 we can see that the initial interval radii are extremely small, which leads to a nice process time for each iteration. We point out that almost all real root checks are done by just *one* Krawczyk iteration, and hardly any overlap is found after all the Krawczyk iteration processes due to the small initial intervals that we give. All of these save a great deal of executing time of our program.

Table 5 Detail data for each iteration, unit:s

Problem	avg. rad. of ans.	avg. rad. of init.	avg. time of iter.	max time of iter.
barry	3.552714e−015	1.377800e−014	0.054600	0.062400
cyclic5	1.614703e−009	7.142857e−007	0.113881	0.140401
cyclic6	4.440892e−016	2.137195e−015	0.183951	0.234002
des18_3	3.768247e−007	9.737288e−007	0.241802	0.296402
eco7	1.998401e−015	1.483754e−013	0.122851	0.156001
eco8	2.109424e−015	3.283379e−013	0.183301	0.218401
geneig	2.664535e−016	5.721530e−014	0.315122	0.436803
kinema	1.998401e−015	6.784427e−011	0.157951	0.218401
reimer4	1.110223e−016	1.258465e−014	0.122851	0.156001
reimer5	1.110223e−016	4.754080e−014	0.195001	0.421203
virasoro	9.472120e−009	2.265625e−006	0.387844	0.624004

5 Conclusion

In this paper, a new algorithm based on hybrid computation is provided for real root isolation of zero-dimensional nonsingular square polynomial systems. The algorithm first applies homotopy continuation to obtain all the approximate roots of the system. For each approximate root, an initial interval which contains the corresponding accurate root is constructed. Then the Krawczyk operator is called to verify all the initial intervals so as to get all the real root isolation boxes. Some necessary check and refinement work are done after that to ensure the boxes are pairwise disjoint and meet width requirement.

In the construction of initial intervals, we give a rigorous radius error bound based on a corollary of the Kantorovich theorem. Some constructive algorithms are presented for both real and complex approximate roots. Meanwhile, we introduce an empirical estimate radius, which has a nice performance in numerical experiments.

In the modification and implementation of the Krawczyk iteration algorithm, some problems of interval arithmetic are also discussed in this paper.

At last we utilize some existing tools to implement our algorithm under Matlab environment. Many benchmarks have been checked along with comparison and analysis.

We also mention some possible future work here. The construction of initial intervals is still too complicated and further optimization should be studied. Also the empirical estimate with more efficiency and accuracy is a considerable question. The strategy for dividing intervals in the Krawczyk iteration could also be improved, which may be helpful in the high dimension cases. It is important to study the cases that the systems are positive dimensional or singular.

In the aspect of implementation, replacing the Matlab implementation with C/C++ codes may improve the performance of our method. Parallel computation can be another technique to speed-up the computation since the verification of all initial intervals can be obviously parallelled.

Acknowledgments The work is partly supported by the ANR-NSFC project EXACTA (ANR-09-BLAN-0371-01/60911130369), NSFC-11001040, NSFC-11271034 and the project SYSKF1207 from ISCAS. The authors especially thank professor Dongming Wang for the early discussion on this topic in 2010 and also thank professor T.Y. Li for his helpful suggestions and his team's work on Hom4ps2-Matlab interface. Thanks also go to Ting Gan who computed the 15 systems in Table 3 and provided us suggestion on possible improvements on our program. Thank Zhenyi Ji who shared with us his insights on our hybrid method. Thank the referees for their valuable constructive comments which help improve the presentation greatly.

References

1. Beltran, C.: A continuation method to solve polynomial systems, and its complexity. Numerische Mathematik. Online rst. doi:10.1007/s00211-010-0334-3
2. Beltran, C., Leykin, A.: Certified numerical homotopy tracking. Exp. Math. **21**(1), 69–83 (2012)
3. Beltran, C., Leykin, A.: Robust certified numerical homotopy tracking. Found. Comput. Math. **13**(2), 253–295 (2013)
4. Benchmarks: http://hom4ps.math.msu.edu/HOM4PS_soft_files/equations.zip
5. Blum, L., Cucker, F., Shub, M., Smale, S.: Complexity and Real Computation. Springer, New York (1997)
6. Boulier, F., Chen, C., Lemaire, F., Moreno Maza, M.: Real root isolation of regular chains. In: Proceedings of ASCM'2009, pp. 15–29 (2009)
7. Cheng, J.-S., Gao, X.-S., Guo, L.-L.: Root isolation of zero-dimensional polynomial systems with linear univariate representation. J. Symbolic Comput. **47**(7), 843–858 (2012)
8. Cheng, J.-S., Gao, X.-S., Li, J.: Root isolation for bivariate polynomial systems with local generic position method. In: Procedings of ISSAC'2009, pp. 103–110
9. Cheng, J.-S., Gao, X.-S., Yap, C.-K.: Complete numerical isolation of real zeros in zero-dimensional triangular systems. In: Proceedings of ISSAC'2007, pp. 92–99 (2007)
10. Collins, G.E., Akritas, A.G.: Polynomial real root isolation using Descartes' rule of signs. In: Proceedings of SYMSAC, pp. 272–275 (1976)
11. Collins, G.E., Loos, R.: Real zeros of polynomials. In: Buchberger, B., Collins, G.E., Loos, R. (eds.) Computer Algebra: Symbolic and Algebraic Computation, pp. 83–94. Springer, New York (1982)
12. Dayton, B., Li, T.Y., Zeng, Z.G.: Multiple zeros of nonlinear systems. Math. Comp. **80**, 2143–2168 (2011)
13. Emiris, I.Z., Mourrain, B., Tsigaridas, E.P.: The DMM bound: multivariate (aggregate) separation bounds. In: Proceedings of ISSAC'2010, pp. 243–250
14. Gragg, G.W., Tapia, R.A.: Optimal error bounds for the Newton-Kantorovich theorem. SIAM J. Numer. Anal. **11**(1), 10–13 (1974)
15. Hauenstein, J.D., Sottile, F.: Algorithm 921: alphaCertified: certifying solutions to polynomial systems. ACM Trans. Math. Softw.(TOMS) **38**(4), 28 (2012)
16. Leykin, A., Verschelde, J., Zhao, A.L.: Newton's method with deflation for isolated singularities of polynomial systems. Theoret. Comput. Sci. **359**, 111–122 (2006)
17. Leykin, A., Verschelde, J., Zhao, A.L.: Higher-order deflation for polynomial systems with isolated singular solutions. In: Dickenstein, A., Schreyer, F., Sommese, A. (eds.) Algorithms in Algeraic Geometry, pp. 79–97. Springer, New York (2008)
18. Li, T.Y.: Numerical solution of multivariate polynomial systems by homotopy continuation methods. Acta Numerica **6**, 399–436 (1997)

19. Li, T. Y.: HOM4PS-2.0. http://hom4ps.math.msu.edu/HOM4PS_soft.htm (2008)
20. Li, T.Y., Wang, X.S.: Solving real polynomial systems with real homotopies. Math. Comput. **60**(202), 669–680 (1993)
21. Mantzaflaris, A., Mourrain, B., Tsigaridas, E.P.: On continued fraction expansion of real roots of polynomial systems, complexity and condition numbers. Theor. Comput. Sci. (TCS) **412**(22), 2312–2330 (2011)
22. Moore, R.E., Kearfott, R.B., Cloud, M.J.: Introduction to Interval Analysis. Society for Industrial and Applied Mathematics, Philadelphia (2009)
23. Mujica, J.: Complex Analysis in Banach Spaces. North-Holland Mathematics Studies. Elsevier, Amsterdam (1986)
24. Rouillier, F.: Solving zero-dimensional systems through the rational univariate representation. Appl. Algebra Eng. Commun. Comput. **9**, 433–461 (1999)
25. Rump, S.M.: INTLAB—INTerval LABoratory. In: Csendes, T. (ed.) Developments in Reliable Computing, pp. 77–104. Kluwer Academic Publishers. http://www.ti3.tu-harburg.de/rump/ (1999)
26. Rump, S.M.: Verification methods: rigorous results using floating-point arithmetic. Acta Numerica **19**, 287–449 (2010)
27. Shen, F.: The real roots isolation of polynomial system based on hybrid computation. Master degree thesis, Peking University (April 2012)
28. Sommese, A., Wampler, C.: The Numerical Solution of Systems of Polynomials: Arising in Engineering and Science. World Scientific, Singapore (2005)
29. Strzebonski, A.W., Tsigaridas, E.P.: Univariate real root isolation in multiple extension fields. In: Proceedings of ISSAC'2012, pp. 343–350
30. Verschelde, J.: PHCpack. http://homepages.math.uic.edu/jan/PHCpack/phcpack.html (1999)
31. Wampler, C.: HomLab. http://nd.edu/cwample1/HomLab/main.html
32. Wu, X., Zhi, L.: Computing the multiplicity structure from geometric involutive form. In: Proceedings of ISSAC'2008, pp. 325–332 (2008)
33. Xia, B.: DISCOVERER: a tool for solving semi-algebraic systems. ACM Commun. Comput. Algebra **41**(3), 102–103 (2007)
34. Xia, B., Zhang, T.: Real solution isolation using interval arithmetic. Comput. Math. Appl. **52**, 853–860 (2006)
35. Yang, L., Xia, B.: An algorithm for isolating the real solutions of semi-algebraic systems. J. Symbolic Comput. **34**, 461–477 (2002)
36. Yang, Z., Zhi, L., Zhu, Y.: Verified error bounds for real solutions of positive-dimensional polynomial systems. In: Proceedings of ISSAC'2013 (2013)
37. Zhang, T.: Isolating real roots of nonlinear polynomial. Master degree thesis, Peking University (2004)
38. Zhang, T., Xiao, R., Xia, B.: Real solution isolation based on interval Krawczyk operator. In: Sung-il, P., Park, H. (eds.) Proceedings of ASCM'2005, pp. 235–237 (2005)

Overview of the Mathemagix Type System

Joris van der Hoeven

Abstract The goal of the MATHEMAGIX project is to develop a new and free software for computer algebra and computer analysis, based on a strongly typed and compiled language. In this paper, we focus on the underlying type system of this language, which allows for heavy overloading, including parameterized overloading with parameters in so called "categories." The exposition is informal and aims at giving the reader an overview of the main concepts, ideas and differences with existing languages. In a forthcoming paper, we intend to describe the formal semantics of the type system in more detail.

Keywords Mathemagix · Type system · Overloading · Parametric polymorphism · Language design · Computer algebra

1 Introduction

1.1 Motivation for a New Language

Until the mid-1990s, the development of computer algebra systems tended to exploit advances in the area of programming languages, and sometimes even influenced the design of new languages. The FORMAC system [2] was developed shortly after the introduction of FORTRAN. Symbolic algebra was an important branch of the artificial intelligence project at MIT during the 1960s. During a while, the MACSYMA system [26, 29, 35] was the largest program written in LISP, and motivated the development of better LISP compilers.

The SCRATCHPAD system [15, 21] was at the origin of yet another interesting family of computer algebra systems, especially after the introduction of

Note This work has been supported by the ANR-09-JCJC-0098-01 MaGiX project, as well as the Digiteo 2009-36HD grant and Région Ile-de-France.

J. van der Hoeven (✉)
LIX, CNRS, École polytechnique, 91128 Palaiseau Cedex, France
e-mail: vdhoeven@lix.polytechnique.fr

© Springer-Verlag Berlin Heidelberg 2014
R. Feng et al. (eds.), *Computer Mathematics*, DOI 10.1007/978-3-662-43799-5_27

domains and categories as function values and dependent types in MODLISP and SCRATCHPAD II [20, 22, 33]. These developments were at the forefront of language design and type theory [12, 27, 28]. SCRATCHPAD later evolved into the AXIOM system [19, 34]. In the A# project [38, 39], later renamed into ALDOR, the language and compiler were redesigned from scratch and further purified.

After this initial period, computer algebra systems have been less keen on exploiting new ideas in language design. One important reason is that a good language for computer algebra is more important for developers than for end users. Indeed, typical end users tend to use computer algebra systems as enhanced pocket calculators, and rarely write programs of substantial complexity themselves. Another reason is specific to the family of systems that grew out of SCRATCHPAD: after IBM's decision to no longer support the development, there has been a long period of uncertainty for developers and users on how the system would evolve. This has discouraged many of the programmers who did care about the novel programming language concepts in these systems.

In our opinion, this has lead to an unpleasant current situation in computer algebra: there is a dramatic lack of a modern, sound and fast general purpose programming language. The major systems MATHEMATICA [40] and MAPLE [11] are both interpreted, weakly typed (even the classical concept of a closure has been introduced only recently in MAPLE!), besides being proprietary and very expensive. The SAGE system [31] relies on PYTHON and merely contents itself to glue together various existing libraries and other software components.

The absence of modern languages for computer algebra is even more critical whenever performance is required. Nowadays, many important computer algebra libraries (such as GMP [13], MPFR [16], FLINT [17], FGB [7], etc.) are directly written in C or C++. Performance issues are also important whenever computer algebra is used in combination with numerical algorithms. We would like to emphasize that high level ideas can be important even for traditionally low level applications. For instance, in a suitable high level language it should be easy to operate on SIMD vectors of, say, 256 bit floating point numbers. Unfortunately, MPFR would have to be completely redesigned in order to make such a thing possible.

For these reasons, we have started the design of a new software, MATHEMAGIX [36, 37], based on a compiled and strongly typed language, featuring signatures, dependent types, and overloading. MATHEMAGIX is intended as a general purpose language, which supports both functional and imperative programming styles. Although the design has greatly been influenced by SCRATCHPAD II and its successors AXIOM and ALDOR, there are several important differences, as we will see. MATHEMAGIX is also a free software, which can be downloaded from `www.mathemagix.org`.

In this paper, we will focus on the underlying type system. We present an informal overview of this system and highlight in which respect it differs from existing systems. We plan to provide a more detailed formal description of the type system in a future paper.

1.2 Main Philosophy Behind the Type System

The central idea behind the design of the MATHEMAGIX language is that the declaration of a function is analogous to the statement of a mathematical theorem, whereas the implementation of the function is analogous to giving a proof. Of course, this idea is also central in the area of automated proof assistants, such as COQ [3, 4] or ISABELLE/HOL [30]. However, MATHEMAGIX is intended to be a general purpose programming language rather than an automated theorem prover. Therefore, we only insist on very detailed declarations, whereas the actual implementations do not need to be formally proven.

One consequence of this design philosophy is that program interfaces admit very detailed specifications: although the actual implementations are not formally proven, the combination of various components is sound as long as each of the components fulfills its specification. By contrast, MAPLE, MATHEMATICA or SAGE functions can only be specified in quite vague manners, thereby introducing a big risk of errors when combining several libraries.

Another consequence of the MATHEMAGIX design is that it allows for massively overloaded notations. This point is crucial for computer algebra and also the main reason why mainstream strongly typed functional programming languages, such as HASKELL [18, 23] or OCAML [25], are not fully suitable for our applications. To go short, we insist on very detailed and unambiguous function declarations, but provide a lot of flexibility at the level of function applications. On the contrary, languages such as OCAML require unambiguous function applications, but excel at assigning types to function declarations in which no types are specified for the arguments.

The MATHEMAGIX type system also allows for a very flat design of large libraries: every function comes with the hypotheses under which it is correct, and can almost be regarded as a module on its own. This is a major difference with respect to AXIOM and ALDOR, where functionality is usually part of a class or a module. In such more hierarchical systems, it is not always clear where to put a given function. For instance, should a converter between lists and vectors be part of the list or the vector class?

1.3 Overview of this Paper

In order to make the above discussion about the main design philosophy more concrete, we will consider the very simple example of computing the square of an element x in a monoid. In Sect. 2, we will show how such a function would typically be written in various existing languages, and compare with what we would do in MATHEMAGIX. For a somewhat older, but more detailed comparison from a similar perspective, we refer to [10].

In Sect. 3, we will continue with a more complete description of the primitives of the type system which have currently been implemented in the compiler. We will also discuss various difficulties that we have encountered and some plans for extensions.

As stated before, we have chosen to remain quite informal in this paper. Nevertheless, in Sect. 4, we will outline the formal semantics of the type system. The main difficulty is to carefully explain what are the possible meanings of expressions based on heavily overloaded notations, and to design a compiler which can determine these meanings automatically.

Given the declarative power of the language, it should be noticed that the compiler will not always be able to find all possible meanings of a program. However, this is not necessarily a problem as long as the compiler never assigns a wrong meaning to an expression. Indeed, given an expression whose meanings are particularly hard to detect, it is not absurd to raise an error, or even to loop forever. Indeed, in such cases, it will always be possible to make the expression easier to understand by adding some explicit casts. Fortunately, most natural mathematical notations also have a semantics which is usually easy to determine: otherwise, mathematicians would have a hard job to understand each other at the first place!

2 Comparison on an Example

We will consider a very simple example in order to illustrate the most essential differences between MATHEMAGIX and various existing programming languages: the computation of the square of an element x of a monoid. Here we recall that a monoid is simply a set M together with an associative multiplication $\cdot : M^2 \to M$. Although our example may seem trivial, we will see that the programming languages that we have investigated fail to treat this example in a completely satisfactory way from our perspective.

2.1 Mathemagix

In MATHEMAGIX, we would start by a formal declaration of the concept of a monoid. As in the case of ALDOR, this is done by introducing the monoid *category*:

```
category Monoid == {
  infix *: (This, This) -> This;
}
```

The built-in type `This` stands for the *carrier* of the monoid. In other words, it will be possible to regard any type `T` with a function `infix *: (T, T) -> T` as a monoid. We may now define the square of an element x of a monoid by

```
forall (M: Monoid)
square (x: M): M == x * x;
```

Given an instance `x` of any type `T` with a multiplication `infix *: (T, T) -> T`, we may thus compute the square of `x` using `square x`. For instance, after inclusion of the standard library file `integer.mmx`, the literal constant `111` has type `Integer` and a multiplication `infix *: (Integer, Integer) -> Integer` is provided. Hence, `square 111` will yield the square of the integer `111`.

In our definition of a monoid, we notice that we did *not* specify the multiplication to be associative. For instance, we might consider an extension of the current language with a keyword `assert`, which would allow us to define

```
category Monoid == {
  infix *: (This, This) -> This;
  assert (forall (x: This, y: This, z: This) x*(y*z) =
                (x*y)*z);
}
```

Nevertheless, nothing withholds the user from replacing the definition by

```
category Monoid == {
  infix *: (This, This) -> This;
  associative: This -> Void;
}
```

At least, this allows the user to *indicate* that the multiplication on a type `T` is associative, by implementing the "dummy" function `associative` for `T`.

Of course, one might consider adding an `assert` primitive to MATHEMAGIX which would really behave as an annotation similar to the `associative` function. However, if we want to take advantage of the mathematical semantics of associativity, then we should also be able to automatically *prove* associativity during type conversions. We regard this as an issue for automatic theorem provers which is beyond the current design goals of MATHEMAGIX. Notice nevertheless that it would be quite natural to extend the language in this direction in the further future.

2.2 Aldor

As stated in the introduction, a lot of the inspiration for MATHEMAGIX comes from the ALDOR system and its predecessors. In ALDOR, the category `Monoid` would be defined using

```
define Monoid : Category == with {
  *: (%, %) -> %;
}
```

However, the `forall` primitive inside MATHEMAGIX for the definition of templates does not have an analogue inside ALDOR. In ALDOR, one would rather define a parameterized class which exports the template. For instance:

```
Squarer (M: Monoid): with {
  square: M -> M;
} == add {
  square (x: M): M == x * x;
}
```

In order to use the template for a particular class, say `Integer`, one has to explicitly import the instantiation of the template for that particular class:

```
import from Squarer (Integer);
```

The necessity to encapsulate templates inside classes makes the class hierarchy in ALDOR rather rigid. It also forces the user to think more than necessary about where to put various functions and templates. This is in particular the case for routines which involve various types in a natural way. For instance, where should we put a converter from vectors to lists? Together with other routines on vectors? With other routines on lists? Or in an entirely separate module?

2.3 C++

The C++ language [32] does provide support for the definition of templates:

```
template<typename M>
square (const M& x) {
  return x * x;
}
```

However, as we see on this example, the current language does not provide a means for requiring `M` to be a monoid, at least in the weak sense from Sect. 2.1 that there exists a multiplication `M operator * (const M&, const M&)`. Several C++ extensions with "signatures" [1] or "concepts" [5] have been proposed in order to add such requirements. C++ also imposes a lot of restrictions on how templates can be used. Most importantly, the template arguments should be known statically, at compile time. Also, instances of user defined types (such as an explicit matrix) cannot be used as template arguments.

In the above piece of code, we also notice that the argument x is of type `const M&` instead of M. This kind of interference of low level details with the type system is at the source of many problems when writing large computer algebras libraries in C++. Although MATHEMAGIX also gives access to various low level details, we decided to follow a quite different strategy in order to achieve this goal. However, these considerations fall outside the main scope of this paper.

2.4 Ocaml

Mainstream strongly typed functional programming languages, such as OCAML and HASKELL, do not provide direct support for operator overloading. Let us first examine the consequences of this point of our view in the case of OCAML. In order to make the types which are associated to expressions by the compiler explicit, the examples in this section will be presented in the form of an interactive session.

First of all, multiplication does not carry the same name for different numeric types. For instance:

```
# let square x = x * x;;
val square: int -> int = <fun>

# let float_square x = x *. x;;
val float_square: float -> float = <fun>
```

At any rate, this means that we somehow have to specify the monoid in which we want to take a square when applying the square function of our example. Nevertheless, modulo acceptance of this additional disambiguation constraint, it is possible to define the analogue of the `Monoid` category and the routine `square`:

```
# module type Monoid =
    sig
      type t
      val mul : t -> t -> t
    end;;

# module Squarer =
   functor (El: Monoid) ->
    struct
      let square x = El.mul x x
    end;;
```

As in the case of ALDOR, we need to encapsulate the `square` function in a special module `Squarer`. Moreover, additional efforts are required in order to instantiate this module for a specific type, such as `int`:

```
# module Int_Monoid =
    struct
      type t = int
      let mul x y = x * y
    end;;

# module Int_Squarer = Squarer (Int_Monoid);;

# Int_Squarer.square 11111;;
- : int = 123454321
```

In our definition of `Int_Monoid`, it should be noticed that we need to specify the multiplication on `int` explicitly. On the one hand, this gives a greater flexibility: for instance, it is straightforward to construct another integer monoid where the addition is used as the monoid multiplication. However, we think that this kind flexibility is rarely useful in the area of computer algebra. In fact, mathematicians rather tend to use a unique notation for multiplication, so mathematical programming languages should rather focus on transposing this habit directly into the language. If one *really* wants to use addition as a multiplication, then it is not hard to define a wrapper class which does precisely this.

2.5 Haskell

In many regards, HASKELL is similar in spirit to OCAML, but HASKELL type classes come closer to overloading than OCAML modules. For instance, the square function can be defined in a quite compact way as follows:

```
class Monoid  a where
    (*) :: a -> a -> a
square x = x * x
```

In order to enable the square function for a particular type, one has to create an instance of the monoid for this particular type. For instance, we may endow `String` with the structure of a monoid by using concatenation as our multiplication:

```
instance Monoid [Char] where
    x * y = x ++ y
```

After this instantiation, we may square the string `"hello"` using `square "hello"` (in practice, the example needs to be slightly tweeked since the operator `*` is already reserved for standard multiplication of numbers; one also has to use the `-XFlexibleInstances` compilation option in order to allow for the instantiation of the string type).

The nice thing of the above mechanism is that we may instantiate other types as monoids as well and share the name `*` of the multiplication operator among all these instantiations. HASKELL style polymorphism thereby features several of the advantages of operator overloading. However, there are some important differences. First of all, it is not allowed to use the same name `*` inside another type class, such as `Ring`, except when the other type class is explicitly derived from `Monoid`. Secondly, the user still has to explicitly instantiate the type classes for specific types: in MATHEMAGIX, the type `String` can automatically be regarded as a `Monoid` as soon as the operator `*` is defined on strings.

2.6 Discussion

Essentially, the difference between MATHEMAGIX and classical strongly typed functional languages such as OCAML and HASKELL is explained by the following observation: if we want to be able to declare the square function simply by writing

```
square x = x * x
```

and *without* specifying the type of `x`, then the symbol `*` should not be too heavily overloaded in order to allow the type system to determine the type of `square`. In other words, no sound strongly typed system can be designed which allows both for highly ambiguous function declarations *and* highly ambiguous function applications.

Whether the user prefers a type system which allows for more freedom at the level of function declarations or function applications is a matter of personal taste. We may regard OCAML and HASKELL as prominent members of the family of strongly typed languages which accomodate a large amount of flexibility at the declaration side. But if we are rather looking for high expressiveness at the function application side, and insist on the possibility to heavily overload notations, then we hope that the MATHEMAGIX type system will be a convenient choice.

We finally notice that signatures are now implemented under various names in a variety of languages. For instance, in JAVA, one may use the concept of an interface. Nevertheless, to the best of our knowledge, the current section describes the main lines along which signatures are conceived in current languages.

3 Overview of the Language

3.1 Ordinary Variables and Functions

There are three main kinds of objects inside the MATHEMAGIX type system: ordinary variables (including functions), classes and categories. Ordinary variables are defined using the following syntax:

```
test?: Boolean == pred? x;   // constant
flag?: Boolean := false;     // mutable
```

In this example, `test?` is a constant, whereas `flag?` is a mutable variable which can be given new values using the assignment operator `:=`. The actual type of the mutable variable `flag?` is `Alias Boolean`. Functions can be declared using a similar syntax:

```
foo (x: Int): Int == x * x;
```

MATHEMAGIX is a fully functional language, so that functions can both be used as arguments and as return values:

```
shift (x: Int) (y: Int): Int == x + y;
iterate (foo: Int -> Int, n: Int)
        (x: Int): Int ==
  if n = 0 then x
  else iterate (foo, n-1) (foo x);
```

The return type and the types of part of the function arguments are allowed to depend on the arguments themselves. For instance:

```
square (x: M, M: Monoid): M == x * x;
```

MATHEMAGIX does not allow for mutually dependent arguments, but dependent arguments can be specified in an arbitrary order. For instance, in the above example, we were allowed to introduce `M: Monoid` *after* the declaration `x: M`. However, the following declaration with mutually dependent arguments is incorrect:

```
nonsense (x: Foo y, y: Bar x): Void == {}
```

3.2 Classes

New classes are defined using the `class` primitive, as in the following example:

```
class Point == {
  mutable {
    x: Double;
    y: Double;
  }
  constructor point (x2: Int, y2: Int) == {
    x == x2;
    y == y2;
  }
}
```

We usually use similar names for constructors as for the class itself, but the user is free to pick other names. The `mutable` keyword specifies that we have both read and read-write accessors `postfix .x` and `postfix .y` for the fields `x` and `y`. Contrary to C++, new accessors can be defined outside the class itself:

```
postfix .length (p: Point): Double ==
  sqrt (p.x * p.x + p.y * p.y);
```

As in the case of functions, classes are allowed to depend on parameters, which may be either type parameters or ordinary values. Again, there may be dependencies among the parameters. One simple example of a class definition with parameters is:

```
class Num_Vec (n: Int) == {
  mutable v: Vector Double;
  constructor num_vec (c: Double) == {
    v == [ c | i: Int in 0..n ];
  }
}
```

3.3 Categories

Categories are the central concept for achieving genericity. We have already seen an example of the definition of a category in Sect. 2.1. Again, categories may take parameters, with possible dependencies among them. For instance:

```
category Module (R: Ring) == {
  This: Abelian_Group;
  infix *: (R, This) -> This;
}
```

The `This` type can occur in the category fields in many ways. In the above example, the line `This: Abelian_Group` means that `Module R` in particular includes all fields of `Abelian_Group`. More generally, `This` can be part of function argument types, of return types, or part of the declaration of an ordinary variable. For instance, the category `To T` below formalizes the concept of a type with an implicit converter to `T`.

```
category Type == {}

category To (T: Type) == {
  convert: This -> T;
}
```

Given an ordinary type `T`, we write `x: T` if `x` is an instance of `T`. In the case of a category `Cat`, we write `T`: Cat if a type `T` *satisfies* the category, that is, if all category fields are defined in the current context, when replacing `This` by `T`. Contrary to OCAML or HASKELL, it follows that MATHEMAGIX is very name sensitive: if we want a type `T` to be a monoid, then we need a multiplication on `T` with the exact name `infix *`. Of course, wrappers can easily be defined if we want different names, but one of the design goals of MATHEMAGIX is that it should be particularly easy to consistently use standard names.

The natural analogues of categories in OCAML and HASKELL are modules and type classes. In the case of MATHEMAGIX, there is only one *carrier* `This`, but the above examples show that it easy to mimick multiple carriers (or "multi-sorted signatures") using parameterized categories. Apart from this difference, MATHEMAGIX categories, OCAML modules and and HASKELL type classes provide a similar degree of expressivity.

3.4 Discrete Overloading

The main strength of the MATHEMAGIX type system is that it allows for heavy though fully type safe overloading. Similarly as in C++ or ALDOR, discrete overloading of a symbol is achieved by declaring it several times with different types:

```
infix * (c: Double, p: Point): Point ==
  point (c * p.x, c * p.y);
infix * (p: Point, c: Double): Point ==
  point (p.x * c, p.y * c);
```

Contrary to C++, non function variables and return values of functions can also be overloaded:

```
bar: Int == 11111;
bar: String == "Hello";
mmout << bar * bar << lf;
mmout << bar >< " John!" << lf;
```

Internally, the MATHEMAGIX type system associates a special *intersection* type `And (Int, String)` to the overloaded variable `bar`. During function applications, MATHEMAGIX consistently takes into account all possible meanings of the arguments and returns a possibly overloaded value which corresponds to all possible meanings of the function application. For instance, consider the overloaded function

```
foo (x: Int): Int == x + x;
foo (s: String): String == reverse s;
```

Then the expression `foo bar` will be assigned the type `And (Int, String)`. An example of a truly ambiguous expression would be `bar = bar`, since it is unclear whether we want to compare the integers `11111` or the strings `"Hello"`. True ambiguities will provoke compile time errors.

3.5 Parametric Overloading

The second kind of parametric overloading relies on the `forall` keyword. The syntax is similar to template declarations in C++, with the difference that all template parameters should be rigourously typed:

```
forall (M: Monoid)
fourth_power (x: M): M == x * x * x * x;
```

Internally, the MATHEMAGIX type system associates a special *universally quantified* type `Forall (M: Monoid, M -> M)` to the overloaded function `fourth_power`. In a similar way, values themselves can be parametrically overloaded. The main challenge for the MATHEMAGIX type system is to compute consistently with intersection types and universally quantified types. For instance, we may define the notation `[ 1, 2, 3 ]` for vectors using

```
forall (T: Type)
operator [] (t: Tuple T): Vector T == vector t;
```

This notation in particular defines the empty vector `[]` which admits the universally quantified type `Forall (T: Type, Vector T)`. In particular, and contrary to what would have been the case in C++, it is not necessary to make the type of `[]` explicit as soon as we perform the template instantiation. Thus, writing

```
v: Vector Int == [];
w: Vector Int == [] >< []; // concatenation
```

would typically be all right. On the other hand, the expression `#[]` (size of the empty vector) is an example of a genuine and parametric ambiguity.

In comparison with C++, it should be noticed in addition that parametric overloading is fully dynamic and that there are no restrictions on the use of ordinary variables as template parameters. Again, there may be dependencies between template arguments. MATHEMAGIX also implements the mechanism of partial specialization. For instance, if we have a fast routine `square` for double precision numbers, then we may define

```
fourth_power (x: Double): Double ==
  square square x;
```

Contrary to C++, partial specialization of a function takes into account both the argument types *and* the return type. This make it more natural to use the partial specialization mechanism for functions for which not all template parameters occur in the argument types:

```
forall (R: Number_Type) pi (): R == ...;
pi (): Double == ...;
```

3.6 Implicit Conversions

One major difference between ALDOR and AXIOM is that ALDOR does *not* contain any mechanism for implicit conversions. Indeed, in AXIOM, the mechanism of implicit conversions [33] partially depends on heuristics, which makes its behaviour quite unpredictable in non trivial situations. We have done a lot of experimentation with the introduction of implicit conversions in the MATHEMAGIX type system, and decided to ban them from the core language. Indeed, systematic implicit conversions introduce too many kinds of ambiguities, which are sometimes of a very subtle nature.

Nevertheless, the parametric overloading facility makes it easy to *emulate* implicit conversions, with the additional benefit that it can be made precise when exactly implicit conversions are permitted. Indeed, we have already introduced the `To T` category, defined by

```
category To (T: Type) == {
  convert: This -> T;
}
```

Here `convert` is the standard operator for type conversions in MATHEMAGIX. Using this category, we may define scalar multiplication for vectors by

```
forall (M: Monoid, C: To M)
infix * (c: C, v: Vector M): Vector M ==
  [ (c :> M) * x | x: M in v ];
```

Here `c :> M` stands for the application of `convert` to `c` and retaining only the results of type `M` (recall that `c` might have several meanings due to overloading). This kind of emulated "implicit" conversions are so common that MATHEMAGIX defines a special notation for them:

```
forall (M: Monoid)
infix * (c :> M, v: Vector M): Vector M ==
  [ c * x | x: M in v ];
```

In particular, this mechanism can be used in order to define converters with various kinds of transitivity:

```
convert (x :> Integer): Rational == x / 1;
convert (cp: Colored_Point) :> Point == cp.p;
```

The first example is also called an *upgrader* and provides a simple way for the construction of instances of more complex types from instances of simpler types. The second example is called a *downgrader* and can be used in order to customize type inheritance, in a way which is unrelated to the actual representation types in memory.

The elimination of genuine implicit converters also allows for several optimizations in the compiler. Indeed, certain operations such as multiplication can be overloaded hundreds of times in non trivial applications. In the above example of scalar multiplication, the MATHEMAGIX compiler takes advantage of the fact that at least one of the two arguments must really be a vector. This is done using a special table lookup mechanism for retaining only those few overloaded values which really have a chance of succeeding when applying a function to concrete arguments.

3.7 Union Types, Structures and Symbolic Expressions

In OCAML and HASKELL, the `type` and `data` keywords allow for the definition of unions and more general data types whose instances are freely built up from a finite number of explicitly given constructors. These languages also provide a powerful system of pattern matching in order to efficiently process instances of such types. In MATHEMAGIX, *structures* offer a similar functionality. For instance, we may define lists using

```
structure List (T: Type) == {
  null ();
  cons (head: T, tail: List T);
}
```

This declaration automatically introduces corresponding predicates `null?` and `cons?`, as well as accessors `.head` and `.tail`. For instance, the length of a list can be computing using

```
forall (T: Type)
prefix # (l: List T): Int ==
  if null? l then 0 else #l.tail + 1;
```

Alternatively, one may use pattern matching as in OCAML:

```
forall (T: Type)
prefix # (l: List T): Int ==
  match l with {
    case null () do return 0;
    case cons (_, t: List T) do return #t + 1;
  }
```

The fact that this code is slightly more verbose than its OCAML analogue stems from the fact that, in accordance with the general design of MATHEMAGIX, all implicit declarations occurring in the patterns (such as `t: List T`) should be explicitly typed by the user. MATHEMAGIX also supports the following HASKELL style of pattern matching:

```
forall (T: Type) {
  prefix # (l: List T): Int := 0;
  prefix # (cons (_, t: List T)): Int := #t + 1;
}
```

MATHEMAGIX structures are in particular very useful for the definition of symbolic expressions. Indeed, such expressions are typically finite unions of special types of expressions, such as literal names, compound expressions, scalar expressions (integers, rationals, etc.), sums, products, polynomials, matrices, etc. For this reason, MATHEMAGIX provides a few useful extensions of the above mechanisms with respect to OCAML and HASKELL:

1. The possibility to define open structures, whose set of basic constructors can be extended *a posteriori*. This typically allows the user to enrich the `Symbolic` type with a user defined kind of "symbolic skew polynomials".
2. The possibility to add new user defined patterns, besides the patterns which directly correspond to the constructors of the structure.
3. Some syntactic sugar for efficient dispatching of routines based on the kind of structure we are dealing with (e.g. there are two kinds of lists: `null` lists and `cons` lists).

We plan to give a full description of these features and how to exploit them in another paper.

3.8 Syntactic Sugar

Functions with several arguments use a classical tuple notation. It would have been possible to follow the OCAML and HASKELL conventions, which rely on currying, and rather regard a binary function $f : T^2 \to T$ as a function of type $T \to (T \to T)$. Although this convention is more systematic and eases the implementation of a compiler, it is also non standard in mainstream mathematics; in MATHEMAGIX, we have chosen to keep syntax as close as possible to classical mathematics. Furthermore, currying may be a source of ambiguities in combination with overloading. For instance, the expression `- 1` might be interpreted as the unary negation applied to 1, or as the operator $x \mapsto 1 - x$.

In order to accomodate for functions with an arbitrary number of arguments and lazy streams of arguments, MATHEMAGIX uses a limited amount of syntactic sugar. Given a type `T`, the type `Tuple T` stands for an arbitrary tuple of arguments of type `T`, and `Generator T` stands for a lazy stream of arguments of type `T`. For instance, `(1, 2)` would be a typical tuple of type `Tuple Int` and `0..10` a typical generator of type `Generator Int`. For instance, the prototype of a function which evaluates a multivariate polynomial at a tuple of points might be

```
forall (R: Ring)
eval (P: MVPol R, p: Tuple R): R == ...;
```

Given a polynomial `P: MVPol Int`, this would allow us to write `eval (p, 1, 2, 3)`. If we had used vectors or lists instead of tuples, then `eval` would always take exactly two arguments, and we would be forced to write `eval (p, cons (1, cons (2, cons (3, null ())))))`. The syntactic sugar takes care of the necessary conversions between tuples and generators. For instance, given a polynomial `P: MVPol Int`, the following would be valid evaluations:

```
eval (P, 1, 2..8, (9, 10), 11..20);
eval (P, (i^2 | i: Int in 0..100));
```

Notice that the notation of function application (or evaluation) can be overloaded itself:

```
postfix .() (fs: Vector (Int -> Int),
             x: Int): Vector Int ==
  [ f x | f: Int -> Int in fs ];
```

3.9 Future Extensions

There are various natural and planned extensions of the current type system.

One of the most annoying problems that we are currently working on concerns literal integers: the expression `1` can naturally be interpreted as a machine `Int` or as a long `Integer`. Consequently, it is natural to consider `1` to be of type `And (Int, Integer)`. For efficiency reasons, it is also natural to implement each of the following operations:

```
infix =: (Int, Int) -> Boolean;
infix =: (Integer, Integer) -> Boolean;
infix =: (Int, Integer) -> Boolean;
infix =: (Integer, Int) -> Boolean;
```

This makes an expression such as `1 = 1` highly ambiguous. Our current solution permits the user to prefer certain operations or types over others. For instance, we would typically prefer the type `Integer` over `Int`, since `Int` arithmetic might overflow. However, we still might prefer `infix =: (Int, Int) -> Boolean` over `infix =: (Int, Integer) -> Boolean`. Indeed, given `i: Int`, we would like the test `i = 0` to be executed fast.

One rather straightforward extension of the type system is to consider other "logical types". Logical implication is already implemented using the `assume` primitive:

```
forall (R: Ring) {
  ...
  assume (R: Ordered)
  sign (P: Polynomial R): Int ==
    if P = 0 then 0 else sign P[deg P];
  ...
}
```

Equivalently, we might have used a quantification `forall (R: Ordered_Ring)` for the declaration of `sign`, where `Ordered_Ring` is the "join" of the categories `Ring` and `Ordered`.

Similarly, the implementation of *existentially quantified* types will allow us to write routines such as

```
forall (K: Field)
exists (L: Algebraic_Extension K)
roots (p: Polynomial K): Vector L == ...;
```

Internally, an instance `x` of a type of the form `exists (C: Cat) F(C)` would be represented by a pair `(C: Cat, x: F(C))`.

We also plan to extend the syntactic sugar. For instance, given two aliases `i, j: Alias Int`, we would like to be able to write `(i, j) := (j, i)` or `(i, j) += (1, 1)`. A macro facility should also be included, comparable to the one that can be found in SCHEME. Some further syntactic features might be added for specific areas. For instance, in the MACAULAY2 system [6, 14], one may use the declaration

```
R = ZZ[x,y]
```

for the simultaneous introduction of the polynomial ring $\mathbb{Z}[x, y]$ *and* the two coordinate functions $x, y : \mathbb{Z}[x, y]$.

In the longer future, we would like to be able to formally describe mathematical properties of categories and algorithms, and provide suitable language constructs for supplying partial or complete correctness proofs.

4 Semantics and Compilation

In this section we will briefly sketch the semantics of the MATHEMAGIX language. For this purpose it is convenient to schematize the language constructs from the previous section by remaining as close as possible to more conventional typed λ-calculus, but with special notations for categories and overloading. Actually, MATHEMAGIX involves two main typed languages:

1. The *source language* contains constructs for building ambiguous expressions and their types. Such overloaded *source programs* always carry explicit types.
2. The *target language* extends the source language with notations for the explicit disambiguation of overloaded expressions. All valid *target programs* are unambiguous.

The job of the compiler is to transform source programs into target programs. The programs in the target language can be interpreted quite naturally in the language of untyped λ-calculus. Hence, in order to describe the semantics of MATHEMAGIX, it suffices to specify how the compiler transforms source program into target programs, and how target programs should be interpreted in classical untyped λ-calculus.

We will use a few notational conventions. For the sake of brevity, we will now use superscripts for specifying types. For instance, $\lambda x^{\text{Integer}} \cdot (x \times x)^{\text{Integer}}$ denotes the function $x \in \mathbb{Z} \mapsto x^2$. For the sake of readability, we will also denote types T, Int, etc. using capitalized identifiers and categories $\boldsymbol{C}$, **Ring**, etc. using bold capitalized identifiers. Similarly, we will use the terms "type expressions" and "category expressions" whenever an expression should be considered as a type or category. Notice however that this terminology is not formally enforced by the language itself.

4.1 Source Language

The *source language* contains three main components:

4.1.1 Typed Lambda Expressions

The first component consists of ordinary typed λ-expressions, and notations for their types:

1. Given expressions f and x, we denote function application by $f(x)$, $(f)x$, or $f x$.
2. Given a variable x, an expression y and type expressions T and U, we denote by $\lambda x^T.y^U$ the lambda expression which sends x of type T to y of type U.
3. We will denote by $T \to U$ the type of the above λ-expression. In the case when U depends on x, we will rather write $T \to U_x$ for this type.

Hence, all lambda expressions are typed and there are no syntactic constraints on the types T and U. However, "badly typed" expressions such as $\lambda x^{\text{Int}} \cdot x^{\text{Boolean}}$ will have no correct interpretation in the section below.

4.1.2 Declarations

The second part of the language concerns declarations of recursive functions, classes and categories.

1. Given variables $x_1, \ldots, x_n$, type expressions $T_1, \ldots, T_n$ and expressions $y_1, \ldots, y_n, z$, we may form the expression $\left(x_1^{T_1} \equiv y_1, \ldots, x_n^{T_n} \equiv y_n\right).z$. The informal meaning is: the expression z, with mutually recursive bindings $x_1^{T_1} \equiv y_1, \ldots, x_n^{T_n} \equiv y_n$.
2. Given variables $x_1, \ldots, x_n$ and type expressions $T_1, \ldots, T_n$, we may form the data type class $\left\langle x_1^{T_1}, \ldots, x_n^{T_n}\right\rangle$. For instance, a list of integers might be declared using $(\mathrm{List} \equiv \mathrm{class}\,\langle \mathrm{nil}^{\mathrm{List}}, \mathrm{cons}^{\mathrm{Int} \to \mathrm{List} \to \mathrm{List}}\rangle).z$. We also introduce a special variable **Class** which will be the type of class $\left\langle x_1^{T_1}, \ldots, x_n^{T_n}\right\rangle$.
3. Given variables $x_1, \ldots, x_n, y$ and type expressions $T_1, \ldots, T_n, U$, we may form the category $y^U \left\langle x_1^{T_1}, \ldots, x_n^{T_n}\right\rangle$. For instance, we might introduce the **Monoid** category using $(\mathbf{Monoid} \equiv \mathrm{This}^{\mathbf{Class}}\,\langle \times^{\mathrm{This} \to \mathrm{This} \to \mathrm{This}}\rangle).z$.

4.1.3 Overloaded Expressions

The last part of the language includes explicit constructs for overloaded expressions and their types:

1. Given two expressions x and y, we may form the overloaded expression $x \wedge y$.
2. Given type expressions T and U, we may form the intersection type $T \cap U$.
3. Given a variable x, a type expression T and an expression y, we may form the parametrically overloaded expression $\bigwedge_{x^T} y$.
4. Given a variable x, a type expression T and a type expression U, we may form the universally quantified type expression $\bigwedge_{x^T} U$.

In the last two cases, the variable x is often (but not necessarily) a type variable A and its type T a category C.

4.2 Target Language

The source language allows us to define an overloaded function such as

$$\mathrm{foo}^{\mathrm{Int} \to \mathrm{Int} \,\cap\, \mathrm{String} \to \mathrm{String}} \equiv \left(\lambda x^{\mathrm{Int}} \cdot (x \times x)^{\mathrm{Int}}\right) \wedge \left(\lambda x^{\mathrm{String}} \cdot (x \bowtie x)^{\mathrm{String}}\right) \quad (1)$$

In a context where 1 is of type Int, it is the job of the compiler to recognize that foo should be interpreted as a function of type Int $\to$ Int in the expression foo(1).

In order to do so, we first extend the source language with a few additional constructs in order to disambiguate overloaded expressions. The extended language will be called the *target language*. In a given context $\mathcal{C}$, we next specify when a source expression x can be interpreted as a non ambiguous expression $\hat{x}$ in the target language. In that case, we will write $\mathcal{C} \models x \rightsquigarrow \hat{x}$ and the expression $\hat{x}$ will always admit a unique type.

For instance, for foo as above, we introduce operators π_1 and π_2 for accessing the two possible meanings, so that

$$\left\{\text{foo}^{\text{Int}\to\text{Int}\,\cap\,\text{String}\to\text{String}}, 1^{\text{Int}}\right\} \models \text{foo}(1) \rightsquigarrow \pi_1\,(\text{foo})\,(1).$$

For increased clarity, we will freely annotate target expressions by their types when appropriate. For instance, we might have written $\pi_1\,(\text{foo})^{\text{Int}\to\text{Int}}\left(1^{\text{Int}}\right)^{\text{Int}}$ instead of $\pi_1\,(\text{foo})\,(1)$.

4.2.1 Disambiguation Operators

In the target language, the following notations will be used for disambiguating overloaded expressions:

1. Given an expression x, we may form the expressions $\pi_1(x)$ and $\pi_2(x)$.
2. Given expressions x and y, we may form the expression $x[y]$. Here x should be regarded as a template and $x[y]$ as its specialization at y.

There are many rules for specifying how to interpret expressions. We list a few of them:

$$\frac{\mathcal{C} \models x \rightsquigarrow \hat{x}^{T \cap U}}{\mathcal{C} \models x \rightsquigarrow \pi_1(\hat{x})^T} \qquad \frac{(\mathcal{C} \models x \rightsquigarrow \hat{x}^T) \wedge (\mathcal{C} \models y \rightsquigarrow \hat{y}^U)}{\mathcal{C} \models (x \wedge y) \rightsquigarrow (\hat{x} \wedge \hat{y})^{T \cap U}}$$

$$\frac{\left(\mathcal{C} \models x \rightsquigarrow \hat{x}^{\bigcap_{y^T} U}\right) \wedge (\mathcal{C} \models z \rightsquigarrow \hat{z}^T)}{\mathcal{C} \models x \rightsquigarrow \hat{x}[\hat{z}]^{U[\hat{z}/y]}}$$

$$\frac{(\mathcal{C} \models T \rightsquigarrow \widehat{T}) \wedge (\mathcal{C} \cup \{x^{\widehat{T}}\} \models y \rightsquigarrow \hat{y}) \wedge (\mathcal{C} \models U \rightsquigarrow \widehat{U})}{\mathcal{C} \models (\lambda x^T \cdot y^U) \rightsquigarrow (\lambda x^{\widehat{T}} \cdot \hat{y}^{\widehat{U}})^{\widehat{T} \to \widehat{U}_x}}$$

Here $U[\hat{z}/y]$ stands for the substitution of $\hat{z}$ for y in U. For instance, given an ambiguous expression x which may be interpreted as $\hat{x}^{T \cap U}$ in the context $\mathcal{C}$, the first rule specifies that we may disambiguate and interpret it as $\pi_1(\hat{x})^T$ as well (the other disambiguazation being $\pi_2\left(\hat{x}^U\right)$. For example,

$$\left\{\mathrm{foo}^{\mathrm{Int}\Rightarrow\mathrm{Int}\cap\mathrm{String}\to\mathrm{String}}, 1^{\mathrm{Int}}\right\} \models \mathrm{foo} \rightsquigarrow \pi_1\,(\mathrm{foo})^{\mathrm{Int}\to\mathrm{Int}}.$$

4.2.2 Category Matching

The second kind of extensions in the target language concern notations for specifying how types match categories:

1. Given expressions T, $f_1, \ldots, f_n$ and C, we may form the expression $T\langle f_1, \ldots, f_n\rangle \Uparrow C$. The informal meaning of this expression is "the type T considered as an instance of C, through specification of the structure $f_1, \ldots, f_n$".
2. Given an expression T, we may form $T \Downarrow$, meaning "forget the category of T".
3. Given expressions x and T, we may form the expression $x \uparrow T$, which allows us to cast to a type T of the form $T = U\langle f_1, \ldots, f_n\rangle \Uparrow C$.
4. Given an expression x, we may form $x \downarrow$.

In order to cast a given type T^{B} to a given category $C = \mathrm{This}^{B}\left\langle x_1^{X_1}, \ldots, x_n^{X_n}\right\rangle$, all fields of the category should admit an interpretation in the current context:

$$\frac{\forall i,\ \left(\mathcal{C} \models X_i\,[T/\mathrm{This}] \rightsquigarrow \widehat{X}_i\right) \wedge \left(\mathcal{C} \models x_i \rightsquigarrow \hat{x}_i^{\widehat{X}_i}\right)}{\mathcal{C} \models T \rightsquigarrow T\langle \hat{x}_1, \ldots, \hat{x}_n\rangle \Uparrow C}.$$

Assuming in addition that $\mathcal{C} \models y \rightsquigarrow \hat{y}^{T}$, we also have $\mathcal{C} \models y \rightsquigarrow \hat{y} \uparrow \left(T\langle \hat{x}_1, \ldots, \hat{x}_n\rangle \Uparrow C\right)$. There are further rules for casting down.

4.3 Compilation

4.3.1 Schematic Behaviour

A target expression x^{T} is said to be reduced if its type T is not of the form $U \cap V$, $\bigcap_{y^{Y}} U$, or $U \Uparrow C$ or $U \Downarrow$. The task of the compiler is to recursively determine all reduced interpretations of all subexpressions of a source program. Since each subexpression x may have several interpretations, we systematically try to represent the set of all possible reduced interpretations by a conjunction $\tilde{x}$ of universally quantified expressions. In case of success, this target expression $\tilde{x}$ will be the result of the compilation in the relevant context $\mathcal{C}$, and we will write $\mathcal{C} \models x \rightsquigarrow^{*} \tilde{x}$.

Let us illustrate this idea on two examples. With foo as in (1) and $c^{\mathrm{String}\cap\mathrm{Int}}$, there are two reduced interpretations of foo (c):

$$\left\{\mathrm{foo}^{\mathrm{Int}\to\mathrm{Int}\cap\mathrm{String}\to\mathrm{String}}, c^{\mathrm{String}\cap\mathrm{Int}}\right\}$$
$$\models \mathrm{foo}\,(c) \rightsquigarrow \pi_1\,(\mathrm{foo})\,(\pi_2\,(c))^{\mathrm{Int}},$$

$$\left\{\text{foo}^{\text{Int}\to\text{Int}\,\cap\,\text{String}\to\text{String}}, c^{\text{String}\,\cap\,\text{Int}}\right\}$$
$$\models \text{foo}\,(c) \rightsquigarrow \pi_2\,(\text{foo})\,(\pi_1\,(c))^{\text{String}}.$$

Hence, the result of the compilation of foo (c) is given by

$$\left\{\text{foo}^{\text{Int}\to\text{Int}\,\cap\,\text{String}\to\text{String}}, c^{\text{String}\,\cap\,\text{Int}}\right\}$$
$$\models \text{foo}\,(c) \rightsquigarrow^* (\pi_1\,(\text{foo})\,(\pi_2\,(c)) \wedge \pi_2\,(\text{foo})\,(\pi_1\,(c)))^{\text{Int}\,\cap\,\text{String}}.$$

In a similar way, the result of compilation may be a parametrically overloaded expression:

$$\{\text{bar}^{\bigcap_{T^C}\,\text{Int}\to\text{List}\,T}, 1^{\text{Int}}\} \models \text{bar}(1) \rightsquigarrow^* \bigwedge_{T^C} \text{bar}\,[T]\,(1)^{\text{List}\,T}.$$

4.3.2 Resolution of Ambiguities

Sometimes, the result $\tilde{x}$ of the compilation of x is a conjunction which contains at least two expressions of the same type. In that case, x is truly ambiguous, so the compiler should return an error message, unless we can somehow resolve the ambiguity. In order to do this, the idea is to define a partial preference relation $\preccurlyeq$ on target expressions and to keep only those expressions in the conjunction $\tilde{x}$ which are maximal for this relation.

For instance, assume that we have a function square of type $(\bigcap_{M^{\text{Monoid}}} M \to M) \cap \text{Int} \to \text{Int}$ and the constant 2012 of type Int. In Sect. 3.5, we have seen that MATHEMAGIX supports partial specialization. Now π_2 (square) is a partial specialization of π_1 (square), but not the inverse. Consequently, we should strictly prefer π_2 (square) over π_1 (square), and π_2 (square) (2012) over π_2 (square) $[I]\,(2012 \uparrow I)\downarrow$, where $I = \text{Int}\left\langle \times^{\text{Int}\to\text{Int}\to\text{Int}} \right\rangle \Uparrow \mathbf{Monoid}$.

As indicated in Sect. 3.9, we are currently investigating further extensions of the preference relation $\preccurlyeq$ *via* user provided preference rules.

4.3.3 Implementation Issues

In absence of universal quantification, the search process for all reduced interpretations can in principle be designed to be finite and complete. The most important implementation challenge for MATHEMAGIX compilers therefore concerns universal quantification.

The main idea behind the current implementation is that all pattern matching is done in two stages: at the first stage, we propose possible matches for free variables introduced during unification of quantified expressions. At a second stage, we verify that the proposed matches satisfy the necessary categorical constraints, and we rerun the pattern matching routines for the actual matches. When proceeding this way, it is guaranteed that casts of a type to a category never involve free variables.

Let us illustrate the idea on the simple example of computing a square. So assume that we have the function *square* of type $\bigcap_{M^{\mathbf{Monoid}}} M \rightarrow M$ in our context, as well as a multiplication $\times$: Int $\rightarrow$ Int $\rightarrow$ Int. In order to compile the expression square (2012^{Int}), the algorithm will attempt to match Int $\rightarrow \mathcal{F}_1$ with $\bigcap_{M^{\mathbf{Monoid}}} M \rightarrow M$ for some free variable $\mathcal{F}_1$. At a first stage, we introduce a new free variable $\mathcal{F}_2^{\mathbf{Monoid}}$ and match $\mathcal{F}_2^{\mathbf{Monoid}} \rightarrow \mathcal{F}_2^{\mathbf{Monoid}}$ against Int $\rightarrow \mathcal{F}_1$. This check succeeds with the bindings $\mathcal{F}_2^{\mathbf{Monoid}} :=$ Int and $\mathcal{F}_1 :=$ Int, but without performing any type checking for these bindings. At a second stage, we have to resolve the innermost binding $\mathcal{F}_2^{\mathbf{Monoid}} :=$ Int and cast Int to **Monoid**. This results in the correct proposal $\mathcal{F}_2^{\mathbf{Monoid}} := I$ for the free variable, where $I \equiv \mathrm{Int}\langle \times^{\mathrm{Int} \rightarrow \mathrm{Int} \rightarrow \mathrm{Int}} \rangle \Uparrow \mathbf{Monoid}$. We finally rematch $I \rightarrow I$ with Int $\rightarrow \mathcal{F}_1$ and find the return type $\mathcal{F}_1 := I$.

In practice the above idea works very well. Apart from more pathological theoretical problems that will be discussed below, the only practically important problem that we do not treat currently, is finding a "smallest common supertype" with respect to type conversions (see also [33]).

For instance, let f be a function of type $\bigcap_{R^{\mathbf{Class}()}} \bigcap_{T^{\mathbf{Into} R}} T \rightarrow T \rightarrow R$. What should be the type of $f x y$, where $x^{X^{\mathbf{Class}()}}$ and $y^{Y^{\mathbf{Class}()}}$ are such that X and Y are different? Theoretically speaking, this should be the type $\bigcap_{R^{\boldsymbol{C}}} R$, where $\boldsymbol{C}$ is the category $T^{\mathbf{Class}}\langle \mathrm{convert}^{X \rightarrow T}, \mathrm{convert}^{Y \rightarrow T} \rangle$. However, the current pattern matching mechanism in the MATHEMAGIX compiler will not find this type.

4.3.4 Theoretical Problems

It is easy to write programs which make the compiler fail or loop forever. For instance, given a context with the category $\mathbf{In}(T) \equiv \mathrm{This}^{\mathbf{Class}} \langle \mathrm{convert}^{\mathrm{This} \rightarrow T} \rangle$ and functions convert and f of types $\bigcap_{T^{\mathbf{Class}}} F(T) \rightarrow T$ and $\bigcap_{T^{\mathbf{In}(\mathrm{Int})}} T \rightarrow T$, the compilation of $f(x^{\mathrm{String}})$ will loop. Indeed, the compiler will successively search for converters String $\rightarrow$ Int, F (String) $\rightarrow$ Int, F (F (String)) $\rightarrow$ Int, etc. Currently, some safeguards have been integrated which will make the compiler abort with an error message when entering this kind of loops.

The expressiveness of the type system actually makes it possible to encode any first order theory directly in the system. For instance, given a binary predicate P and function symbols f, g, the statement $\forall x, P(f(x), g(x)) \Rightarrow P(g(g(x)), f(x))$ might be encoded by the declaration of a function $\bar{P}$ of type $\bigcap_{\bar{x}^{\boldsymbol{C}}} \bar{g}(\bar{g}(\bar{x})) \rightarrow \bar{f}(\bar{x}) \rightarrow$ Boolean, where $\boldsymbol{C} = T^{\mathbf{Class}} \langle \bar{P}^{\bar{f}(T) \rightarrow \bar{g}(T) \rightarrow \mathrm{Boolean}} \rangle$.

These negative remarks are counterbalanced by the fact that the type system is not intended to prove mathematical theorems, but rather to make sense out of commonly used overloaded mathematical notations. It relies upon the shoulders of the user to use the type system in order to define such common notations and not misuse it in order to prove general first order statements. Since notations are intended to be easily understandable at the first place, they can usually be given a sense by following simple formal procedures. We believe that our type system is powerful enough to cover most standard notations in this sense.

The above discussion shows that we do not aim completeness for the MATHEMAGIX system. So what about soundness? The rules for interpretation are designed in such a way that all interpretations are necessarily correct. The only possible problems which can therefore occur are that the compiler loops forever or that it is not powerful enough to automatically find certain non trivial interpretations.

We also notice that failure of the compiler to find the intended meaning does not necessarily mean that we will get an error message or that the compiler does not terminate. Indeed, theoretically speaking, we might obtain a correct interpretation, even though the intended interpretation should be preferred. In particular, it is important to use the overloading facility in such a way that all possible interpretations are always correct, even though some of them may be preferred.

Finally, an interesting research question is to investigate which sublanguages of MATHEMAGIX do admit a complete type system. For instance, if we exclude parametric overloading, then the type system becomes complete. Similarly, if parameteric overloading is subject to additional constraints, then the type system might still be complete. For instance, what if the category only contains functions of the type $T_1 \to \cdots \to T_n \to U$, where at least one of the types T_i involves This, and where each of the types $T_1, \ldots, T_n, U$ is either equal to This or free from $This$? Another natural kind of requirement in the case of upgraders would be to insist on always upgrading "simpler" types (such as R) into more "complex" types (such as Polynomial (R)), and never the other way around. Similarly, downgraders should always downgrade more complex types into simpler types.

4.4 Execution

Given an expression x on which the compilation process succeeds, we finally have to show what it means to evaluate x. So let $\tilde{x}$ with $\varnothing \models x \rightsquigarrow^* \tilde{x}$ be the expression in the target language which is produced by the compiler. The target language has the property that it is quite easy to "downgrade" $\tilde{x}$ into an expression of classical untyped λ-calculus. This reduces the evaluation semantics of MATHEMAGIX to the one of this calculus.

Some of the most prominent rules for rewriting $\tilde{x}$ into a term of classical untyped λ-calculus are the following:

1. Overloaded expressions $x \wedge y$ are rewritten as pairs $\lambda f \cdot f x y$.
2. The projections π_1 and π_2 are simply true : $\lambda x.\lambda y \cdot x$ and false : $\lambda x \cdot \lambda y \cdot y$.
3. Template expressions $\bigwedge_{x^T} y$ are rewritten as λ-expressions $\lambda x \cdot y$.
4. Template instantiation $x[y]$ is rewritten into function application $x(y)$.
5. Instances $T\left\langle x_1^{U_1}, \ldots, x_n^{U_n}\right\rangle \Uparrow C$ of categories are implemented as n-tuples $\lambda f \cdot T x_1 \cdots x_n$.

For instance, consider the template $\bigwedge_{M^{\text{Monoid}}} \lambda x^M \cdot (x \times x)^M$. After compilation, this template is transformed into the expression $\lambda M \cdot \lambda x. \left(\pi_1^1 M\right) xx$, where $\pi_i^n = \lambda x_1. \dots .\lambda x_n \cdot x_i$.

One of the aims of the actual MATHEMAGIX compiler is to be compatible with existing C libraries and C++ template libraries. For this reason, the backend of MATHEMATICS really transforms expressions in the target language into C++ programs instead of terms of untyped λ-calculus.

5 Ongoing Work and Perspectives

The current MATHEMAGIX compiler available from www.mathemagix.org implements the type system described in this paper. In addition, the language contains features for using C++ template libraries in a generic way. Until recently, we only relied on C++ for the development of our libraries, so the MATHEMAGIX system provides a range of C++ template libraries for mathematical computations [24, 37].

Now that we have completed the implementation of a basic version of the compiler, our first major challenge is to rewrite the most important C++ template libraries directly in MATHEMAGIX. This process will involve further finetuning of the language and various extensions which will control how to optimize things. As soon as we have gained more practical experience, we plan to give a more detailed account on the advantages of the MATHEMAGIX language for the implementation of computer algebra libraries.

Concerning efficiency and code optimization, we aim to give the user access to very low level details, such as SIMD instructions or the layout of long integers in memory, while providing powerful mechanisms for abstraction. The language will also provide fine grained control over when a template will be instantiated for particular parameters and when to execute the generic code. One of our aims is that the compiler will be able to generate code of a quality comparable to the "codelets" in FFTW3 [8, 9], but for a wide range of problems, and without the need to switch between various languages (we recall that FFTW3 is a C library which relies on OCAML for the generation of codelets).

Plans for the longer future include support for various types of parallellism and interoperability with other general purpose languages and computer algebra systems. We also hope to interface MATHEMAGIX with one or more automatic theorem provers, which would allow us to specify the semantics of programs in an even more detailed way.

References

1. Baumgartner, G., Russo, V.F.: Implementing signatures for C++. ACM Trans. Program. Lang. Syst. **19**(1), 153–187 (1997)

2. Bond, E., Auslander, M., Grisoff, S., Kenney, R., Myszewski, M., Sammet, J., Tobey, R., Zilles, S.: FORMAC an experimental formula manipulation compiler. In: Proceedings of the 1964 19th ACM National Conference, ACM'64, pp. 112–101. ACM, New York (1964)
3. Coquand, T., et al.: The coq proof assistant. http://coq.inria.fr/ (1984)
4. Coquand, T., Huet, G.: The calculus of constructions. Inf. Comput. **76**(2–3), 95–120 (1988)
5. Dos Reis, G., Stroustrup, B.: Specifying C++ concepts. SIGPLAN Not. **41**(1), 295–308 (2006)
6. Eisenbud, D., Grayson, D.R., Stillman, M.E., Sturmfels, B. (eds.): Computations in Algebraic Geometry With Macaulay 2. Springer, London (2002)
7. Faugère., J.-C., FGb: a library for computing gröbner bases. In: Fukuda, K., Hoeven, J., Joswig, M., Takayama, N. (eds.) Mathematical Software - ICMS'10, Volume 6327 of Lecture Notes in Computer Science, pp. 84–87. Springer, Heidelberg (2010)
8. Frigo, M., Johnson, S.G.: The design and implementation of FFTW3. Proc. IEEE, **93**(2), 216–231 (2005) (Special issue on "Program Generation, Optimization, and Platform Adaptation")
9. Frigo, M.: A fast Fourier transform compiler. In: Proceedings of the 1999 ACM SIGPLAN Conference on Programming Language Design and Implementation, vol. 34, pp. 169–180. ACM (1999)
10. Garcia, R., Järvi, J., Lumsdaine, A., Siek, J.G., Willcock, J.: A comparative study of language support for generic programming. In: Proceedings of the 2003 ACM SIGPLAN Conference on Object-Oriented Programming, Systems, Languages, and AppLications (OOPSLA'03) (2003)
11. Geddes, K., Gonnet, G.: Maplesoft. Maple. http://www.maplesoft.com/products/maple/ (1980)
12. Girard, J.Y.: Une extension de l'interprétation de Gödel à l'analyse, et son application à l'élimination de coupures dans l'analyse et la théorie des types. In: Fenstad, J.E. (eds.) Proceedings of the Second Scandinavian Logic Symposium, pp. 63–92. North-Holland Publishing Co. (1971)
13. Granlund, T., et al.: GMP, the GNU multiple precision arithmetic library. http://www.swox.com/gmp (1991)
14. Grayson, D.R., Stillman, M.E.: Macaulay2, a software system for research in algebraic geometry. http://www.math.uiuc.edu/Macaulay2/
15. Griesmer, J.H., Jenks, R.D., Yun, D.Y.Y.: SCRATCHPAD User's Manual. Computer Science Department Monograph Series, IBM Research Division (1975)
16. Hanrot, G., Lefèvre, V., Ryde K., Zimmermann, P.: MPFR, a C library for multiple-precision floating-point computations with exact rounding. http://www.mpfr.org (2000)
17. Hart, W.: An introduction to Flint. In: Fukuda, K., van der Hoeven, J., Joswig, M., Takayama, N. (eds.) Mathematical Software - ICMS'10, Third International Congress on Mathematical Software, Kobe, Japan, 13–17 Sept, 2010, Volume 6327 of Lecture Notes in Computer Science, pp. 88–91. Springer (2010)
18. Hudak, P., Hughes, J., Jones S.P., Wadler, P.: A history of haskell: being lazy with class. In: Proceedings of the Third ACM SIGPLAN Conference on History of Programming Languages, HOPL III, pp. 12–1. ACM, New York (2007)
19. Jenks, R.D., Sutor, R.: AXIOM: The Scientific Computation System. Springer, New York (1992)
20. Jenks, R.D.: Modlisp - an introduction (invited). In: Proceedings of the International Symposiumon on Symbolic and Algebraic Computation, EUROSAM'79, pp. 466–480. Springer, London (1979)
21. Jenks, R.D., et al.: The SCRATCHPAD language. SIGPLAN Not. **9**(4), 101–111 (1974)
22. Jenks, R.D.: A language for computational algebra. SIGPLAN Not. **16**(11), 22–29 (1981)
23. Jones, S.P., et al.: The Haskell 98 language and libraries: the revised report. J. Funct. Program. **13**(1), 0–255 (2003) (http://www.haskell.org/definition/)
24. Lecerf, G.: Mathemagix: towards large scale programming for symbolic and certified numeric computations. In: Fukuda, K., van der Hoeven, J., Joswig, M., Takayama, N., (eds.) Mathematical Software - ICMS'10, Third International Congress on Mathematical Software, Kobe, Japan, 13–17 Sept 2010, Volume 6327 of Lecture Notes in Computer Science, pp. 329–332. Springer, 2010
25. Leroy, X., et al.: OCaml. http://caml.inria.fr/ocaml/ (1996)

26. Martin, W.A., Fateman, R.J.: The MACSYMA system. In: Proceedings of the Second ACM Symposium on Symbolic And Algebraic Manipulation, SYMSAC'71, pp. 59–75. ACM, New York (1971)
27. Martin-Löf, P.: Constructive mathematics and computer programming. Log. Methodol. Philos. Sci. **VI**, 153–175 (1979)
28. Milner, R.: A theory of type polymorphism in programming. J. Comput. Syst. Sci. **17**, 348–375 (1978)
29. Moses, J.: Macsyma: A personal history. J. Symb. Comput. **47**(2), 123–130 (2012)
30. Nipkow, T., Paulson, L., Wenzel, M.: Isabelle/Hol. http://www.cl.cam.ac.uk/research/hvg/Isabelle/ (1993)
31. Stein, W.A., et al.: Sage mathematics software. The sage development team, 2004. http://www.sagemath.org
32. Stroustrup, B.: The C++ Programming Language, 2nd edn. Addison-Wesley, Reading (1995)
33. Sutor, R.S., Jenks, R.D.: The type inference and coercion facilities in the scratchpad ii interpreter. SIGPLAN Not. **22**(7), 56–63 (1987)
34. The Axiom computer algebra system. http://wiki.axiom-developer.org/FrontPage
35. The Maxima computer algebra system (free version). http://maxima.sourceforge.net/ (1998)
36. van der Hoeven, J., Lecerf, G., Mourain, B., et al.: Mathemagix. http://www.mathemagix.org (2002)
37. van der Hoeven, J., Lecerf, G., Mourrain, B., Trébuchet, P., Berthomieu, J., Diatta, D., Manzaflaris, A.: Mathemagix, the quest of modularity and efficiency for symbolic and certified numeric computation. ACM Commun. Comput. Algebra **45**(3/4), 186–188 (2012)
38. Watt,S., Broadbery, P.A., Dooley, S.S., Iglio, P., Morrison, S.C., Steinbach, J.M., Sutor, R.S.: A first report on the A# compiler. In: Proceedings of the International Symposium on Symbolic and Algebraic Computation, ISSAC'94, pp. 25–31. ACM, New York (1994)
39. Watt, S., et al.: Aldor programming language. http://www.aldor.org/ (1994)
40. Wolfram Research. Mathematica. http://www.wolfram.com/mathematica/ (1988)

Resultant-Free Computation of Indefinite Hyperexponential Integrals

Xiaoli Wu

Abstract In this note, we describe a special structure of differential Gosper forms of rational functions, which allows us to design a new and simple algorithm for constructing differential Gosper forms without the resultant computation and integer-root finding. Moreover, we present an algorithm for computing a universal denominator of the first-order linear differential equation which the Almkvist–Zeilberger algorithm solves.

1 Introduction

The Gosper algorithm [1] and its continuous analogue [2] have played an important role in symbolic integration and summation. Zeilberger developed his fast algorithms [2, 3] for creative telescoping based on the parametrization of the Gosper algorithm and its continuous analogue. These algorithms allow us to verify many combinatorial identities in a systematic way [4].

In this note, we focus on the discussion of the Almkvist–Zeilberger algorithm [2], which is a differential analogue of the Gosper algorithm. Throughout, we let k be a field of characteristic zero, $k[x]$ be the ring of polynomials in x over k, and $k(x)$ be the field of rational functions in x over k. Let $'$ denote the usual derivation d/dx on the field $k(x)$. For a polynomial $p \in k[x]$, let $\text{lc}(p)$ denote the leading coefficient of p. A field E is called a *differential extension* of $k(x)$ if it contains $k(x)$ and $'$ extends as a derivation on E.

This work was partially supported by the Tian Yuan Special Funds of the National Natural Science Foundation of China (Grant No. 11126089).

X. Wu (✉)
The School of Science, Hangzhou Dianzi University, Hangzhou 310018, Zhejiang, China
e-mail: xlwu@amss.ac.cn

Key Laboratory of Mathematics Mechanization, Chinese Academy of Sciences, Beijing 100190, China

© Springer-Verlag Berlin Heidelberg 2014
R. Feng et al. (eds.), *Computer Mathematics*, DOI 10.1007/978-3-662-43799-5_28

Definition 1 Let E be a differential extension of $k(x)$. A nonzero element $h \in E$ is said to be *hyperexponential* over $k(x)$ if the logarithmic derivative h'/h is in $k(x)$. The rational quotient h'/h is called the *certificate* of h.

A hyperexponential function h over $k(x)$ is said to be *hyperexponential integrable* if there exists another hyperexponential function g such that $g' = h$. The Almkvist–Zeilberger algorithm decides whether a given hyperexponential function is hyperexponential integrable or not. The crucial step is the computation of differential Gosper forms of rational functions. After this, the integrability problem is reduced to finding polynomial solutions of a first-order linear differential equation with polynomial coefficients. The first result of this note is a special property of differential Gosper forms, which leads to a new way to compute differential Gosper forms by combining an idea in [5, Sect. 3] such that the computation of resultants and integer-root finding are avoided.

Instead of the Almkvist–Zeilberger algorithm, one can solve the integrability problem for hyperexponential functions using Abramov's algorithm [6]. To decide whether h is hyperexponential integrable or not is equivalent to find rational solutions of the equation

$$y' + fy = 1, \tag{1}$$

where f is the certificate of h. Abramov's algorithm first computes a polynomial $d \in k[x]$ such that the denominator of any rational solution divides d. This polynomial is called a *universal denominator* for Eq. (1). With universal denominators, we reduce the problem to finding polynomial solutions of a related first-order linear differential equation. The second result of this note is to give a special way to find a universal denominator for Eq. (1).

2 The Almkvist–Zeilberger Algorithm

Let h be a nonzero hyperexponential function over $k(x)$ and let $f \in k(x)$ be the certificate of h. If h is hyperexponential integrable, i.e., $h = g'$ for some hyperexponential function g over $k(x)$, then $g = ah$ for some $a \in k(x)$, since $g'/g \in k(x)$. Moveover, the rational function a is a rational solution of Eq. (1). The Almkvist–Zeilberger algorithm is a special method for solving this first-order differential equation.

Definition 2 For a rational function $f \in k(x)$, we call the triple $(p, q, r) \in k[x]^3$ a *differential Gosper form* of f if

$$f = \frac{p'}{p} + \frac{q}{r}, \quad \text{where } \gcd(r, q - ir') = 1 \quad \text{for all integer } i \geq 0. \tag{2}$$

We first recall the algorithm for computing differential Gosper forms in [2, Sect. 5]. The algorithm relies on the following lemma.

Lemma 1 *Let $f = u/v \in k(x)$ be such that $u, v \in k[x]$ and* $\gcd(u, v) = 1$ *and let $j \in \mathbb{N}$. If $w := \gcd(v, u - jv')$ is a polynomial of positive degree, then the pair $(\bar{u}, \bar{v}) \in k[x]^2$ with $\bar{v} = v/w$ and $\bar{u} = (u - jw'\bar{v})/w$ satisfies* $\gcd(\bar{v}, \bar{u} - j\bar{v}') = 1$.

Proof First, we show that $w \mid (u - jw'\bar{v})$. It is easy to see that $j \neq 0$ because $\gcd(u, v)$ is 1. Since $v = w\bar{v}$, we have $u - jv' = u - jw'\bar{v} - jw\bar{v}'$. Therefore $w|(u - jv')$ implies that $w|(u - jw'\bar{v})$. By definition of w, we have $\gcd(v/w, (u - jv')/w) = 1$. A direct calculation yields that

$$\bar{u} - j\bar{v}' = \frac{u - jw'\bar{v}}{w} - j\bar{v}' = \frac{u - jw'\bar{v} - jw\bar{v}'}{w} = \frac{u - jv'}{w}.$$

So we get $\gcd(\bar{v}, \bar{u} - j\bar{v}') = 1$. □

To decide whether there exists a nonnegative integer j such that $\gcd(v, u - jv')$ is of positive degree, one way is to compute the nonnegative integer roots of the resultant $R := \operatorname{resultant}_x(v, u - zv') \in k[z]$. The process of computing differential Gosper forms is starting from $p := 1, q := u$ and $r := v$. If $w := \gcd(v, u - jv')$ is of positive degree for some nonnegative integer j, then we update

$$p \leftarrow pw^j, \quad r \leftarrow r/w, \quad q \leftarrow \frac{q - jw'(r/w)}{w}.$$

This process is repeated until $\gcd(r, q - ir') = 1$ for all $i \in \mathbb{N}$. The proposition below shows that any irreducible factor of p computed by the process above is a factor of v with multiplicity one, which is a special case of Fact 8.31 in [7]. For completeness, we provide a more direct proof of this fact.

Proposition 1 *Let $u, v \in k[x]$ be such that* $\gcd(u, v) = 1$ *and let* $v = v_1 v_2^2 \cdots v_m^m$ *be the squarefree factorization of v. Let w be an irreducible polynomial in $k[x]$. If $w| \gcd(v, u - jv')$ for some $j \in \mathbb{N}$, then $w|v_1$.*

Proof Suppose that $w|v_i$ for some $i \in \{2, \ldots, m\}$. Then $w|v'$ since $\gcd(v, v') = v_2 v_3^2 \cdots v_m^{m-1}$. Since $w|(u - jv')$, we have $w|u$, which implies that $w| \gcd(u, v)$. This is a contradiction with the assumption that $\gcd(u, v) = 1$. □

Remark 1 The polynomial v_1 in Proposition 1 can be computed as follows: (i). Compute $\tilde{v} = \gcd(v, v')$ and $v^* = v/\tilde{v}$. (ii). Compute $\bar{v} = \gcd(\tilde{v}, \tilde{v}')$ and $v^{**} = \tilde{v}/\bar{v}$. (iii). $v_1 = v^*/v^{**}$.

Combining Proposition 1 with an idea in [5, Sect. 3], we can compute differential Gosper forms of the polynomial p without resultant calculation and integer-root finding.

Algorithm 1 Input: A rational function $f = u/v \in k(x)$
Output: A differential Gosper form (p, q, r) of f.

1. Compute the product v_1 of multiplicity-one factors of v as in Remark 1;

2. Factor $v_1 = p_1 \cdots p_n$ into irreducible factors.
3. Set $p := 1$. For i from 1 to n, test whether there exists $\lambda_i \in \mathbb{N}$ such that

$$p_i \mid (u - \lambda_i p_i'(v/p_i)). \tag{3}$$

If such a nonnegative integer λ_i exists, then update $p := p_i^{\lambda_i} p$.
4. Set $q/r := f - p'/p$, and return (p, q, r).

In Step 3, we can observe that there is either no such λ_i or a unique one, and there cannot be more than one such number.

Remark 2 For an irreducible polynomial p_i, the test in Eq. (3) is very simple. It suffices to decide whether the ratio $\text{rem}(u, p_i)/\text{rem}(p_i'(v/p_i), p_i)$ is a nonnegative integer or not. As noted in [5, Sect. 3], the algorithm above, even though involving the irreducible factorization in Step 2, is less time-consuming than that of resultant calculation and integer-root finding in practice.

After obtaining a differential Gosper form (p, q, r) of f, the problem of finding rational solutions of $y' + fy = 1$ is reduced to finding polynomial solutions of the equation

$$rz' + (q + r')z = p.$$

If this equation has a polynomial solution $s \in k[x]$, then h is hyperexponential integrable with $g = srh/p$. One can observe that the computation of differential Gosper forms not only predicts the denominator of the rational solutions but also computes part of the numerator of the rational solutions. This advantage allows the Almkvist–Zeilberger algorithm to outperform other methods in practice.

It is easy to see that a rational function $f \in k(x)$ is hyperexponential integrable if and only if $f = g'$ for some $g \in k(x)$. For rational functions, there are many classical methods that solve the integrability problem [8–10]. For this special class of inputs, we present a necessary condition on the integrability, which is used to show the irrationality of certain integrals of rational functions.

Lemma 2 *Let $f = s/t \in k(x)$ with $s, t \in k[x]$ and $\gcd(s, t) = 1$. Let $t = t_1 t_2^2 \cdots t_m^m$ be the squarefree factorization of t over k. If $f = g'$ for some $g \in k(x)$, then $t_1 \in k$.*

Proof Suppose that $f = g'$ for some $g \in k(x)$. Write $g = a/b$ with $a, b \in k[x]$ and $\gcd(a, b) = 1$. The equality $f = g'$ implies that

$$sb^2 = t(a'b - ab'). \tag{4}$$

Suppose that $p \in k[x]$ is an irreducible factor of t but $p^2 \nmid t$. Then $t = p\bar{t}$ with $p \nmid \bar{t}$. Substituting this into Eq. (4) yields

$$sb^2 = p\bar{t}(a'b - ab'). \tag{5}$$

Since $p \mid sb^2$ and $p \nmid s$, we have $p \mid b$. Write $b = p^n \bar{b}$ with $p \nmid \bar{b}$, where $n \in \mathbb{N}$. Then

$$sp^{2n}\bar{b}^2 = p\bar{t}(a'p^n\bar{b} - anp'p^{n-1}\bar{b} - ap^n\bar{b}').$$

Note that the left hand-side is divisible by p^{n+1}. Thus we get $p \mid nap'\bar{t}\bar{b}$, which is a contradiction. So the polynomial t has no multiplicity-one factor, i.e., $t_1 \in k$. □

Example 1 By Lemma 2, the rational function $1/x \neq r'(x)$ for any rational function $r \in k(x)$, which is equivalent to that $\int 1/x\, dx = \log(x)$ is not a rational function.

3 Universal Denominators

Abramov's algorithm [6] is used to find rational solutions of linear differential equations with polynomial coefficients of any order. We can specialize Abramov's algorithm to find rational solutions of the equation

$$y' + fy = 1. \tag{6}$$

The following proposition shows the structure of the universal denominators of Eq. (6).

Proposition 2 *Let $f = u/v \in k(x)$ be such that* $\gcd(u, v) = 1$ *and let $y = a/b \in k(x)$ with* $\gcd(a, b) = 1$ *be a rational solution of Eq.* (6). *Let v_1 be the product of multiplicity-one factors of v. If $p \in k[x]$ is an irreducible factor of b with multiplicity λ, then $p \mid v_1$ and* $\lambda = \mathrm{rem}(u, p)/\mathrm{rem}(p'v/p, p) \in \mathbb{N}$.

Proof Substituting $f = u/v$ and $y = a/b$ into Eq. (6) yields

$$(a'v + ua)b - avb' = b^2 v. \tag{7}$$

Assume that p is an irreducible factor of b with multiplicity λ. Write $b = p^\lambda \bar{b}$ with $p \nmid \bar{b}$. Replacing b by $p^\lambda \bar{b}$ in Eq. (7), we get

$$(a'v + ua)p^\lambda \bar{b} - av(\lambda p' p^{\lambda-1}\bar{b} + p^\lambda \bar{b}') = p^{2\lambda}\bar{b}^2 v. \tag{8}$$

The right side of Eq. (8) is divisible by p^λ, so is the left one. Thence we obtain $p \mid \lambda avp'\bar{b}$. Since p is coprime to a, p', and $\bar{b}$, we have $p \mid v$. Write $v = p\bar{v}$. Substituting this into Eq. (8) and dividing both sides by p^λ yield

$$a\bar{b}(u - \lambda\bar{v}p') + p\bar{v}(a'\bar{b} - a\bar{b}') = p^{\lambda+1}\bar{b}^2\bar{v}. \tag{9}$$

The right side of Eq. (9) is divisible by p, so is the left one. Hence $p \mid (u - \lambda\bar{v}p')$. Since both u and p' are not divisible by p, so $\bar{v}$ is not divisible by p, which implies that $p \mid v_1$. Moreover, the divisibility $p \mid (u - \lambda \bar{v} p')$ implies that $\lambda = \text{rem}(u, p)/\text{rem}(p'\bar{v}, p) \in \mathbb{N}$. □

Remark 3 A more systematic way to show results as Proposition 2 is via order estimates (c.f. see [11, Chap. 4]). The Proposition above can be shown by observing that the order of the derivative of f at some pole of f differs from the order of f at this pole by exactly one. Nevertheless, we elaborate those ideas in the proof above.

We present an algorithm for computing a universal denominator of Eq. (6) by using Proposition 2.

Algorithm 2 Input: A rational function $f = u/v \in k(x)$
Output: A polynomial $b \in k[x]$ such that b is a universal denominator of Eq. (6).

1. Compute the product v_1 of multiplicity-one factors of v as in Remark 1;
2. Factor $v_1 = p_1 \cdots p_n$ into irreducible factors.
3. Set $b := 1$. For i from 1 to n, test whether there exists $\lambda_i \in \mathbb{N}$ such that

$$p_i \mid (u - \lambda_i p_i'(v/p_i)). \tag{10}$$

   If such a nonnegative integer λ_i exists, then update $b := p_i^{\lambda_i} b$.
4. Return b.

Remark 4 One can see that the process above is almost the same as that in Algorithm 1. It is quite a mysterious step in the Almkvist–Zeilberger algorithm that passing from rational solutions to polynomial solutions via differential Gosper forms. However, we give a more direct way via Proposition 2. A similar result in the discrete case for hypergeometric terms was shown by Chen et al. in [12].

All that remains is to look for a polynomial $a \in k[x]$ satisfying Eq. (7). To do this, we need an upper bound μ on the degree of a. The degree bound estimation will proceed in three cases.

Case 1. Assume that $\deg(u) > \deg(v) - 1$. Then the highest degree of all terms on the left hand-side of Eq. (7) is $\deg(u) + \deg(b) + \deg(a)$, which leads to

$$\mu := \deg(v) + \deg(b) - \deg(u).$$

Case 2. Assume that $\deg(u) = \deg(v) - 1$. Let

$$u = \text{lc}(u)x^{\deg(v)-1} + \text{lower terms} \quad \text{and} \quad v = \text{lc}(v)x^{\deg(v)} + \text{lower terms}.$$

2.1. The following equation has no nonnegative integer solution for μ

$$\mu \text{lc}(v) - \deg(b)\text{lc}(v) + \text{lc}(u) = 0. \tag{11}$$

Equating degrees on both sides of Eq. (7) leads to

$$\mu := \deg(b) + 1.$$

2.2. Equation (11) has a nonnegative integer solution μ_0 for μ. Then

$$\mu = \max\{\mu_0, \deg(b) + 1\}.$$

Case 3. Assume that $\deg(u) < \deg(v) - 1$. In this case, $\mu_0 = \deg(b)$ is a nonnegative integer solution of

$$\mu - \deg(b) = 0,$$

which is derived from computing the leading coefficient of $ba' - ab'$. So we can take

$$\mu = \max\{\mu_0, \deg(b) + 1\} = \deg(b) + 1.$$

Remark 5 In Case 2.2, we claim that Eq. (7) has no polynomial solution if $\mu_0 = \deg(b) + 1$. Suppose there exists a polynomial solution $a \in k[x]$. Note that the leading coefficient of the left hand-side of Eq. (7) is

$$\mathrm{lc}(a)\mathrm{lc}(b)(\deg(a)\mathrm{lc}(v) - \deg(b)\mathrm{lc}(v) + \mathrm{lc}(u)).$$

Suppose that $\deg(a) \neq \mu_0$. Then the degree of the left hand-side of Eq. (7) is not equal to $2\deg(b) + \deg(v)$, a contradiction. Suppose that $\deg(a) = \mu_0$. Then the degree of the left hand-side of Eq. (7) is less than $2\deg(b) + \deg(v)$, since μ_0 is a root of

$$\mu\mathrm{lc}(b)(\deg(a)\mathrm{lc}(v) - \deg(b)\mathrm{lc}(v) + \mathrm{lc}(u)).$$

This also leads to a contradiction. So the claim holds.

The procedure on bound estimation is similar to the one in [2]. However, in Remark 5, we give a necessary condition that can decide whether there exists a polynomial solution or not quickly. In a nutshell, the degree bound is essential to the coding of our algorithm. Now our algorithm is summarized as follows.

Algorithm 3 Input: A hyperexponential function h with certificate $f \in k(x)$. Output: A hyperexponential function g such that $g' = h$ if h is hyperexponential integrable. Otherwise, return NULL.

1. Write $f = u/v$ with $u, v \in k[x]$ and $\gcd(u, v) = 1$;
2. Apply Algorithm 2 to compute a universal denominator b of Eq. (6);
3. Compute a polynomial solution a of Eq. (7);
4. Return $g = ah/b$ if Step 3 finds a polynomial solution. Otherwise, return NULL.

We illustrate our algorithm with an example.

Example 2 Given a hyperexponential function $h = (1+2x)\exp(x)/(2\sqrt{x})$. First,

$$\frac{h'}{h} = \frac{4x^2+4x-1}{2x(2x+1)},$$

which gives $u = 4x^2+4x-1$ and $v = 2x(2x+1)$. Secondly, we will consider the irreducible factors of $v_1 = v$, which are x and $2x+1$. Then

$$\mathrm{rem}(4x^2+4x-1-4\lambda x, 2x+1) = 0$$

yields $\lambda = 1$ while

$$\mathrm{rem}(4x^2+4x-1-2\lambda(2x+1), x) = 0$$

yields $\lambda = 1/2$. So the universal denominator b is $2x+1$. Finally, we try to find a polynomial solution of the equation

$$(4x^2-1)a + 2(2x^2+x)a' = 2x(4x^2+4x+1),$$

and the upper bound on the degree of polynomial solutions is $\deg(v)+\deg(b)-\deg(u) = 1$ because $\deg(u) = 2 > \deg(v)-1 = 1$. Using the method of undetermined coefficients we get $a = 2x$. Hence h is hyperexponential integrable and $g = \sqrt{x}\exp(x)$.

4 Conclusion

We have implemented our algorithms in **Maple** 13. According to the experiments, we find that the Almkvist-Zeilberger algorithm together with our improvements outperforms the method in Sect. 3. The reason is that the previous method predicts part of the numerator of rational solutions via differential Gosper forms. By this prediction, the linear system solved for polynomial solutions is much smaller than that in the second method. This motivates the question for future work whether one can refine Abramov's algorithm to predict part of the numerator when finding rational solutions of a linear differential equation with polynomial coefficients of any order.

Acknowledgments The author would like to thank the anonymous referees for their constructive and helpful comments.

References

1. Gosper Jr, R.W.: Decision procedure for indefinite hypergeometric summation. Proc. Nat. Acad. Sci. U.S.A. **75**(1), 40–42 (1978)
2. Almkvist, G., Zeilberger, D.: The method of differentiating under the integral sign. J. Symbolic Comput. **10**, 571–591 (1990)

3. Zeilberger, D.: The method of creative telescoping. J. Symbolic Comput. **11**(3), 195–204 (1991)
4. Petkovšek, M., Wilf, H.S., Zeilberger, D.: $A = B$. With a Foreword by Donald E. Knuth. A K Peters Ltd., Wellesley (1996)
5. Geddes, K.O., Le, H.Q., Li, Z.: Differential rational normal forms and a reduction algorithm for hyperexponential functions. In: ISSAC'04: Proceedings of the 2004 International Symposium on Symbolic and Algebraic Computation, pp. 183–190. ACM, New York (2004)
6. Abramov, S.A.: Rational solutions of linear differential and difference equations with polynomial coefficients. USSR Comput. Math. Math. Phys. **29**(6), 7–12 (1989)
7. Gerhard, J.: Modular Algorithms in Symbolic Summation and Symbolic Integration (LNCS). Springer, Berlin (2004)
8. Hermite, C.: Sur l'intégration des fractions rationnelles. Ann. Sci. École Norm. Sup. **2**(1), 215–218 (1872)
9. Horowitz, E.: Algorithms for partial fraction decomposition and rational function integration. In: Proceedings ofSYMSAC'71, pp. 441–457. ACM, New York (1971)
10. Ostrogradskiĭ, M.V.: De l'intégration des fractions rationnelles. Bull. de la classe physico-mathématique de l'Acad. Impériale des Sciences de Saint-Pétersbourg **4**(145–167), 286–300 (1845)
11. Bronstein, M.: Symbolic Integration I: Transcendental Functions, Volume 1 of Algorithms and Computation in Mathematics, 2nd edn. Springer, Berlin (2005)
12. Chen, W.Y.C., Paule, P., Saad, H.L.: Converging to Gosper's algorithm. Adv. Appl. Math. **41**(3), 351–364 (2008)

ImUp: A Maple Package for Uniformity-Improved Reparameterization of Plane Curves

Jing Yang, Dongming Wang and Hoon Hong

Abstract We present a software package for computing piecewise rational reparameterizations of parametric plane curves, which have improved uniformities of angular speed. The package ImUp is implemented in Maple on the basis of some recently developed algorithms of reparameterization using piecewise Möbius transformations. We discuss some implementation issues and illustrate the capability and performance of the public functions of ImUp with examples and experiments. It is shown that the quality of plots of plane curves may be effectively improved by means of reparameterization using ImUp.

Keywords Parametric plane curve · Angular speed uniformity · Möbius transformation · ImUp package · Piecewise reparameterization

1 Introduction

This paper presents the main functions of a Maple package for computing reparameterizations of given rational parametric plane curves with improved uniformities. Each computed reparameterization may be composed of one or several C^0 or C^1 continuous pieces of rational parameterizations and its uniformity is either

---

J. Yang (✉)
LMIB – School of Mathematics and Systems Science, Beihang University,
Beijing 100191, China
e-mail: yangjing@smss.buaa.edu.cn

D. Wang
Laboratoire D'Informatique de Paris 6, CNRS – Université Pierre et Marie Curie,
4 Place Jussieu, BP 169, 75252 Paris Cedex 05, France
e-mail: Dongming.Wang@lip6.fr

H. Hong
Department of Mathematics, North Carolina State University, Box 8205,
Raleigh 27695, USA
e-mail: hong@ncsu.edu

© Springer-Verlag Berlin Heidelberg 2014

R. Feng et al. (eds.), *Computer Mathematics*, DOI 10.1007/978-3-662-43799-5_29

optimal or close to a given value. The package named ImUp (standing for **Im**proving **U**niformities of **p**arameterizations and pronounced as "I'm Up") is implemented on the basis of the algorithms described in the three papers by the authors [1–3].

Parameterizations of curves and surfaces are commonly used for plotting (e.g., in computer-aided geometric design) and their uniformities directly affect the quality of the plots. For example, a parameterization of a plane curve will generate plotting points with a natural distribution (i.e., the more the curve bends, the bigger the density of points is, and vice versa) if the angular speed of the parameterization is uniform. It is thus necessary to compute reparameterizations with desired uniformities from given parameterizations. This computational problem has been studied by a number of researchers using various techniques (see [1–8] and references therein). However, we are not aware of any software tools or packages, which are publicly available for the problem of uniform reparameterization of curves and/or surfaces.

Given a parameterization p of a rational plane curve with angular speed uniformity u, the ImUp package is capable of computing a piecewise rational reparameterization p^* with angular speed uniformity u^* satisfying two kinds of requirements: (1) p^* has a given number of pieces with C^0 continuity and u^* is optimal; (2) p^* is C^1 continuous and u^* is close to a given objective uniformity as much as possible. The reparameterization is standard if it is composed of only one piece. The underlying algorithms used for computing the reparameterization, based on standard, C^0 and C^1 piecewise Möbius transformations and some (constrained) optimization techniques, are rather efficient and can improve the uniformity considerably in most cases. The package also contains functions for computing the angular speed, uniformity and optimal Möbius transformation of any parameterization of a rational plane curve and for plotting the curve using the (piecewise) parameterization.

The rest of the paper is structured as follows. Section 2 provides a brief review of the methods and techniques of reparameterization implemented in the package ImUp. We discuss some implementation issues in Sect. 3 and illustrate the usage of ImUp's public functions in Sect. 4. Examples and experimental results are provided in Sect. 5 to show the improvement of uniformity and quality of plots by reparameterization.

2 Problems and Methods

In this section, we formulate the main problems of uniformity-improved reparameterization which the package ImUp is developed to address and explain the methods and techniques implemented in the package. We start with some concepts and notations.

2.1 Terminology and Problems

For any rational parametric curve

$$p = (x(t), y(t)), \quad t \in [0, 1],$$

we recall its angle, angular speed, average angular speed, and uniformity of angular speed defined respectively as follows:

$$\theta_p = \arctan \frac{y'}{x'}, \quad \omega_p = |\theta'_p|, \quad \mu_p = \int_0^1 \omega_p \, dt, \quad u_p = \frac{\mu_p^2}{\mu_p^2 + \int_0^1 (\omega_p - \mu_p)^2 \, dt}.$$

The uniformity u_p ranges over $(0, 1]$. When $u_p = 1$, p is called an *arc-angle parameterization.*

For a proper parameter transformation $r(s)$, $p \circ r$ is a reparameterization of p with angular speed

$$\omega_{p \circ r} = (\omega_p \circ r)(s) \cdot r'(s) \tag{1}$$

and uniformity

$$u_{p \circ r} = \mu_p^2 \Big/ \int_0^1 \frac{\omega_p^2}{(r^{-1})'}(t) \, dt,$$

where r^{-1} is the inverse function of r. Through a proper parameter transformation r_p satisfying

$$(r_p)^{-1} = \int_0^t \omega_p(\gamma) \, d\gamma \Big/ \mu_p,$$

one can convert p into an arc-angle parameterization. Except in the case when p is a straight line, r_p is always irrational. Therefore, the problem of finding a rational reparameterization of p via parameter transformation is essentially to look for a rational approximation of r_p.

The method proposed in [1–3] allows us to compute an approximation of r_p using (piecewise) Möbius transformations. A *piecewise Möbius transformation* is defined as a map m of the following form

$$m(s) = \begin{cases} \vdots \\ m_i(s), & \text{if } s \in [s_i, s_{i+1}]; \\ \vdots \end{cases} \tag{2}$$

where

$$\begin{aligned} m_i(s) &= t_i + \Delta t_i \frac{(1 - \alpha_i)\tilde{s}_i}{(1 - \alpha_i)\tilde{s}_i + (1 - \tilde{s}_i)\alpha_i}, \\ \tilde{s}_i &= (s - s_i)/\Delta s_i, \quad \Delta t_i = t_{i+1} - t_i, \quad \Delta s_i = s_{i+1} - s_i, \quad 0 \le i \le N-1; \\ 0 &= t_0 < \cdots < t_N = 1, \quad 0 = s_0 < \cdots < s_N = 1, \quad 0 < \alpha_0, \ldots, \alpha_{N-1} < 1. \end{aligned} \tag{3}$$

The piecewise Möbius transformation m defined above is C^0 continuous. When $N = 1$, it degenerates to an *α-Möbius transformation* [3]. If m satisfies

$$m_i'(s_{i+1}) = m_{i+1}'(s_{i+1})$$

for $0 \le i \le N - 2$, it becomes a *C^1 piecewise Möbius transformation.*

The problems which the ImUp package can solve may be formulated as follows.

(1) Given a rational parameterization p of a plane curve and a positive integer N, compute an optimal C^0 piecewise reparameterization of p with N pieces.
(2) Given a rational parameterization p of a plane curve and an objective uniformity $\bar{u}$, compute a C^1 piecewise reparameterization of p whose uniformity is close to $\bar{u}$.

In the following subsections, we shall explain how to solve these two problems using the reparameterization methods by piecewise Möbius transformations described in [1, 2]. The methods work for $N > 1$ under the condition that ω_p is nonzero on [0, 1], so this condition is assumed.

2.2 *Optimal C^0 Piecewise Reparameterization with Fixed Number of Pieces*

For any given p and integer $N \geq 1$ as before, the method presented in [1] can compute a locally optimal partition $T = (t_0, \ldots, t_N)$ and globally optimal $S = (s_0, \ldots, s_N)$ and $\alpha = (\alpha_0, \ldots, \alpha_{N-1})$ for constructing an optimal C^0 piecewise Möbius transformation m (which has $3N - 2$ free parameters). The computed triple (T, S, α) has to maximize u_{pom}.

As μ_p is a constant, maximizing u_{pom} is equivalent to minimizing

$$\eta_{p,m} = \int_0^1 \frac{\omega_p^2}{(m^{-1})'}(t)\,dt$$

over (T, S, α). It is shown in [1] that if T is arbitrary but fixed, then there is a unique critical point (T, S_T, α_T) at which $\eta_{p,m}$ reaches its global minimum η_{p,m_T}; in this case, S_T and α_T are optimal and can be expressed in terms of T, and so can η_{p,m_T}. Therefore, the problem of finding an optimal triple (T, S, α) can be reduced to that of optimizing η_{p,m_T} over T. The latter is a problem of optimization under linear inequality constraints and can thus be solved by using Zoutendijk's method of feasible directions [9].

An optimal C^0 piecewise Möbius transformation can be easily constructed from the computed optimal (T, S, α) according to (2) and (3).

2.3 C^1 Piecewise Reparameterization with Objective Uniformity

It is difficult to find an exact optimal solution to the problem of minimizing $\eta_{p,m}$ over (T, S, α) for C^1 piecewise reparameterization because the reformulated objective function has a highly nonlinear dependency on the involved variables. A method is proposed in [2] for computing a triple (T, S, α) which provides an optimal solution for the problem that is not exact, but is close to the exact one. This method consists of two main steps.

1. Construct a C^1 piecewise Möbius transformation as follows.
   1.1. Compute a partition T of the unit interval [0, 1] by solving $\omega_p'(t) = 0$ for t over (0, 1).
   1.2. Choose $s_i = \int_0^{t_i} \omega_p \, dt \Big/ \mu_p$ $(1 \le i \le N-1)$ to obtain S (then T and S are almost optimal).
   1.3. Compute the exact optimal α using T and S.
   1.4. Construct a C^1 piecewise Möbius transformation from (T, S, α).
2. Use step 1 to construct a C^1 piecewise Möbius transformation m_1 for p, then m_2 for $p \circ m_1$, and so on. This process leads to a series $m_1, \ldots, m_n$ of C^1 piecewise Möbius transformations such that

$$p_n = p_{n-1} \circ m_n = p \circ m_1 \circ \cdots \circ m_n$$

   has its uniformity close to the given objective uniformity.

The method explained above involves computations with floating-point numbers, so the results of computation are not always guaranteed to be complete. To avoid using floating-point numbers for some computations, an alternative approach is proposed in [2] to refine the partition T through an iteration process. The iteration terminates when the objective uniformity is reached, or only minor improvement on the uniformity can be made. From (T, S, α) computed with the refined partition, a C^1 piecewise reparameterization of p whose uniformity is close to the objective uniformity can be easily constructed as well according to step 1.

3 Implementation

The package ImUp has been implemented in the computer algebra system Maple[1] (and it could also be done in other popular systems like Mathematica[2] and Matlab[3]). In this section, we discuss some of the implementation issues.

---

[1] http://www.maplesoft.com/products/maple/.

[2] http://www.wolfram.com/mathematica/.

[3] http://www.mathworks.com/products/matlab/.

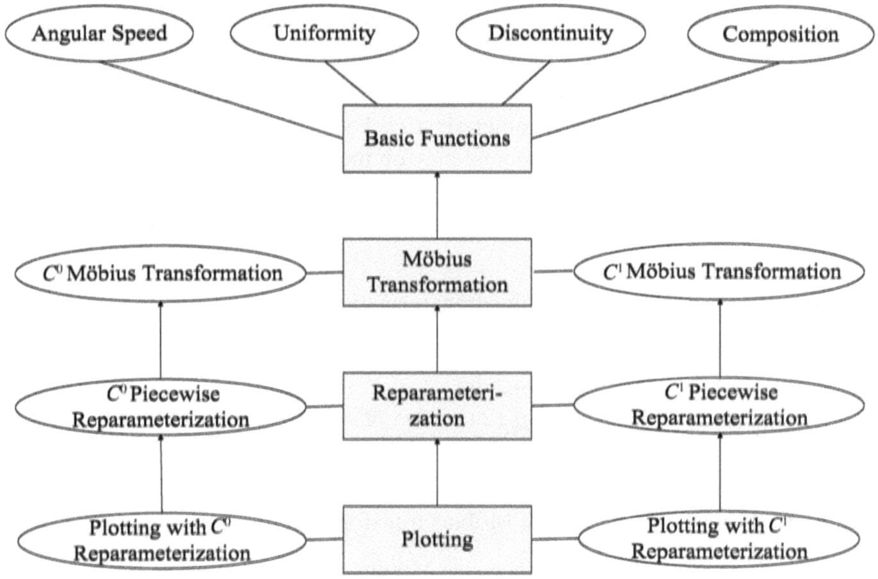

Fig. 1 ImUp structure

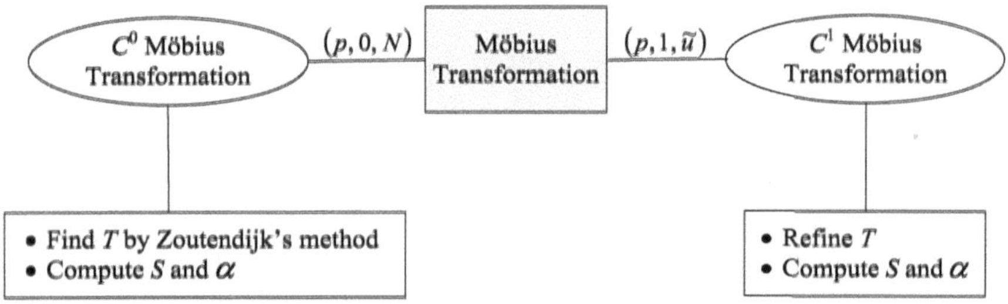

Fig. 2 Module of Möbius transformation

3.1 *Architecture of* ImUp

Figure 1 shows the block-diagram structure of the package ImUp which consists of four modules: basic functions, Möbius transformation, reparameterization, and plotting (of which the Möbius transformation module is essential).

There are four basic functions which compute the composition $p \circ m$, ω_p/ω_{pom}, u_p/u_{pom}, and the discontinuity of $p \circ m$ when m is piecewise, for any given parameterization p and Möbius transformation m. The Möbius transformation module allows one to compute piecewise Möbius transformations in different ways (see Fig. 2). Once a piecewise Möbius transformation is computed, one can construct a reparameterization of p using the reparameterization module. The plotting module may be used to plot the curve using its reparameterization.

3.2 Data Structure

In the programming, we have used the following simple data structure

$$[[t_0, \ldots, t_N], [s_0, \ldots, t_N], [\alpha_0, \ldots, \alpha_{N-1}]],$$

instead of a piecewise linear fractional function, to represent a piecewise Möbius transformation m. This is because the parameters in the transformation are frequently used during the computation and directly providing these parameters (instead of extracting them from the fractional-function representation of m) may help reduce computing time; whereas the fractional-function representation of m, as well as the reparameterization using m, can be easily constructed from these parameters.

3.3 Technical Considerations

We implemented the reparameterization algorithms initially in a straightforward and naive manner, but the resulting program did not perform well. We then examined the program, with attention to efficiency and reliability, and enhanced its performance after having resolved a few causes identified. In what follows, we discuss these causes and their resolution.

3.3.1 Angular Speed

As defined before, $\omega_p \equiv |\theta_p'|$. Since $\omega_p \neq 0$ over [0, 1], $\omega_p = \theta_p'$ if $\theta_p' > 0$, or $\omega_p = -\theta_p'$ if $\theta_p' < 0$. Hence, for integration and computing the average angular speed, ω_p may be replaced by θ_p' because the terms in the integrands are ω_p^2 and

$$\mu_p = \left| \int_0^1 \theta_p' \, dt \right| = \int_0^1 |\theta_p'| \, dt.$$

Therefore, θ_p' can be used instead of $|\theta_p'|$. By getting rid of the absolute value symbol in this way, one can avoid solving nonpolynomial equations during the refinement of interval partition for the C^1 piecewise reparameterization.

3.3.2 Zoutendijk's Method

For the C^0 piecewise reparameterization, the main computational problem is to search for the locally optimal partition of [0, 1] using the Zoutendijk's method of feasible directions. This optimization problem may be stated precisely as follows.

Given $\phi = \sum_{k=0}^{N-1} \sqrt{\Delta t_k (2\sqrt{A_k C_k} + B_k)}$,
where

$$A_k = \int_{t_k}^{t_{k+1}} \omega_p^2 \cdot (1-\tilde{t})^2 \, dt, \quad B_k = \int_{t_k}^{t_{k+1}} \omega_p^2 \cdot 2t \cdot (1-\tilde{t}) \, dt,$$

$$C_k = \int_{t_k}^{t_{k+1}} \omega_p^2 \cdot \tilde{t}^2 \, dt \quad \text{and} \quad \tilde{t} = (t - t_k)/\Delta t_k,$$

find $T^* = (t_0^*, \ldots, t_N^*)$ such that ϕ is locally optimal with respect to the sequence T^*, subject to the linear constraints $0 = t_0 < t_1 < \cdots < t_{N-1} < t_N = 1$.

The constraints in the above problem are different from those in typical optimization problems which Zoutendijk's method is used to solve: the feasible region is an open set for our problem, while it is closed for others. The openness of the feasible set may cause computational overflow when the objective function ϕ is evaluated on the boundary of the feasible set. This issue can be resolved by modifying Zoutendijk's method in the following way.

One main step in Zoutendijk's algorithm is to find λ^* such that

$$\min_{0 \le \lambda \le \lambda_{\max}} \phi(T^{(k)} + \lambda T_d^{(k)})$$

is reached at $\lambda = \lambda^*$, where $T^{(k)}$ is the initial T, $T_d^{(k)}$ is the feasible direction T_d at the kth iteration and $\lambda_{\max}$ is the maximum admissible step length. Due to the openness of the feasible set, $\lambda_{\max}$ cannot be reached, so λ ranges over $[0, \lambda_{\max})$. The one-dimensional search we introduced can always find an almost optimal step length λ. It is different from the one-dimensional search used in Zoutendijk's method and may be described as follows. First divide $[0, \lambda_{\max})$ into N_λ equidistant partitions with list of nodes

$$[i\,\lambda_{\max}/N_\lambda : \ 1 \le i \le N_\lambda - 1]$$

and then evaluate the objective function ϕ at each node. Once $T^{(k)} + \lambda T_d^{(k)}$ goes out of the feasible set, the evaluation process terminates. The optimal ϕ and λ can be chosen from the values of ϕ at the partition nodes. If $\lambda = \lambda_{\max}/N_\lambda$ (the smallest node) cannot keep $T^{(k)} + \lambda T_d^{(k)}$ in the feasible set, then increase N_λ to reduce the step length until the step length becomes admissible.

Linear inequality constraints may be expressed as $AT \ge b$. Another main step in Zoutendijk's method is to decompose the matrix A and the vector b for the corresponding constraints into $[A_1, A_2]$ and $[b_1, b_2]$ such that $A_1 T^{(k)} = b_1$ and $A_2 T^{(k)} > b_2$, where $T^{(k)}$ is the initial T at the kth iteration. This step may be skipped for our problem because the feasible set is open and $AT^{(k)} > b$ is always guaranteed in the iteration. It is easy to see that A_1 is null and its nullity can also be used to simplify the constraints

$$A_1 T_d \geq 0, \quad -1 \leq T_d^{\langle i \rangle} \leq 1, \quad 1 \leq i \leq N-1$$

for $\min \nabla\phi(T^{(k)})^T T_d$ in the next step, where $T_d^{\langle i \rangle}$ denotes the ith element of T_d.

Using a modified version of Zoutendijk's method, we may compute a locally optimal T and calculate the globally optimal S and α in the C^0 piecewise reparameterization.

3.3.3 Discontinuity of Angular Speed

For any piecewise function $f(x)$ over $[0, 1]$ with partition nodes $x_i, 0 = x_0 < \cdots < x_N = 1$, we define the local discontinuity of $f(x)$ at $x = x_i$ $(1 \leq i \leq N-1)$ to be

$$D_i(f) = 2\,\frac{|f(x_i^+) - f(x_i^-)|}{|f(x_i^+)| + |f(x_i^-)|}$$

and the (global) discontinuity of $f(x)$ over $[0, 1]$ to be

$$D(f) = \sum_{i=1}^{N-1} D_i(f)/(N-1).$$

Let p, m and t_i, s_i be as before. As ω_p is assumed to be nonzero, $\omega_{p\circ m}$ is also nonzero over $[0, 1]$. By (1) and the relation $\omega_p(t_i^+) = \omega_p(t_i^-) = \omega_p(t_i)$, the local discontinuity for the angular speed of $q = p \circ m$ at $s = s_i$ is

$$\begin{aligned} D_i(\omega_q) &= 2\,\frac{|\omega_q(s_i^+) - \omega_q(s_i^-)|}{\omega_q(s_i^+) + \omega_q(s_i^-)} = 2\,\frac{|\omega_p(t_i^+)m'(s_i^+) - \omega_p(t_i^-)m'(s_i^-)|}{\omega_p(t_i^+)m'(s_i^+) + \omega_p(t_i^-)m'(s_i^-)} \\ &= 2\,\frac{|m'(s_i^+) - m'(s_i^-)|}{m'(s_i^+) + m'(s_i^-)} = D_i(m'). \end{aligned}$$

Therefore,

$$D(\omega_q) = \sum_{i=1}^{N-1} D_i(\omega_q)/(N-1) = \sum_{i=1}^{N-1} D_i(m')/(N-1) = D(m').$$

This shows that the discontinuity of ω_q is equal to that of m', so computing $D(\omega_q)$ may be reduced to computing $D(m')$ (the latter is more efficient).

4 Public Interface

This section presents the public functions of the ImUp package and their usage. It serves as a quick reference manual for the users of the package. For simplicity, we use N*-piecewise* to indicate that the number of pieces in the piecewise Möbius transformation or reparameterization is N. In what follows, p is a given rational parameterization of a plane curve.

4.1 *AngularSpeed*

The calling sequence is

$$\text{AngularSpeed}(p, m)$$

where the second argument is optional. It computes the angular speed of p or a reparameterization $p \circ m$, where m is a piecewise Möbius transformation.

4.2 *Uniformity*

The calling sequence is

$$\text{Uniformity}(p, m)$$

where the second argument is optional. It computes the uniformity of the angular speed of p or a reparameterization $p \circ m$, where m is a piecewise Möbius transformation.

4.3 *MoebiusTransformation*

The calling sequence is

$$\text{MoebiusTransformation}(p, \text{opt}, N|\bar{u})$$

where the second argument is 0 or 1 for C^0 or C^1 piecewise Möbius transformation respectively. The third argument depends on the second: it is N when opt $= 0$, or $\bar{u}$ when opt $= 1$. It computes an optimal C^0 N-piecewise Möbius transformation, or a C^1 piecewise Möbius transformation from which the constructed reparameterization of p has uniformity close (as much as possible) to the objective uniformity $\bar{u}$.

4.4 ReparameterizationN

The calling sequence is

$$\text{ReparameterizationN}(p, N).$$

It computes an optimal C^0 N-piecewise reparameterization of p.

4.5 ReparameterizationU

The calling sequence is

$$\text{ReparameterizationU}(p, \bar{u}).$$

It computes a C^1 piecewise reparameterization of p whose uniformity is close (as much as possible) to the objective uniformity $\bar{u}$.

4.6 ImUpPlotN

The calling sequence is

$$\text{ImUpPlotN}(p, N, N_{\text{pt}}).$$

It plots the curve p using an optimal C^0 N-piecewise reparameterization of p with N_{pt} points.

4.7 ImUpPlotU

The calling sequence is

$$\text{ImUpPlotU}(p, \bar{u}, N_{\text{pt}}).$$

It plots the curve p using a C^1 piecewise reparameterization of p whose uniformity is close (as much as possible) to the objective uniformity $\bar{u}$ with N_{pt} points.

The ImUp package will be available as http://DongmingWang.org/ImUp.zip for download.

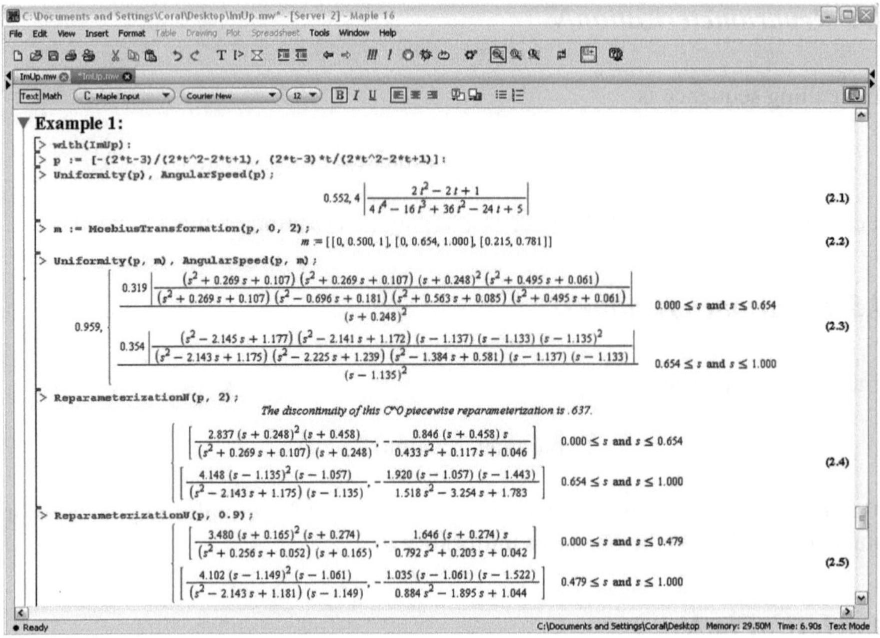

Fig. 3 Example session for the usage of ImUp functions

5 Examples and Experiments

In this section, we illustrate the usage of ImUp functions by means of example sessions and provide some experimental results to show the performance of the package.

5.1 Examples

Figure 3 is a Maple session which shows the usage of some ImUp functions. Figure 4 displays three plots of a curve segment using three plotting functions and with the same number of plotting points. The plot generated by ImUpPlotU is the best and the one generated by ImUpPlotN looks better than the one generated by Maple's built-in function. The uniformity of the piecewise reparameterization, which is composed of six pieces and computed by ReparameterizationU and whose plot is shown on the right-hand side, is 0.992, while the uniformity of the piecewise reparameterization, which has only two pieces and is computed by ReparameterizationN and whose plot is shown in the middle, is 0.737.

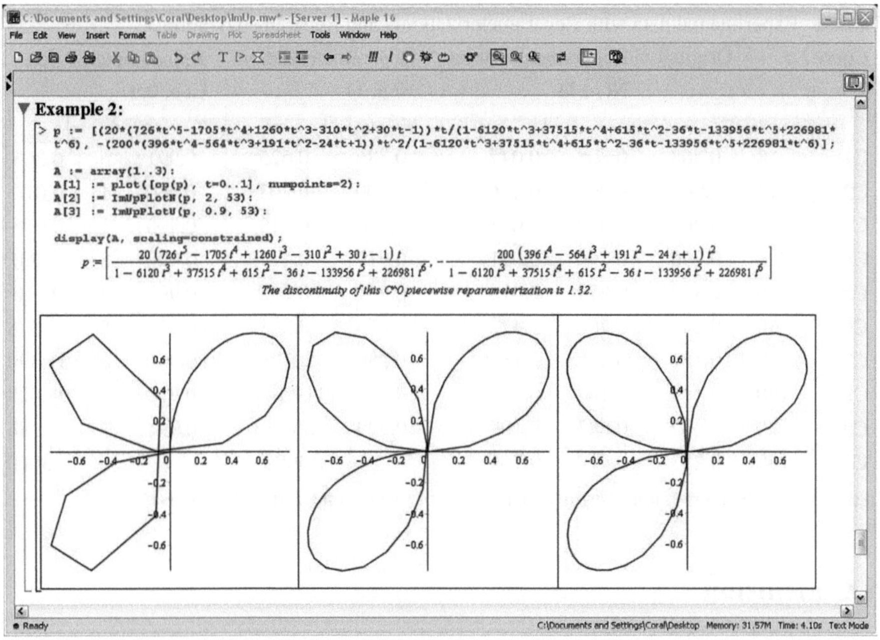

Fig. 4 Plots of a curve segment using Maple's built-in function plot (left), ImUpPlotN (middle) and ImUpPlotU (right). Note that "numpoints" in the function plot is *not the exact* number of points generated for the plot. It specifies the minimum number of points to be generated and the plotting function often generates more points. For this example with numpoints = 2, the number of points generated for the plot is 53

5.2 Experiments

We have tested our implementation in Maple 16 on a PC Intel(R) Core(TM)2 Duo CPU P8400 @2.26 GHz with 2 G of RAM using a number of plane curves. Some experimental results for the functions ReparameterizationN and ReparameterizationU with 10 curves are given in Table 1, where the curves 1–5 are taken from [10] with specialized values of parameters and the curves 6–10 are randomly generated by Maple with fixed degree.

From Table 1, it may be seen that both of the reparameterization functions can improve the uniformity of angular speed to a certain extent and the numbers of pieces are also acceptable. These two functions can deal with curves of degree as high as 100. They should be able to meet the need of applications.

Table 1 Experimental data with ReparameterizationN and ReparameterizationU

Curve	d	Original u	Reparameterization N(p, 1)		Reparameterization N(p, 3)		Reparameterization U(p, 0.9)		
			u	T	u	T	u	N	T
C1	8	0.808	0.808	0.047	0.886	0.625	0.991	6	0.265
C2	12	0.960	0.960	0.016	1.000	0.203	0.960	1	0.016
C3	6	0.906	0.919	0.031	1.000	0.219	0.906	1	0.016
C4	6	0.796	0.796	0.031	0.911	0.437	0.996	8	0.250
C5	6	0.879	0.879	0.015	0.961	0.328	0.997	4	0.110
C6	50	0.647	0.706	1.422	0.973	8.203	0.971	5	10.578
C7	70	0.181	0.418	3.157	0.550	20.204	0.960	6	31.390
C8	100	0.184	0.184	6.188	0.989	119.141	0.970	2	95.828
C9	120	0.682	0.683	9.109	0.999	66.813	0.994	2	37.188
C10	150	0.253	0.479	29.765	–	>3000	–	–	>3000

In the table, d is degree, u uniformity, N number of pieces, and T time (seconds)

6 Conclusion

We have presented a software package ImUp implemented in Maple for computing piecewise rational reparameterizations of plane curves. The ImUp public functions are described, their usage is illustrated by examples and some implementation issues are discussed. Plots generated by ImUp functions appear better than their corresponding plots generated by Maple's built-in function in terms of smoothness. Experimental results show that the reparameterization functions in ImUp have good performance. The package is being extended by implementing the general framework developed in [11] on angular speed for rational parameterizations of space curves.

Acknowledgments This work was supported partially by the Open Fund of SKLSDE (Grant No. SKLSDE-2011KF-02) and ANR-NSFC Project EXACTA (Grant No. ANR-09-BLAN-0371-01/60911130369).

References

1. Yang, J., Wang, D., Hong, H.: Improving angular speed uniformity by optimal C^0 piecewise reparameterization. In: Gerdt, V.P., Koepf, W., Mayr, E.W., Vorozhtsov E.V. (eds.) Proceedings of the 14th International Workshop on Computer Algebra in Scientific Computing (Maribor, Slovenia, September 3–6, 2012), pp. 349–360, LNCS 7442, Springer, Heidelberg (2012)
2. Yang, J., Wang, D., Hong, H.: Improving angular speed uniformity by C^1 piecewise reparameterization. In: Ida, T., Fleuriot, J. (eds.) Proceedings of the 9th International Workshop on Automated Deduction in Geometry (Edinburgh, UK, September 17–19, 2012), pp. 33–47. Springer, Berlin/Heidelberg (2013)
3. Yang, J., Wang, D., Hong, H.: Improving angular speed uniformity by reparameterization. Comput. Aided Geom. Des. **30**(7), 636–652 (2013)

4. Costantini, P., Farouki, R., Manni, C., Sestini, A.: Computation of optimal composite reparameterizations. Comput. Aided Geom. Des. **18**(9), 875–897 (2001)
5. Farouki, R.: Optimal parameterizations. Comput. Aided Geom. Des. **14**(2), 153–168 (1997)
6. Jüttler, B.: A vegetarian approach to optimal parameterizations. Comput. Aided Geom. Des. **14**(9), 887–890 (1997)
7. Liang, X., Zhang, C., Zhong, L., Liu, Y.: C^1 continuous rational reparameterization using monotonic parametric speed partition. In: Proceedings of the 9th International Conference on Computer-Aided Design and Computer Graphics (Hong Kong, December 7–10, 2005), pp. 16–21, IEEE Computer Society (2005)
8. Patterson, R., Bajaj, C.: Curvature adjusted parameterization of curves. Computer Science Technical Report CSD-TR-907, Paper 773, Purdue University, USA (1989)
9. Zoutendijk, G.: Methods of Feasible Directions: A Study in Linear and Nonlinear Programming. Elsevier Publishing Company, Amsterdam/New York (1960)
10. Curves of high degree: http://www.geometryatlas.com/categories/Curves-Curves_of_High_Degree/all (2013). Accessed 11 September 2013
11. Hong, H., Wang, D., Yang, J.: A framework for improving uniformity of parameterizations of curves. Sci. China Inf. Sci. **56**(10), 108101:1–108101:22 (2013).

The Diagonal Reduction Algorithm Using Fast Givens

Wen Zhang, Sanzheng Qiao and Yimin Wei

Abstract Recently, a new lattice basis reduction notion, called diagonal reduction, was proposed for lattice-reduction-aided detection (LRAD) of multiinput multioutput (MIMO) systems. In this paper, we improve the efficiency of the diagonal reduction algorithm by using the fast Givens transformations. The technique of the fast Givens is applicable to a family of LLL-type lattice reduction methods to improve efficiency. Also, in this paper, we investigate dual diagonal reduction and derive an upper bound of the proximity factors for a family of dual reduction aided successive interference cancelation (SIC) decoding. Our upper bound not only extends an existing bound for dual LLL reduction to a family of dual reduction methods, but also improves the existing bound.

Keywords Lattice reduction · Diagonal reduction · Fast givens · Dual reduction · MIMO detection

---

973 Program Project Under Grant 2010CB327900.
National Natural Science Foundation of China Under Grant 11271084.

W. Zhang
School of Mathematics and Physics, Qingdao University of Science and Technology, Qingdao 266000, People's Republic of China
e-mail: zhangwen9801@gmail.com

S. Qiao
Shanghai Key Laboratory of Contemporary Applied Mathematics, Department of Computing and Software, McMaster University, Hamilton, ON L8S 4K1, Canada
e-mail: qiao@cas.mcmaster.ca

Y. Wei (✉)
Shanghai Key Laboratory of Contemporary Applied Mathematics, School of Mathematical Sciences, Fudan University, Shanghai 200433, People's Republic of China
e-mail: ymwei@fudan.edu.cn

© Springer-Verlag Berlin Heidelberg 2014
R. Feng et al. (eds.), *Computer Mathematics*, DOI 10.1007/978-3-662-43799-5_30

1 Introduction

Lattice basis reduction plays an important role in the detection of wireless multiple-input multiple-output (MIMO) systems. For detection problems of lattice type, the optimal maximum-likelihood (ML) decoding can be modeled as the closest vector problem (CVP) [1, 16], which has been proved to be NP-hard [2]. Although many algorithms, like the sphere decoding algorithm [5, 14], can solve CVP exactly, the complexity of these algorithms increases exponentially with the number of transmit antennas [1, 5, 6]. Thus, such optimal solvers are infeasible for real-time systems, where timing is critical. To satisfy the time constraint, many suboptimal solvers with polynomial complexity, like the successive interference cancelation (SIC) decoding, have been proposed [3, 12]. However, the suboptimal detectors may suffer from heavy performance loss at a low signal-to-noise ratio (SNR). It has been found that lattice reduction, used as an efficient preprocessor, has the potential to achieve high performance for suboptimal decoding algorithms. Recently, many reduction algorithms, such as the Lenstra-Lenstra-Lovász (LLL) algorithm [7], effective LLL algorithm [10], partial LLL algorithm [11, 17], and diagonal reduction algorithm [19], have been proposed for SIC decoding. It is proved in [9, 18] that SIC decoding aided by the above reduction notions can achieve the same receive diversity order as the infinite lattice decoding (ILD).

Of all the aforementioned lattice reduction algorithms, the diagonal reduction algorithm is the most efficient one. From our observation [19], the total computation of the diagonal reduction is dominated by the computation of the Givens rotations. Thus, in this paper, we propose to improve the efficiency of the diagonal reduction by replacing the Givens rotation with the more efficient and mathematically equivalent fast Givens transformation [4, p. 218]. The improvement is achieved by substantially reducing the number of multiplication operations required, because two entries of the 2-by-2 fast Givens matrix equal 1. Moreover, the fast Givens technique is general in that it can be incorporated into all the LLL-type lattice reduction methods to enhance performance.

Also, we investigate the basis reduction for dual lattices. In [9], the LLL and effective LLL algorithms for dual lattices are presented. In this paper, we investigate the diagonal reduction for dual lattices and prove that the dual basis of a diagonal reduced basis is also diagonal reduced. In addition, we derive an upper bound for the proximity factors of a family of dual LLL-type reduction aided SIC decoding. Our upper bound not only extends an existing bound for LLL reduction in [9] to a family of reduction methods, but also improves the existing one.

The rest of the paper is organized as follows. In Sect. 2, we briefly introduce the systems model and review the diagonal reduction algorithm. The new algorithm using the fast Givens is given in Sect. 3. Section 4 presents the diagonal reduction for dual lattices and our new upper bound for the proximity factors. In Sect. 5, we demonstrate our simulation results.

Notations: $\mathbf{B}^{\mathrm{T}}$, $\mathbf{B}^{\dagger}$, and $\det(\mathbf{B})$ denote the transpose, the Moore-Penrose inverse, and the determinant of a matrix $\mathbf{B}$ respectively, $\Re(z)$ and $\Im(z)$ the real and imaginary parts of a complex number z, $\lfloor a \rceil$ the integer nearest to a real number a.

2 Lattice Basis Reduction

2.1 System Model

Consider a MIMO system consisting of n_T transmit antennas and m_T receive antennas. The relationship between the $n_T \times 1$ transmitted signal vector $\mathbf{x}$ and the $m_T \times 1$ received signal vector $\mathbf{y}$ is given by

$$\mathbf{y} = \mathbf{H}\mathbf{x} + \mathbf{n}, \tag{1}$$

where $\mathbf{H}$, $\mathbf{y}$, $\mathbf{n}$ represent the channel matrix, the received and additive noise signals, respectively. In general, the entries of both $\mathbf{H}$ and $\mathbf{n}$ are assumed to be complex-valued independently and identically distributed (i.i.d.) Gaussian variables. Treating the real and imaginary parts of (1) separately, an equivalent real-valued system of doubled size can be obtained:

$$\mathbf{y} = \mathbf{B}\mathbf{x} + \mathbf{n}, \tag{2}$$

where

$$\mathbf{y} = \begin{bmatrix} \Re(\mathbf{y}) \\ \Im(\mathbf{y}) \end{bmatrix}, \quad \mathbf{n} = \begin{bmatrix} \Re(\mathbf{n}) \\ \Im(\mathbf{n}) \end{bmatrix}, \quad \mathbf{B} = \begin{bmatrix} \Re(\mathbf{H}) & -\Im(\mathbf{H}) \\ \Im(\mathbf{H}) & \Re(\mathbf{H}) \end{bmatrix}.$$

Given a MIMO system modeled as (2), the optimum ML decoding is equivalent to the following CVP:

$$\min_{\mathbf{x} \in \mathscr{A}} \|\mathbf{y} - \mathbf{B}\mathbf{x}\|_2. \tag{3}$$

where the constellation $\mathscr{A}$ is of lattice type. Unfortunately, CVP has been proved to be NP-hard [2], and all existing algorithms for solving (3) have an exponential complexity with the lattice dimension n [5, 6]. Recently, lattice-reduction-aided SIC decoding turned out to be extremely promising, since its bit-error-rate (BER) performance can effectively approximate the ML decoding with a complexity of only $O(n^3)$ operations [9, 15].

2.2 Diagonal Reduction Algorithm

In this section, we first introduce some concepts of lattices and the SIC decoding, then we describe the diagonal reduction method [19].

Given a matrix $\mathbf{B} \in \mathbb{R}^{m \times n}$ $(n \leq m)$ of full column rank, then a *lattice* generated by $\mathbf{B}$ is defined by $L(\mathbf{B}) = \{\mathbf{Bz} : \mathbf{z} \in \mathbb{Z}^n\}$. The columns of $\mathbf{B}$ form a *basis* for the lattice $L(\mathbf{B})$. An integer matrix $\mathbf{Z} \in \mathbb{Z}^{n \times n}$ is called *unimodular* if $|\det(\mathbf{Z})| = 1$. The columns of a matrix $\mathbf{B}'$ can form a basis for $L(\mathbf{B})$ if and only if there exists a unimodular matrix $\mathbf{Z}$ such that $\mathbf{B}' = \mathbf{BZ}$. The *volume* of $L(\mathbf{B})$ is defined as $\mathrm{vol}(L(\mathbf{B})) = \sqrt{\det(\mathbf{B}^{\mathrm{T}}\mathbf{B})}$, which is independent of the choice of basis. Let $\lambda(L)$ be the Euclidean length of the shortest nonzero vector in a lattice L, then it is well known that $\lambda(L)/\mathrm{vol}(L)^{1/n}$ is upper bounded over all n-dimension lattices L, and the *Hermite's constant* γ_n is defined as the supremum of $\lambda(L)^2/\mathrm{vol}(L)^{2/n}$ over all n-dimension lattices. Finding the exact value of γ_n is very difficult. The exact value of γ_n is only known for $1 \leq n \leq 8$ and $n = 24$ [13, p. 33]. For an arbitrary dimension n, an upper bound of the Hermite's constant is given in [13, p. 35]:

$$\gamma_n \leq 1 + \frac{n}{4}, \quad \text{for all } n \geq 1. \tag{4}$$

A *lattice reduction algorithm* finds a unimodular matrix $\mathbf{Z}$ for a given $\mathbf{B}$ such that the columns of $\mathbf{BZ}$ are reasonably short. Lattice reduction has now become a powerful tool for enhancing the performance of suboptimal MIMO detectors, since it can significantly improve the orthogonality of the channel matrix.

Given a lattice generator matrix $\mathbf{B} \in \mathbb{R}^{m \times n}$ and its QR decomposition $\mathbf{B} = \mathbf{QR}$, where $\mathbf{Q} \in \mathbb{R}^{m \times n}$ has orthonormal columns and $\mathbf{R} \in \mathbb{R}^{n \times n}$ is upper triangular. From [8, 17], the efficiency of sphere decoding and the performance of SIC decoding is determined by the arrangement of the diagonal elements of $\mathbf{R}$. Based on this fact, various reduction notions, such as the LLL reduction [7], effective LLL reduction [10], partial LLL reduction [11, 17], and diagonal reduction [19], have been proposed. Among all the aforementioned reduction notions, the diagonal reduction is the weakest, consequently, the least computationally demanding.

Definition 1 (*Diagonal reduction* [19]) A basis matrix $\mathbf{B} \in \mathbb{R}^{m \times n}$ is said to be diagonal reduced with the parameter ω $(1/4 < \omega < 1)$, if the entries $r_{i,j}$ of the upper triangular factor $\mathbf{R}$ in its QR decomposition $\mathbf{B} = \mathbf{QR}$ satisfy

$$(r_{k-1,k} - \mu_k r_{k-1,k-1})^2 + r_{k,k}^2 \geq \omega r_{k-1,k-1}^2, \tag{5}$$

for all $1 < k \leq n$, where $\mu_k = \lfloor r_{k-1,k}/r_{k-1,k-1} \rceil$.

From the above definition, diagonal reduction only imposes one simple constraint on the diagonal entries of $\mathbf{R}$. However, it is proved in [19] that diagonal-reduction-aided SIC decoding has identical performance as LLL-reduction-aided SIC decoding. A generic implementation of diagonal reduction can be found in Fig. 1.

Input: $\mathbf{Q} \in \mathbb{R}^{m\times n}$, $\mathbf{R} \in \mathbb{R}^{n\times n}$, ω
Output: Updated $\mathbf{Q}$ and updated $\mathbf{R}$ that is diagonal reduced with the parameter ω and a unimodular $\mathbf{Z}$ that reduces $\mathbf{R}$

1: Initialization $\mathbf{Z} \leftarrow \mathbf{I}_n$
2: $k \leftarrow 2$
3: **while** $k \leq n$ **do**
4:     $\mu_k \leftarrow \lfloor \mathbf{R}(k-1,k)/\mathbf{R}(k-1,k-1) \rceil$
5:     **if** the condition (5) is not satisfied **then**
6:         **if** $\mu_k \neq 0$ **then**
7:             $\mathbf{R}(1:k-1,k) \leftarrow \mathbf{R}(1:k-1,k) - \mu_k \mathbf{R}(1:k-1,k-1)$
8:             $\mathbf{Z}(:,k) \leftarrow \mathbf{Z}(:,k) - \mu_k \mathbf{Z}(:,k-1)$
9:         **end if**
10:         swap columns $k-1$ and k in $\mathbf{R}$ and $\mathbf{Z}$
11:         find a Givens rotation $\mathbf{G}$ to restore the upper triangular structure of $\mathbf{R}$
12:         $\mathbf{R}(k-1:k,k-1:n) \leftarrow \mathbf{G}\mathbf{R}(k-1:k,k-1:n)$
13:         $\mathbf{Q}(:,k-1:k) \leftarrow \mathbf{Q}(:,k-1:k)\mathbf{G}^{\mathrm{T}}$
14:         $k \leftarrow \max(k-1,2)$
15:     **else**
16:         $k \leftarrow k+1$
17:     **end if**
18: **end while**

Fig. 1 Diagonal reduction algorithm (DR) [19]

3 Diagonal Reduction Using Fast Givens

From Fig. 1, the computational cost of the diagonal reduction algorithm includes two parts: the size-reduction (lines 7–8) and the Givens rotation (lines 11–13). The simulation results in [19] indicate that the overall complexity of the algorithm is dominated by the Givens rotations as the lattice dimension n increases. Thus, we propose the use of the fast Givens transformation in place of the Givens rotations to speed up the diagonal reduction algorithm.

Like the Givens rotation, the fast Given can be used to introduce zeros into selected positions. Specifically, given a lattice generator matrix $\mathbf{B}$, the fast Givens transformation is based on the following decomposition:

$$\mathbf{B} = \mathbf{F}\mathbf{D}^{-1}\mathbf{R}, \tag{6}$$

where $\mathbf{D} = \mathrm{diag}(d_i)$ is a positive diagonal matrix, $\mathbf{F}\mathbf{D}^{-1/2}$ represents the orthogonal factor in the QR decomposition of $\mathbf{B}$, and $\mathbf{D}^{-1/2}\mathbf{R}$ represents the upper triangular factor.

How can the fast Givens introduce zeros? In the 2-by-2 case, given $\mathbf{x} = [x_1, x_2]^{\mathrm{T}}$ and the corresponding diagonal elements $d_1, d_2 > 0$, we first compute

$$\alpha = -x_1/x_2, \quad \beta = -\alpha d_2/d_1, \quad \text{and } \gamma = -\alpha\beta.$$

When $\gamma \leq 1$, we have the type 1 fast Givens:

$$\mathbf{F} = \begin{bmatrix} \beta & 1 \\ 1 & \alpha \end{bmatrix} \tag{7}$$

and update d_1 and d_2:

$$\widehat{d}_1 \leftarrow (1+\gamma)d_2 \quad \text{and} \quad \widehat{d}_2 \leftarrow (1+\gamma)d_1. \tag{8}$$

When $\gamma > 1$, setting

$$\alpha \leftarrow 1/\alpha, \quad \beta \leftarrow 1/\beta, \quad \text{and}\ \gamma \leftarrow 1/\gamma,$$

we have the type 2 fast Givens:

$$\mathbf{F} = \begin{bmatrix} 1 & \beta \\ \alpha & 1 \end{bmatrix} \tag{9}$$

and update d_1 and d_2:

$$\widehat{d}_1 \leftarrow (1+\gamma)d_1 \quad \text{and} \quad \widehat{d}_2 \leftarrow (1+\gamma)d_2. \tag{10}$$

Then it can be verified that

$$\mathbf{F}\begin{bmatrix} x_1 \\ x_2 \end{bmatrix} = \begin{bmatrix} \times \\ 0 \end{bmatrix}$$

and

$$\begin{bmatrix} \widehat{d}_1 & 0 \\ 0 & \widehat{d}_2 \end{bmatrix}^{-1/2} \mathbf{F} \begin{bmatrix} d_1 & 0 \\ 0 & d_2 \end{bmatrix}^{1/2}$$

is orthogonal.

In our fast Givens-based diagonal reduction algorithm, all the transformations are based on the decomposition (6). In the beginning, we compute the QR decomposition $\mathbf{B} = \mathbf{QR}$ and set $\mathbf{F} = \mathbf{Q}$ and $\mathbf{D} = \mathbf{I}_n$. Thus, in this case, the size-reduction in each iteration is the same as lines 7–8 of Fig. 1. But the diagonal reduction condition (5) becomes

$$d_{k-1}^{-1}(r_{k-1,k} - \mu_k r_{k-1,k-1})^2 + d_k^{-1} r_{k,k}^2 \geq \omega d_{k-1}^{-1} r_{k-1,k-1}^2, \tag{11}$$

for $1 < k \leq n$. The diagonal reduction algorithm using fast Givens (DRFG) is summarized in Fig. 2.

In comparison with the original diagonal reduction algorithm, DRFG saves a substantial number of multiplication operations, since two entries of the 2-by-2 fast Givens matrix are equal to 1. However, DRFG introduces overhead, such as the

Input: $\mathbf{Q} \in \mathbb{R}^{m\times n}$, $\mathbf{R} \in \mathbb{R}^{n\times n}$, ω
Output: Updated $\mathbf{Q}$ and updated $\mathbf{R}$ that is diagonal reduced with the parameter ω and a unimodular $\mathbf{Z}$ that reduces $\mathbf{R}$

1: $\mathbf{F} \leftarrow \mathbf{Q}$, $\mathbf{D} \leftarrow \mathbf{I}_n$, $\mathbf{Z} \leftarrow \mathbf{I}_n$
2: $k \leftarrow 2$
3: **while** $k \leq n$ **do**
4: $\quad \mu_k \leftarrow \lfloor \mathbf{R}(k-1,k)/\mathbf{R}(k-1,k-1) \rceil$
5: $\quad$ **if** the condition (11) is not satisfied **then**
6: $\quad\quad$ **if** $\mu_k \neq 0$ **then**
7: $\quad\quad\quad \mathbf{R}(1:k-1,k) \leftarrow \mathbf{R}(1:k-1,k) - \mu_k \mathbf{R}(1:k-1,k-1)$
8: $\quad\quad\quad \mathbf{Z}(:,k) \leftarrow \mathbf{Z}(:,k) - \mu_k \mathbf{Z}(:,k-1)$
9: $\quad\quad$ **end if**
10: $\quad\quad$ swap columns $k-1$ and k in $\mathbf{R}$ and $\mathbf{Z}$
11: $\quad\quad$ construct fast Givens matrix $\mathbf{F}$ of type (7) or type (9) such that $\mathbf{F}\begin{bmatrix} r_{k-1,k-1} \\ r_{k,k-1} \end{bmatrix} = \begin{bmatrix} \times \\ 0 \end{bmatrix}$
12: $\quad\quad \mathbf{R}(k-1:k,k-1:n) \leftarrow \mathbf{F}\mathbf{R}(k-1:k,k-1:n)$
13: $\quad\quad \mathbf{F}(:,k-1:k) \leftarrow \mathbf{F}(:,k-1:k)\mathbf{F}^{\mathrm{T}}$
14: $\quad\quad$ using (8) or (10) to update d_{k-1} and d_k
15: $\quad\quad k \leftarrow \max(k-1,2)$
16: $\quad$ **else**
17: $\quad\quad k \leftarrow k+1$
18: $\quad$ **end if**
19: **end while**
20: $\mathbf{Q} \leftarrow \mathbf{F}\mathbf{D}^{-1/2}$, $\mathbf{R} \leftarrow \mathbf{D}^{-1/2}\mathbf{R}$

Fig. 2 Diagonal reduction algorithm using fast givens (DRFG)

computations in line 14 and line 20. Our simulation results presented in Sect. 5 show that overall DRFG is more efficient than DR.

4 Dual Diagonal Reduction

In this section, after a brief introduction to dual basis, we first investigate diagonal reduction of dual basis and prove that if a primal basis is diagonal reduced, then its dual basis is also diagonal reduced. Then we derive an upper bound of proximity factor of SIC decoding, which not only improves an existing bound for the dual LLL reduction in [9], but also extends it to a family of dual LLL-type reductions.

4.1 Dual Lattice Reduction

Let L be an n-dimensional lattice in $\mathbb{R}^m$, then the dual lattice L^* of L is defined as the set

$$L^* = \{\mathbf{u} \mid \langle \mathbf{u}, \mathbf{v} \rangle \in \mathbb{Z}, \text{ for all } \mathbf{v} \in L\}, \tag{12}$$

where $\langle \mathbf{u}, \mathbf{v} \rangle$ is the inner product of $\mathbf{u}$ and $\mathbf{v}$. Suppose that $\mathbf{B}$ is a primal basis matrix of L, then it is obvious that the columns of $\mathbf{B}^{\dagger\mathrm{T}}$ form a basis for its dual lattice L^*. In this paper, we adopt the definition of the dual basis $\mathbf{B}^* \triangleq \mathbf{B}^{\dagger\mathrm{T}}\mathbf{J}$ [9], where

$$\mathbf{J} \triangleq \begin{bmatrix} 0 \cdots 0 \; 1 \\ 0 \cdots 1 \; 0 \\ \vdots \cdots \vdots \; \vdots \\ 1 \cdots 0 \; 0 \end{bmatrix}$$

A dual lattice is closely related to its corresponding primal lattice. For instance, we have $L^{**} = L$ and $\det(L^*) = 1/\det(L)$.

Given a primal basis matrix $\mathbf{B}$, then the dual basis reduction is to perform a lattice reduction algorithm on its dual basis $\mathbf{B}^*$. Like the primal basis reduction, dual basis reduction can also return a well reduced basis of the primal lattice. Suppose that $\mathbf{Z}^*$ is the unimodular matrix that reduces the dual basis $\mathbf{B}^*$, then the corresponding reduced primal basis is given by

$$\mathbf{B}' = (\mathbf{B}^{\dagger\mathrm{T}}\mathbf{J}\mathbf{Z}^*)^{\dagger\mathrm{T}}\mathbf{J} = \mathbf{B}\mathbf{J}(\mathbf{Z}^*)^{\dagger\mathrm{T}}\mathbf{J},$$

where $\mathbf{J}(\mathbf{Z}^*)^{\dagger\mathrm{T}}\mathbf{J}$ is the unimodular matrix associated with the primal lattice.

To study the reduction properties of diagonal reduction on dual lattices, the following result is essential.

Lemma 1 *Let* $\mathbf{B} = \mathbf{QR}$ *and* $\mathbf{B}^* = \mathbf{Q}^*\mathbf{R}^*$ *be the QR decompositions of the primal basis* $\mathbf{B}$ *and its dual basis* $\mathbf{B}^*$*, respectively. Then*

$$\mathbf{Q}^* = \mathbf{QJ}, \quad \mathbf{R}^* = \mathbf{J}\mathbf{R}^{-\mathrm{T}}\mathbf{J}. \tag{13}$$

Proof It is easy to verify that $\mathbf{B}^\dagger = \mathbf{R}^{-1}\mathbf{Q}^\mathrm{T}$. Thus, we have

$$\begin{aligned} \mathbf{B}^* &= \mathbf{B}^{\dagger\mathrm{T}}\mathbf{J} = (\mathbf{R}^{-1}\mathbf{Q}^\mathrm{T})^\mathrm{T}\mathbf{J} = \mathbf{Q}\mathbf{R}^{-\mathrm{T}}\mathbf{J} \\ &= (\mathbf{QJ}) \cdot (\mathbf{J}\mathbf{R}^{-\mathrm{T}}\mathbf{J}). \end{aligned} \tag{14}$$

Obviously, $\mathbf{QJ}$ has orthonormal columns and $\mathbf{J}\mathbf{R}^{-\mathrm{T}}\mathbf{J}$ is an upper triangular matrix, thus the proof is completed.

Based on the above lemma, we can obtain the following result.

Proposition 1 *If the lattice basis matrix* $\mathbf{B}$ *is diagonal reduced, then its dual basis* $\mathbf{B}^*$ *is also diagonal reduced.*

Proof Let $\mathbf{R} = [r_{i,j}]$ and $\mathbf{R}^*$ be the upper triangular factors of $\mathbf{B}$ and $\mathbf{B}^*$, respectively. Then from Lemma 1,

$$\mathbf{R}^* = \mathbf{J}\mathbf{R}^{-\mathrm{T}}\mathbf{J}$$
$$= \begin{bmatrix} \frac{1}{r_{n,n}} & -\frac{r_{n-1,n}}{r_{n-1,n-1}r_{n,n}} & \times & & \times \\ & \frac{1}{r_{n-1,n-1}} & -\frac{r_{n-2,n-1}}{r_{n-2,n-2}r_{n-1,n-1}} & & \times \\ & & \frac{1}{r_{n-2,n-2}} & \ddots & \vdots \\ & & & \ddots & -\frac{r_{1,2}}{r_{1,1}r_{2,2}} \\ & & & & \frac{1}{r_{1,1}} \end{bmatrix} \tag{15}$$

Since $\mathbf{B}$ is diagonal reduced, we then have

$$\left(r_{k-1,k} - \left\lfloor \frac{r_{k-1,k}}{r_{k-1,k-1}} \right\rceil \cdot r_{k-1,k-1}\right)^2 + r_{k,k}^2 \geq \omega r_{k-1,k-1}^2, \tag{16}$$

for all $1 < k \leq n$. Multiplying the both sides of (16) with $\frac{1}{(r_{k-1,k-1}r_{k,k})^2}$, we obtain

$$\left(\frac{r_{k-1,k}}{r_{k-1,k-1}r_{k,k}} - \left\lfloor \frac{r_{k-1,k}}{r_{k-1,k-1}} \right\rceil \cdot \frac{1}{r_{k,k}}\right)^2 + \left(\frac{1}{r_{k-1,k-1}}\right)^2 \geq \omega \left(\frac{1}{r_{k,k}}\right)^2,$$

which implies that $\mathbf{R}^*$ is also diagonal reduced.

4.2 Proximity Factor

To characterize the performance gap between suboptimal decoding and ILD, a proximity factor was defined in [8] and further discussed in [9, 18]. Given a lattice generator matrix $\mathbf{B} = [\mathbf{b}_1, ..., \mathbf{b}_n]$, denote ϕ_i the acute angle between $\mathbf{b}_i$ and the linear space spanned by the previous $i-1$ basis vectors, then the proximity factor of SIC decoding is defined as:

$$\rho_i \triangleq \sup_{\mathbf{B}\in\mathscr{B}_{\text{Red}}} \frac{\lambda^2(L(\mathbf{B}))}{\|\mathbf{b}_i\|_2^2 \sin^2\phi_i}, \tag{17}$$

where the supremum is taken over the set $\mathscr{B}_{\text{Red}}$ of bases satisfying a certain reduction notion for any n-dim lattice L. We further define $\rho \triangleq \max_i\{\rho_i\}$. From [9], the average error probability of SIC decoding can be bounded by

$$P_{e,\text{SIC}}(\text{SNR}) \leq \sum_{i=1}^{n} P_{e,\text{ILD}}\left(\frac{\text{SNR}}{\rho_i}\right) \leq n P_{e,\text{ILD}}\left(\frac{\text{SNR}}{\rho}\right)$$

for arbitrary SNR.

Denote ρ_{LLL}, ρ_{DLLL}, ρ_{DR}, and ρ_{DDR} the proximity factors of SIC decoding aided by LLL reduction, dual LLL reduction, diagonal reduction, and dual diagonal reduction, respectively. An upper bound of ρ_{LLL} is given in [18]:

$$\rho_{\text{LLL}} \leq \gamma_n \cdot \beta^{\frac{n-1}{2}} \leq \left(1 + \frac{n}{4}\right) \beta^{\frac{n-1}{2}}, \tag{18}$$

where $\beta = 1/(\omega - 1/4) \geq 4/3$. Following the argument in [19], it is easy to prove that

$$\rho_{i,\text{DR}} = \rho_{i,\text{LLL}} \leq \gamma_i \cdot \beta^{\frac{i-1}{2}}. \tag{19}$$

Thus,

$$\rho_{\text{DR}} = \rho_{n,\text{DR}} \leq \gamma_n \cdot \beta^{\frac{n-1}{2}}. \tag{20}$$

For dual reduction, an upper bound of ρ_{DLLL} is given in [9]:

$$\rho_{\text{DLLL}} \leq \beta^{n-1}. \tag{21}$$

In the following, we improve the upper bound (21). From (19) and Proposition 1, we can obtain that

$$\rho_{i,\text{DDR}} = \sup_{\mathbf{B}^* \in \mathscr{B}_{\text{DR}}} \frac{\lambda^2(L(\mathbf{B}))}{r_{i,i}^2} = \sup_{\mathbf{B} \in \mathscr{B}_{\text{DR}}} \frac{\lambda^2(L(\mathbf{B}))}{r_{i,i}^2} \leq \gamma_i \cdot \beta^{\frac{i-1}{2}}. \tag{22}$$

Thus,

$$\rho_{\text{DDR}} = \rho_{n,\text{DDR}} \leq \gamma_n \cdot \beta^{\frac{n-1}{2}}. \tag{23}$$

Following the above argument, it is easy to prove that the proximity factors of SIC decoding aided by all dual LLL-type reduction notions, such as dual LLL reduction, dual effective LLL reduction, and dual partial LLL reduction, can be upper bounded by the right-hand side of (23). Comparing (23) with (20), SIC decoding aided by primal and dual diagonal reductions are expected to have the same performance. This shall be confirmed by the simulation results presented in Sect. 5.

5 Simulation Results

In this section, we present our simulation results on comparing the efficiency of the proposed algorithm DRFG with the original algorithm DR. All experiments were performed on matrices with random entries, drawn from an i.i.d. zero-mean, unit variance Gaussian distribution. Without loss of generality, all testing matrices were

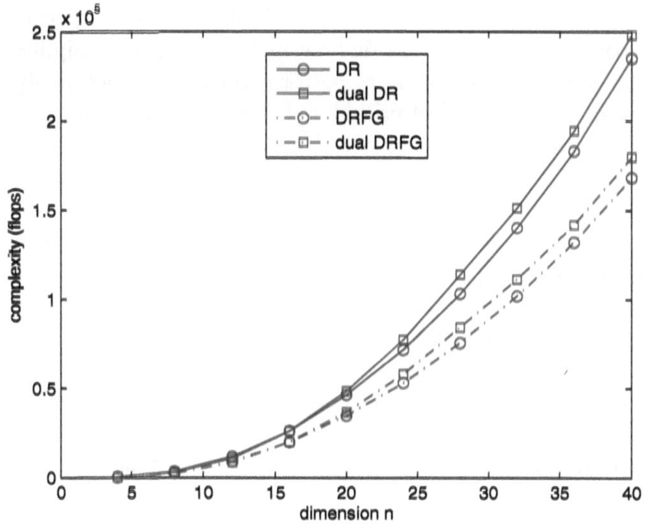

Fig. 3 The average complexity (in flops) of the diagonal reduction (DR), dual diagonal reduction (dual DR), DRFG, and dual DRFG algorithms

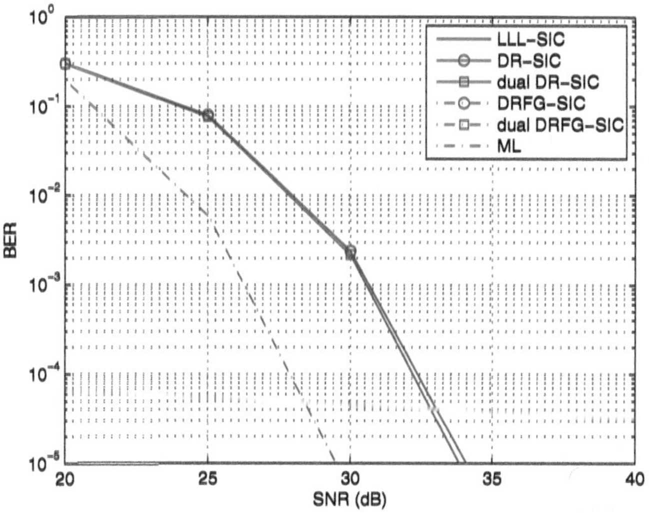

Fig. 4 Simulated BER of SIC aided by the LLL, DR, dual DR, DRFG, and the dual DRFG for 64-QAM over an 8×8 uncoded MIMO fading channel

set to square matrices. For each size, we generated 1,000 random matrices and took an average. The parameter ω in the reduction algorithms was set to 0.99.

Although the new algorithm DRFG is expected to be faster than the original algorithm DR, the computations in line 14 and line 20 of Fig. 2 introduce overhead.

To compare the overall complexity of the algorithms, we experimented on the floating-point operations (flops)[1] carried out by the algorithms. Figure 3 depicts our results on the average numbers of flops performed by the reduction algorithms. The figure shows that in both cases of primal and dual lattice reduction, DRFG is more efficient than DR, and the performance gap between them widens quickly as the dimension increases. This indicates that the overhead introduced by the fast Givens is insignificant. Also note that the DR (DRFG) algorithm is slightly faster than its dual counter part dual DR (dual DRFG) algorithm. This is due to the additional computation, for instance, the calculation of $\mathbf{B}^{\dagger}$, required by the dual reduction.

We also investigated the reduction quality of different reduction algorithms measured by the BER performance of the SIC decoding. Specifically, using a 64-QAM constellation, Fig. 4 depicts the simulated BER curves of lattice-reduction-aided SIC over an 8×8 uncoded MIMO fading channel. We have found that the SIC aided by the four diagonal reduction algorithms have identical BER performance to that aided by the LLL algorithm. This is consistent with the theoretical analysis presented in Sect. 4.2.

Acknowledgments We would like to thank Professor Lihong Zhi and referees for their useful comments.

References

1. Agrell, E., Eriksson, T., Vardy, A., Zeger, K.: Closest point search in lattices. IEEE Trans. Inf. Theory **48**, 2201–2214 (2002)
2. Arora, S., Babai, L., Stern, J.: The hardness of approximate optima in lattices, codes, and systems of linear equations. J. Comput. Syst. Sci. **54**, 317–331 (1997)
3. Babai, L.: On Lovász's lattice reduction and the nearest lattice point problem. Combinatorica **6**, 1–13 (1986)
4. Golub, G.H., Van Loan, C.F.: Matrix Computations. The Johns Hopkins University Press, Baltimore (1996)
5. Hassibi, B., Vikalo, H.: On the sphere-decoding algorithm I: Expected complexity. IEEE Trans. Signal Process. **53**, 2806–2818 (2005)
6. Jaldén, J., Ottersen, B.: On the complexity of sphere decoding in digital communications. IEEE Trans. Signal Process. **53**, 1474–1484 (2005)
7. Lenstra, A.K., Lenstra, H.W., Lovász, L.: Factorizing polynomials with rational coefficients. Math. Ann. **261**, 515–534 (1982)
8. Ling, C.: Towards characterizing the performance of approximate lattice decoding. Proc. Int. Symp. Turbo Codes/Int. Conf. Source Channel Coding '06. Munich, Germany (2006).
9. Ling, C.: On the proximity factors of lattice reduction-aided decoding. IEEE Trans. Signal Process. **59**, 2795–2808 (2011)
10. Ling, C., Howgrave-Graham, N.: Effective LLL reduction for lattice decoding. In: Proceedings IEEE Int. Symp. Inf. Theory (ISIT). Nice, France (2007).
11. Ling, C., Mow, W.H., Gan, L.: Dual-lattice ordering and partial lattice reduction for SIC-based MIMO detection. IEEE J. Sel. Topics Signal Process. **3**, 975–985 (2009)

[1] Flop count: addition/multiplication/division/max/rounding, 1 flop.

12. Mow, W.H.: Universal lattice decoding: Principle and recent advances. Wireless Wirel. Commun. Mobile Comput. **3**, 553–569 (2003). (Special Issue on Coding and Its Applications Wireless CDMA System)
13. Nguyen, P.Q., Vallée, B.: The LLL Algorithm: Survey and Applications. Springer, Berlin (2009)
14. Schnorr, C.P., Euchner, M.: Lattice basis reduction: Improved practical algorithms and solving subset sum problems. Math. Progr. **66**, 181–199 (1994)
15. Taherzadeh, M., Mobasher, A., Khandani, A.K.: LLL reduction achieves the receive diversity in MIMO decoding. IEEE Trans. Inf. Theory **53**, 4801–4805 (2007)
16. Wübben, D., Seethaler, D., Jaldén, J., Marz, G.: Lattice reduction: a survey with applications in wireless communications. IEEE Signal Process. Mag. **28**, 70–91 (2011)
17. Xie, X., Chang, X.W., Borno, M.A.: Partial LLL reduction. In: Proceedings of IEEE GLOBECOM (2011)
18. Zhang, W., Qiao, S., Wei, Y.: HKZ and Minkowski reduction algorithms for lattice-reduction-aided MIMO detection. IEEE Trans. Signal Process. **60**, 5963–5976 (2012)
19. Zhang, W., Qiao, S., Wei, Y.: A diagonal lattice reduction algorithm for MIMO detection. IEEE Signal Process. Lett. **19**, 311–314 (2012)

Constructing Generalized Bent Functions from Trace Forms of Galois Rings

Xiaoming Zhang, Baofeng Wu, Qingfang Jin and Zhuojun Liu

Abstract Quaternary constant-amplitude codes (codes over $\mathbb{Z}_4$) of length 2^m exist for every positive integer m, and every codeword of such a code corresponds to a function from the binary m-tuples to $\mathbb{Z}_4$ having the bent property, called a generalized bent function. In this chapter, we extend previous constructions and propose a general approach which can lead to more generalized bent functions.

1 Introduction

Multicode code-division multiple access (MC-CDMA) is a simple scheme to implement rate adaption in CDMA systems. The principal drawback of this technique is that the transmitted signals can have a high peak-to-average power ratio (PAPR). Paterson showed codewords with low PAPR were exactly those which were far from the first-order Reed–Muller code [6]. The code consisting of codewords with PAPR $= 1$ is called constant-amplitude code. But binary constant-amplitude codes cannot exist for lengths 2^m when m is odd. Schmidt studied quaternary constant-

This research is supported by National Basic Research Program of China (Grant No. 2011CB302400).

X. Zhang (✉) · B. Wu (✉) · Q. Jin (✉) · Z. Liu (✉)
Key Laboratory of Mathematics Mechanization, AMSS,
Chinese Academy of Science, Beijing, China
e-mail: xmzhang@amss.ac.cn

B. Wu
e-mail: wubaofeng@amss.ac.cn

Q. Jin
e-mail: qfjin@amss.ac.cn

Z. Liu
e-mail: zliu@mmrc.iss.ac.cn

© Springer-Verlag Berlin Heidelberg 2014
R. Feng et al. (eds.), *Computer Mathematics*, DOI 10.1007/978-3-662-43799-5_31

amplitude codes (codes over $\mathbb{Z}_4$) of length 2^m and these codes exist for every positive integer m [7]. He reduced the problem of constructing constant-amplitude codes to the construction of generalized bent function and presented one construction of quaternary generalized bent function by exploiting the connection with algebraic codes over $\mathbb{Z}_4$, in particular quaternary Reed–Muller, Kerdock and Delsarte-Goethals codes. For more details, we refer to [6, 7, 9].

The remainder of this chapter is organized as follows. In Sect. 2, some notations and basic facts are introduced which will be used several times in the following content. In Sect. 3, we construct a family of generalized bent functions which generalize Schmidt's construction. In Sect. 4, we consider a more general case which can lead to generalized bent functions and we also consider how to construct those generalized bent functions using algorithms and their complexity over finite fields or skew-polynomial rings.

2 Preliminary

A generalized Boolean function is defined as a mapping

$$f : \mathbb{Z}_2^m \longrightarrow \mathbb{Z}_{2^h},$$

where h is a positive integer. Every f can be written uniquely in 2-adic expansion

$$f(x_0, \ldots, x_{m-1}) = \sum_{j=0}^{h-1} f_j(x_0, \ldots, x_{m-1})2^j,$$

where each f_j is a mapping from $\mathbb{Z}_2^m$ to $\mathbb{Z}_2$ for $j = 0, 1, \ldots, h-1$.

The Fourier transform of $f : \mathbb{Z}_2^m \longrightarrow \mathbb{Z}_{2^h}$ is given by $\widehat{f} : \mathbb{Z}_2^m \longrightarrow \mathbb{C}$ with

$$\widehat{f}(u) = \sum_{x \in \mathbb{Z}_2^m} \omega^{f(x)}(-1)^{x \cdot u}$$

where "·" denotes the scalar product in $\mathbb{Z}_2^m$ and ω is a primitive 2^h-th root of unity in $\mathbb{C}$.

Definition 2.1 A function $f : \mathbb{Z}_2^m \longrightarrow \mathbb{Z}_{2^h}$ is called a generalized bent function if $|\widehat{f}(u)| = 2^{m/2}$ for all $u \in \mathbb{Z}_2^m$.

Now we briefly provide some facts about Galois rings and associated fields, for details we refer to [5, 7]. In this chapter, we consider Galois rings with characteristic equal to a power of 2.

Define $\mu : \mathbb{Z}_{2^h} \longrightarrow \mathbb{F}_2$ to be the map:

$$\begin{aligned} \mu : \mathbb{Z}_{2^h} &\longrightarrow \mathbb{F}_2, \\ \sum_{i=0}^{h-1} a_i 2^i &\longmapsto a_0 \end{aligned}$$

This map can be extended to act on polynomials:

$$\begin{aligned} \mu : \mathbb{Z}_{2^h}[x] &\longrightarrow \mathbb{F}_2[x] \\ \sum_{i=0}^{m} a_i x^i &\longmapsto \sum_{i=0}^{m} \mu(a_i) x^i \end{aligned}$$

A polynomial $p(x) \in \mathbb{Z}_{2^h}[x]$ is called monic basic irreducible if $p(x)$ is monic and $\mu(p(x))$ is irreducible over $\mathbb{F}_2$. The Galois ring $\mathscr{R}_{h,m}$ is defined by $\mathscr{R}_{h,m} \cong \mathbb{Z}_{2^h}[x]/(p(x))$, where $p(x)$ is a monic basic irreducible polynomial over $\mathbb{F}_2$ of degree m.

Let $\xi \in \mathscr{R}_{h,m}$ be a root of a monic basic primitive polynomial $p(x)$ with degree m, then ξ generates a cyclic group $\mathscr{T}^*_{h,m}$ of order $2^m - 1$. The set

$$\mathscr{T}_{h,m} := \{0\} \cup \mathscr{T}^*_{h,m}$$

is called the Teichmüller set of $\mathscr{R}_{h,m}$. Every element $z \in \mathscr{R}_{h,m}$ can be uniquely written as

$$z = \sum_{i=0}^{m-1} z_i \xi^i, \ z_i \in \mathbb{Z}_{2^h}$$

which is called the additive representation of z. Moreover, z can be uniquely expressed in the form

$$z = \sum_{i=0}^{h-1} z_i 2^i, \ z_i \in \mathscr{T}_{h,m}$$

called the 2-adic representation of z.

Lemma 1 (see [7]) *Let $z \in \mathscr{R}_{h,m}$, we define the map $\eta : \mathscr{R}_{h,m} \longrightarrow \mathscr{T}_{h,m}$ by $\eta(z) = z^{2^{m(h-1)}}$. For any $x, y \in \mathscr{T}_{h,m}$, we define the operation $\oplus$ as*

$$x \oplus y = \eta(x + y).$$

Then $< \mathscr{T}_{h,m}, \oplus >$ is an abelian group. Hence, $< \mathscr{T}_{h,m}, \oplus, \cdot > \cong < \mathbb{F}_{2^m}, +, \cdot >$.

Let $z \in \mathscr{R}_{h,m}$ have 2-adic expansion

$$z = \sum_{i=0}^{h-1} z_i 2^i, z_i \in \mathscr{T}_{h,m},$$

then the map $\sigma : \mathscr{R}_{h,m} \longrightarrow \mathscr{R}_{h,m}$ defined by

$$\sigma(z) = \sum_{i=0}^{h-1} z_i^2 2^i$$

is the Frobenius automorphism on $\mathscr{R}_{h,m}$ with respect to the ground ring $\mathbb{Z}_{2^h}$. The absolute trace function $\mathrm{Tr}_m : \mathscr{R}_{h,m} \longrightarrow \mathbb{Z}_{2^h}$ is then defined to be

$$\mathrm{Tr}_m(z) = \sum_{i=0}^{m-1} \sigma^i(z).$$

From now on, we consider operations in $\mathscr{R}_{2,m}$ and use tr to denote the trace function of finite fields, and use Tr to denote the trace function of Galois rings. Note that $2\mathrm{Tr}(r) = 2\mathrm{tr}(\mu(\mathrm{r}))$ for any $r \in \mathscr{R}_{2,m}$.

3 Construction of Generalized Bent Function

Schmidt constructed generalized bent functions as follows:

Theorem 1 (see [7]) *Suppose $m \geq 3$ and let $f : \mathscr{T}_{2,m} \longrightarrow \mathbb{Z}_4$ be given by*

$$f(x) = \varepsilon + \mathrm{Tr}(ax + 2bx^3),\ \varepsilon \in \mathbb{Z}_4, a \in \mathscr{R}_{2,m}, b \in \mathscr{T}_{2,m}^*.$$

Then $f(x)$ is a generalized bent function if either $\mu(a) = 0$ and the equation $x^3 + \frac{1}{\mu(b)} = 0$ has no solution in $\mathbb{F}_{2^m}$, or $\mu(a) \neq 0$ and $x^3 + x + \frac{\mu(b)^2}{\mu(a)^6} = 0$ has no solution in $\mathbb{F}_{2^m}$. Here, μ is the modulo-2 reduction map.

In this section, we extend the construction in Theorem 1, replacing 3 by $1 + 2^k$, where k is a positive integer. We consider the following functions:

$$f(x) = \varepsilon + \mathrm{Tr}(ax + 2bx^{1+2^k}), \varepsilon \in \mathbb{Z}_4, a \in \mathscr{R}_{2,m}, b \in \mathscr{T}_{2,m}^*.$$

Next we give the conditions under which $f(x)$ is a generalized bent function, and these conditions can be described more precisely.

Theorem 2 *Suppose $m \geq 5$ and let $f(x) = \varepsilon + \mathrm{Tr}(ax + 2bx^{1+2^k})$, where $\varepsilon \in \mathbb{Z}_4, a \in \mathscr{R}_{2,m}, b \in \mathscr{T}_{2,m}^*$. Then $f(x)$ is a generalized bent function if:*

(i) *$\mu(a) = 0$ and the equation $x^{2^{2k}-1} + \frac{1}{\mu(b)^{2^k-1}} = 0$ has no solution in $\mathbb{F}_{2^m}$; or*

(ii) *$\mu(a) \neq 0$ and the equation $\mu(b)^{2^k} x^{2^{2k}-1} + \mu(a)^{2^{k+1}} x^{2^k-1} + \mu(b) = 0$ has no solution in $\mathbb{F}_{2^m}$.*

Proof Consider the exponential sum

$$\rho(a,b) = \sum_{x \in \mathscr{T}_{2,m}} \omega^{\mathrm{Tr}(ax+2bx^{2^k+1})}.$$

We have

$$\rho^2(a,b) = \sum_{x,y \in \mathscr{T}_{2,m}} \omega^{\mathrm{Tr}[a(x+y)+2b(x^{1+2^k}+y^{1+2^k})]}.$$

Setting

$$z = x \oplus y,$$

with the definition of the operation $\oplus$ in Sect. 2, we have

$$\begin{aligned} z &= (x+y)^{2^m} = [(x+y)^2]^{2^{m-1}} \\ &= [(x^2+y^2+2xy)^2]^{2^{m-2}} \\ &= \cdots \\ &= x^{2^m} + y^{2^m} + 2(xy)^{2^{m-1}} \end{aligned}$$

which together with $y \equiv x + z \pmod 2$ implies

$$x + y = z + 2\sqrt{xy} = z + 2x + 2\sqrt{xz}.$$

Next we compute $\mathrm{Tr}(a[x+y])$ and $2\mathrm{Tr}(b[x^{1+2^k} + y^{1+2^k})])$.

$$\begin{aligned} \mathrm{Tr}(a[x+y]) &= \mathrm{Tr}(a[z+2x+2\sqrt{xz}]) \\ &= \mathrm{Tr}(az) + 2\mathrm{Tr}(a^{2^k}x^{2^k}) + 2\mathrm{Tr}(a^{2^{k+1}}x^{2^k}z^{2^k}), \end{aligned}$$

$$\begin{aligned} 2\mathrm{Tr}[b(x^{1+2^k} + y^{1+2^k})] &= 2\mathrm{Tr}[b(x^{1+2^k} + (x+z+2\sqrt{xz})^{1+2^k})] \\ &= 2\mathrm{Tr}[b(x^{2^k}z + xz^{2^k} + z^{1+2^k})] \\ &= 2\mathrm{Tr}(bz^{1+2^k}) + 2\mathrm{Tr}(bzx^{2^k}) + 2\mathrm{Tr}(b^{2^k}x^{2^k}z^{2^{2k}}) \\ &= 2\mathrm{Tr}(bz^{1+2^k}) + 2\mathrm{Tr}[x^{2^k}(bz + b^{2^k}z^{2^{2k}})]. \end{aligned}$$

Since $\langle \mathscr{T}_{2,m}, \oplus \rangle$ is an abelian group, z takes on every value in $\mathscr{T}_{2,m}$ when x is fixed and y ranges over $\mathscr{T}_{2,m}$. Hence we have

$$\rho^2(a,b) = \sum_{z\in\mathcal{T}_{2,m}} \omega^{\mathrm{Tr}(az+2bz^{1+2^k})} \sum_{x\in\mathcal{T}_{2,m}} \omega^{2\mathrm{Tr}(x^{2^k}h(z))}$$

where

$$h(z) = b^{2^k}z^{2^{2k}} + a^{2^{k+1}}z^{2^k} + bz + a^{2^k}.$$

The inner sum

$$\begin{aligned}\sum_{x\in\mathcal{T}_{2,m}} \omega^{2\mathrm{Tr}(x^{2^k}h(z))} &= \sum_{x\in\mathcal{T}_{2,m}} (-1)^{\mathrm{tr}(\mu(x^{2^k}h(z)))}\\ &= \sum_{x\in\mathcal{T}_{2,m}} (-1)^{\mathrm{tr}(\mu(x\,\sqrt[2^k]{h(z)}))}\\ &= \sum_{x\in\mathcal{T}_{2,m}} (-1)^{\mathrm{tr}(\mu(x)\,\sqrt[2^k]{\widetilde{h}(z)})}\end{aligned}$$

where $\widetilde{h}(z) = \mu(h(z))$. By the property of the trace function,

$$\sum_{x\in\mathcal{T}_{2,m}} \omega^{2\mathrm{Tr}(x^{2^k}h(z))} = \begin{cases} 0, & \text{if } \widetilde{h}(z) \neq 0,\\ 2^m, & \text{if } \widetilde{h}(z) = 0.\end{cases}$$

Therefore, if $\widetilde{h}(z)$ has exactly one root, the absolute value of the Fourier transform of f

$$\widehat{f}(u) = \omega^{\varepsilon} \sum_{z\in\mathcal{T}_{2,m}} \omega^{\mathrm{Tr}([a+2u]z+2bz^{1+2^k})}$$

is $2^{\frac{m}{2}}$ for each $u \in \mathcal{T}_{2,m}$, and f is a generalized bent function.

Now we aim at characterizing the cases where $\widetilde{h}(z)$ has exactly one root. Write $\alpha = \mu(a)$ and $\beta = \mu(b)$. There are two cases, depending on whether $\alpha = \mu(a) = 0$ or not.

Case 1: When $\alpha = 0$, $\widetilde{h}(z) = \beta^{2^k}z^{2^{2k}} + \beta z = z(\beta^{2^k}z^{2^{2k}-1} + \beta)$. If $\beta^{2^k}z^{2^{2k}-1} + \beta = 0$ has no solution in $\mathbb{F}_{2^m}$, $f(x)$ is a generalized bent function, which corresponds (i).

Case 2: When $\alpha \neq 0$ and the equation $\beta^{2^k}x^{2^{2k}-1} + \alpha^{2^{k+1}}x^{2^k-1} + \beta = 0$ has no solution in $\mathbb{F}_{2^m}$, then $\widetilde{h}(z) = 0$ has one solution in $\mathbb{F}_{2^m}$ and $f(x)$ is generalized bent function.

Remark For any positive $k \in \mathbb{Z}$, there always exist $a \in \mathcal{R}_{2,m}$ and $b \in \mathcal{T}_{2,m}^*$ such that the function we construct is a generalized bent function.

Proof Let γ be a primitive element of $\mathbb{F}_{2^m}$, and let $\alpha = \mu(a)$, $\beta = \mu(b)$.

Condition (i) in the theorem is equivalent to $\alpha = 0$ and $\beta \notin \langle \gamma^{\frac{2^{(2k,m)}-1}{2^{(k,m)}-1}} \rangle$;

Suppose $(\alpha, \beta) \in \mathbb{F}_{2^m}^* \times \mathbb{F}_{2^m}^*$ and $\beta^{2^k} x^{2^{2k}-1} + \alpha^{2^{k+1}} x^{2^k-1} + \beta = 0$ has a root in $\mathbb{F}_{2^m}$. Let $y = x^{2^k-1}$ and since $y = 0$ is not a root, we have

$$\beta^{2^k} y^{2^k} + \frac{\beta}{y} = \alpha^{2^{k+1}}$$

Then $(\alpha, \beta) \in \mathbb{F}_{2^m}^* \times \mathbb{F}_{2^m}^*$ satisfies

$$\alpha = G_\beta(y)$$

where $G_\beta(y) = (\beta^{2^k} y^{2^k} + \frac{\beta}{y})^{\frac{1}{2^{k+1}}}$. Thus condition (ii) in the theorem is equivalent to

$$\bigcup_{\beta \in \mathbb{F}_{2^m}^*} G(\langle \gamma^{2^k-1} \rangle) \times \{\beta\} \gneqq \mathbb{F}_{2^m}^* \times \mathbb{F}_{2^m}^* = \bigcup_{\beta \in \mathbb{F}_{2^m}^*} \mathbb{F}_{2^m}^* \times \{\beta\}.$$

This holds since $G_\beta(y)$ will never be a permutation polynomial over $\mathbb{F}_{2^m}$ [3]. So in both cases, there always exist $a \in \mathscr{R}_{2,m}$ and $b \in \mathscr{T}_{2,m}^*$ such that $f(x) = \varepsilon + \mathrm{Tr}(ax + 2bx^{1+2^k})$ is a generalized bent function.

4 General Construction of Generalized Bent Function

In Sect. 3, both constructions use the quadratic forms over finite fields. In fact, we can consider a more general case in this section.

Theorem 3 *Suppose $m \geq 5$ and $f(x) = \varepsilon + \mathrm{Tr}(ax + 2bxL(x))$, where $L(x) = \sum_{i=0}^{m-1} a_i x^{2^i} \in \mathscr{T}_{2,m}[x]$, $\varepsilon \in \mathbb{Z}_4$, $a \in \mathscr{R}_{2,m}$ and $b \in \mathscr{T}_{2,m}^*$. Let $\alpha = \mu(a), \beta = \mu(b), \alpha_i = \mu(a_i)$. Then $f(x)$ is a generalized bent function if*

$$\sum_{i=0}^{m-1} (\beta\alpha_i z^{2^i} + (\beta\alpha_i)^{2^{m-i}} z^{2^{m-i}}) + \alpha^2 z$$

is a linearized permutation polynomial over $\mathbb{F}_{2^m}$.

Proof Denote

$$\rho(a, b) = \sum_{x \in \mathscr{T}_{2,m}} \omega^{\mathrm{Tr}(ax+2bxL(x))}$$

We consider the following sum:

$$\rho^2(a, b) = \sum_{x,y \in \mathscr{T}_{2,m}} \omega^{\mathrm{Tr}(a[x+y]+2b[xL(x)+yL(y)])}$$

We have

$$\mathrm{Tr}(a[x+y]) = \mathrm{Tr}(a[z+2x+2\sqrt{xz}]) = \mathrm{Tr}(az) + 2\mathrm{Tr}(ax) + 2\mathrm{Tr}(a^2xz);$$

$$\begin{aligned}
2\mathrm{Tr}(b[xL(x)+yL(y)]) &= 2\mathrm{Tr}(b[xL(x) + (z+x+2\sqrt{xz})L(z+x+2\sqrt{xz})]) \\
&= 2\mathrm{Tr}(b[xL(x)+xL(x)+xL(z)+zL(x)+zL(z)]) \\
&= 2\mathrm{Tr}(b[xL(z)+zL(x)]) + 2\mathrm{Tr}(bzL(z)) \\
&= 2\mathrm{tr}(x\overline{L_b}(z) + x\tilde{\tilde{L}}_b(z)) + 2\mathrm{Tr}(zL(z)),
\end{aligned}$$

where $\overline{L_b}(z) = \sum_{i=0}^{m-1} \beta\alpha_i z^{2^i}$ and $\tilde{\tilde{L}}_b(z) = \sum_{i=0}^{m-1} (\beta\alpha_i)^{2^{m-i}} z^{2^{m-i}}$.

Then

$$\begin{aligned}
\rho^2(a,b) &= \sum_{x,y\in\mathcal{T}_{2,m}} \omega^{\mathrm{Tr}(a[x+y]+2b[xL(x)+yL(y)])} \\
&= \sum_{x,z\in\mathcal{T}_{2,m}} \omega^{\mathrm{Tr}(az+2zL(z))+2\mathrm{tr}(x[\overline{L_b}(z)+\tilde{\tilde{L}}_b(z)+\alpha^2 z+\alpha])} \\
&= \sum_{z\in\mathcal{T}_{2,m}} \omega^{\mathrm{Tr}(az+2zL(z))} \sum_{x\in\mathbb{F}_{2^m}} (-1)^{\mathrm{tr}(xh(z))}
\end{aligned}$$

where

$$\begin{aligned}
h(z) &= \overline{L_b}(z) + \tilde{\tilde{L}}_b(z) + \alpha^2 z + \alpha \\
&= \sum_{i=0}^{m-1} (\beta\alpha_i z^{2^i} + (\beta\alpha_i)^{2^{m-i}} z^{2^{m-i}}) + \alpha^2 z + \alpha.
\end{aligned}$$

If $\sum_{i=0}^{m-1}(\beta\alpha_i z^{2^i} + (\beta\alpha_i)^{2^{m-i}} z^{2^{m-i}}) + \alpha^2 z$ is a permutation polynomial over $\mathbb{F}_{2^m}$, then $h(z) = 0$ has one solution in $\mathbb{F}_{2^m}$ and $|\rho^2(a,b)| = 2^m$, which implies $f(x)$ is a generalized bent function.

Now we talk about how to test whether it will promise a generalized bent function for $L(x) = \sum_{i=0}^{m-1} a_i x^{2^i} \in \mathcal{R}_{2,m}[x]$ from an algorithmic aspect. When discussing algorithmic problems of the finite field $\mathbb{F}_{q^m}$, we always assume a basis $\{\theta_i\}_{i=0}^{m-1}$ for the field extension over $\mathbb{F}_q$ is given and all conjugates of the basis elements, i.e., $\theta_i^{q^j}$, $0 \le i, j \le m-1$ are precomputed. By the complexity of an algorithm, we mean the number of operations needed to be done in ground field $\mathbb{F}_q$. Generally speaking, for an operation in $\mathbb{F}_{q^m}$, the complexity is $\mathcal{O}(M(m))$, where $M(m) = mlogmloglogm$. For convenience, we use the notation $\tilde{\mathcal{O}}(h) = \mathcal{O}(h(logh)^k)$ for some constant k.

Recall that $f(x) = \varepsilon + \mathrm{Tr}(ax + 2bxL(x))$ is a generalized bent function if

$$\begin{aligned} h(z) &= \overline{L_b}(z) + \overline{\overline{L}}_b(z) + \alpha^2 z + \alpha \\ &= \sum_{i=0}^{m-1} (\beta\alpha_i z^{2^i} + (\beta\alpha_i)^{2^{m-i}} z^{2^{m-i}}) + \alpha^2 z \in \mathbb{F}_{2^m}[x] \end{aligned}$$

is a permutation polynomial. Generally, to test whether a given polynomial over a finite field is a permutation polynomial is very hard: for $f(x) \in \mathbb{F}_q[x]$ of degree m, the complexity is $\widetilde{\mathcal{O}}(mlogq)$ by von zur Gathen's random polynomial time algorithm [1], and $\widetilde{\mathcal{O}}((mq)^{\frac{6}{7}})$ by Shparlinski's deterministic polynomial time algorithm [8]. But for a linearized polynomial, it is not necessary to adapt these algorithms due to its special form.

Let $B(x) = \sum_{i=0}^{m-1} b_i x^{q^i} \in \mathbb{F}_{q^m}[x]$, the following result in [4] is well known:

Theorem 4 (see [4]) *$B(x)$ is a permutation polynomial if and only if matrix*

$$D(B) = \begin{pmatrix} b_0 & b_1 & \cdots & b_{m-1} \\ b_{m-1}^q & b_0^q & \cdots & b_{m-2}^q \\ \cdots & \cdots & \cdots & \cdots \\ b_1^{q^{m-1}} & b_2^{q^{m-1}} & \cdots & b_0^{q^{m-1}} \end{pmatrix}$$

is nonsingular.

Now we assume

$$\begin{aligned} h(x) &\equiv \sum_{i=0}^{m-1} (\beta\alpha_i z^{2^i} + (\beta\alpha_i)^{2^{m-i}} z^{2^{m-i}}) + \alpha^2 z \\ &\equiv \sum_{i=0}^{m-1} h_i x^{2^i} \in \mathbb{F}_{2^m}[x], \end{aligned}$$

then $f(x)$ is a generalized bent function if the matrix $D(h)$ is nonsingular over $\mathbb{F}_{2^m}$. To get $D(h)$, we need to compute $h_i^{2^j}$, $0 \le i, j \le m-1$, where $h_0 = \alpha^2$ and $h_i = \beta\alpha_i + \beta\alpha_{m-i}^{2^i}$ for $1 \le i \le m-1$. The total complexity is $\mathcal{O}(m^2)M(m)logm = \mathcal{O}(m^3)$. To test singularity of $D(h)$, there still needs $\mathcal{O}(m^3)$ operations on $\mathbb{F}_{2^m}$ by Gaussian elimination, so the complexity is at least $\mathcal{O}(m^3 M(m)) = \widetilde{\mathcal{O}}(m^4)$.

Another method to test whether $h(x)$ is a permutation polynomial is to compute greatest common right divisors (GCRD) of polynomials in $\mathbb{F}_{2^m}[x;\sigma]$. Here σ is the Frobenius automorphism of $\mathbb{F}_{2^m}$ over $\mathbb{F}_2$ and $\mathbb{F}_{2^m}[x;\sigma]$ is the skew-polynomial ring over $\mathbb{F}_{2^m}$, which consists of ordinary polynomials over $\mathbb{F}_{2^m}$ but with a noncommutative multiplication $x \cdot a = \sigma(a)x$ for any $a \in \mathbb{F}_{2^m}$. A well-known result of Ore shows that $\mathbb{F}_{2^m}[x;\sigma] \cong \mathcal{L}_m$, where $\mathcal{L}_m$ is the ring consisting of all linearized polynomials over $\mathbb{F}_{2^m}$ with ordinary addition and polynomial composition as the ring operations.

Theorem 5 (see [10]) $h(x) = \sum_{i=0}^{m-1} h_i x^i \in \mathbb{F}_{2^m}[x;\sigma]$ *is a linearized permutation polynomial if and only if* $\mathrm{GCRD}(\sum_{i=0}^{m-1} h_i x^i, x^m - 1) = 1$.

Due to this result, the problem of testing whether $f(x)$ is a generalized bent function can be reduced to computing GCRD($\sum_{i=0}^{m-1} h_i x^i$, $x^m - 1$) in $\mathbb{F}_{2^m}[x;\sigma]$. The complexity of the GCRD algotithm in $\mathbb{F}_{2^m}[x;\sigma]$ is $\widetilde{\mathcal{O}}(m^4)$ according to [2], which is the same as former method. Fortunately, when $h(x) \in \mathbb{F}_2[x]$, the GCRD of $h(x)$ and $x^m - 1$ degenerates to the greatest common divisor of $h(x)$ and $x^m - 1$ in $\mathbb{F}_2[x]$, the complexity of which is $\mathcal{O}(m^3)$ using Euclidean algorithm or $\mathcal{O}(m^2 logm)$ using fast Euclidean algorithm [1].

Suppose $\overline{L}(x) = \mu(L(x)) \in \mathbb{F}_2[x]$, $a = b = 1$. In this case, we can replace above conditions by computing greatest common divisor of two polynomials over finite fields, which can be completed efficiently in polynomial time using Euclidean algorithm. More precisely,

Corollary 1 *Let* $f(x) = \varepsilon + \mathrm{Tr}(x + 2xL(x))$, *where* $\varepsilon \in \mathbb{Z}_4$, $x \in \mathcal{T}_{2,m}$, $L(x) = \sum_{i=0}^{m-1} a_i x^{2^i} \in \mathcal{T}_{2,m}[x]$ *and* $\alpha_i = \mu(a_i) \in \mathbb{F}_2$ *for* $i = 0, 1, \ldots, m-1$. *Then* $f(x)$ *is a generalized bent function if* $gcd(\sum_{i=0}^{m-1} \beta_i x^i, x^m - 1) = 1$ *where* $\beta_0 = 1$, $\beta_i = \alpha_i + \alpha_{m-i}^{2^i}$ *for* $i = 0, 1, \ldots, m-1$.

Proof Using Theorem 3, we have

$$\rho^2(1,1) = \sum_{z \in \mathcal{T}} \omega^{\mathrm{Tr}(z+2zL(z))} \sum_{x \in \mathcal{T}} \omega^{2\mathrm{tr}[x(\overline{L}(z)+\widetilde{\overline{L}}(z)+z+1)]}$$

$$= \sum_{z \in \mathcal{T}} \omega^{\mathrm{Tr}(z+2zL(z))} \sum_{x \in \mathbb{F}_{2^m}} (-1)^{\mathrm{tr}(xh(z))}$$

where $\overline{L}(z) = \alpha_0 z + \sum_{i=1}^{m-1} \alpha_i^{2^{m-i}} z^{2^{m-i}}$, $h(z) = \overline{L}(z) + \widetilde{\overline{L}}(z) + z + 1$.
Denote

$$L_1(z) = h(z) - 1 = \sum_{i=0}^{m-1} \beta_i z^{2^i}$$

where $\beta_0 = 1$ and $\beta_i = \alpha_i + \alpha_{m-1}^{2^i}$ for $i = 0, 1, \ldots, m-1$. Since $L_1(z) \in \mathbb{F}_2[z]$ is a linearized polynomial, $L_1(z)$ is a permutation polynomial if and only if $gcd(\sum_{i=0}^{m-1} \beta_i x^i, x^m - 1) = 1$. In this case, $h(z) = L_1(z) + 1 = 0$ has only one solution in $\mathbb{F}_{2^m}$ and $|\rho(1,1)|^2 = 2^m$, so $f(x)$ is a generalized bent function.

5 Conclusion

In this chapter, we extend Schmidt's construction of generalized bent functions. Using the tools of quadratic forms and linearized polynomials over finite fields, we also consider a general case of generalized bent functions with the similar form.

When $\overline{L}(x) = \mu(L(x)) \in \mathbb{F}_2[x]$, $a = b = 1$, we can construct a family of generalized bent functions by computing greatest common divisor of two polynomials over finite fields, which can be completed in polynomial time using Euclidean algorithm. All above generalized bent functions can be used to construct quaternary constant-amplitude codes.

References

1. Gathen, J.V.Z.: Tests for permutation polynomials. SIAM J. Comput. **20**, 591–602 (1991)
2. Giesbrecht, M.: Factoring in skew-polynomial rings over finite fields. J. Symb. Comput. **26**, 463–486 (1998)
3. Li, Y., Wang, M.: Permutation polynomials EA-equivalent to the inverse function over GF(2^n). Cryptogr. Commun. **3**(3), 175–186 (2011)
4. Lidl, R., Niederreiter, H.: Finite Fields. Cambridge University Press, Cambridge (1997)
5. McDonald, B.R.: Finite Rings with Identity. Marcel Dekker, New York (1974)
6. Paterson, K.G.: Generalized Reed-Muller codes and power control in OFDM modulation. IEEE Trans. Inf. Theory **46**, 104–120 (2000)
7. Schmidt, K.U.: Quaternary constant-amplitude codes for multicode CDMA. IEEE Trans. Inf. Theory **55**, 1824–1832 (2009)
8. Shparlinski, I.E.: A deterministic test for permutation polynomials. Comput. Complex. **2**, 129–132 (1992)
9. Solé P., Tokareva N.: Connections between quaternary and binary bent functions. In: Cryptology ePrint Archives http://www.eprint.iacr.org/2009/544 (2009)
10. Wu, B., Liu, Z.: Linearized polynomials over finite fields revisited. Finite Fields Appl. **2**, 79–100 (2013)

Matrix Formulae of Differential Resultant for First Order Generic Ordinary Differential Polynomials

Zhi-Yong Zhang, Chun-Ming Yuan and Xiao-Shan Gao

Abstract In this paper, a matrix representation for the differential resultant of two generic ordinary differential polynomials f_1 and f_2 in the differential indeterminate y with order one and arbitrary degree is given. That is, a nonsingular matrix is constructed such that its determinant contains the differential resultant as a factor. Furthermore, the algebraic sparse resultant of f_1, f_2, δf_1 and δf_2 treated as polynomials in y, y', y'' is shown to be a nonzero multiple of the differential resultant of f_1 and f_2. Although very special, this seems to be the first matrix representation for a class of nonlinear generic differential polynomials.

Keywords Matrix formula · Differential resultant · Sparse resultant · Macaulay resultant

1 Introduction

Multivariate resultant, which gives a necessary condition for a set of $n+1$ polynomials in n variables to have common solutions, is an important tool in elimination theory. One of the major issues in the resultant theory is to give a matrix representation for the resultant, which allows fast computation of the resultant using existing methods of determinant computation. By a matrix representation of the resultant, we mean a nonsingular square matrix whose determinant contains the resultant as a factor. There exist stronger forms of matrix representations. For instance, in the case of two univariate polynomials in one variable, there exist matrix formulae named after Sylvester and Bézout, whose determinants equal the resultant. Unfortunately, such

Z.-Y. Zhang · C.-M. Yuan (✉) · X.-S. Gao
KLMM, Academy of Mathematics and Systems Science,
Chinese Academy of Sciences, Beijing, People's Republic of China
e-mail: cmyuan@mmrc.iss.ac.cn

Z.-Y. Zhang
e-mail: zhiyong-2008@163.com

X.-S. Gao
e-mail: xgao@mmrc.iss.ac.cn

© Springer-Verlag Berlin Heidelberg 2014
R. Feng et al. (eds.), *Computer Mathematics*, DOI 10.1007/978-3-662-43799-5_32

determinant formulae do not generally exist for multivariate resultants. Macaulay showed that the multivariate resultant can be represented as a ratio of two determinants of certain Macaulay matrixes [16]. D'Andrea established a similar result for the multivariate sparse resultant [8] based on the pioneering work on sparse resultant [2, 3, 11, 20]. This paper will study matrix representations for differential resultants.

Using the analog between ordinary differential operators and univariate polynomials, the differential resultant for two linear ordinary differential operators was studied by Berkovich and Tsirulik [1] using Sylvester style matrices. The subresultant theory was first studied by Chardin [6] for two differential operators and then by Li [15] and Hong [12] for the more general Ore polynomials.

For nonlinear differential polynomials, the differential resultant is more difficult to define and study. The differential resultant for two nonlinear differential polynomials in one variable was defined by Ritt in [17, p. 47]. In [23, p. 46], Zwillinger proposed to define the differential resultant of two differential polynomials as the determinant of a matrix following the idea of algebraic multivariate resultant, but did not give details. General differential resultants were defined by Carrà Ferro using Macaulay's definition of algebraic resultants [5]. But, the treatment in [5] is not complete, as will be shown in Sect. 2.2 of this paper. In [22], Yang et al. used the idea of algebraic Dixon resultant to compute the differential resultant. Although very efficient, this approach is not complete and does not provide a matrix representation for the differential resultant. Differential resultants for linear ordinary differential polynomials were studied by Rueda-Sendra [18]. In [19], Rueda gave a matrix representation for a generic sparse linear system. In [9], the first rigorous definition for the differential resultant of $n+1$ differential polynomials in n differential indeterminates was given and its properties were proved. In [13, 14], the sparse resultant for differential Laurent polynomials was defined and a single exponential time algorithm to compute the sparse resultant was given. Note that an ideal approach is used in [9, 13, 14], and whether the multivariate differential resultant admits a matrix representation is left as an open issue.

In this paper, based on the idea of algebraic sparse resultants and Macaulay resultants, a matrix representation for the differential resultant of two generic ordinary differential polynomials f_1, f_2 in the differential indeterminate y with order one and arbitrary degree is given. The constructed square matrix has entries equal to the coefficients of f_1, f_2, their derivatives, or zero, whose determinant is a nonzero multiple of the differential resultant. Furthermore, we prove that the sparse resultant of $f_1, f_2, \delta f_1$ and δf_2 treated as polynomials in y, y', y'' is not zero and contains the differential resultant of f_1 and f_2 as a factor. Although very special, this seems to be the first matrix representation for a class of nonlinear generic differential polynomials.

The rest of the paper is organized as follows. In Sect. 2, the method of Carrà Ferro is briefly introduced and the differential resultant is defined following [9]. In Sect. 3, a matrix representation for the differential resultant of two differential polynomials with order one and arbitrary degree is given. In Sect. 4, it is shown that the differential resultant can be computed as a factor of a special algebraic sparse resultant. In Sect. 5, the conclusion is presented and a conjecture is proposed.

2 Preliminaries

To motivate what we do, we first briefly recall Carrà Ferro's definition for differential resultant and then give a counter example to show the incompleteness of Carrà Ferro's method when dealing with nonlinear generic differential polynomials. Finally, definition for differential resultant given in [9] is introduced.

2.1 A Sketch of Carrà Ferro's Definition

Let K be an ordinary differential field of characteristic zero with δ as a derivation operator. $K\{y\} = K[\delta^n y, n \in \mathbb{N}]$ is the differential ring of the differential polynomials in the differential indeterminate y with coefficients in K. Let p_1 (respectively p_2) be a differential polynomial of order m and degree d_1 (respectively of order n and degree d_2) in $K\{y\}$. According to Carrà Ferro [5], the differential resultant of p_1 and p_2, denoted by $\delta\mathrm{R}(p_1, p_2)$, is defined to be the Macaulay's algebraic resultant of $m + n + 2$ differential polynomials

$$\mathscr{P}(p_1, p_2) = \{\delta^n p_1, \delta^{n-1} p_1, \ldots, p_1, \delta^m p_2, \delta^{m-1} p_2, \ldots, p_2\}$$

in the polynomial ring $S_{m+n} = K[y, \delta y, \ldots, \delta^{m+n} y]$ in $m + n + 1$ variables.

Specifically, let

$$D = 1 + (n + 1)(d_1 - 1) + (m + 1)(d_2 - 1), \quad L = \binom{m + n + 1 + D}{m + n + 1}.$$

Let $y_i = \delta^i y$ for all $i = 0, 1, \ldots, m + n$. For each $a = (a_0, \ldots, a_{m+n}) \in \mathbb{N}^{m+n+1}$, $Y^a = y_0^{a_0} \cdots y_{m+n}^{a_{m+n}}$ is a power product in S_{m+n}. M^D_{m+n+1} stands for the set of all power products in S_{m+n} of degree less than or equal to D. Obviously, the cardinality of M^D_{m+n+1} equals L. In a similar way it is possible to define $M^{D-d_1}_{m+n+1}$ which has $L_1 = \binom{m+n+1+D-d_1}{m+n+1}$ monomials, and $M^{D-d_2}_{m+n+1}$ which has $L_2 = \binom{m+n+1+D-d_2}{m+n+1}$ monomials. The monomials in M^D_{m+n+1}, $M^{D-d_1}_{m+n+1}$ and $M^{D-d_2}_{m+n+1}$ are totally ordered using first the degree and then the lexicographic order derived from $y_0 < y_1 < \cdots < y_{m+n}$.

Definition 1 The $(((n + 1)L_1 + (m + 1)L_2) \times L)$-matrix

$$M(\delta, n, m) = M(\delta^n p_1, \ldots, \delta p_1, p_1, \delta^m p_2, \ldots, \delta p_2, p_2),$$

is defined in the following way: for each i such that $(j - 1)L_1 < i \leq jL_1$ the coefficients of the polynomial $Y^a \delta^{n+1-j} p_1$ are the entries of the ith row for each $Y^a \in M^{D-d_1}_{m+n+1}$ and each $j = 1, \ldots, n + 1$, while for each i such that

$$(n + 1)L_1 + (j - n - 2)L_2 < i \leq (n + 1)L_1 + (j - n - 1)L_2$$

the coefficients of the polynomial $Y^a \delta^{m+n+2-j} p_2$ are the entries of the ith row for each $Y^a \in M_{m+n+1}^{D-d_2}$ and each $j = n+2, \ldots, m+n+2$, that are written with respect to the power products in M_{m+n+1}^{D} in decreasing order.

Definition 2 The differential resultant of p_1 and p_2 is defined to be

$$\gcd(\det(P) : P \text{ is an } (L \times L)\text{-submatrix of } M(\delta, n, m)).$$

2.2 A Counter Example and Definition of Differential Resultant

In this subsection, we use Carrà Ferro's method [5] to construct matrix formula of two nonlinear generic ordinary differential polynomials with order one and degree two,

$$\begin{aligned} g_1 &= a_0 y_1^2 + a_1 y_1 y + a_2 y^2 + a_3 y_1 + a_4 y + a_5, \\ g_2 &= b_0 y_1^2 + b_1 y_1 y + b_2 y^2 + b_3 y_1 + b_4 y + b_5, \end{aligned} \tag{1}$$

where, hereinafter, $y_1 = \delta y$ and a_i, b_i with $i = 0, \ldots, 5$ are generic differential indeterminates.

For differential polynomials in (1), we have $d_1 = d_2 = 2, m = n = 1, D = 5, L = 56$, and $L_1 = L_2 = 20$. The set of column monomials is

$$M_3^5 = \{y_2^5, y_2^4 B_3^1, y_2^3 B_3^2, y_2^2 B_3^3, y_2 B_3^4, B_3^5\},$$

where, and throughout the paper, $B = \{1, y, y_1, y_2\}$ and B_i^j denotes all monomials of total degree less than or equal to j in the first i elements of B. For example, $B_2^2 = \{1, y, y^2\}$ and $B_3^2 = \{1, y, y_1, y^2, yy_1, y_1^2\}$. Note that the monomials of M_3^3 are $M_3^3 = \{y_2^3, y_2^2 B_3^1, y_2 B_3^2, B_3^3\} = B_4^3$.

According to Definition 1, $M(\delta, 1, 1)$ is an 80×56 matrix

$$\begin{array}{c} \begin{matrix} y_2^5 & y_2^4 y_1 & \cdots & y_2^3 & \cdots & y & 1 \end{matrix} \\ \left(\begin{matrix} 0 & d_1 a_0 & \cdots & \delta a_5 & \cdots & 0 & 0 \\ & & & \cdots\cdots & & & \\ 0 & 0 & \cdots & 0 & \cdots & \delta a_4 & \delta a_5 \\ 0 & 0 & \cdots & a_0 & \cdots & 0 & 0 \\ & & & \cdots\cdots & & & \\ 0 & 0 & \cdots & 0 & \cdots & a_4 & a_5 \\ 0 & d_2 b_0 & \cdots & \delta b_5 & \cdots & 0 & 0 \\ & & & \cdots\cdots & & & \\ 0 & 0 & \cdots & 0 & \cdots & \delta b_4 & \delta b_5 \\ 0 & 0 & \cdots & b_0 & \cdots & 0 & 0 \\ & & & \cdots\cdots & & & \\ 0 & 0 & \cdots & 0 & \cdots & b_4 & b_5 \end{matrix}\right) \begin{matrix} \left.\right\} B_3^3 \delta g_1 \\ \left.\right\} B_3^3 g_1 \\ \left.\right\} B_3^3 \delta g_2 \\ \left.\right\} B_3^3 g_2 \end{matrix} \end{array}$$

Obviously, the entries of the first column are all zero in $M(\delta, 1, 1)$, since the monomial y_2^5 never appears in any row polynomial $Y * f$, where the monomial $Y \in M_3^3$ and $f \in \{\delta g_1, g_1, \delta g_2, g_2\}$. Consequently, the differential resultant of g_1 and g_2 is identically zero according to this definition.

Actually, the differential resultant is defined using an ideal approach for two generic differential polynomials in one differential indeterminate in [17] and $n+1$ generic differential polynomials in n differential indeterminates in [9]. f is said to be a *generic differential polynomial* in differential indeterminates $\mathbb{Y} = \{y_1, \ldots, y_n\}$ with order s and degree m if f contains all the monomials of degree up to m in $y_1, \ldots, y_n$ and their derivatives up to order s. Furthermore, the coefficients of f are also differential indeterminates. For instance, g_1 and g_2 in (1) are two generic differential polynomials.

Theorem 3 [9] *Let $p_0, p_1, \ldots, p_n$ be generic differential polynomials with order s_i and coefficient sets $\mathbf{u}_i$ respectively. Then $[p_0, p_1, \ldots, p_n]$ is a prime differential ideal in $\mathbb{Q}\{\mathbb{Y}, \mathbf{u}_0, \ldots, \mathbf{u}_n\}$. And*

$$[p_0, p_1, \ldots, p_n] \cap \mathbb{Q}\{\mathbf{u}_0, \ldots, \mathbf{u}_n\} = \mathrm{sat}(\mathbf{R}(\mathbf{u}_0, \ldots, \mathbf{u}_n)) \tag{2}$$

is a prime differential ideal of codimension one, where $\mathbf{R}$ is defined to be the differential sparse resultant of $p_0, p_1, \ldots, p_n$, which has the following properties

(a) $\mathbf{R}(\mathbf{u}_0, \mathbf{u}_1, \ldots, \mathbf{u}_n)$ is an irreducible polynomial and differentially homogeneous in each $\mathbf{u}_i$.
(b) $\mathbf{R}(\mathbf{u}_0, \mathbf{u}_1, \ldots, \mathbf{u}_n)$ is of order $h_i = s - s_i$ in $\mathbf{u}_i$ $(i = 0, \ldots, n)$ with $s = \sum_{l=0}^{n} s_l$.
(c) The differential resultant can be written as a linear combination of p_i and the derivatives of p_i up to the order $s - s_i$. Precisely, we have

$$\mathbf{R}(\mathbf{u}_0, \mathbf{u}_1, \ldots, \mathbf{u}_n) = \sum_{i=0}^{n} \sum_{j=0}^{s-s_i} h_{ij} p_i^{(j)}$$

where $h_{ij} \in \mathbb{Q}[\mathbb{Y}, \mathbb{Y}^{(1)} \ldots, \mathbb{Y}^{(s)}, \mathbf{u}_0^{(s-s_0)}, \ldots, \mathbf{u}_n^{(s-s_n)}]$.

3 Matrix Formula for Differential Polynomials

In this section, we will give a matrix representation for the following generic differential polynomials in y

$$\begin{aligned} f_1 &= a_{y_1^{d_1}} y_1^{d_1} + a_{y_1^{d_1-1}y} y_1^{d_1-1} y + \cdots + a_0, \\ f_2 &= b_{y_1^{d_2}} y_1^{d_2} + b_{y_1^{d_2-1}y} y_1^{d_2-1} y + \cdots + b_0, \end{aligned} \tag{3}$$

where $y_1 = \delta y$, $1 \leq d_1 \leq d_2$, and $a_{y_1^{d_1}}, \ldots, a_0, b_{y_1^{d_2}}, \ldots, b_0$ are differential indeterminates.

3.1 Matrix Construction

In this subsection, we show that when choosing a proper column monomial set, a square matrix can be constructed following Macaulay's idea [16].

By (c) of Theorem 3, the differential resultant for f_1, f_2 can be written as a linear combination of $f_1, f_2, \delta f_1$ and δf_2 which are treated as polynomials in variables $y, y_1, y_2 = \delta y_1$. So, we will try to construct a matrix representation for the differential resultant of f_1 and f_2 from these four polynomials.

From Sect. 2, it is easy to see that the main problem in Carrà Ferro's definition is that the matrix $M(\delta, n, m)$ contains too many columns. Or equivalently, the monomial set M^D_{m+n+1} used to represent the columns is too large.

Consider the monomial set

$$\mathscr{E} = B_3^D \cup y_2 B_3^{D-1} \tag{4}$$

with $D = 2d_1 + 2d_2 - 3$. We will show that if using $\mathscr{E}$ as the column monomial set, a nonsingular square matrix can be constructed.

Define the *main monomial* of polynomials $p_1 = \delta f_1, p_2 = \delta f_2, p_3 = f_1, p_4 = f_2$ to be

$$\mathrm{mm}(p_1) = y_2\, y_1^{d_1-1}, \quad \mathrm{mm}(p_2) = y_1^{d_2}, \quad \mathrm{mm}(p_3) = y^{d_1}, \quad \mathrm{mm}(p_4) = 1. \tag{5}$$

Then, we can divide $\mathscr{E}$ into four mutually disjoint sets $\mathscr{E} = \mathscr{S}_1 \cup \mathscr{S}_2 \cup \mathscr{S}_3 \cup \mathscr{S}_4$, where

$$\begin{aligned}
\mathscr{S}_1 &= \{Y^\alpha \in \mathscr{E} : \mathrm{mm}(p_1) \text{ divides } Y^\alpha\},\\
\mathscr{S}_2 &= \{Y^\alpha \in \mathscr{E} : \mathrm{mm}(p_1) \text{ does not divide } Y^\alpha \text{ but } \mathrm{mm}(p_2) \text{ does}\},\\
\mathscr{S}_3 &= \{Y^\alpha \in \mathscr{E} : \mathrm{mm}(p_1), \mathrm{mm}(p_2) \text{ do not divide } Y^\alpha \text{ but } \mathrm{mm}(p_3) \text{ does}\},\\
\mathscr{S}_4 &= \{Y^\alpha \in \mathscr{E} : \mathrm{mm}(p_1), \mathrm{mm}(p_2), \mathrm{mm}(p_3) \text{ do not divide } Y^\alpha\}.
\end{aligned} \tag{6}$$

As a consequence, we can write down a system of equations:

$$\begin{aligned}
Y^\alpha/\mathrm{mm}(p_1) * p_1 &= 0, \quad \text{for } Y^\alpha \in \mathscr{S}_1,\\
Y^\alpha/\mathrm{mm}(p_2) * p_2 &= 0, \quad \text{for } Y^\alpha \in \mathscr{S}_2,\\
Y^\alpha/\mathrm{mm}(p_3) * p_3 &= 0, \quad \text{for } Y^\alpha \in \mathscr{S}_3,\\
Y^\alpha/\mathrm{mm}(p_4) * p_4 &= 0, \quad \text{for } Y^\alpha \in \mathscr{S}_4.
\end{aligned} \tag{7}$$

Observe that the total number of equations is the number of elements in $\mathscr{E}$ and denoted by $N = (D+1)^2$.

Regarding the monomials in (7) as unknowns, we obtain a system of N linear equations in these monomial unknowns. Denote the coefficient matrix of the system of linear equations (7) by D_{d_1,d_2} whose entries are zero or the coefficients of f_i and $\delta f_i, i = 1, 2$.

Note that the main monomials of the polynomials are not the maximal monomials in the sense of Macaulay [16], so the monomials on the left hand side of (7) may not be contained in $\mathscr{E}$. Next, we prove that this does not occur for our main monomials.

Lemma 1 *The coefficient matrix D_{d_1,d_2} of system* (7) *is square.*

Proof The coefficient matrix of system (7) has $N = |\mathscr{E}|$ rows. In order to prove the lemma, it suffices to demonstrate that, for each $Y^\alpha \in \mathscr{S}_i, i = 1, \ldots, 4$, all monomials in $[Y^\alpha/\mathrm{mm}(p_i)] * p_i$ are contained in $\mathscr{E}$. Recall that $\mathscr{E} = B_3^D \cup y_2 B_3^{D-1}$. Then by (6), one has

$$\begin{aligned}
\mathscr{S}_1 &= B_3^{D-d_1} * \mathrm{mm}(p_1) = B_3^{D-d_1} * \mathrm{mm}(\delta f_1),\\
\mathscr{S}_2 &= B_3^{D-d_2} * \mathrm{mm}(p_2) = B_3^{D-d_2} * \mathrm{mm}(\delta f_2),\\
\mathscr{S}_3 &= T_1 * \mathrm{mm}(p_3) = T_1 * \mathrm{mm}(f_1),\\
\mathscr{S}_4 &= T_2 * \mathrm{mm}(p_4) = T_2 * \mathrm{mm}(f_2),
\end{aligned} \tag{8}$$

where

$$\begin{aligned}
T_1 &= \left\{\left(\bigcup_{i=0}^{d_2-1} y_1^i B_2^{D-d_1-i}\right) \bigcup \left(y_2 \bigcup_{i=0}^{d_1-2} y_1^i B_2^{D-d_1-1-i}\right)\right\},\\
T_2 &= \left\{\left(\bigcup_{i=0}^{d_2-1} y_1^i B_2^{d_1-1}\right) \bigcup \left(y_2 \bigcup_{i=0}^{d_1-2} y_1^i B_2^{d_1-1}\right)\right\}.
\end{aligned} \tag{9}$$

Note that the representation of $\mathscr{S}_2$ in (8) is obtained with the help of the condition $d_1 \le d_2$.

Hence, Eq. (7) become

$$\begin{aligned}
B_3^{D-d_1} * \delta f_1 &= 0,\\
B_3^{D-d_2} * \delta f_2 &= 0,\\
T_1 * f_1 &= 0,\\
T_2 * f_2 &= 0.
\end{aligned} \tag{10}$$

Since the monomial set of δf_1 is $B_3^{d_1} \cup y_2 * B_3^{d_1-1}$, the monomial set of $B_3^{D-d_1} * \delta f_1$ is $B_3^{D-d_1} * (B_3^{d_1} \cup y_2 * B_3^{d_1-1}) = B_3^D \cup y_2 B_3^{D-1} = \mathscr{E}$. So monomials in the first set of equations in (10) are in $\mathscr{E}$. Since the monomial set of f_1 is $B_3^{d_1} = \cup_{l=0}^{d_1} y_1^l B_2^{d_1-l}$, the monomial set of $T_1 * f_1$ is $T_{11} \cup y_2 T_{12}$, where $T_{11} = \bigcup_{k=0}^{d_1+d_2-1} y_1^k B_2^{D-k}$ and $T_{12} = \bigcup_{k=0}^{2d_1-2} y_1^k B_2^{D-k-1}$. Since $d_1 \ge 1$ and $d_2 \ge 1$, we have $d_1 + d_2 - 1 \le$

$D = 2d_1 + 2d_2 - 3$ and hence $T_{11} \subset B_3^D$. Since $d_1 \geq 1$ and $d_2 \geq 1$, we have $2d_1 - 2 \leq D - 1 = 2d_1 + 2d_2 - 4$ and hence $T_{12} \subset B_3^{D-1}$. As a consequence, $T_{11} \cup y_2 T_{12} \subset \mathcal{E}$ and the monomials in the third set of equations in (10) are in $\mathcal{E}$. Other cases can be proved similarly. Thus all monomials in the left hand side of (10) are in $\mathcal{E}$. This proves the lemma. □

It is worthy to say that, due to the decrease of the number of monomials in $\mathcal{E}$ compared with the method by Carrà Ferro, the size of the matrix D_{d_1,d_2} decreases significantly. Precisely, when we choose $m = n = 1$, the size of the matrix D_{d_1,d_2} is $(D+1)^2 \times (D+1)^2$ and one can see that the size of the matrix given by Carrà Ferro is $2\binom{3+D-d_2}{3} + 2\binom{3+D-d_1}{3} \times \binom{3+D}{3}$ where $D = 2d_1 + 2d_2 - 3$. One can show that $2\binom{3+D-d_2}{3} + 2\binom{3+D-d_1}{3} \geq \binom{3+D}{3} \geq (D+1)^2$ for $d_1, d_2 \geq 1$ and the last equality holds only for $d_1 = d_2 = 1$.

3.2 Matrix Representation for Differential Resultant

In this section, we show that $\det(D_{d_1,d_2})$ is not identically equal to zero and contains the differential resultant as a factor.

Lemma 2 $\det(D_{d_1,d_2})$ *is not identically equal to zero.*

Proof It suffices to show that there exists a unique monomial in the sense that it is different from all other monomials in the expansion of $\det(D_{d_1,d_2})$.

The coefficients of the main monomials in δf_1, δf_2, f_1 and f_2 are respectively

$$\begin{array}{lll} \delta f_1 : a_{y_1^{d_1}} & \text{the coefficient of } \mathrm{mm}(\delta f_1) = y_2 y_1^{d_1-1}, & \\ \delta f_2 : \delta b_{y_1^{d_2}} + b_{y_1^{d_2-1}y} & \text{the coefficient of } \mathrm{mm}(\delta f_2) = y_1^{d_2}, & \\ f_1 : a_{y^{d_1}} & \text{the coefficient of } \mathrm{mm}(f_1) = y^{d_1}, & \\ f_2 : b_0 & \text{the coefficient of } \mathrm{mm}(f_2) = 1. & \end{array} \quad (11)$$

We will show that the monomial $(a_{y_1^{d_1}})^{n_1}(\delta b_{y_1^{d_2}})^{n_2}(a_{y^{d_1}})^{n_3}(b_0)^{n_4}$ is a unique one by the following four steps, where n_i is the number of elements in $\mathcal{S}_i$ with $i = 1, \ldots, 4$. From (10), $n_1 = |B_3^{D-d_1}|$, $n_2 = |B_3^{D-d_2}|$, $n_3 = |T_1|$, $n_4 = |T_2|$.

Step 1. Observe that, in δf_1, $a_{y^{d_1}}$ only occurs in the coefficient of $y_1 y^{d_1-1}$ with the form $a_{y^{d_1}} + \delta a_{y_1 y^{d_1-1}}$. Furthermore, $\delta a_{y_1 y^{d_1-1}}$ only occurs in this term given by δf_1 and no other places of D_{d_1,d_2}. So using the transformation

$$\delta a_{y_1 y^{d_1-1}} = c_{y_1 y^{d_1-1}} - d_1 a_{y^{d_1}}, \quad \text{with other coefficients unchanged,} \quad (12)$$

where $c_{y_1 y^{d_1-1}}$ is a new differential indeterminate, D_{d_1,d_2} is transformed to a new matrix which is singular if and only if the original one is singular. Still denote the matrix by D_{d_1,d_2}.

From (10), for a monomial $M \in T_1$, $a_{y^{d_1}}$ is the coefficient of the monomial My^{d_1} in each polynomial $T_1 * f_1$ and hence in each corresponding row of D_{d_1,d_2}. Then $a_{y^{d_1}}$ is in different rows and columns of D_{d_1,d_2}, and this gives the factor $(a_{y^{d_1}})^{n_3}$. Delete those rows and columns of D_{d_1,d_2} containing $a_{y^{d_1}}$ and denote the remaining matrix by $D^{(1)}_{d_1,d_2}$. From (10), the columns deleted are represented by monomials $y^{d_1}T_1$. So, $D^{(1)}_{d_1,d_2}$ is still a square matrix.

Step 2. Let $M \in B_3^{D-d_1}$. The term $a_{y_1^{d_1}}$ occurs in $M * \delta f_1$ as the coefficient of the monomial $y_2 y_1^{d_1-1}M$, or equivalently it occurs in the columns represented by $y_2 y_1^{d_1-1}M$. This gives the factor $(a_{y_1^{d_1}})^{n_1}$. It is easy to check that $a_{y_1^{d_1}}$ does not occur in other places of $D^{(1)}_{d_1,d_2}$. From the definition for T_1 (9), the columns deleted in case 1 correspond those columns represented by monomials of the form $y_2^{k_2} y_1^{k_1} y^{d_1}$ where either $k_2 = 0$ and $k_1 < d_2$ or $k_2 = 1$ and $k_1 < d_1 - 1$. Then $\{y^{d_1}T_1\} \cap \{y_2 y_1^{d_1-1} B_3^{D-d_1}\} = \emptyset$, or equivalently those columns of D_{d_1,d_2} containing $a_{y_1^{d_1}}$ are still in $D^{(1)}_{d_1,d_2}$. Similar to case 1, one can delete those rows and columns of $D^{(1)}_{d_1,d_2}$ containing $a_{y_1^{d_1}}$ and denote the remaining matrix by $D^{(2)}_{d_1,d_2}$ which is still a square matrix. From (10), the columns deleted are represented by monomials $y_2 y_1^{d_1-1} B_3^{D-d_1}$.

Step 3. At the moment, $D^{(2)}_{d_1,d_2}$ only contains coefficients of f_2 and δf_2. Observe that b_0 only occurs in the rows corresponding to $T_2 * f_2$, where T_2 is defined in (9). Note that δb_0 instead of b_0 occurs in δf_2. Since $\{y^{d_1}T_1\} \cap T_2 = \emptyset$ and $\{y_2 y_1^{d_1-1} B_3^{D-d_1}\} \cap T_2 = \emptyset$, the columns of D_{d_1,d_2} containing b_0, represented by T_2, are not deleted in case 1 and case 2. Then, we have the factor $(b_0)^{n_4}$. Similarly, delete those rows and columns of $D^{(2)}_{d_1,d_2}$ containing b_0 and denote the remaining matrix by $D^{(3)}_{d_1,d_2}$ which is still a square matrix. From (10), the columns deleted are represented by monomials T_2.

Step 4. From (10), the rows of $D^{(3)}_{d_1,d_2}$ are from coefficients of $B_3^{D-d_2} * \delta f_2$. The term $\delta b_{y_1^{d_2}}$ is in the coefficient of the monomial $M * y_1^{d_2}$ in $M * \delta f_2$ for $M \in B_3^{D-d_2}$, and $\delta b_{y_1^{d_2}}$ does not occur in other places of $M * \delta f_2$. Furthermore, since $\{y^{d_1}T_1\} \cap \{y_1^{d_2} B_3^{D-d_2}\} = \emptyset$, $\{y_2 y_1^{d_1-1} B_3^{D-d_1}\} \cap \{y_1^{d_2} B_3^{D-d_2}\} = \emptyset$, and $T_2 \cap \{y_1^{d_2} B_3^{D-d_2}\}\} = \emptyset$, the columns containing the term $\delta b_{y_1^{d_2}}$ are not deleted in the first three cases. Then, we have the factor $(\delta b_{y_1^{d_2}})^{n_2}$.

Following the above procedures step by step, the coefficients of choosing main monomials of the polynomials f_1, f_2, δf_1 and δf_2 occur in each row and each column of D_{d_1,d_2} and only once, and the monomial $\left(a_{y_1^{d_1}}\right)^{n_1} \left(\delta b_{y_1^{d_2}}\right)^{n_2} \left(a_{y^{d_1}}\right)^{n_3} \left(b_0\right)^{n_4}$ is a unique one in the expansion of the determinant of D_{d_1,d_2}. So the lemma follows. □

Note that the selection of main monomials in above algorithm is not unique, thus there may exist other ways to construct matrix formula for system (3).

Corollary 4 *Following the above notations, for any* $Y^\alpha \in \mathscr{S}_i$, *if all monomials of* $[Y^\alpha/\mathrm{mm}(p_j)] * p_j$ *are contained in* $\mathscr{E}(j \neq i)$, *then the rearranged matrix, which is obtained by replacing the row polynomials* $[Y^\alpha/\mathrm{mm}(p_i)] * p_i$ *by* $[Y^\alpha/\mathrm{mm}(p_j)] * p_j$, *is not identically equal to zero.*

Corollary 4 follows from the fact that the proof of Lemma 2 is independent of the number of elements in $\mathscr{S}_i$ as long as the main monomials are the same.

The relation between $\det(D_{d_1,d_2})$ and differential resultant of f_1 and f_2, denoted by $\mathbf{R}$, is stated as the following theorem.

Theorem 5 $\det(D_{d_1,d_2})$ *is a nonzero multiple of* $\mathbf{R}$.

Proof From Lemma 2, $\det(D_{d_1,d_2})$ is nonzero. In the matrix D_{d_1,d_2}, multiply a column monomial $M \neq 1$ in $\mathscr{E}$ to the corresponding column and add the result to the constant column corresponding to the monomial 1. Then the constant column becomes $Y^\alpha * p_i$ with $p_1 = \delta f_1, p_2 = \delta f_2, p_3 = f_1, p_4 = f_2$ and $Y^\alpha \in \mathscr{S}_i/\mathrm{mm}(p_i), i = 1, \ldots, 4$. Since a determinant is multilinear on the columns, expanding the matrix by the constant column, we obtain

$$\det(D_{d_1,d_2}) = h_1 f_1 + h_2 \delta f_1 + h_3 f_2 + h_4 \delta f_2, \tag{13}$$

where h_j are differential polynomials. From (2), $\det(D_{d_1,d_2}) \in \mathrm{sat}(\mathbf{R})$. On the other hand, from Theorem 3, $\mathbf{R}$ is irreducible and the order of $\mathbf{R}$ about the coefficients of f_1, f_2 is one. Therefore, $\mathbf{R}$ must divide $\det(D_{d_1,d_2})$. □

From Theorem 5, we can easily deduce a degree bound $N = 4(d_1 + d_2 - 1)^2$ for the differential resultant of f_1 and f_2. The main advantage to represent the differential resultant as a factor of the determinant of a matrix is that we can use fast algorithms of matrix computation to compute the differential resultant as did in the algebraic case [4].

Suppose that $\det(D_{d_1,d_2})$ is expanded as a polynomial. Then the differential resultant can be found by the following result.

Corollary 6 *Suppose* $\det(D_{d_1,d_2}) = \prod_{i=1}^{s} P_i^{e_i}$ *is an irreducible factorization of* $\det(D_{d_1,d_2})$ *in* $\mathbb{Q}[C_{f_1}, C_{f_2}]$, *where* $C_{f_i}, i = 1, 2$ *are the sets of coefficients of* f_i. *Then there exists a unique factor, say* P_1, *which is in* $[f_1, f_2]$ *and is the differential resultant of* f_1 *and* f_2.

Proof From (c) of Theorems 3 and 5, $\mathbf{R} \in [f_1, f_2]$ and is an irreducible factor of $\det(D_{d_1,d_2})$. Suppose $\det(D_{d_1,d_2})$ contains another factor, say P_2, which is also in $[f_1, f_2]$. Then $P_2 \in \mathrm{sat}(\mathbf{R})$ by (2). Since $\mathbf{R}$ is irreducible with order one and P_2 is of order no more than one, P_2 must equal $\mathbf{R}$, which contradicts to the hypothesis. □

3.3 Example (1) Revisited

In this section, we apply the method just proposed to construct a matrix representation for the differential resultant of the system (1).

Following the method given in the proceeding section, for system (1), we have $D = 2d_1 + 2d_2 - 3 = 5$ and select the main monomials of $\delta g_1, \delta g_2, g_1, g_2$ are $y_2 y_1, y_1^2, y^2, 1$, respectively. Then $\mathscr{E} = y_2 B_3^4 \cup B_3^5$ is divided into the following four disjoint sets

$$
\begin{aligned}
\mathscr{S}_1 &= y_2 y_1 B_3^3, \\
\mathscr{S}_2 &= y_1^2 B_3^3, \\
\mathscr{S}_3 &= y^2 \left[B_2^3 \cup y_1 B_2^2 \cup y_2 B_2^2 \right], \\
\mathscr{S}_4 &= B_2^1 \cup y_1 B_2^1 \cup y_2 B_2^1.
\end{aligned} \tag{14}
$$

Using (7) and regarding the monomials in $\mathscr{E}$ as variables, we obtain the matrix $D_{2,2}$, which is a 36×36 square matrix in the following form.

$$
\begin{array}{c}
\begin{array}{ccccccc}
y_2 y_1^4 & y_2 y_1^3 y & \cdots & y_2 y_1^2 y^2 & \cdots & y & 1
\end{array} \\
\left(
\begin{array}{ccccccc}
2a_0 & a_1 & \cdots & 0 & \cdots & 0 & 0 \\
 & & & \cdots\cdots & & & \\
0 & 0 & \cdots & 0 & \cdots & \delta a_4 & \delta a_5 \\
0 & 0 & \cdots & a_0 & \cdots & 0 & 0 \\
 & & & \cdots\cdots & & & \\
0 & 0 & \cdots & 0 & \cdots & a_4 & a_5 \\
2b_0 & b_1 & \cdots & 0 & \cdots & 0 & 0 \\
 & & & \cdots\cdots & & & \\
0 & 0 & \cdots & 0 & \cdots & \delta b_4 & \delta b_5 \\
0 & 0 & \cdots & b_0 & \cdots & 0 & 0 \\
 & & & \cdots\cdots & & & \\
0 & 0 & \cdots & 0 & \cdots & b_4 & b_5
\end{array}
\right)
\begin{array}{l}
\left.\begin{array}{c} \\ \\ \\ \end{array}\right\} B_3^3 \delta g_1 \\
\left.\begin{array}{c} \\ \\ \\ \end{array}\right\} (B_2^3 \cup y_1 B_2^2 \cup y_2 B_2^2) g_1 \\
\left.\begin{array}{c} \\ \\ \\ \end{array}\right\} B_3^3 \delta g_2 \\
\left.\begin{array}{c} \\ \\ \\ \end{array}\right\} (B_2^1 \cup y_1 B_2^1 \cup y_2 B_2^1) g_2
\end{array}
\end{array}
$$

As shown in the proof of Lemma 2, $(a_0)^{10}(a_2)^{10}(b_5)^6(\delta b_0)^{10}$ is a unique monomial in the expansion of the determinant of $D_{2,2}$. Hence, the differential resultant of g_1 and g_2 is a factor of $\det(D_{2,2})$. Note that in Carrà Ferro's construction for g_1, g_2, $M(\delta, 1, 1)$ is an 80×56 matrix, which is larger than $D_{2,2}$.

In particular, suppose $a_0 = b_0 = 1$ and a_i, b_i are differential constants, i.e., $\delta a_i = \delta b_i = 0, i = 1, \ldots, 5$. Then $D_{2,2}$ can be expanded as a polynomial and the differential resultant of g_1 and g_2 can be found with Corollary 6, which is a polynomial of degree 12 and contains 3,210 terms. This is the same as the result obtained in [22].

4 Differential Resultant as the Algebraic Sparse Resultant

In this section, we show that differential resultant of f_1 and f_2 is a factor of the algebraic sparse resultant of the system $\{f_1, f_2, \delta f_1$ and $\delta f_2\}$.

4.1 Results About Algebraic Sparse Resultant

In this subsection, notions of algebraic sparse resultants are introduced. Details can be found in [4, 7, 11, 20].

A set S in $\mathbb{R}^n$ is said to be convex if it contains the line segment connecting any two points in S. If a set is not itself convex, its convex hull is the smallest convex set containing it and denoted by Conv(S). A set $V = \{a_1, \ldots, a_m\}$ is called a vertex set of a convex set Q if each point $q \in Q$ can be expressed as:

$$q = \sum_{j=1}^{m} \lambda_j a_j, \quad \text{with} \sum_{j=1}^{m} \lambda_j = 1 \text{ and } \lambda_j \geq 0,$$

and each a_j is called a *vertex* of Q.

Consider $n+1$ generic sparse polynomials in the algebraic indeterminates $x_1, \ldots, x_n$:

$$p_i = u_{i0} + u_{i1}M_{i1} + \cdots + u_{il_i}M_{il_i}, \quad i = 1, \ldots, n+1,$$

where u_{ij} are indeterminates and $M_{ik} = \prod_{s=1}^{n} x_s^{e^{ik_s}}$ are monomials in $\mathbb{Q}[x_1, \ldots, x_n]$ with exponent vectors $a_{ik} = (e^{ik_1}, \ldots, e^{ik_n}) \in \mathbb{Z}^n$. Note that we assume each p_i contains a constant term u_{i0}. For $a = (e^1, \ldots, e^n) \in \mathbb{Z}^n$, the corresponding monomial is denoted as $M(a) = \prod_{s=1}^{n} x_s^{e^s}$.

The finite set $\mathscr{A}_i \subset \mathbb{Z}^n$ of all monomial exponents appearing in p_i is called the *support* of p_i, denoted by supp(p_i). Its cardinality is $l_i = |\mathscr{A}_i|$. The *Newton polytope* $Q_i \subset \mathbb{R}^n$ of p_i is the *convex hull* of $\mathscr{A}_i$, denoted by $Q_i = \text{Conv}(\mathscr{A}_i)$. Since Q_i is the convex hull for a finite set of points, it must have a vertex set. For simplicity, we assume that each $\mathscr{A}_i$ is of dimension n as did in [11, p. 252]. Let $\mathbf{u}$ be the set of coefficients of $p_i, i = 0, \ldots, n$. Then, the ideal

$$(p_1, p_2, \ldots, p_{n+1}) \cap \mathbb{Q}[\mathbf{u}] = (\mathscr{R}(\mathbf{u})) \tag{15}$$

is principal and the generator $\mathscr{R}$ is defined to be the sparse resultant of $p_1, \ldots, p_{n+1}$ [11, p. 252]. When the coefficients $\mathbf{u}$ of p_i are specialized to certain values $\mathbf{v}$, the sparse resultant for the specialized polynomials is defined to be $\mathscr{R}(\mathbf{v})$. The matrix representation of $\mathscr{R}$ is associated with the decomposition of the Minkowski sum of the Newton polytopes Q_i.

The *Minkowski sum* of the convex polytopes Q_i

$$Q = Q_1 + \cdots + Q_{n+1} = \{q_1 + \cdots + q_{n+1} | q_i \in Q_i\}.$$

is still convex and of dimension n.

Choose sufficiently small numbers $\delta_i > 0$ and let $\delta = (\delta_1, \ldots, \delta_n) \in \mathbb{R}^n$ be a perturbed vector. Then the points which lie in the interior of the perturbed district $\mathcal{E} = \mathbb{Z}^n \cap (Q + \delta)$ are chosen as the *column monomial set* [4] to construct the matrix for the sparse resultant.

Choose $n + 1$ sufficiently generic linear lifting functions $l_1, \ldots, l_{n+1} \in \mathbb{Z}[x_1, \ldots, x_n]$ and define the lifted Newton polytopes $\widehat{Q}_i = \{\widehat{q}_i = (q_i, l_i(q_i)) : q_i \in Q_i\} \subset \mathbb{R}^{n+1}$. Let

$$\widehat{Q} = \sum_{i=1}^{n+1} \widehat{Q}_i \subset \mathbb{R}^{n+1}$$

which is an $(n + 1)$-dimensional convex polytope. The *lower envelope* of $\widehat{Q}$ with respect to vector $(0, \ldots, 0, 1) \in \mathbb{R}^{n+1}$ is the union of all the n-dimensional faces of $\widehat{Q}$, whose inner normal vector has positive last component.

Let $\pi : (q_1, \ldots, q_{n+1}) \mapsto (q_1, \ldots, q_n)$ be the projection to the first n coordinates from $\mathbb{R}^{n+1}$ to $\mathbb{R}^n$. Then π is a one to one map between the lower envelope of $\widehat{Q}$ and Q [4]. The genericity requirements on l_i assure that every point $\widehat{q}$ on the lower envelope can be uniquely expressed as $\widehat{q} = \widehat{q}_1 + \cdots + \widehat{q}_{n+1}$ with $\widehat{q}_i \in \widehat{Q}_i$, such that the sum of the projections under π of these points leads to a unique sum of $q = q_1 + \cdots + q_{n+1} \in Q \subset \mathbb{R}^n$ with $q_i \in Q_i$, which is called the *optimal (Minkowski) sum* of q. For $\mathcal{F}_i \subset Q_i$, $R = \sum_{i=1}^{n+1} \mathcal{F}_i$ is called an optimal sum, if each element of R can be written as a unique optimal sum $\sum_{i=0}^{n} p_i$ for $p_i \in \mathcal{F}_i$.

A polyhedral *subdivision* of an n-dimensional polytope Q consists of finitely many n-dimensional polytopes $R_1, \ldots, R_s$, called the cells of the subdivision, such that $Q = R_1 \cup \cdots \cup R_s$ and for $i \neq j$ and $R_i \cap R_j$ is either empty or a face of both R_i and R_j. A polyhedral subdivision is called a *mixed subdivision* if each cell R_l can be written as an optimal sum $R_l = \sum_{i=1}^{n+1} \mathcal{F}_i$, where each $\mathcal{F}_i$ is a face of Q_i and $n = \sum_{i=1}^{n+1} \dim(\mathcal{F}_i)$. Furthermore, if $R_j = \sum_{i=1}^{n+1} \mathcal{F}'_i$ is another cell in the subdivision, then $R_l \cap R_j = \sum_{i=1}^{n+1} (\mathcal{F}_i \cap \mathcal{F}'_i)$. A cell $R_l = \sum_{i=1}^{n+1} \mathcal{F}_i$ is called *mixed* if $\dim(\mathcal{F}_i) \leq 1$ for all i; otherwise, it is a *nonmixed cell.* As a result of $n = \sum_{i=1}^{n+1} \dim(\mathcal{F}_i)$, a mixed cell has one unique vertex, which satisfies $\dim(\mathcal{F}_{i_0}) = 0$, while a nonmixed cell has at least two vertices.

Recall $\delta = (\delta_1, \delta_2, \ldots, \delta_n)$, where $0 < \delta_i < 1$. If $Q = R_1 \cup \cdots \cup R_s$ is a subdivision of Q, then $\delta + Q = (\delta + R_1) \cup \cdots \cup (\delta + R_s)$ is a subdivision of $\delta + Q$.

Let $q \in \mathbb{Z}^n \cap (Q + \delta)$ lie in the interior of a cell $\delta + \mathcal{F}_1 + \cdots + \mathcal{F}_{n+1}$ of a mixed subdivision for $Q + \delta$, where $\mathcal{F}_i$ is a face of Q_i. The *row content function* of q is defined as the largest integer such that $\mathcal{F}_i$ is a vertex. In fact, all the vertices in the optimal sum of p can be selected as the row content functions. Hence, we define *generalized row content functions* (GRC for brief) as one of the integers, not necessary largest, such that $\mathcal{F}_i$ is a vertex.

Suppose that we have a mixed subdivision of Q. With a fixed GRC, a sparse resultant matrix can be constructed as follows. For each $i = 1, \ldots, n+1$, define the subset S_i of $\mathcal{E}$ as follows:

$$S_i = \{q \in \mathcal{E} \mid \mathrm{GRC}(q) = (i, j_0)\}, \tag{16}$$

where $j_0 \in \{1, \ldots, m_i\}$, m_i is the number of vertexes of Q_i, and we obtain a disjoint union for $\mathcal{E}$:

$$\mathcal{E} = S_1 \cup \cdots \cup S_{n+1}. \tag{17}$$

For $q \in S_i$, let $q = q_1 + \cdots + q_{n+1} \in Q$ be an optimal sum of q. Then, q_i is a vertex of Q_i and the corresponding monomial $M(q_i)$ is called the *main monomial* of p_i and denoted by $\mathrm{mm}(p_i)$, similar to what we did in Sect. 3. Main monomials have the following important property [7, p. 350].

Lemma 3 *If $q \in S_i$, then the monomials in $(M(q)/\mathrm{mm}(p_i))p_i$ are contained in $\mathcal{E}$.*

Now consider the following equation systems

$$(M(q)/\mathrm{mm}(p_i))p_i, \quad q \in S_i, i = 1, \ldots, n+1. \tag{18}$$

Treating the monomials in $\mathcal{E}$ as variables, by Lemma 3, the coefficient matrix for the equations in (18) is an $|\mathcal{E}| \times |\mathcal{E}|$ square matrix, called the *sparse resultant matrix*. The sparse resultant of $p_i, i = 1, \ldots, n+1$ is a factor of the determinant of this matrix.

In [3, 4], Canny and Emiris used linear programming algorithms to find the row content functions and to construct S_i. We briefly describe this procedure below.

Now assume Q_i has the vertex set $V_i = \{a_{i1}, \ldots, a_{i\,m_i}\}$. A point $q \in \mathbb{Z}^n \cap (Q+\delta)$ implies that $q \in \sigma + \delta$ with a cell $\sigma \in Q$. In order to obtain the generalized row content functions of q, we wish to find the optimal sum of $q - \delta$ in terms of the vertexes in V_i. Introducing variables $\lambda_{ij}, i = 1, \ldots, n+1, j = 1, \ldots, m_i$, one has

$$q - \delta = \sum_{i=1}^{n+1} q_i = \sum_{i=1}^{n+1}\sum_{j=1}^{m_i} \lambda_{ij}\, a_{ij}, \quad \text{with} \sum_{j=1}^{m_i} \lambda_{ij} = 1 \text{ and } \lambda_{ij} \geq 0. \tag{19}$$

On the other hand, in order to make the lifted points lie on the lower envelope of $\widehat{Q}$, one must force the "height" of the listed points minimal, thus requiring to find λ_{ij} such that

$$\sum_{i=1}^{n+1}\sum_{j=1}^{m_i} \lambda_{ij}\, l_i(a_{ij}) \text{ to be minimized} \tag{20}$$

under the linear constraint conditions (19), where $l_i(a_{ij})$ is a random linear function in a_{ij}.

For $q \in \mathscr{E}$, let λ^*_{ij} be an optimal solution for the linear programming problem (20). Then $q - \delta = \sum_{i=1}^{n+1} q_i^*$ where $q_i^* = \sum_{j=1}^{m_i} \lambda^*_{ij} l_i(a_{ij})$. $a^*_{ij_0}$ is a vertex of Q_i if and only if there exists a j_0 such that $\lambda^*_{ij_0} = 1$ and $\lambda^*_{ij} = 0$ for $j \neq j_0$. In this case, the generalized row content function of q is (i, j_0) and $\mathrm{mm}(p_i)$ is $M(a^*_{ij_0})$. It is shown that when the lift functions l_i are general enough, all S_i can be computed in the above way [4].

In order to study the linear programming problem (20), we need to recall a lemma about the optimality criterion for the general linear programming problem

$$\begin{aligned} &\min_{x} && z = c^T x \\ &\text{subject to} && Ax = b, \quad \text{with } l \leq x \leq u, \end{aligned} \tag{21}$$

where A is an $m \times n$ rectangular matrix, b is a column vector of dimension m, c and x are column vectors of dimension n, and the superscript T stands for transpose. In order for the linear programming problem to be meaningful, the row rank of A must be less than the column rank of A. We thus can assume A to be row full rank. Let $n_1, \ldots, n_m$ be linear independent columns of A. Then the corresponding $x_{n_1}, \ldots, x_{n_m}$ are called *basic variables of x*. Let B be the matrix consisting of the $n_1, \ldots, n_m$ columns of A. Then B is an $m \times m$ invertible matrix. Lemma 4 below gives an optimality criterion for the linear programming problem (21).

Lemma 4 [10] *Let x_B be a basic variables set of x, where B is the corresponding coefficient matrix of x_B. If the corresponding basic feasible solution $x_B = B^{-1}b \geq 0$ and the conditions $c_B B^{-1} A - c \leq 0$ hold, where c_B is the row vector obtained by listing the coefficients of x_B in the object function, then an optimal solution for the linear programming problem (21) can be given as $x_B = B^{-1}b$ and all other x_i equals zero, which is called the optimal solution determined by the basic variables x_B.*

4.2 Algebraic Sparse Resultant Matrix

In this subsection, we show that the sparse resultant for $f_1, f_2, \delta f_1$ and δf_2 is nonzero and contains the differential resultant of f_1 and f_2 as a factor.

For the differential polynomials f_1 and f_2 given in (3), consider the $p_1 = \delta f_1, p_2 = \delta f_2, p_3 = f_1, p_4 = f_2$ as algebraic polynomials in y, y_1, y_2. The monomial sets of $\delta f_1, \delta f_2, f_1$ and f_2 are $B_3^{d_1} \cup y_2 * B_3^{d_1 - 1}$, $B_3^{d_2} \cup y_2 * B_3^{d_2 - 1}$, $B_3^{d_1}$, and $B_3^{d_2}$ respectively. For convenience, we will not distinguish a monomial M and its exponential vector when there exists no confusion. Then the Newton polytopes for $\delta f_1, \delta f_2, f_1$ and f_2 are respectively,

$$
\begin{aligned}
Q_1 &= \mathrm{Conv}(\sup(\delta f_1)) = \mathrm{Conv}(B_3^{d_1} \cup y_2 * B_3^{d_1-1}) \subset \mathbb{R}^3,\\
Q_2 &= \mathrm{Conv}(\sup(\delta f_2)) = \mathrm{Conv}(B_3^{d_2} \cup y_2 * B_3^{d_2-1}) \subset \mathbb{R}^3,\\
Q_3 &= \mathrm{Conv}(\sup(f_1)) = \mathrm{Conv}(B_3^{d_1}) \subset \mathbb{R}^3,\\
Q_4 &= \mathrm{Conv}(\sup(f_2)) = \mathrm{Conv}(B_3^{d_2}) \subset \mathbb{R}^3,
\end{aligned} \tag{22}
$$

The Newton polytopes Q_1 and Q_3 are shown in Fig. 1 (for $d_1 = 5$) while Q_2 and Q_4 have similar polytopes as Q_1 and Q_3 but with different sizes respectively.

Let the Minkowski sum $Q = Q_1 + Q_2 + Q_3 + Q_4$. In order to compute the column monomial set, we choose a perturbed vector $\delta = (\delta_1, \delta_2, \delta_3)$ with $0 < \delta_i < 1$ with $i = 1, 2, 3$. Then the points in $\mathbb{Z}^3 \cap (Q + \delta)$ is easily shown to be $yy_1y_2\mathscr{E}$ where $\mathscr{E}$ is given in (4). Note that using $\mathscr{E}$ or $yy_1y_2\mathscr{E}$ as the column monomial set will lead to the same matrix.

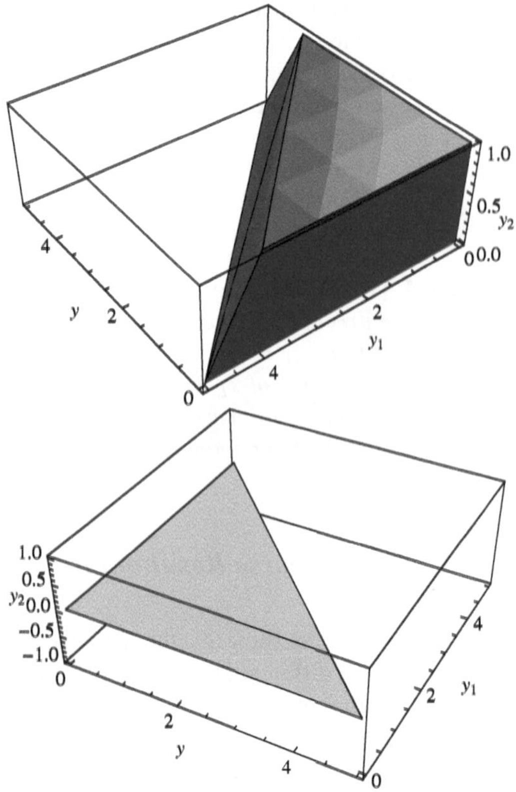

Fig. 1 The Newton polytopes Q_1 and Q_3

The vertex sets of Q_i, denoted by V_i, are respectively

$$
\begin{aligned}
V_1 &:= \{(0,0,0), (0,0,1), (0,d_1-1,1), (0,d_1,0), (d_1-1,0,1), (d_1,0,0)\},\\
V_2 &:= \{(0,0,0), (0,0,1), (0,d_2-1,1), (0,d_2,0), (d_2-1,0,1), (d_2,0,0)\},\\
V_3 &:= \{(0,0,0), (0,d_1,0), (d_1,0,0)\},\\
V_4 &:= \{(0,0,0), (0,d_2,0), (d_2,0,0)\}.
\end{aligned}
$$

Let the lifting functions be $l_i = (L_{i1}, L_{i2}, L_{i3}), i = 1, \dots, 4$, where L_{ij} are parameters to be determined later. From (20), the object function of the linear programming problem to be solved is

$$
\begin{aligned}
\min_{\lambda_{ij}}(&\lambda_{12}L_{13} + \lambda_{13}[L_{12}(d_1-1) + L_{13}] + \lambda_{14}L_{12}d_1 + \lambda_{15}[L_{11}(d_1-1) + L_{13}]\\
&+\lambda_{16}L_{11}d_1 + \lambda_{22}L_{23} + \lambda_{23}[L_{22}(d_2-1) + L_{23}] + \lambda_{24}L_{22}d_2\\
&+\lambda_{25}[L_{21}(d_2-1) + L_{23}] + \lambda_{26}L_{21}d_2 + \lambda_{32}L_{32}d_1 + \lambda_{33}L_{31}d_1\\
&+\lambda_{42}L_{42}d_2 + \lambda_{43}L_{41}d_2)
\end{aligned}
\tag{23}
$$

under the constraints

$$
\begin{aligned}
&A_1 = \lambda_{15}(d_1-1) + \lambda_{16}d_1 + \lambda_{25}(d_2-1) + \lambda_{26}d_2 + \lambda_{33}d_1 + \lambda_{43}d_2,\\
&A_2 = \lambda_{13}(d_1-1) + \lambda_{14}d_1 + \lambda_{23}(d_2-1) + \lambda_{24}d_2 + \lambda_{32}d_1 + \lambda_{42}d_2,\\
&A_3 = \lambda_{12} + \lambda_{13} + \lambda_{15} + \lambda_{22} + \lambda_{23} + \lambda_{25},\\
&\sum_{j=1}^{m_i} \lambda_{ij} = 1,\ i = 1, \dots, 4,\\
&\lambda_{ij} \geq 0, i = 1, \dots, 4, j = 1, \dots, m_i \text{ with } m_1 = m_2 = 6, m_3 = m_4 = 3,
\end{aligned}
\tag{24}
$$

where $A_1 = \varepsilon_1 - \delta_1, A_2 = \varepsilon_2 - \delta_2, A_3 = \varepsilon_3 - \delta_3$ with $(\varepsilon_1, \varepsilon_2, \varepsilon_3) \in \mathbb{Z}^3 \cap (Q + \delta)$. According to the procedure given in Sect. 4.1, we need to solve the linear programming problem (23) for each $(\varepsilon_1, \varepsilon_2, \varepsilon_3) \in \mathbb{Z}^n \cap (Q + \delta)$. Note that L_{ij} are parameters. What we need to do is to show that there exist L_{ij} such that the solutions of (23) make the corresponding main monomials to be the ones selected by us in (5). More precisely, we need to determine L_{ij} such that for each $q \in \mathbb{Z}^n \cap (Q + \delta)$, the optimal solution for the linear programming problem (23) consists one of the following cases:

$\lambda_{13} = 1$ implies $\mathrm{GRC}(q) = (1, 3)$, the vertex is $(0, d_1 - 1, 1)$, and $\mathrm{mm}(\delta f_1) = y_2 y_1^{d_1 - 1}$,

$\lambda_{24} = 1$ implies $\mathrm{GRC}(q) = (2, 4)$, the vertex is $(0, d_2, 0)$, and $\mathrm{mm}(\delta f_2) = y_1^{d_2}$,

$\lambda_{33} = 1$ implies $\mathrm{GRC}(q) = (3, 3)$, the vertex is $(d_1, 0, 0)$, and $\mathrm{mm}(f_1) = y^{d_1}$,

$\lambda_{41} = 1$ implies $\mathrm{GRC}(q) = (4, 1)$, the vertex is $(0, 0, 0)$, and $\mathrm{mm}(f_2) = 1$.

The following lemma proves that the above statement is valid.

Lemma 5 *There exist L_{ij} such that the optimal solution of the corresponding linear programming problem* (23) *can be chosen such that the corresponding main monomials for $f_1, f_2, \delta f_1$ and δf_2 are* $\mathrm{mm}(f_1) = y^{d_1}$, $\mathrm{mm}(f_2) = 1$, $\mathrm{mm}(\delta f_1) = y_2 y_1^{d_1 - 1}$, $\mathrm{mm}(\delta f_2) = y_1^{d_2}$ *respectively and $\mathscr{E}$ can be written as a disjoint union $\mathscr{E} = S_1 \cup S_2 \cup S_3 \cup S_4$, where S_i is defined in* (16).

Proof We write the linear programming problem as the standard form (21). It is easy to see

$$\begin{aligned} c = \ & (0, L_{13}, (d_1 - 1)L_{12} + L_{13}, d_1 L_{12}, (d_1 - 1)L_{11} + L_{13}, d_1 L_{11}, \\ & 0, L_{23}, (d_2 - 1)L_{22} + L_{23}, d_2 L_{22}, (d_2 - 1)L_{21} + L_{23}, d_2 L_{21}, \\ & 0, d_1 L_{32}, d_1 L_{31}, 0, d_2 L_{42}, d_2 L_{41}). \end{aligned}$$

Let $\delta = (\delta_1, \delta_2, \delta_3)$ be a sufficiently small vector in sufficiently generic position. Then

$$A = \begin{pmatrix} 0 & 0 & 0 & 0 & \widetilde{d}_1 & d_1 & 0 & 0 & 0 & 0 & \widetilde{d}_2 & d_2 & 0 & 0 & d_1 & 0 & 0 & d_2 \\ 0 & 0 & \widetilde{d}_1 & d_1 & 0 & 0 & 0 & 0 & \widetilde{d}_2 & d_2 & 0 & 0 & 0 & d_1 & 0 & 0 & d_2 & 0 \\ 0 & 1 & 1 & 0 & 1 & 0 & 0 & 1 & 1 & 0 & 1 & 0 & 0 & 0 & 0 & 0 & 0 & 0 \\ 1 & 1 & 1 & 1 & 1 & 1 & 0 & 0 & 0 & 0 & 0 & 0 & 0 & 0 & 0 & 0 & 0 & 0 \\ 0 & 0 & 0 & 0 & 0 & 0 & 1 & 1 & 1 & 1 & 1 & 1 & 0 & 0 & 0 & 0 & 0 & 0 \\ 0 & 0 & 0 & 0 & 0 & 0 & 0 & 0 & 0 & 0 & 0 & 0 & 1 & 1 & 1 & 0 & 0 & 0 \\ 0 & 0 & 0 & 0 & 0 & 0 & 0 & 0 & 0 & 0 & 0 & 0 & 0 & 0 & 0 & 1 & 1 & 1 \end{pmatrix},$$

where $\widetilde{d}_1 = d_1 - 1, \widetilde{d}_2 = d_2 - 1$, which is a 7×18 matrix and $b = (A_1, A_2, A_3, 1, 1, 1, 1)$. It is easy to see that the rank of A is 7, since $d_1 \geq 1$.

From (4), we have $\mathscr{E} = \mathbb{Z}^3 \cap (Q + \delta) = y y_1 y_2 (B_3^D \cup y_2 B_3^{D-1})$, where $D = 2d_1 + 2d_2 - 3$. We will construct a disjoint union $\mathscr{E} = S_1 \cup S_2 \cup S_3 \cup S_4$ like (17) such that the corresponding main monomials are respectively $\mathrm{mm}(\delta f_1) = y_2 y_1^{d_1 - 1}$, $\mathrm{mm}(\delta f_2) = y_1^{d_1}$, $\mathrm{mm}(f_1) = y^{d_1}$, $\mathrm{mm}(f_2) = 1$.

Four cases will be considered.

Case 1. We will give the conditions about L_{ij} under which $\mathrm{mm}(\delta f_1) = y_2 y_1^{d_1 - 1}$, or equivalently, the linear programming problem (23) has an optimal solution where $\lambda_{13} = 1$. As a consequence, S_1 will also be constructed.

As shown by Lemma 4, an optimal solution for a linear programming problem can be uniquely determined by a set of basic variables. We will construct the required optimal solutions by choosing different sets of basic variables. Four subcases are considered.

1.1. Selecting basic variables as $\mathrm{vet}_{11} = \{\lambda_{13}, \lambda_{23}, \lambda_{24}, \lambda_{32}, \lambda_{33}, \lambda_{41}, \lambda_{43}\}$ while other variables are nonbasic variables and equal to zero. Due the constraint $\lambda_{11} + \lambda_{12} + \cdots + \lambda_{16} = 1$, we have $\lambda_{13} = 1$. Then for any such an optimal solution of the linear programming problem (23), in the optimal sum of any element

$q = q_1 + q_2 + q_3 + q_4, q_1 = (0, d_1 - 1, 1)$ is a vertex of Q_1, $\text{mm}(\delta f_1) = y_2 y_1^{d_1-1}$, and the corresponding q belongs to S_1 as defined in (16).

We claim that the basic feasible solutions in vet_{11} must be nondegenerate meaning that all basic variables are positive, that is, $x_B = B^{-1}b > 0$. A mixed cell $R = \sum_{i=1}^{4} \mathscr{F}_i$, where $\mathscr{F}_i$ is a face of Q_i, must satisfy the dimension constraint $\sum_{i=1}^{4} \dim(\mathscr{F}_i) = 3$. In the cell corresponding to the basic variables vet_{11}, $\mathscr{F}_1 = (0, d_1 - 1, 1)$ is a vertex of Q_1, $\mathscr{F}_4$ is a one dimensional face of Q_4 of the form $\lambda_{41}V_{41} + \lambda_{43}V_{43}$, where $V_{41} = (0, 0, 0)$, $V_{43} = (d_2, 0, 0)$, and $\lambda_{41} + \lambda_{43} = 1$. $\mathscr{F}_2$ and $\mathscr{F}_3$ are one dimensional faces of Q_2 and Q_3 respectively. In order for the dimension constraint $\sum_{i=1}^{4} \dim(\mathscr{F}_i) = 3$ to be valid, the claim must be true. For otherwise, one of the variables in vet_{11} must be zero, say $\lambda_{41} = 0$. Then $\lambda_{43} = 1$ and $\mathscr{F}_4$ becomes a vertex, which implies $\sum_{i=1}^{4} \dim(\mathscr{F}_i) < 3$, a contradiction.

From Lemma 4, the coefficient matrix of basic variables in (24) is

$$B_{11} = \begin{pmatrix} 0 & 0 & 0 & 0 & d_1 & 0 & d_2 \\ d_1 - 1 & d_2 - 1 & d_2 & d_1 & 0 & 0 & 0 \\ 1 & 1 & 0 & 0 & 0 & 0 & 0 \\ 1 & 0 & 0 & 0 & 0 & 0 & 0 \\ 0 & 1 & 1 & 0 & 0 & 0 & 0 \\ 0 & 0 & 0 & 1 & 1 & 0 & 0 \\ 0 & 0 & 0 & 0 & 0 & 1 & 1 \end{pmatrix},$$

with $\text{rank}(B_{11}) = 7$. For all (A_1, A_2, A_3) and $b = (A_1, A_2, A_3, 1, 1, 1, 1)$, the requirement $B_{11}^{-1}b > 0$ in Lemma 4 gives

$$1 < A_3 < 2, d_1 + d_2 < A_2 + A_3 < 2d_1 + d_2,$$
$$2d_1 + d_2 < A_1 + A_2 + A_3 < 2d_1 + 2d_2.$$

Substituting $A_1 = \varepsilon_1 - \delta_1, A_2 = \varepsilon_2 - \delta_2, A_3 = \varepsilon_3 - \delta_3$ into the above inequalities and considering that $(\varepsilon_1, \varepsilon_2, \varepsilon_3)$ are integer points, we have

$$\varepsilon_3 = 2, \ \varepsilon_2 = d_1 + d_2 - 1, \ldots, 2d_1 + d_2 - 2,$$
$$\varepsilon_1 + \varepsilon_2 = 2d_1 + d_2 - 1, \ldots, 2d_1 + 2d_2 - 2. \tag{25}$$

On the other hand, $c_{B_{11}} = ((d_1-1)L_{12}+L_{13}, (d_2-1)L_{22}+L_{23}, d_2L_{22}, d_1L_{32}, d_1 L_{31}, 0, d_2L_{41})$. After simplification and rearrangement, the condition $c_{B_{11}} B_{11}^{-1} A - c \le 0$ in Lemma 4 becomes

$$\{L_{12} - L_{11} + L_{31} - L_{32}, L_{12} + L_{31} - L_{32} - L_{41},$$
$$L_{13} - L_{12} + L_{22} - L_{23}, L_{22} - L_{21} + L_{31} - L_{32}, \tag{26}$$
$$L_{22} + L_{31} - L_{32} - L_{41}, L_{31} - L_{41}, L_{32} - L_{31} + L_{41} - L_{42}\} \le 0$$

where, hereinafter, $\{w_1, \ldots, w_s\} \le 0$ means $w_i \le 0$ for $i = 1, \ldots, s$.

By Lemma 4, if (25) and (26) are valid, we obtain an optimal solution of the linear programming problem (23) which is determined by the basic variables vet_{11}. Hence, if (26) is valid, the corresponding $q = (\varepsilon_1, \varepsilon_2, \varepsilon_3)$ in and (25) are in S_1, since in the optimal decomposition of $q = q_1 + q_2 + q_3 + q_4$, $q_1 = (0, d_1 - 1, 1)$ is a vertex.

1.2. Similarly, choosing the basic variables as $\text{vet}_{12} = \{\lambda_{13}, \lambda_{23}, \lambda_{24}, \lambda_{31}, \lambda_{32}, \lambda_{33}, \lambda_{41}\}$, which generates a new basic matrix B_{12}, and from $B_{12}^{-1}b > 0$, we obtain

$$0 < A_1, d_1 + d_2 < A_2 + A_3, 1 < A_3 < 2, A_1 + A_2 + A_3 < 2d_1 + d_2,$$

which in turn lead to the following values for $\varepsilon_1, \varepsilon_2, \varepsilon_3$

$$\begin{aligned} &\varepsilon_3 = 2, \quad \varepsilon_1 = 1, \ldots, \quad \varepsilon_2 = d_1 + d_2 - 1, \ldots \\ &\varepsilon_1 + \varepsilon_2 = d_1 + d_2, \ldots, 2d_1 + d_2 - 2. \end{aligned} \tag{27}$$

The condition $c_{B_{12}} B_{12}^{-1} A - c \leq 0$ leads to the following constraints on L_{ij}, $i = 1, \ldots, 4$, $j = 1, 2, 3$,

$$\begin{aligned} &\{L_{12} - L_{11} + L_{31} - L_{32}, L_{12} - L_{32}, \\ &L_{13} - L_{12} + L_{22} - L_{23}, L_{22} - L_{32}, \\ &L_{22} - L_{21} + L_{31} - L_{32}, L_{31} - L_{41}, L_{32} - L_{42}\} \leq 0. \end{aligned} \tag{28}$$

1.3. Similarly, the basic variables $\text{vet}_{13} = \{\lambda_{13}, \lambda_{23}, \lambda_{24}, \lambda_{32}, \lambda_{33}, \lambda_{41}, \lambda_{42}\}$ lead to

$$\begin{aligned} &\varepsilon_3 = 2, \quad \varepsilon_1 = 1, \ldots, d_1, \\ &\varepsilon_1 + \varepsilon_2 = 2d_1 + d_2 - 1, \ldots, 2d_1 + 2d_2 - 2, \end{aligned} \tag{29}$$

and

$$\begin{aligned} &\{L_{12} - L_{42}, L_{12} - L_{11} + L_{31} - L_{32}, \\ &L_{13} - L_{12} + L_{22} - L_{23}, L_{22} - L_{21} + L_{31} - L_{32}, \\ &L_{22} - L_{42}, L_{31} - L_{32} - L_{41} + L_{42}, L_{32} - L_{42}\} \leq 0. \end{aligned} \tag{30}$$

1.4. Similarly, the basic variables $\text{vet}_{14} = \{\lambda_{13}, \lambda_{23}, \lambda_{24}, \lambda_{33}, \lambda_{41}, \lambda_{42}, \lambda_{43}\}$ lead to

$$\begin{aligned} &\varepsilon_3 = 2, \quad \varepsilon_1 = d_1 + 1, \ldots, \quad \varepsilon_2 = d_1 + d_2 - 1, \ldots \\ &\varepsilon_1 + \varepsilon_2 = 2d_1 + d_2, \ldots, 2d_1 + 2d_2 - 2. \end{aligned} \tag{31}$$

and

$$\{L_{12} - L_{11} + L_{41} - L_{42}, L_{12} - L_{42},\\ L_{13} - L_{12} + L_{22} - L_{23}, L_{22} - L_{21} + L_{41} - L_{42},\\ L_{22} - L_{42}, L_{31} - L_{41}, L_{31} - L_{32} - L_{41} + L_{42}\} \leq 0. \tag{32}$$

Let S_1 be the set $(\varepsilon_1, \varepsilon_2, \varepsilon_3)$ defined in (25), (27), (29), and (31). Then for $\eta \in S_1$ and an optimal sum of $\eta = q_1 + q_2 + q_3 + q_4$, since $\lambda_{13} = 1$, q_1 must be the vertex $(0, d_1 - 1, 1)$ of Q_1. Therefore, $\mathrm{mm}(\delta f_1) = y_2 y_1^{d_1 - 1}$. Of course, in order for this statement to be valid, L_{ij} must satisfy constraints (26), (28), (30), and (32). We will show later that these constraints indeed have common solutions.

The following three cases can be treated similarly, and we only list the conditions for $\varepsilon_1, \varepsilon_2, \varepsilon_3$ while the concrete requirements for L_{ij} are listed at the end of the proof.

Case 2. In order for $\mathrm{mm}(\delta f_2) = y_1^{d_1}$, we choose the basic variables

$$\begin{aligned}
\mathrm{vet}_{21} &= \{\lambda_{13}, \lambda_{14}, \lambda_{24}, \lambda_{31}, \lambda_{32}, \lambda_{33}, \lambda_{41}\},\\
\mathrm{vet}_{22} &= \{\lambda_{13}, \lambda_{14}, \lambda_{24}, \lambda_{33}, \lambda_{41}, \lambda_{42}, \lambda_{43}\},\\
\mathrm{vet}_{23} &= \{\lambda_{13}, \lambda_{14}, \lambda_{24}, \lambda_{32}, \lambda_{33}, \lambda_{41}, \lambda_{43}\},\\
\mathrm{vet}_{24} &= \{\lambda_{13}, \lambda_{14}, \lambda_{24}, \lambda_{32}, \lambda_{33}, \lambda_{41}, \lambda_{42}\},\\
\mathrm{vet}_{25} &= \{\lambda_{11}, \lambda_{12}, \lambda_{13}, \lambda_{24}, \lambda_{31}, \lambda_{33}, \lambda_{41}\},\\
\mathrm{vet}_{26} &= \{\lambda_{13}, \lambda_{14}, \lambda_{15}, \lambda_{24}, \lambda_{33}, \lambda_{41}, \lambda_{43}\},\\
\mathrm{vet}_{27} &= \{\lambda_{12}, \lambda_{13}, \lambda_{14}, \lambda_{15}, \lambda_{24}, \lambda_{33}, \lambda_{41}\},
\end{aligned}$$

which lead to the following elements of S_2

$$\begin{aligned}
&\varepsilon_3 = 1,\ 0 < \varepsilon_1,\ d_1 + d_2 - 1 < \varepsilon_2, \varepsilon_1 + \varepsilon_2 = d_1 + d_2 + 1, \ldots, 2d_1 + d_2 - 1;\\
&\varepsilon_3 = 1,\ d_1 < \varepsilon_1,\ d_1 + d_2 - 1 < \varepsilon_2, \varepsilon_1 + \varepsilon_2 = 2d_1 + d_2 + 1, \ldots, 2d_1 + 2d_2 - 1;\\
&\varepsilon_3 = 1,\ \varepsilon_2 = d_1 + d_2, \ldots, 2d_1 + d_2 - 1,\ \varepsilon_1 + \varepsilon_2 = 2d_1 + d_2, \ldots, 2d_1 + 2d_2 - 1;\\
&\varepsilon_3 = 1,\ \varepsilon_1 = 1, \ldots, d_1,\ \varepsilon_1 + \varepsilon_2 = 2d_1 + d_2, \ldots, 2d_1 + 2d_2 - 1;\\
&\varepsilon_3 = 1,\ \varepsilon_1 = 1, \ldots, d_1,\ \varepsilon_2 = d_2 + 1, \ldots, d_1 + d_2 - 1;\\
&\varepsilon_3 = 1,\ \varepsilon_2 = d_2 + 1, \ldots, d_1 + d_2 - 1,\ \varepsilon_2 = 2d_1 + d_2, \ldots, 2d_1 + 2d_2 - 1;\\
&\varepsilon_3 = 1,\ \varepsilon_1 = d_1 + 1, \ldots,\ \varepsilon_2 = d_2 + 1, \ldots, \varepsilon_1 + \varepsilon_2 = d_1 + d_2 + 2, \ldots, 2d_1 + d_2 - 1.
\end{aligned}$$

Case 3. In order for $\mathrm{mm}(f_1) = y_1^{d_1}$, we choose the basic variables

$$\begin{aligned}
\mathrm{vet}_{31} &= \{\lambda_{15}, \lambda_{16}, \lambda_{24}, \lambda_{26}, \lambda_{33}, \lambda_{41}, \lambda_{43}\},\\
\mathrm{vet}_{32} &= \{\lambda_{13}, \lambda_{15}, \lambda_{23}, \lambda_{24}, \lambda_{33}, \lambda_{41}, \lambda_{43}\},\\
\mathrm{vet}_{33} &= \{\lambda_{15}, \lambda_{23}, \lambda_{24}, \lambda_{25}, \lambda_{33}, \lambda_{41}, \lambda_{43}\},\\
\mathrm{vet}_{34} &= \{\lambda_{12}, \lambda_{13}, \lambda_{15}, \lambda_{23}, \lambda_{24}, \lambda_{33}, \lambda_{41}\},
\end{aligned}$$

which lead to the following results about $\varepsilon_1, \varepsilon_2, \varepsilon_3$ in S_3

$$\begin{aligned}
&\varepsilon_3 = 1,\ \ \varepsilon_2 = 1, \ldots, d_2,\ \ \varepsilon_1 + \varepsilon_2 = 2d_1 + d_2, \ldots, 2d_1 + 2d_2 - 1;\\
&\varepsilon_3 = 2,\ \ \varepsilon_2 = d_2, \ldots, d_1 + d_2 - 2, \varepsilon_1 + \varepsilon_2 = 2d_1 + d_2 - 1, \ldots, 2d_1 + 2d_2 - 2;\\
&\varepsilon_3 = 2,\ \ \varepsilon_2 = 1, \ldots, d_2 - 1,\ \ \varepsilon_1 + \varepsilon_2 = 2d_1 + d_2 - 1, \ldots, 2d_1 + 2d_2 - 2;\\
&\varepsilon_3 = 2,\ \ \varepsilon_1 = d_1 + 1, \ldots,\ \ \varepsilon_2 = d_2, \ldots, \varepsilon_1 + \varepsilon_2 = d_1 + d_2 + 1, \ldots, 2d_1 + d_2 - 2.
\end{aligned} \tag{33}$$

Case 4. In order for $\mathrm{mm}(f_2) = 1$, we choose the following basic variables

$$\begin{aligned}
\mathrm{vet}_{41} &= \{\lambda_{11}, \lambda_{12}, \lambda_{21}, \lambda_{24}, \lambda_{26}, \lambda_{31}, \lambda_{41}\},\\
\mathrm{vet}_{42} &= \{\lambda_{11}, \lambda_{12}, \lambda_{24}, \lambda_{26}, \lambda_{31}, \lambda_{33}, \lambda_{41}\},\\
\mathrm{vet}_{43} &= \{\lambda_{11}, \lambda_{12}, \lambda_{15}, \lambda_{24}, \lambda_{26}, \lambda_{33}, \lambda_{41}\},\\
\mathrm{vet}_{44} &= \{\lambda_{12}, \lambda_{13}, \lambda_{23}, \lambda_{24}, \lambda_{31}, \lambda_{33}, \lambda_{41}\},\\
\mathrm{vet}_{45} &= \{\lambda_{12}, \lambda_{23}, \lambda_{24}, \lambda_{25}, \lambda_{31}, \lambda_{33}, \lambda_{41}\},\\
\mathrm{vet}_{46} &= \{\lambda_{12}, \lambda_{21}, \lambda_{22}, \lambda_{23}, \lambda_{25}, \lambda_{31}, \lambda_{41}\},\\
\mathrm{vet}_{47} &= \{\lambda_{12}, \lambda_{15}, \lambda_{23}, \lambda_{25}, \lambda_{26}, \lambda_{33}, \lambda_{41}\},
\end{aligned}$$

which correspond to the elements in S_4

$$\begin{aligned}
&\varepsilon_3 = 1,\ \ 0 < \varepsilon_1,\ \ 0 < \varepsilon_2,\ \ \varepsilon_1 + \varepsilon_2 = 2, \ldots, d_2;\\
&\varepsilon_3 = 1,\ \ \varepsilon_2 = 1, \ldots, d_2,\ \ \varepsilon_1 + \varepsilon_2 = d_2 + 1, \ldots, d_1 + d_2;\\
&\varepsilon_3 = 1,\ \ \varepsilon_2 = 1, \ldots, d_2,\ \ \varepsilon_1 + \varepsilon_2 = d_1 + d_2 + 1, \ldots, 2d_1 + d_2 - 1;\\
&\varepsilon_3 = 2,\ \ \varepsilon_1 = 1, \ldots, d_1,\ \ \varepsilon_2 = d_2, \ldots, d_1 + d_2 - 2;\\
&\varepsilon_3 = 2,\ \ \varepsilon_2 = 1, \ldots, d_2 - 1,\ \ \varepsilon_1 + \varepsilon_2 = d_2, \ldots, d_1 + d_2 - 1;\\
&\varepsilon_3 = 2,\ \ \varepsilon_1 = 1, \ldots,\ \ \varepsilon_2 = 1, \ldots, d_2 - 2,\ \ \varepsilon_1 + \varepsilon_2 = 2, \ldots, d_2 - 1;\\
&\varepsilon_3 = 1,\ \ \varepsilon_2 = 1, \ldots, d_2 - 1, \varepsilon_1 + \varepsilon_2 = d_1 + d_2 - 1, \ldots, 2d_1 + d_2 - 2.
\end{aligned}$$

Merge all the constraints for L_{ij}, we obtain

$$\begin{aligned}
&L_{11} - L_{12} - L_{21} + L_{22} \le 0,\\
&L_{13} \le L_{23}, L_{21} \le L_{31} \le L_{11} \le L_{41},\\
&L_{22} \le L_{12} \le L_{32} \le L_{42}, L_{31} = L_{32} + L_{41} - L_{42}.
\end{aligned} \tag{34}$$

The solution set for system (34) is nonempty. For example, $l_1 = (7, -4, -5)$, $l_2 = (5, -9, 5)$, $l_3 = (6, 2, 1)$, $l_4 = (8, 4, 7)$, which will be used for example (1), satisfy the conditions in (34).

We can also check that $\mathscr{E} = S_1 \cup S_2 \cup S_3 \cup S_4$ is a disjoint union for $\mathscr{E}$. The lemma is proved. □

We now have the main result of this section.

Theorem 7 *The sparse resultant of $f_1, f_2, \delta f_1, \delta f_2$ as polynomials in y, y_1, y_2 is not identically zero and contains the differential resultant of f_1 and f_2 as a factor.*

Proof Note that $a_0, b_0, \delta a_0, \delta b_0$, which are the zero degree terms of $f_1, f_2, \delta f_1, \delta f_2$ respectively, are algebraic indeterminates. As a consequence,

$$J_1 = (f_1, f_2, \delta f_1, \delta f_2)$$

is a prime ideal in $\mathbb{Q}[\mathbf{u}, y, y_1, y_2]$, where $\mathbf{u}$ is the set of the coefficients of f_1, f_2 are their first order derivatives. Let

$$J_2 = J_1 \cap \mathbb{Q}[\mathbf{u}].$$

Then J_2 is also a prime ideal. We claim that

$$J_2 = (\mathbf{R}) \tag{35}$$

where $\mathbf{R}$ is the differential resultant of f_1 and f_2. From (c) of Theorem 3, $\mathbf{R} \in J_2$. Let $T \in J_2$. Then $T \in J_1 \subset [f_1, f_2]$. From (2), the pseudo remainder of T with respect to $\mathbf{R}$ is zero. Also note that the order of T in a_i, b_i is less than or equal to 1. From (a) and (b) of Theorem 3, $\mathbf{R}$ must be a factor of T, which proves (35).

From Lemma 5, the main monomials for $f_1, f_2, \delta f_1, \delta f_2$ are the same as those used to construct $\mathscr{S}_1, \mathscr{S}_2, \mathscr{S}_3, \mathscr{S}_4$ in (5). As a consequence, we have $S_1 \subset \mathscr{S}_1$. For $q \in \mathscr{S}_1 \setminus S_1$, q must be in some S_i, say $q \in S_2$. Then from Lemma 5, the monomials in $(M(q)/\mathrm{mm}(\delta f_2))\delta f_2$ are contained in $\mathscr{E}$. By Corollary 4, the sparse resultant matrix of $f_1, f_2, \delta f_1, \delta f_2$ obtained after move q from S_2 to S_1 is still nonsingular. Doing such movements repeatedly will lead to $\mathscr{S}_1 = S_1, \mathscr{S}_2 = S_2, \mathscr{S}_3 = S_3, \mathscr{S}_4 = S_4$. As a consequence, the sparse resultant is not identically zero.

From (15), we have $\mathscr{R} \in J_1$ which implies $\mathscr{R} \in J_2$. Since $\mathbf{R}$ is irreducible, $\mathbf{R}$ must be a factor of $\mathscr{R}$. □

4.3 Example (1) Revisited

We show how to construct a nonsingular algebraic sparse resultant matrix of the system $\{g_1, g_2, \delta g_1, \delta g_2\}$, where g_1, g_2 are from (1).

Using the algorithm for sparse resultant in [3, 4], we choose perturbed vector $\delta = (0.01, 0.01, 0.01)$ and the lifting functions $l_1 = (7, -4, -5), l_2 = (5, -9, 5), l_3 = (6, 2, 1), l_4 = (8, 4, 7)$, where l_i corresponds to Q_i defined in (22) with $d_1 = d_2 = 2$. These lift functions satisfy the conditions (34).

By Lemma 5, the main monomials for $g_1, g_2, \delta g_1, \delta g_2$ are identical with those given in Sect. 3.3. Let S_1, S_2, S_3, S_4 be those constructed as in the proof of Lemma 5. After the following changes

$$\text{move } \{y_2y_1y^3, y_2y_1y^2\} \text{ in } S_3 \text{ to } S_1,$$
$$\text{move } \{y_2y^2, y_1y^2, y^3, y^2\} \text{ in } S_4 \text{ to } S_3,$$
$$\text{move } \{y_2y_1y, y_2y_1\} \text{ in } S_4 \text{ to } S_1,$$

we have $\mathscr{S}_i = S_i, i = 1, \ldots, 4$. Then by Corollary 4, the sparse resultant matrix constructed with the original S_1, S_2, S_3, S_4 is nonsingular and contains the differential resultant as a factor.

5 Conclusion and Discussion

In this paper, a matrix representation for two first order nonlinear generic ordinary differential polynomials f_1, f_2 is given. That is, a nonsingular matrix is constructed such that its determinant contains the differential resultant as a factor. The constructed matrix is further shown to be an algebraic sparse matrix of $f_1, f_2, \delta f_1$ and δf_2 when certain special lift functions are used. Combining the two results, we show that the sparse resultant of $f_1, f_2, \delta f_1$ and δf_2 is not zero and conatins the differential resultant of f_1 and f_2 as a factor.

It can be seen that to give a matrix representation for $n + 1$ generic polynomials in n variables is far from solved, even in the case of $n = 1$. Based on what is proved in this paper, we propose the following conjecture.

Conjecture Let $\mathscr{P} = \{f_1, f_2, \ldots, f_{n+1}\}$ be $n+1$ generic differential polynomials in n indeterminates, $\text{ord}(f_i) = s_i$, and $s = \sum_{i=0}^{n} s_i$.

Then the sparse resultant of the algebraic polynomial system

$$f_1, \delta f_1, \ldots \delta^{s-s_0} f_1, \ldots, f_{n+1}, \delta f_{n+1}, \ldots \delta^{s-s_n} f_{n+1} \tag{36}$$

is not zero and contains the differential resultant of $\mathscr{P}$ as a factor.

Acknowledgments This article is partially supported by a National Key Basic Research Project of China (2011CB302400) and by grants from NSFC (60821002, 11101411). Z.Y. Zhang acknowledges the support of China Postdoctoral Science Foundation funded project (20110490620).

References

1. Berkovich, L.M., Tsirulik, V.G.: Differential resultants and some of their applications. Differ. Equ. **22**, 750–757 (1986)
2. Canny, J.F., Emiris, I.Z.: An efficient algorithm for the sparse mixed resultant. In: Cohen, G., Mora, T., Moreno, O. (eds.) Proceedings of 10th International Symposium on Applied Algebra, Algebraic Algorithms and Error-Correcting Codes, May 1993. Lecture Notes in Computer Science, vol. 263, pp. 89–104. Springer, Heidelberg (1993)
3. Canny, J.F., Emiris, I.Z.: Efficient incremental algorithms for the sparse resultant and the mixed volume. J. Symb. Comput. **20**(2), 117–149 (1995)

4. Canny, J.F., Emiris, I.Z.: A subdivision-based algorithm for the sparse resultant. J. ACM **47**(3), 417–451 (2000)
5. Carrà-Ferro, G.: A resultant theory for the systems of two ordinary algebraic differential equations. Appl. Algebra Eng. Commun. Comput. **8**, 539–560 (1997)
6. Chardin, M.: Differential resultants and subresultants. Fundamentals of Computation Theory. Lecture Notes in Computer Science, vol. 529, pp. 180–189. Springer, Berlin (1991)
7. Cox, D., Little, J., O'Shea, D.: Using Algebraic Geometry. Springer, New York (1998)
8. D'Andrea, C.: Macaulay style formulas for the sparse resultant. Trans. Amer. Math. Soc. **354**, 2595–2629 (2002)
9. Gao, X.S., Li, W., Yuan, C.M.: Intersection theory in differential algebraic geometry: generic intersections and the differential chow form. Trans. Amer. Math. Soc. **365**, 4575–4632 (2013)
10. Gass, S.I.: Linear Programming: Methods and Applications, 5th edn. McGraw-Hill, Inc., New York (1984)
11. Gelfand, I.M., Kapranov, M., Zelevinsky, A.V.: Discriminants, Resultants and Multidimensional Determinants. Birkhauser, Boston (1994)
12. Hong, H.: Ore subresultant coefficients in solutions. Appl. Algebra Eng. Commun. Comput. **12**(5), 421–428 (2001)
13. Li, W., Gao, X.S., Yuan, C.M.: Sparse differential resultant. In: Proceedings of ISSAC 2011, San Jose, CA, USA, pp. 225–232, ACM Press, New York (2011)
14. Li, W., Yuan, C.M., Gao, X.S.: Sparse differential resultant for Laurent differential polynomials. arXiv:1111.1084v3 (2012)
15. Li. Z.: A Subresultant Theory for Linear Differential, Linear Difference and Ore Polynomials, with Applications. PhD thesis, Johannes Kepler University (1996)
16. Macaulay, F.S.: The Algebraic Theory of Modular Systems. Proc. Cambridge Univ. Press, Cambridge (1916)
17. Ritt, J.F.: Differential Equations from the Algebraic Standpoint. American Mathematical Society, New York (1932)
18. Rueda, S.L., Sendra, J.R.: Linear complete differential resultants and the implicitization of linear DPPEs. J. Symb. Comput. **45**, 324–341 (2010)
19. Rueda, S.L.: Linear sparse differential resultant formulas. Linear Algebra Appl. **438**, 4296–4321 (2013)
20. Sturmfels, B.: Sparse elimination theory. In: Eisenbud, D., Robbiano, L. (eds.) Computation Algebraic Geometry and Commutative Algebra, Proceedings, Cortona, June 1991, pp. 264–298. Cambridge University Press, Cambridge (1991)
21. Sturmfels, B.: On the newton polytope of the resultant. J. Algebraic Comb. **3**, 207–236 (1994)
22. Yang, L., Zeng, Z., Zhang, W.: Differential elimination with dixon resultants. Abstract in Proceedings of ACA2009, June 2009, Montreal, Canada. Preprint No. 2, SKLTC, East China Normal University, July, 2011
23. Zwillinger, D.: Handbook of Differential Equations. Academic Press, San Diego (1998)

MIX
Papier aus verantwortungsvollen Quellen
Paper from responsible sources
FSC® C105338

If you have any concerns about our products,
you can contact us on
ProductSafety@springernature.com

In case Publisher is established outside the EU,
the EU authorized representative is:
Springer Nature Customer Service Center GmbH
Europaplatz 3, 69115 Heidelberg, Germany

Printed by Libri Plureos GmbH
in Hamburg, Germany